D0871522

WHAT IS MATHEMATICS?

WHAT IS
Mathematics?

AN ELEMENTARY APPROACH TO IDEAS AND METHODS

BY

RICHARD COURANT
HEAD OF THE DEPARTMENT OF MATHEMATICS
NEW YORK UNIVERSITY

AND

HERBERT ROBBINS
DEPARTMENT OF MATHEMATICS
NEW YORK UNIVERSITY

Oxford University Press

LONDON NEW YORK TORONTO

Ninth Printing, 1958

Tenth Printing, 1960

PRINTED IN THE UNITED STATES OF AMERICA

DEDICATED TO

ERNEST, GERTRUDE, HANS

AND LEONORE COURANT

PREFACE TO THE FIRST EDITION

For more than two thousand years some familiarity with mathematics has been regarded as an indispensable part of the intellectual equipment of every cultured person. Today the traditional place of mathematics in education is in grave danger. Unfortunately, professional representatives of mathematics share in the responsibility. The teaching of mathematics has sometimes degenerated into empty drill in problem solving, which may develop formal ability but does not lead to real understanding or to greater intellectual independence. Mathematical research has shown a tendency toward overspecialization and overemphasis on abstraction. Applications and connections with other fields have been neglected. However, such conditions do not in the least justify a policy of retrenchment. On the contrary, the opposite reaction must and does arise from those who are aware of the value of intellectual discipline. Teachers, students, and the educated public demand constructive reform, not resignation along the line of least resistance. The goal is genuine comprehension of mathematics as an organic whole and as a basis for scientific thinking and acting.

Some splendid books on biography and history and some provocative popular writings have stimulated the latent general interest. But knowledge cannot be attained by indirect means alone. Understanding of mathematics cannot be transmitted by painless entertainment any more than education in music can be brought by the most brilliant journalism to those who never have listened intensively. Actual contact with the *content* of living mathematics is necessary. Nevertheless technicalities and detours should be avoided, and the presentation of mathematics should be just as free from emphasis on routine as from forbidding dogmatism which refuses to disclose motive or goal and which is an unfair obstacle to honest effort. It is possible to proceed on a straight road from the very elements to vantage points from which the substance and driving forces of modern mathematics can be surveyed.

The present book is an attempt in this direction. Inasmuch as it presupposes only knowledge that a good high school course could impart, it may be regarded as popular. But it is not a concession to the dangerous tendency toward dodging all exertion. It requires a certain degree of intellectual maturity and a willingness to do some thinking on one's own. The book is written for beginners and scholars, for students and teachers, for philosophers and engineers, for class rooms and libraries. Perhaps this is too ambitious an intention. Under the pressure of other work some compromise had to be made in publishing the book after many years of preparation, yet before it was really finished. Criticism and suggestions will be welcomed.

At any rate, it is hoped that the book may serve a useful purpose as a contribution to American higher education by one who is profoundly grateful for the opportunity offered him in this country. While responsibility for the plan and philosophy of this publication rests with the undersigned, any credit for merits it may have must be shared with Herbert Robbins. Ever since he became associated with the task, he has unselfishly made it his own cause, and his collaboration has played a decisive part in completing the work in its present form.

Grateful acknowledgment is due to the help of many friends. Discussions with Niels Bohr, Kurt Friedrichs, and Otto Neugebauer have influenced the philosophical and historical attitude; Edna Kramer has given much constructive criticism from the standpoint of the teacher; David Gilbarg prepared the first lecture notes from which the book originated; Ernest Courant, Norman Davids, Charles de Prima, Alfred Horn, Herbert Mintzer, Wolfgang Wasow, and others helped in the endless task of writing and rewriting the manuscript, and contributed much in improving details; Donald Flanders made many valuable suggestions and scrutinized the manuscript for the printer; John Knudsen, Hertha von Gumppenberg, Irving Ritter, and Otto Neugebauer prepared the drawings; H. Whitney contributed to the collection of exercises in the appendix. The General Education Board of the Rockefeller Foundation generously supported the development of courses and notes which then became the basis of the book. Thanks are also due to the Waverly Press, and in particular Mr. Grover C. Orth, for their extremely competent work; and to the Oxford University Press, in particular Mr. Philip Vaudrin and Mr. W. Oman, for their encouraging initiative and coöperation.

New Rochelle, N. Y., August 22, 1941.

R. Courant.

PREFACE TO THE SECOND, THIRD AND FOURTH EDITIONS

During the last years the force of events has led to an increased demand for mathematical information and training. Now more than ever there exists the danger of frustration and disillusionment unless students and teachers try to look beyond mathematical formalism and manipulation and to grasp the real essence of mathematics. This book was written for such students and teachers, and the response to the first edition encourages the authors in the hope that it will be helpful.

Criticism by many readers has led to numerous corrections and improvements. For generous help with the preparation of the fourth edition cordial thanks are due to Mrs. Natascha Artin.

New Rochelle, N. Y., March 18, 1943.
October 10, 1945.
October 28, 1947.

R. Courant

HOW TO USE THE BOOK

The book is written in a systematic order, but it is by no means necessary for the reader to plow through it page by page and chapter by chapter. For example, the historical and philosophical introduction might best be postponed until the rest of the book has been read. The different chapters are largely independent of one another. Often the beginning of a section will be easy to understand. The path then leads gradually upward, becoming steeper toward the end of a chapter and in the supplements. Thus the reader who wants general information rather than specific knowledge may be content with a selection of material that can be made by avoiding the more detailed discussions.

The student with slight mathematical background will have to make a choice. Asterisks or small print indicate parts that may be omitted at a first reading without seriously impairing the understanding of subsequent parts. Moreover, no harm will be done if the study of the book is confined to those sections or chapters in which the reader is most interested. Most of the exercises are not of a routine nature; the more difficult ones are marked with an asterisk. The reader should not be alarmed if he cannot solve many of these.

High school teachers may find helpful material for clubs or selected groups of students in the chapters on geometrical constructions and on maxima and minima.

It is hoped that the book will serve both college students from freshman to graduate level and professional men who are genuinely interested in science. Moreover, it may serve as a basis for college courses of an unconventional type on the fundamental concepts of mathematics. Chapters III, IV, and V could be used for a course in geometry, while Chapters VI and VIII together form a self-contained presentation of the calculus with emphasis on understanding rather than routine. They could be used as an introductory text by a teacher who is willing to make active contributions in supplementing the material according to specific needs and especially in providing further numerical examples. Numerous exercises scattered throughout the text and an additional collection at the end should facilitate the use of the book in the class room.

It is even hoped that the scholar will find something of interest in details and in certain elementary discussions that contain the germ of a broader development.

CONTENTS

WHAT IS MATHEMATICS?

Mathematics as an expression of the human mind reflects the active will, the contemplative reason, and the desire for aesthetic perfection. Its basic elements are logic and intuition, analysis and construction, generality and individuality. Though different traditions may emphasize different aspects, it is only the interplay of these antithetic forces and the struggle for their synthesis that constitute the life, usefulness, and supreme value of mathematical science.

Without doubt, all mathematical development has its psychological roots in more or less practical requirements. But once started under the pressure of necessary applications, it inevitably gains momentum in itself and transcends the confines of immediate utility. This trend from applied to theoretical science appears in ancient history as well as in many contributions to modern mathematics by engineers and physicists.

Recorded mathematics begins in the Orient, where, about 2000 B.C., the Babylonians collected a great wealth of material that we would classify today under elementary algebra. Yet as a science in the modern sense mathematics only emerges later, on Greek soil, in the fifth and fourth centuries B.C. The ever-increasing contact between the Orient and the Greeks, beginning at the time of the Persian empire and reaching a climax in the period following Alexander's expeditions, made the Greeks familiar with the achievements of Babylonian mathematics and astronomy. Mathematics was soon subjected to the philosophical discussion that flourished in the Greek city states. Thus Greek thinkers became conscious of the great difficulties inherent in the mathematical concepts of continuity, motion, and infinity, and in the problem of measuring arbitrary quantities by given units. In an admirable effort the challenge was met, and the result, Eudoxus' theory of the geometrical continuum, is an achievement that was only paralleled more than two thousand years later by the modern theory of irrational numbers. The deductive-postulational trend in mathematics originated at the time of Eudoxus and was crystallized in Euclid's *Elements*.

However, while the theoretical and postulational tendency of Greek mathematics remains one of its important characteristics and has exercised an enormous influence, it cannot be emphasized too strongly that application and connection with physical reality played just as important a part in the mathematics of antiquity, and that a manner of presentation less rigid than Euclid's was very often preferred.

It may be that the early discovery of the difficulties connected with "incommensurable" quantities deterred the Greeks from developing the art of numerical reckoning achieved before in the Orient. Instead they forced their way through the thicket of pure axiomatic geometry. Thus one of the strange detours of the history of science began, and perhaps a great opportunity was missed. For almost two thousand years the weight of Greek geometrical tradition retarded the inevitable evolution of the number concept and of algebraic manipulation, which later formed the basis of modern science.

After a period of slow preparation, the revolution in mathematics and science began its vigorous phase in the seventeenth century with analytic geometry and the differential and integral calculus. While Greek geometry retained an important place, the Greek ideal of axiomatic crystallization and systematic deduction disappeared in the seventeenth and eighteenth centuries. Logically precise reasoning, starting from clear definitions and non-contradictory, "evident" axioms, seemed immaterial to the new pioneers of mathematical science. In a veritable orgy of intuitive guesswork, of cogent reasoning interwoven with nonsensical mysticism, with a blind confidence in the superhuman power of formal procedure, they conquered a mathematical world of immense riches. Gradually the ecstasy of progress gave way to a spirit of critical self-control. In the nineteenth century the immanent need for consolidation and the desire for more security in the extension of higher learning that was prompted by the French revolution, inevitably led back to a revision of the foundations of the new mathematics, in particular of the differential and integral calculus and the underlying concept of limit. Thus the nineteenth century not only became a period of new advances, but was also characterized by a successful return to the classical ideal of precision and rigorous proof. In this respect it even surpassed the model of Greek science. Once more the pendulum swung toward the side of logical purity and abstraction. At present we still seem to be in this period, although it is to be hoped that the resulting unfortunate separation between pure mathematics and the vital applications, perhaps inevitable in times of critical revision, will be followed by an era of closer unity. The regained internal strength and, above all, the enormous simplification attained on the basis of clearer comprehension make it possible today to master the mathematical theory without losing sight of applications. To establish once again an organic union between pure and applied science and a sound balance between abstract generality and colorful individuality may well be the paramount task of mathematics in the immediate future.

This is not the place for a detailed philosophical or psychological analysis of mathematics. Only a few points should be stressed. There seems to be a great danger in the prevailing overemphasis on the deductive-postulational character of mathematics. True, the element of constructive invention, of directing and motivating intuition, is apt to elude a simple philosophical formulation; but it remains the core of any mathematical achievement, even in the most abstract fields. If the crystallized deductive form is the goal, intuition and construction are at least the driving forces. A serious threat to the very life of science is implied in the assertion that mathematics is nothing but a system of conclusions drawn from definitions and postulates that must be consistent but otherwise may be created by the free will of the mathematician. If this description were accurate, mathematics could not attract any intelligent person. It would be a game with definitions, rules, and syllogisms, without motive or goal. The notion that the intellect can create meaningful postulational systems at its whim is a deceptive half-truth. Only under the discipline of responsibility to the organic whole, only guided by intrinsic necessity, can the free mind achieve results of scientific value.

While the contemplative trend of logical analysis does not represent all of mathematics, it has led to a more profound understanding of mathematical facts and their interdependence, and to a clearer comprehension of the essence of mathematical concepts. From it has evolved a modern point of view in mathematics that is typical of a universal scientific attitude.

Whatever our philosophical standpoint may be, for all purposes of scientific observation an object exhausts itself in the totality of possible relations to the perceiving subject or instrument. Of course, mere perception does not constitute knowledge and insight; it must be coordinated and interpreted by reference to some underlying entity, a "thing in itself," which is not an object of direct physical observation, but belongs to metaphysics. Yet for scientific procedure it is important to discard elements of metaphysical character and to consider observable facts always as the ultimate source of notions and constructions. To renounce the goal of comprehending the "thing in itself," of knowing the "ultimate truth," of unraveling the innermost essence of the world, may be a psychological hardship for naive enthusiasts, but in fact it was one of the most fruitful turns in modern thinking.

Some of the greatest achievements in physics have come as a reward for courageous adherence to the principle of eliminating metaphysics. When Einstein tried to reduce the notion of "simultaneous events occurring at different places" to observable phenomena, when he unmasked as a metaphysical prejudice the belief that this concept must have a scientific meaning in itself, he had found the key to his theory of relativity. When Niels Bohr and his pupils analyzed the fact that any physical observation must be accompanied by an effect of the observing instrument on the observed object, it became clear that the sharp simultaneous fixation of position and velocity of a particle is not possible in the sense of physics. The far-reaching consequences of this discovery, embodied in the modern theory of quantum mechanics, are now familiar to every physicist. In the nineteenth century the idea prevailed that mechanical forces and motions of particles in space are things in themselves, while electricity, light, and magnetism should be reduced to or "explained" as mechanical phenomena, just as had been done with heat. The "ether" was invented as a hypothetical medium capable of not entirely explained mechanical motions that appear to us as light or electricity. Slowly it was realized that the ether is of necessity unobservable; that it belongs to metaphysics and not to physics. With sorrow in some quarters, with relief in others, the mechanical explanations of light and electricity, and with them the ether, were finally abandoned.

A similar situation, even more accentuated, exists in mathematics. Throughout the ages mathematicians have considered their objects, such as numbers, points, etc., as substantial things in themselves. Since these entities had always defied attempts at an adequate description, it slowly dawned on the mathematicians of the nineteenth century that the question of the meaning of these objects as substantial things does not make sense within mathematics, if at all. The only relevant assertions concerning them do not refer to substantial reality; they state only the interrelations between mathematically "undefined objects" and the rules governing operations with them. What points, lines, numbers "actually" *are* cannot and need not be discussed in mathematical science. What matters and what corresponds to "verifiable" fact is structure and relationship, that two points determine a line, that numbers combine according to certain rules to form other numbers, etc. A clear insight into the necessity of a dissubstantiation of elementary mathematical concepts has been one of the most important and fruitful results of the modern postulational development.

Fortunately, creative minds forget dogmatic philosophical beliefs whenever adherence to them would impede constructive achievement. For scholars and layman alike it is not philosophy but active experience in mathematics itself that alone can answer the question: What is mathematics?

CHAPTER I

THE NATURAL NUMBERS

INTRODUCTION

Number is the basis of modern mathematics. But what is number? What does it mean to say that $\frac{1}{2} + \frac{1}{2} = 1$, $\frac{1}{2} \cdot \frac{1}{2} = \frac{1}{4}$, and $(-1)(-1) = 1$? We learn in school the mechanics of handling fractions and negative numbers, but for a real understanding of the number system we must go back to simpler elements. While the Greeks chose the geometrical concepts of point and line as the basis of their mathematics, it has become the modern guiding principle that all mathematical statements should be reducible ultimately to statements about the *natural numbers*, 1, 2, 3, $\cdots$. "God created the natural numbers; everything else is man's handiwork." In these words Leopold Kronecker (1823–1891) pointed out the safe ground on which the structure of mathematics can be built.

Created by the human mind to count the objects in various assemblages, numbers have no reference to the individual characteristics of the objects counted. The number six is an abstraction from all actual collections containing six things; it does not depend on any specific qualities of these things or on the symbols used. Only at a rather advanced stage of intellectual development does the abstract character of the idea of number become clear. To children, numbers always remain connected with tangible objects such as fingers or beads, and primitive languages display a concrete number sense by providing different sets of number words for different types of objects.

Fortunately, the mathematician as such need not be concerned with the philosophical nature of the transition from collections of concrete objects to the abstract number concept. We shall therefore accept the natural numbers as given, together with the two fundamental operations, addition and multiplication, by which they may be combined.

§1. CALCULATION WITH INTEGERS

1. Laws of Arithmetic

The mathematical theory of the natural numbers or *positive integers* is known as *arithmetic*. It is based on the fact that the addition and multiplication of integers are governed by certain laws. In order to state these laws in full generality we cannot use symbols like 1, 2, 3 which refer to specific integers. The statement

$$1 + 2 = 2 + 1$$

is only a particular instance of the general law that the sum of two integers is the same regardless of the order in which they are considered. Hence, when we wish to express the fact that a certain relation between integers is valid irrespective of the values of the particular integers involved, we shall denote integers symbolically by letters $a, b, c, \cdots$. With this agreement we may state five fundamental laws of arithmetic with which the reader is familiar:

$$1)\ a + b = b + a, \qquad 2)\ ab = ba,$$

$$3)\ a + (b + c) = (a + b) + c, \qquad 4)\ a(bc) = (ab)c,$$

$$5)\ a(b + c) = ab + ac.$$

The first two of these, the *commutative* laws of addition and multiplication, state that one may interchange the order of the elements involved in addition or multiplication. The third, the *associative* law of addition, states that addition of three numbers gives the same result whether we add to the first the sum of the second and third, or to the third the sum of the first and second. The fourth is the associative law of multiplication. The last, the *distributive* law, expresses the fact that to multiply a sum by an integer we may multiply each term of the sum by this integer and then add the products.

These laws of arithmetic are very simple, and may seem obvious. But they might not be applicable to entities other than integers. If a and b are symbols not for integers but for chemical substances, and if "addition" is used in a colloquial sense, it is evident that the commutative law will not always hold. For example, if sulphuric acid is added to water, a dilute solution is obtained, while the addition of water to pure sulphuric acid may result in disaster to the experimenter. Similar illustrations will show that in this type of chemical "arithmetic" the associative and distributive laws of addition may also fail. Thus one can imagine types of arithmetic in which one or more of the laws 1)–5) do not hold. Such systems have actually been studied in modern mathematics.

A concrete model for the abstract concept of integer will indicate the intuitive basis on which the laws 1)–5) rest. Instead of using the usual number symbols 1, 2, 3, etc., let us denote the integer that gives the number of objects in a given collection (say the collection of apples on a particular tree) by a set of dots placed in a rectangular box, one dot for each object. By operating with these boxes we may investigate the laws of the arithmetic of integers. To add two integers a and b, we place the corresponding boxes end to end and remove the partition.

Fig. 1. Addition.

To multiply a and b, we arrange the dots in the two boxes in rows, and form a new box with a rows and b columns of dots. The rules 1)–5)

Fig. 2. Multiplication.

will now be seen to correspond to intuitively obvious properties of these operations with boxes.

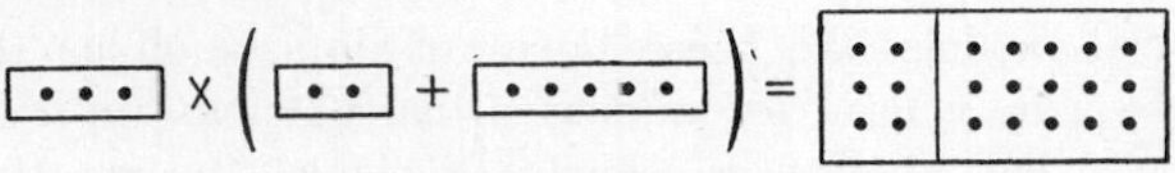

Fig. 3. The Distributive Law.

On the basis of the definition of addition of two integers we may define the relation of *inequality*. Each of the equivalent statements, $a < b$ (read, "a is less than b") and $b > a$ (read, "b is greater than a"), means that box b may be obtained from box a by the addition of a properly chosen third box c, so that $b = a + c$. When this is so we write

$$c = b - a,$$

which defines the operation of *subtraction*.

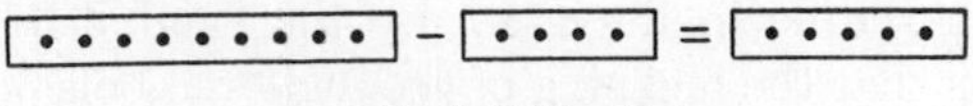

Fig. 4. Subtraction.

Addition and subtraction are said to be *inverse operations*, since if the addition of the integer d to the integer a is followed by the subtraction of the integer d, the result is the original integer a:

$$(a + d) - d = a.$$

It should be noted that the integer $b - a$ has been defined only when $b > a$. The interpretation of the symbol $b - a$ as a *negative integer* when $b < a$ will be discussed later (p. 54 et seq.).

It is often convenient to use one of the notations, $b \geq a$ (read, "b is greater than or equal to a") or $a \leq b$ (read, "a is less than or equal to b"), to express the denial of the statement, $a > b$. Thus, $2 \geq 2$, and $3 \geq 2$.

We may slightly extend the domain of positive integers, represented by boxes of dots, by introducing the integer *zero*, represented by a completely empty box. If we denote the empty box by the usual symbol 0, then, according to our definition of addition and multiplication,

$$a + 0 = a,$$
$$a \cdot 0 = 0,$$

for every integer a. For $a + 0$ denotes the addition of an empty box to the box a, while $a \cdot 0$ denotes a box with no columns; i.e. an empty box. It is then natural to extend the definition of subtraction by setting

$$a - a = 0$$

for every integer a. These are the characteristic arithmetical properties of zero.

Geometrical models like these boxes of dots, such as the ancient abacus, were widely used for numerical calculations until late in the middle ages, when they were slowly displaced by greatly superior symbolic methods based on the decimal system.

2. The Representation of Integers

We must carefully distinguish between an integer and the symbol, 5, V, $\cdots$, etc., used to represent it. In the decimal system the ten digit symbols, 0, 1, 2, 3, $\cdots$, 9, are used for zero and the first nine positive integers. A larger integer, such as "three hundred and seventy-two," can be expressed in the form

$$300 + 70 + 2 = 3 \cdot 10^2 + 7 \cdot 10 + 2,$$

and is denoted in the decimal system by the symbol 372. Here the important point is that the meaning of the digit symbols 3, 7, 2 depends on their *position* in the units, tens, or hundreds place. With this "positional notation" we can represent any integer by using only the ten digit symbols in various combinations. The general rule is to express an integer in the form illustrated by

$$z = a \cdot 10^3 + b \cdot 10^2 + c \cdot 10 + d,$$

where the digits a, b, c, d are integers from zero to nine. The integer z is then represented by the abbreviated symbol

$$abcd.$$

We note in passing that the coefficients d, c, b, a are the remainders left after successive divisions of z by 10. Thus

	Remainder
10)372	
10)37	2
10)3	7
0	3

The particular expression given above for z can only represent integers less than ten thousand, since larger integers will require five or more digit symbols. If z is an integer between ten thousand and one hundred thousand, we can express it in the form

$$z = a \cdot 10^4 + b \cdot 10^3 + c \cdot 10^2 + d \cdot 10 + e,$$

and represent it by the symbol $abcde$. A similar statement holds for integers between one hundred thousand and one million, etc. It is very useful to have a way of indicating the result in perfect generality by a single formula. We may do this if we denote the different coefficients, $e, d, c, \cdots$, by the single letter a with different "subscripts," $a_0, a_1, a_2, a_3, \cdots$; and indicate the fact that the powers of ten may be as large as necessary by denoting the highest power, not by 10^3 or 10^4 as in the examples above, but by 10^n, where n is understood to stand for an arbitrary integer. Then the general method for representing an integer z in the decimal system is to express z in the form

$$(1) \qquad z = a_n \cdot 10^n + a_{n-1} \cdot 10^{n-1} + \cdots + a_1 \cdot 10 + a_0,$$

and to represent it by the symbol

$$a_n a_{n-1} a_{n-2} \cdots a_1 a_0 .$$

As in the special case above, we see that the digits $a_0, a_1, a_2, \cdots, a_n$ are simply the successive remainders when z is divided repeatedly by 10.

In the decimal system the number ten is singled out to serve as a base. The layman may not realize that the selection of ten is not essential, and that any integer greater than one would serve the same purpose. For example, a *septimal* system (base 7) could be used. In such a system, an integer would be expressed as

$$(2) \qquad b_n \cdot 7^n + b_{n-1} \cdot 7^{n-1} + \cdots + b_1 \cdot 7 + b_0,$$

where the b's are digits from zero to six, and denoted by the symbol

$$b_n b_{n-1} \cdots b_1 b_0 .$$

Thus "one hundred and nine" would be denoted in the septimal system by the symbol 214, meaning

$$2 \cdot 7^2 + 1 \cdot 7 + 4.$$

As an exercise the reader may prove that the general rule for passing from the base ten to any other base B is to perform successive divisions of the number z by B; the remainders will be the digits of the number in the system with base B. For example:

$$\begin{array}{rl} 7\underline{)109} & \text{Remainder} \\ 7\underline{)15} & 4 \\ 7\underline{)2} & 1 \\ 0 & 2 \end{array}$$

109 (decimal system) = 214 (septimal system).

It is natural to ask whether any particular choice of base would be most desirable. We shall see that too small a base has disadvantages, while a large base requires the learning of many digit symbols, and an extended multiplication table. The choice of twelve as base has been advocated, since twelve is exactly divisible by two, three, four, and six, and, as a result, work involving division and fractions would often be simplified. To write any integer in terms of the base twelve (duodecimal system), we require two new digit symbols for ten and eleven. Let us write α for ten and β for eleven. Then in the duodecimal system "twelve" would be written 10, "twenty-two" would be 1α, "twenty-three" would be 1β, and "one hundred thirty-one" would be $\alpha\beta$.

The invention of positional notation, attributed to the Sumerians or Babylonians and developed by the Hindus, was of enormous significance for civilization. Early systems of numeration were based on a purely additive principle. In the Roman symbolism, for example, one wrote

CXVIII = one hundred + ten + five + one + one + one.

The Egyptian, Hebrew, and Greek systems of numeration were on the same level. One disadvantage of any purely additive notation is that more and more new symbols are needed as numbers get larger. (Of course, early scientists were not troubled by our modern astronomical or atomic magnitudes.) But the chief fault of ancient systems, such as the Roman, was that computation with numbers was so difficult that only the specialist could handle any but the simplest problems. It is quite different with the Hindu positional system now in use. This was introduced into medieval Europe by the merchants of Italy, who learned it from the Moslems. The positional system has the agreeable property that all numbers, however large or small, can be represented by the use of a small set of different digit symbols (in the decimal system, these are the "Arabic numerals" 0, 1, 2, $\cdots$, 9). Along with this goes the more important advantage of ease of computation. The rules of reckoning with numbers represented in positional notation can be stated in the form of addition and multiplication tables for the digits that can be memorized once and for all. The ancient art of computation, once confined to a few adepts, is now taught in elementary school. There are not many instances where scientific progress has so deeply affected and facilitated everyday life.

3. Computation in Systems Other than the Decimal

The use of ten as a base goes back to the dawn of civilization, and is undoubtedly due to the fact that we have ten fingers on which to count. But the number words of many languages show remnants of the use of other bases, notably twelve and twenty. In English and German the words for 11 and 12 are not constructed on the decimal principle of combining 10 with the digits, as are the "teens," but are linguistically independent of the words for 10. In French the words "vingt" and "quatre-vingt" for 20 and 80 suggest that for some purposes a system with base 20 might have been used. In Danish the word for 70, "halvfirsindstyve," means half-way (from three times) to four times twenty. The Babylonian astronomers had a system of notation that was partly sexagesimal (base 60), and this is believed to account for the customary division of the hour and the angular degree into 60 minutes.

In a system other than the decimal the rules of arithmetic are the same, but one must use different tables for the addition and multiplication of digits. Accustomed to the decimal system and tied to it by the number words of our language, we might at first find this a little confusing. Let us try an example of multiplication in the septimal system. Before proceeding, it is advisable to write down the tables we shall have to use:

Addition

	1	2	3	4	5	6
1	2	3	4	5	6	10
2	3	4	5	6	10	11
3	4	5	6	10	11	12
4	5	6	10	11	12	13
5	6	10	11	12	13	14
6	10	11	12	13	14	15

Multiplication

	1	2	3	4	5	6
1	1	2	3	4	5	6
2	2	4	6	11	13	15
3	3	6	12	15	21	24
4	4	11	15	22	26	33
5	5	13	21	26	34	42
6	6	15	24	33	42	51

Let us now multiply 265 by 24, where these number symbols are written in the septimal system. (In the decimal system this would be equivalent to multiplying 145 by 18.) The rules of multiplication are the same as in the decimal system. We begin by multiplying 5 by 4, which is 26, as the multiplication table shows.

$$\begin{array}{r} 265 \\ 24 \\ \hline 1456 \\ 563 \\ \hline 10416 \end{array}$$

We write down 6 in the units place, "carrying" the 2 to the next place. Then we find $4 \cdot 6 = 33$, and $33 + 2 = 35$. We write down 5, and proceed in this way until everything has been multiplied out. Adding 1,456 + 5,630, we get $6 + 0 = 6$ in the units place, $5 + 3 = 11$ in the sevens place. Again we write down 1 and keep 1 for the forty-nines place, where we have $1 + 6 + 4 = 14$. The final result is $265 \cdot 24 = 10{,}416$.

To check this result we may multiply the same numbers in the decimal system. 10,416 (septimal system) may be written in the decimal system by finding the powers of 7 up to the fourth: $7^2 = 49$, $7^3 = 343$, $7^4 = 2{,}401$. Hence $10{,}416 = 2{,}401 + 4 \cdot 49 + 7 + 6$, this evaluation being in the decimal system. Adding these numbers we find that 10,416 in the septimal system is equal to 2,610 in the decimal system. Now we multiply 145 by 18 in the decimal system; the result is 2,610, so the calculations check.

Exercises: 1) Set up the addition and multiplication tables in the duodecimal system and work some examples of the same sort.

2) Express "thirty" and "one hundred and thirty-three" in the systems with the bases 5, 7, 11, 12.

3) What do the symbols 11111 and 21212 mean in these systems?

4) Form the addition and multiplication tables for the bases 5, 11, 13.

From a theoretical point of view, the positional system with the base 2 is singled out as the one with the smallest possible base. The only digits in this *dyadic system* are 0 and 1; every other number z is represented by a row of these symbols. The addition and multiplication tables consist merely of the rules $1 + 1 = 10$ and $1 \cdot 1 = 1$. But the disadvantage of this system is obvious: long expressions are needed to represent small numbers. Thus seventy-nine, which may be expressed as $1 \cdot 2^6 + 0 \cdot 2^5 + 0 \cdot 2^4 + 1 \cdot 2^3 + 1 \cdot 2^2 + 1 \cdot 2 + 1$, is written in the dyadic system as 1,001,111.

As an illustration of the simplicity of multiplication in the dyadic system, we shall multiply seven and five, which are respectively 111 and 101. Remembering that $1 + 1 = 10$ in this system, we have

$$\begin{array}{r} 111 \\ 101 \\ \hline 111 \\ 111 \\ \hline 100011 \end{array} = 2^5 + 2 + 1,$$

which is thirty-five, as it should be.

Gottfried Wilhelm Leibniz (1646–1716), one of the greatest intellects of his time, was fond of the dyadic system. To quote Laplace: "Leibniz saw in his binary arithmetic the image of creation. He imagined that Unity represented God, and zero the void; that the Supreme Being drew all beings from the void, just as unity and zero express all numbers in his system of numeration."

Exercise: Consider the question of representing integers with the base a. In order to name the integers in this system we need words for the digits $0, 1, \cdots, a - 1$ and for the various powers of a: $a, a^2, a^3, \cdots$. How many different number words are needed to name all numbers from zero to one thousand, for $a = 2, 3, 4, 5, \cdots, 15$? Which base requires the fewest? (Examples: If $a = 10$, we need ten words for the digits, plus words for 10, 100, and 1000, making a total of 13. For $a = 20$, we need twenty words for the digits, plus words for 20 and 400, making a total of 22. If $a = 100$, we need 100 plus 1.)

*§2. THE INFINITUDE OF THE NUMBER SYSTEM. MATHEMATICAL INDUCTION

1. The Principle of Mathematical Induction

There is no end to the sequence of integers $1, 2, 3, 4, \cdots$; for after any integer n has been reached we may write the next integer, $n + 1$. We express this property of the sequence of integers by saying that there are *infinitely many* integers. The sequence of integers represents the simplest and most natural example of the mathematical infinite, which plays a dominant rôle in modern mathematics. Everywhere in this book we shall have to deal with collections or "sets" containing infinitely many mathematical objects, like the set of all points on a line or the set of all triangles in a plane. The infinite sequence of integers is the simplest example of an infinite set.

The step by step procedure of passing from n to $n + 1$ which generates the infinite sequence of integers also forms the basis of one of the most fundamental patterns of mathematical reasoning, the principle of mathematical induction. "Empirical induction" in the natural sciences proceeds from a particular series of observations of a certain phenomenon to the statement of a general law governing all occurrences of this phenomenon. The degree of certainty with which the law is thereby established depends on the number of single observations and confirmations. This sort of inductive reasoning is often entirely convincing; the prediction that the sun will rise tomorrow in the east is as certain as anything can be, but the character of this statement is not the same as that of a theorem proved by strict logical or mathematical reasoning.

In quite a different way *mathematical induction* is used to establish the truth of a mathematical theorem for an infinite sequence of cases, the first, the second, the third, and so on without exception. Let us denote by A a statement that involves an arbitrary integer n. For example, A may be the statement, "The sum of the angles in a convex polygon of $n + 2$ sides is n times 180 degrees." Or A' may be the assertion, "By drawing n lines in a plane we cannot divide the plane into more than 2^n parts." To prove such a theorem for *every* integer n it does not suffice to prove it separately for the first 10 or 100 or even 1000 values of n. This indeed would correspond to the attitude of empirical induction. Instead, we must use a method of strictly mathematical and non-empirical reasoning whose character will be indicated by the following proofs for the special examples A and A'. In the case A, we know that for $n = 1$ the polygon is a triangle, and from elementary geometry the sum of the angles is known to be $1 \cdot 180°$. For a quadrilateral, $n = 2$, we draw a diagonal which divides the quadrilateral into two triangles. This shows immediately that the sum of the angles of the quadrilateral is equal to the sum of the angles in the two triangles, which yields $180° + 180° = 2 \cdot 180°$. Proceeding to the case of a pentagon with 5 edges, $n = 3$, we decompose it into a triangle plus a quadrilateral. Since the latter has the angle sum $2 \cdot 180°$, as we have just proved, and since the triangle has the angle sum $180°$, we obtain $3 \cdot 180$ degrees for the 5-gon. Now it is clear that we can proceed indefinitely in the same way, proving the theorem for $n = 4$, then for $n = 5$, and so on. Each statement follows in the same way from the preceding one, so that the general theorem A can be established for all n.

Similarly we can prove the theorem A'. For $n = 1$ it is obviously true, since a single line divides the plane into 2 parts. Now add a second line. Each of the previous parts will be divided into two new parts, unless the new line is parallel to the first. In either case, for

$n = 2$ we have not more than $4 = 2^2$ parts. Now we add a third line. Each of the previous domains will either be cut into two parts or be left untouched. Thus the sum of parts is not greater than $2^2 \cdot 2 = 2^3$. Knowing this to be true, we can prove the next case in the same way, and so on indefinitely.

The essential idea in the preceding arguments is to establish a general theorem A for all values of n by successively proving a sequence of special cases, $A_1, A_2, \cdots$. The possibility of doing this depends on two things: a) There is a general method for showing that *if* any statement A_r is true then the next statement, A_{r+1}, will *also* be true. b) The first statement A_1 is *known* to be true. That these two conditions are sufficient to establish the truth of *all* the statements $A_1, A_2, A_3, \cdots$ is a logical principle which is as fundamental to mathematics as are the classical rules of Aristotelian logic. We formulate it as follows:

Let us suppose that we wish to establish a whole infinite sequence of mathematical propositions

$$A_1, A_2, A_3, \cdots$$

which together constitute the general proposition A. *Suppose that* a) *by some mathematical argument it is shown that if* r *is any integer and* ***if*** *the assertion* $\mathrm{A_r}$ *is known to be true* **then** *the truth of the assertion* $\mathrm{A_{r+1}}$ *will follow, and that* b) *the first proposition* $\mathrm{A_1}$ ***is*** *known to be true. Then all the propositions of the sequence must be true, and* A *is proved.*

We shall not hesitate to accept this, just as we accept the simple rules of ordinary logic, as a basic principle of mathematical reasoning. For we can establish the truth of every statement A_n, starting from the given assertion b) that A_1 is true, and proceeding by repeated use of the assertion a) to establish successively the truth of A_2, A_3, A_4, and so on until we reach the statement A_n. The principle of mathematical induction thus rests on the fact that after any integer r there is a next, $r + 1$, and that any desired integer n may be reached by a finite number of such steps, starting from the integer 1.

Often the principle of mathematical induction is applied without explicit mention, or is simply indicated by a casual "etc." or "and so on." This is especially frequent in elementary instruction. But the explicit use of an inductive argument is indispensable in more subtle proofs. We shall give a few illustrations of a simple but not quite trivial character.

2. The Arithmetical Progression

For every value of n, *the sum* $1 + 2 + 3 + \cdots + \mathrm{n}$ *of the first* n *integers is equal to* $\frac{\mathrm{n}(\mathrm{n}+1)}{2}$. In order to prove this theorem by mathematical induction we must show that for every n the assertion A_n:

$$1 + 2 + 3 + \cdots + n = \frac{n(n+1)}{2} \tag{1}$$

is true. a) We observe that if r is an integer and if the statement A_r is known to be true, i.e. if it is known that

$$1 + 2 + 3 + \cdots + r = \frac{r(r+1)}{2},$$

then by adding the number $(r + 1)$ to both sides of this equation we obtain the equation

$$1 + 2 + 3 + \cdots + r + (r+1) = \frac{r(r+1)}{2} + (r+1)$$

$$= \frac{r(r+1) + 2(r+1)}{2} = \frac{(r+1)(r+2)}{2},$$

which is precisely the statement A_{r+1}. b) The statement A_1 is obviously true, since $1 = \frac{1 \cdot 2}{2}$. Hence, by the principle of mathematical induction, the statement A_n is true for every n, as was to be proved.

Ordinarily this is shown by writing the sum $1 + 2 + 3 + \cdots + n$ in two forms:

$$S_n = 1 + 2 + \cdots + (n-1) + n$$

and

$$S_n = n + (n-1) + \cdots + 2 + 1.$$

On adding, we see that each pair of numbers in the same column yields the sum $n + 1$, and, since there are n columns in all, it follows that

$$2S_n = n(n+1),$$

which proves the desired result.

From (1) we may immediately derive the formula for the sum of the first $(n + 1)$ terms of any *arithmetical progression*,

$$(2)\quad P_n = a + (a + d) + (a + 2d) + \cdots + (a + nd) = \frac{(n+1)(2a+nd)}{2}.$$

For

$$P_n = (n + 1)a + (1 + 2 + \cdots + n)d = (n + 1)a + \frac{n(n + 1)d}{2}$$

$$= \frac{2(n + 1)a + n(n + 1)d}{2} = \frac{(n + 1)(2a + nd)}{2}.$$

For the case $a = 0$, $d = 1$, this is equivalent to (1).

3. The Geometrical Progression

One may treat the general geometrical progression in a similar way. We shall prove that for every value of n

$$(3)\qquad G_n = a + aq + aq^2 + \cdots + aq^n = a\frac{1 - q^{n+1}}{1 - q}.$$

(We suppose that $q \neq 1$, since otherwise the right side of (3) has no meaning.)

Certainly this assertion is true for $n = 1$, for then it states that

$$G_1 = a + aq = \frac{a(1 - q^2)}{1 - q} = \frac{a(1 + q)(1 - q)}{(1 - q)} = a(1 + q).$$

And *if* we assume that

$$G_r = a + aq + \cdots + aq^r = a\frac{1 - q^{r+1}}{1 - q},$$

then we find as a consequence that

$$G_{r+1} = (a + aq + \cdots + aq^r) + aq^{r+1} = G_r + aq^{r+1} = a\frac{1 - q^{r+1}}{1 - q} + aq^{r+1}$$

$$= a\frac{(1 - q^{r+1}) + q^{r+1}(1 - q)}{1 - q} = a\frac{1 - q^{r+1} + q^{r+1} - q^{r+2}}{1 - q} = a\frac{1 - q^{r+2}}{1 - q}.$$

But this is precisely the assertion (3) for the case $n = r + 1$. This completes the proof.

In elementary textbooks the usual proof proceeds as follows. Set

$$G_n = a + aq + \cdots + aq^n,$$

and multiply both sides of this equation by q, obtaining

$$qG_n = aq + aq^2 + \cdots + aq^{n+1}.$$

Now subtract corresponding sides of this equation from the preceding equation, obtaining

$$G_n - qG_n = a - aq^{n+1},$$

$$(1 - q)G_n = a(1 - q^{n+1}),$$

$$G_n = a\frac{1 - q^{n+1}}{1 - q}.$$

4. The Sum of the First n Squares

A further interesting application of the principle of mathematical induction refers to the sum of the first n squares. By direct trial one finds that, at least for small values of n,

$$(4) \qquad 1^2 + 2^2 + 3^2 + \cdots + n^2 = \frac{n(n+1)(2n+1)}{6},$$

and one might *guess* that this remarkable formula is valid for *all integers* n. To *prove* this, we shall again use the principle of mathematical induction. We begin by observing that *if* the assertion A_n, which in this case is the equation (4), is true for the case $n = r$, so that

$$1^2 + 2^2 + 3^2 + \cdots + r^2 = \frac{r(r+1)(2r+1)}{6},$$

then on adding $(r + 1)^2$ to both sides of this equation we obtain

$$1^2 + 2^2 + 3^2 + \cdots + r^2 + (r+1)^2 = \frac{r(r+1)(2r+1)}{6} + (r+1)^2$$

$$= \frac{r(r+1)(2r+1) + 6(r+1)^2}{6} = \frac{(r+1)[r(2r+1) + 6(r+1)]}{6}$$

$$= \frac{(r+1)(2r^2 + 7r + 6)}{6} = \frac{(r+1)(r+2)(2r+3)}{6},$$

which is precisely the assertion A_{r+1} in this case, since it is obtained by substituting $r + 1$ for n in (4). To complete the proof we need only remark that the assertion A_1, in this case the equation

$$1^2 = \frac{1(1+1)(2+1)}{6},$$

is obviously true. Hence the equation (4) is true for every n.

Formulas of a similar sort may be found for higher powers of the integers, $1^k + 2^k + 3^k + \cdots + n^k$, where k is any positive integer As an exercise, the reader may prove by mathematical induction that

$$(5) \qquad 1^3 + 2^3 + 3^3 + \cdots + n^3 = \left[\frac{n(n+1)}{2}\right]^2.$$

It should be remarked that although the principle of mathematical induction suffices to *prove* the formula (5) once this formula has been written down, the proof gives no indication of how this formula was arrived at in the first place; why precisely the expression $[n(n+1)/2]^2$ should be guessed as an expression for the sum of the first n cubes, rather than $[n(n+1)/3]^2$ or $(19n^2 - 41n + 24)/2$ or any of the infinitely many expressions of a similar type that could have been considered. The fact that the proof of a theorem consists in the application of certain simple rules of logic does not dispose of the creative element in mathematics, which lies in the choice of the possibilities to be examined. The question of the origin of the *hypothesis* (5) belongs to a domain in which no very general rules can be given; experiment, analogy, and constructive intuition play their part here. But once the correct hypothesis is formulated, the principle of mathematical induction is often sufficient to provide the proof. Inasmuch as such a proof does not give a clue to the act of discovery, it might more fittingly be called a *verification.*

*5. An Important Inequality

In a subsequent chapter we shall find use for the inequality

$$(6) \qquad (1 + p)^n \geq 1 + np,$$

which holds for every number $p > -1$ and positive integer n. (For the sake of generality we are anticipating here the use of negative and non-integral numbers by allowing p to be any number greater than -1. The proof for the general case is exactly the same as in the case where p is a positive integer.) Again we use mathematical induction.

a) *If* it is true that $(1 + p)^r \geq 1 + rp$, then on multiplying both sides of this inequality by the positive number $1 + p$, we obtain

$$(1 + p)^{r+1} \geq 1 + rp + p + rp^2.$$

Dropping the positive term rp^2 only strengthens this inequality, so that

$$(1 + p)^{r+1} \geq 1 + (r + 1)p,$$

which shows that the inequality (6) will also hold for the next integer, $r + 1$. b) It is obviously true that $(1 + p)^1 \geq 1 + p$. This completes the proof that (6) is true for every n. The restriction to numbers $p > -1$ is essential. If $p < -1$, then $1 + p$ is negative and the argument in a) breaks down, since if both members of an inequality are multiplied by a negative quantity, the sense of the inequality is reversed. (For example, if we multiply both sides of the inequality $3 > 2$ by -1 we obtain $-3 > -2$, which is false.)

*6. The Binomial Theorem

Frequently it is important to have an explicit expression for the nth power of a binomial, $(a + b)^n$. We find by explicit calculation that

for $n = 1$, $(a + b)^1 = a + b$,

for $n = 2$, $(a + b)^2 = (a + b)(a + b) = a(a + b) + b(a + b)$
$$= a^2 + 2ab + b^2,$$

for $n = 3$, $(a + b)^3 = (a + b)(a + b)^2 = a(a^2 + 2ab + b^2)$
$$+ b(a^2 + 2ab + b^2) = a^3 + 3a^2b + 3ab^2 + b^3,$$

and so on. What general law of formation lies behind the words "and so on"? Let us examine the process by which $(a + b)^2$ was computed. Since $(a + b)^2 = (a + b)(a + b)$, we obtained the expression for $(a + b)^2$ by multiplying each term in the expression $a + b$ by a, then by b, and adding. The same procedure was used to calculate $(a + b)^3 = (a + b)(a + b)^2$. We may continue in the same way to calculate $(a + b)^4$, $(a + b)^5$, and so on indefinitely. The expression for $(a + b)^n$ will be obtained by multiplying each term of the previously obtained expression for $(a + b)^{n-1}$ by a, then by b, and adding. This leads to the following diagram:

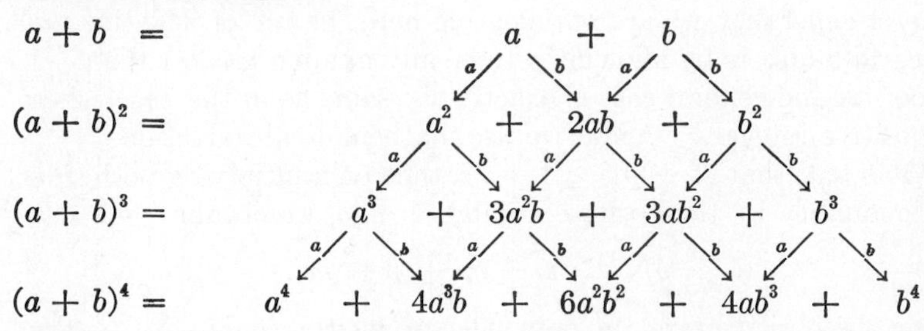

....,

which gives at once the general rule for forming the coefficients in the expansion of $(a + b)^n$. We construct a triangular array of numbers, starting with the coefficients 1, 1 of $a + b$, and such that each number of the triangle is the sum of the two numbers on each side of it in the preceding row. This array is known as *Pascal's Triangle.*

$$
\begin{array}{ccccccccccccccc}
 & & & & & & 1 & & 1 & & & & & & \\
 & & & & & 1 & & 2 & & 1 & & & & & \\
 & & & & 1 & & 3 & & 3 & & 1 & & & & \\
 & & & 1 & & 4 & & 6 & & 4 & & 1 & & & \\
 & & 1 & & 5 & & 10 & & 10 & & 5 & & 1 & & \\
 & 1 & & 6 & & 15 & & 20 & & 15 & & 6 & & 1 & \\
1 & & 7 & & 21 & & 35 & & 35 & & 21 & & 7 & & 1
\end{array}
$$

....

The n*th row of this array gives the coefficients in the expansion of* $(a + b)^n$ *in descending powers of* a *and ascending powers of* b; thus

$$(a + b)^7 = a^7 + 7a^6b + 21a^5b^2 + 35a^4b^3 + 35a^3b^4 + 21a^2b^5 + 7ab^6 + b^7.$$

Using a concise subscript and superscript notation we may denote the numbers in the nth row of Pascal's Triangle by

$$C_0^n = 1, C_1^n, C_2^n, C_3^n, \cdots, C_{n-1}^n, C_n^n = 1.$$

Then the general formula for $(a + b)^n$ may be written

$$(7) \quad (a + b)^n = a^n + C_1^n a^{n-1}b + C_2^n a^{n-2}b^2 + \cdots + C_{n-1}^n ab^{n-1} + b^n.$$

According to the law of formation of Pascal's Triangle, we have

$$(8) \qquad C_i^n = C_{i-1}^{n-1} + C_i^{n-1}.$$

As an exercise, the experienced reader may use this relation, together with the fact that $C_0^1 = C_1^1 = 1$, to show by mathematical induction that

$$(9) \qquad C_i^n = \frac{n(n-1)(n-2)\cdots(n-i+1)}{1\cdot 2\cdot 3\cdots i} = \frac{n!}{i!\,(n-i)!}.$$

(For any positive integer n, the symbol $n!$ (read, "n factorial") denotes the product of the first n integers: $n! = 1\cdot 2\cdot 3 \cdots n$. It is convenient also to define $0! = 1$, so that 9) is valid for $i = 0$ and $i = n$.) This explicit formula for the coefficients in the binomial expansion is sometimes called the *binomial theorem.* (See also p. 475.)

Exercises: Prove by mathematical induction:

1) $\frac{1}{1\cdot 2} + \frac{1}{2\cdot 3} + \cdots + \frac{1}{n(n+1)} = \frac{n}{n+1}$.

2) $\frac{1}{2} + \frac{2}{2^2} + \frac{3}{2^3} + \cdots + \frac{n}{2^n} = 2 - \frac{n+2}{2^n}$.

*3) $1 + 2q + 3q^2 + \cdots + nq^{n-1} = \dfrac{1 - (n+1)q^n + nq^{n+1}}{(1-q)^2}$.

*4) $(1+q)(1+q^2)(1+q^4)\cdots(1+q^{2^n}) = \dfrac{1 - q^{2^{n+1}}}{1-q}$.

Find the sum of the following geometrical progressions:

5) $\dfrac{1}{1+x^2} + \dfrac{1}{(1+x^2)^2} + \cdots + \dfrac{1}{(1+x^2)^n}$.

6) $1 + \dfrac{x}{1+x^2} + \dfrac{x^2}{(1+x^2)^2} + \cdots + \dfrac{x^n}{(1+x^2)^n}$.

7) $\dfrac{x^2-y^2}{x^2+y^2} + \left(\dfrac{x^2-y^2}{x^2+y^2}\right)^2 + \cdots + \left(\dfrac{x^2-y^2}{x^2+y^2}\right)^n$.

Using formulas (4) and (5) prove:

*8) $1^2 + 3^2 + \cdots + (2n+1)^2 = \dfrac{(n+1)(2n+1)(2n+3)}{3}$.

*9) $1^3 + 3^3 + \cdots + (2n+1)^3 = (n+1)^2(2n^2+4n+1)$.

10) Prove the same results directly by mathematical induction.

*7. Further Remarks on Mathematical Induction

The principle of mathematical induction may be generalized slightly to read: "If a sequence of statements A_s, A_{s+1}, A_{s+2}, $\cdots$ is given, where s is some positive integer, and if

a) For every value of $r \geq s$, the truth of A_{r+1} will follow from the truth of A_r, and

b) A_s is known to be true,

then all the statements A_s, A_{s+1}, A_{s+2} $\cdots$ are true; that is to say, A_n is true for all $n \geq s$." Precisely the same reasoning used to establish the truth of the ordinary principle of mathematical induction applies here, with the sequence 1, 2, 3, $\cdots$ replaced by the similar sequence $s, s+1, s+2, s+3 \cdots$. By using the principle in this form we can strengthen somewhat the inequality on page 15 by eliminating the possibility of the "=" sign. We state: *For every* $p \neq 0$ *and* > -1 *and every integer* $n \geq 2$,

$$(1+p)^n > 1 + np. \tag{10}$$

The proof will be left to the reader.

Closely related to the principle of mathematical induction is the "principle of the smallest integer" which states that *every non-empty set* C *of positive integers has a smallest member*. A set is empty if it has no members, e.g., the set of straight circles or the set of integers n such that $n > n$. For obvious reasons we exclude such sets in the statement of the principle. The set C may be finite, like the set 1, 2, 3, 4, 5, or infinite, like the set of all even numbers 2, 4, 6, 8, 10, $\cdots$. Any non-empty set C must contain at least one integer, say n, and the smallest of the integers 1, 2, 3, $\cdots$, n that belongs to C will be the smallest integer in C.

The only way to realize the significance of this principle is to observe that it does *not* apply to every set C of numbers that are not integers; for example, the set of positive fractions $1, \frac{1}{2}, \frac{1}{3}, \frac{1}{4}, \cdots$ does not contain a smallest member.

From the point of view of logic it is interesting to observe that the principle of the smallest integer may be used to *prove* the principle of mathematical induction as a theorem. To this end, let us consider any sequence of statements $A_1, A_2, A_3, \cdots$ such that

a) For any positive integer r the truth of A_{r+1} will follow from that of A_r.

b) A_1 is known to be true.

We shall show the hypothesis that any one of the A's is false to be untenable. For if even one of the A's were false, the set C of *all* positive integers n for which A_n is false would be non-empty. By the principle of the smallest integer, C would contain a smallest integer, p, which must be > 1 because of b). Hence A_p would be false, but A_{p-1} true. This contradicts a).

Once more we emphasize that the principle of mathematical induction is quite distinct from empirical induction in the natural sciences. The confirmation of a general law in any finite number of cases, no matter how large, cannot provide a proof for the law in the rigorous mathematical sense of the word, even if no exception is known at the time. Such a law would remain only a very reasonable *hypothesis*, subject to modification by the results of future experience. In mathematics, a law or a theorem is proved only if it can be shown to be a necessary logical consequence of certain assumptions which are accepted as valid. There are many examples of mathematical statements which have been verified in every particular case considered thus far, but which have not yet been proved to hold in general (for an example see p. 30). One may *suspect* that a theorem is true in all generality by observing its truth in a number of examples; one may then attempt to *prove* it by mathematical induction. If the attempt succeeds the theorem is proved to be true; if the attempt fails, the theorem may be true or false and may some day be proved or disproved by other methods.

In using the principle of mathematical induction one must always be sure that the conditions a) and b) are really satisfied. Neglect of this precaution may lead to absurdities like the following, in which the reader is invited to discover the fallacy. We shall "prove" that *any two positive integers are equal*; for example, that $5 = 10$.

First a definition: If a and b are two unequal positive integers, we define $\max(a, b)$ to be a or b, whichever is greater; if $a = b$ we set $\max(a, b) = a = b$. Thus $\max(3, 5) = \max(5, 3) = 5$, while $\max(4, 4) = 4$. Now let A_n be the statement, "If a and b are any two positive integers such that $\max(a, b) = n$, then $a = b$."

a) Suppose A_r to be true. Let a and b be any two positive integers such that $\max(a, b) = r + 1$. Consider the two integers

$$\alpha = a - 1$$
$$\beta = b - 1;$$

then $\max(\alpha, \beta) = r$. Hence $\alpha = \beta$, for we are assuming A_r to be true. It follows that $a = b$; hence A_{r+1} is true.

b) A_1 is obviously true, for if $\max(a, b) = 1$, then since a and b are by hypothesis positive integers they must both be equal to 1. Therefore, by mathematical induction, A_n is true for every n.

Now if a and b are any two positive integers whatsoever, denote $\max(a, b)$ by r. Since A_n has been shown to be true for every n, in particular A_r is true. Hence $a = b$.

SUPPLEMENT TO CHAPTER I

THE THEORY OF NUMBERS

INTRODUCTION

The integers have gradually lost their association with superstition and mysticism, but their interest for mathematicians has never waned. Euclid (circa 300 B.C.), whose fame rests on the portion of his *Elements* that forms the foundation of geometry studied in high school, seems to have made original contributions to number theory, while his geometry was largely a compilation of previous results. Diophantus of Alexandria (circa 275 A.D.), an early algebraist, left his mark on the theory of numbers. Pierre de Fermat (1601–1665), a jurist of Toulouse, and one of the greatest mathematicians of his time, initiated the modern work in this field. Euler (1707–1783), the most prolific of mathematicians, included much number-theoretical work in his researches. Names prominent in the annals of mathematics—Legendre, Dirichlet, Riemann—can be added to the list. Gauss (1777–1855), the foremost mathematician of modern times, who devoted himself to many different branches of mathematics, is said to have expressed his opinion of number theory in the remark, "Mathematics is the queen of the sciences and the theory of numbers is the queen of mathematics."

§1. THE PRIME NUMBERS

1. Fundamental Facts

Most statements in number theory, as in mathematics as a whole, are concerned not with a single object—the number 5 or the number 32—but with a whole class of objects that have some common property, such as the class of all even integers,

$$2, 4, 6, 8, \cdots,$$

or the class of all integers divisible by 3,

$$3, 6, 9, 12, \cdots,$$

or the class of all squares of integers,

$$1, 4, 9, 16, \cdots,$$

and so on.

Of fundamental importance in number theory is the class of all *primes.* Most integers can be resolved into smaller factors: $10 = 2 \cdot 5$, $111 = 3 \cdot 37$, $144 = 3 \cdot 3 \cdot 2 \cdot 2 \cdot 2 \cdot 2$, etc. Numbers that cannot be so resolved are known as prime numbers or primes. More precisely, *a prime is an integer* p, *greater than one, which has no factors other than itself and one.* (An *integer* a is said to be a *factor* or *divisor* of an *integer* b if there is some *integer* c such that $b = ac$.) The numbers 2, 3, 5, 7, 11, 13, 17, $\cdots$ are primes, while 12, for example, is not, since $12 = 3 \cdot 4$. The importance of the class of primes is due to the fact that *every* integer can be expressed as a *product of primes*: if a number is not itself a prime, it may be successively factored until all the factors are primes; thus $360 = 3 \cdot 120 = 3 \cdot 30 \cdot 4 = 3 \cdot 3 \cdot 10 \cdot 2 \cdot 2 = 3 \cdot 3 \cdot 5 \cdot 2 \cdot 2 \cdot 2 = 2^3 \cdot 3^2 \cdot 5$. An integer (other than 0 or 1) which is not a prime is said to be *composite.*

One of the first questions that arises concerning the class of primes is whether there is only a finite number of different primes or whether the class of primes contains infinitely many members, like the class of all integers, of which it forms a part. The answer is: *There are infinitely many primes.*

The proof of the infinitude of the class of primes as given by Euclid remains a model of mathematical reasoning. It proceeds by the "indirect method". We start with the tentative assumption that the theorem is false. This means that there would be only a finite number of primes, perhaps very many—a billion or so—or, expressed in a general and non-committal way, n. Using the subscript notation we may denote these primes by $p_1, p_2, \cdots, p_n$. Any other number will be composite, and must be divisible by at least one of the primes $p_1, p_2, \cdots, p_n$. We now produce a contradiction by constructing a number A which differs from every one of the primes $p_1, p_2, \cdots, p_n$ because it is larger than any of them, and which nevertheless is not divisible by any of them. This number is

$$A = p_1 p_2 \cdots p_n + 1,$$

i.e. 1 plus the product of what we supposed to be all the primes. A is larger than any of the p's and hence must be composite. But A divided by p_1 or by p_2, etc., always leaves the remainder 1; therefore A has none of the p's as a divisor. Since our initial assumption that there is only a finite number of primes leads to this contradiction, the assumption is seen to be absurd, and hence its contrary must be true. This proves the theorem.

Although this proof is indirect, it can easily be modified to give a method for constructing, at least in theory, an infinite sequence of primes. Starting with any prime number, such as $p_1 = 2$, suppose we have found n primes $p_1, p_2, \cdots, p_n$; we then observe that the number $p_1p_2 \cdots p_n + 1$ either is itself a prime or contains as a factor a prime which differs from those already found. Since this factor can always be found by direct trial, we are sure in any case to find at least one new prime p_{n+1}; proceeding in this way we see that the sequence of constructible primes can never end.

Exercise: Carry out this construction starting with $p_1 = 2$, $p_2 = 3$ and find 5 more primes.

When a number has been expressed as a product of primes, we may arrange these prime factors in any order. A little experience shows that, except for this arbitrariness in the order, the decomposition of a number N into primes is unique: *Every integer* N *greater than* 1 *can be factored into a product of primes in only one way.* This statement seems at first sight to be so obvious that the layman is very much inclined to take it for granted. But it is by no means a triviality, and the proof, though perfectly elementary, requires some subtle reasoning. The classical proof given by Euclid of this "fundamental theorem of arithmetic" is based on a method or "algorithm" for finding the greatest common divisor of two numbers. This will be discussed on page 44. Here we shall give instead a proof of more recent vintage, somewhat shorter and perhaps more sophisticated than Euclid's. It is a typical example of an indirect proof. We shall assume the existence of an integer capable of two essentially different prime decompositions, and from this assumption derive a contradiction. This contradiction will show that the hypothesis that there exists an integer with two essentially different prime decompositions is untenable, and hence that the prime decomposition of every integer is unique.

*If there exists a positive integer capable of decomposition into two essentially different products of primes, there will be a *smallest* such integer (see p. 18),

$$(1) \qquad m = p_1p_2 \cdots p_r = q_1q_2 \cdots q_s ,$$

where the p's and q's are primes. By rearranging the order of the p's and q's if necessary, we may suppose that

$$p_1 \leq p_2 \leq \cdots \leq p_r , \qquad q_1 \leq q_2 \leq \cdots \leq q_s .$$

Now p_1 cannot be equal to q_1, for if it were we could cancel the first factor from each side of equation (1) and obtain two essentially different prime decompositions of an integer smaller than m, contradicting the choice of m as the *smallest* integer for which this is possible. Hence either $p_1 < q_1$ or $q_1 < p_1$. Suppose $p_1 < q_1$. (If $q_1 < p_1$ we simply interchange the letters p and q in what follows.) We form the integer

$$m' = m - (p_1 q_2 q_3 \cdots q_s). \tag{2}$$

By substituting for m the two expressions of equation (1) we may write the integer m' in either of the two forms

$$m' = (p_1 p_2 \cdots p_r) - (p_1 q_2 \cdots q_s) = p_1(p_2 p_3 \cdots p_r - q_2 q_3 \cdots q_s) \tag{3}$$

$$m' = (q_1 q_2 \cdots q_s) - (p_1 q_2 \cdots q_s) = (q_1 - p_1)(q_2 q_3 \cdots q_s) \tag{4}$$

Since $p_1 < q_1$, it follows from (4) that m' is a positive integer, while from (2) it follows that m' is smaller than m. Hence the prime decomposition of m' must be *unique,* aside from the order of the factors. But from (3) it appears that the prime p_1 is a factor of m', hence from (4) p_1 must appear as a factor of either $(q_1 - p_1)$ or $(q_2 q_3 \cdots q_s)$. (This follows from the assumed uniqueness of the prime decomposition of m'; see the reasoning in the next paragraph.) The latter is impossible, since all the q's are larger than p_1. Hence p_1 must be a factor of $q_1 - p_1$, so that for some integer h,

$$q_1 - p_1 = p_1 \cdot h \qquad \text{or} \qquad q_1 = p_1(h + 1).$$

But this shows that p_1 is a factor of q_1, contrary to the fact that q_1 is a prime. This contradiction shows our initial assumption to be untenable and hence completes the proof of the fundamental theorem of arithmetic.

An important corollary of the fundamental theorem is the following: *If a prime* p *is a factor of the product* ab, *then* p *must be a factor of either* a *or* b. For if p were a factor of neither a nor b, then the product of the prime decompositions of a and b would yield a prime decomposition of the integer ab *not containing* p. On the other hand, since p is assumed to be a factor of ab, there exists an integer t such that

$$ab = pt.$$

Hence the product of p by a prime decomposition of t would yield a prime decomposition of the integer ab *containing* p, contrary to the fact that the prime decomposition of ab is unique.

Examples: If one has verified the fact that 13 is a factor of 2652, and the fact that $2652 = 6 \cdot 442$, one may conclude that 13 is a factor of 442. On the other hand, 6 is a factor of 240, and $240 = 15 \cdot 16$, but 6 is not a factor of either 15 or 16. This shows that the assumption that p is *prime* is an essential one.

Exercise: In order to find all the divisors of any number a we need only decompose a into a product

$$a = p_1^{\alpha_1} \cdot p_2^{\alpha_2} \cdots p_r^{\alpha_r},$$

where the p's are distinct primes, each raised to a certain power. *All* the divisors of a are the numbers

$$b = p_1^{\beta_1} \cdot p_2^{\beta_2} \cdots p^{\beta_r},$$

where the β's are any integers satisfying the inequalities

$$0 \leq \beta_1 \leq \alpha_1, \; 0 \leq \beta_2 \leq \alpha_2, \; \cdots, \; 0 \leq \beta_r \leq \alpha_r.$$

Prove this statement. As a consequence, show that the number of different divisors of a (including the divisors a and 1) is given by the product

$$(\alpha_1 + 1)(\alpha_2 + 1) \cdots (\alpha_r + 1).$$

For example,

$$144 = 2^4 \cdot 3^2$$

has $5 \cdot 3$ divisors. They are 1, 2, 4, 8, 16, 3, 6, 12, 24, 48, 9, 18, 36, 72, 144.

2. The Distribution of the Primes

A list of all the primes up to any given integer N may be constructed by writing down in order all the integers less than N, striking out all those which are multiples of 2, then all those remaining which are multiples of 3, and so on until all composite numbers have been eliminated. This process, known as the "sieve of Eratosthenes," will catch in its meshes the primes up to N. Complete tables of primes up to about 10,000,000 have gradually been computed by refinements of this method, and they provide us with a tremendous mass of empirical data concerning the distribution and properties of the primes. On the basis of these tables we can make many highly plausible conjectures (as though number theory were an experimental science) which are often extremely difficult to prove.

a. Formulas Producing Primes

Attempts have been made to find simple arithmetical formulas that yield only primes, even though they may not give all of them. Fermat made the famous conjecture (but not the definite assertion) that all numbers of the form

$$F(n) = 2^{2^n} + 1$$

are primes. Indeed, for $n = 1, 2, 3, 4$ we obtain

$$F(1) = 2^2 + 1 = 5,$$
$$F(2) = 2^{2^2} + 1 = 2^4 + 1 = 17,$$
$$F(3) = 2^{2^3} + 1 = 2^8 + 1 = 257,$$
$$F(4) = 2^{2^4} + 1 = 2^{16} + 1 = 65{,}537,$$

all primes. But in 1732 Euler discovered the factorization $2^{2^5} + 1 = 641 \cdot 6{,}700{,}417$; hence $F(5)$ is not a prime. Later, more of these "Fermat numbers" were found to be composite, deeper number-theoretical methods being required in each case because of the insurmountable difficulty of direct trial. To date it has not even been proved that any of the numbers $F(n)$ is a prime for $n > 4$.

Another remarkable and simple expression which produces many primes is

$$f(n) = n^2 - n + 41.$$

For $n = 1, 2, 3, \cdots, 40$, $f(n)$ is a prime; but for $n = 41$, we have $f(n) = 41^2$, which is no longer a prime.

The expression

$$n^2 - 79n + 1601$$

yields primes for all n up to 79, but fails when $n = 80$. On the whole, it has been a futile task to seek expressions of a simple type which produce only primes. Even less promising is the attempt to find an algebraic formula which shall yield *all* the primes.

b. Primes in Arithmetical Progressions

While it was simple to prove that there are infinitely many primes in the sequence of all integers, 1, 2, 3, 4, $\cdots$, the step to sequences such as 1, 4, 7, 10, 13, $\cdots$ or 3, 7, 11, 15, 19, $\cdots$ or, more generally, to any arithmetical progression, $a, a + d, a + 2d, \cdots a + nd, \cdots$, where a and d have no common factor, was much more difficult. All observations pointed to the fact that *in each such progression there are infinitely many primes*, just as in the simplest one, 1, 2, 3, $\cdots$. It required an enormous effort to prove this general theorem. Lejeune Dirichlet (1805–1859), one of the leading mathematicians of the nineteenth century, obtained full success by applying the most advanced tools of mathematical analysis then known. His original papers on the subject rank even now among the outstanding achievements in mathematics, and after a hundred years the proof has not yet been simplified enough to be within the reach of students who are not well trained in the technique of the calculus and of function theory.

Although we cannot attempt to prove Dirichlet's general theorem, it is easy to generalize Euclid's proof of the infinitude of primes to cover some *special* arithmetical progressions such as $4n + 3$ and $6n + 5$. To treat the first of these, we observe that any prime greater than 2 is odd (since otherwise it would be divisible by 2) and hence is of the form $4n + 1$ or $4n + 3$, for some integer n. Furthermore, the product of two numbers of the form $4n + 1$ is again of that form, since

$$(4a + 1)(4b + 1) = 16ab + 4a + 4b + 1 = 4(4ab + a + b) + 1.$$

Now suppose there were but a finite number of primes, $p_1, p_1, \cdots p_n$, of the form $4n + 3$, and consider the number

$$N = 4(p_1 p_2 \cdots p_n) - 1 = 4(p_1 \cdots p_n - 1) + 3.$$

Either N is itself a prime, or it may be decomposed into a product of primes, none of which can be $p_1, \cdots, p_n$, since these divide N with a remainder -1. Furthermore, all the prime factors of N cannot be of the form $4n + 1$, for N is not of that form and, as we have seen, the product of numbers of the form $4n + 1$ is again of that form. Hence at least one prime factor must be of the form $4n + 3$, which is impossible, since we saw that none of the p's, which we supposed to be *all* the primes of the form $4n + 3$, can be a factor of N. Therefore the assumption that the number of primes of the form $4n + 3$ is finite has led to a contradiction, and hence the number of such primes must be infinite.

Exercise: Prove the corresponding theorem for the progression $6n + 5$.

c. *The Prime Number Theorem*

In the search for a law governing the distribution of the primes, the decisive step was taken when mathematicians gave up futile attempts to find a simple mathematical formula yielding all the primes or giving the exact number of primes contained among the first n integers, and sought instead for information concerning the *average* distribution of the primes among the integers.

For any integer n let us denote by A_n the number of primes among the integers $1, 2, 3, \cdots, n$. If we underline the primes in the sequence consisting of the first few integers: $1\ \underline{2}\ \underline{3}\ 4\ \underline{5}\ 6\ \underline{7}\ 8\ 9\ 10\ \underline{11}\ 12\ \underline{13}\ 14\ 15\ 16\ \underline{17}\ 18\ \underline{19} \cdots$ we can compute the first few values of A_n:

$A_1 = 0, A_2 = 1, A_3 = A_4 = 2, A_5 = A_6 = 3, A_7 = A_8 = A_9 = A_{10} = 4, A_{11} = A_{12} = 5, A_{13} = A_{14} = A_{15} = A_{16} = 6, A_{17} = A_{18} = 7, A_{19} = 8$, etc.

If we now take any sequence of values for n which increases without limit, say

$$n = 10, 10^2, 10^3, 10^4, \cdots,$$

then the corresponding values of A_n,

$$A_{10}, A_{10^2}, A_{10^3}, A_{10^4}, \cdots,$$

will also increase without limit (although more slowly). For we know that there are infinitely many primes, so the values of A_n will sooner or later exceed any finite number. The "density" of the primes among the first n integers is given by the ratio A_n/n, and from a table of primes the values of A_n/n may be computed empirically for fairly large values of n.

n	A_n/n
10^3	0.168
10^6	0.078498
10^9	0.050847478
$\cdots$	$\cdots\cdots\cdots\cdots$

The last entry in this table may be regarded as giving the probability that an integer picked at random from among the first 10^9 integers will be a prime, since there are 10^9 possible choices, of which A_{10^9} are primes.

The distribution of the individual primes among the integers is extremely irregular. But this irregularity "in the small" disappears if we fix our attention on the average distribution of the primes as given by the ratio A_n/n. The simple law that governs the behavior of this ratio is one of the most remarkable discoveries in the whole of mathematics. In order to state the *prime number theorem* we must define the "natural logarithm" of an integer n. To do this we take two perpendicular axes in a plane, and consider the locus of all points in the plane the product of whose distances x and y from these axes is equal to one. In terms of the coördinates x and y this locus, an equilateral hyperbola, is defined by the equation $xy = 1$. We now define log n to be the *area* in Figure 5 bounded by the hyperbola, the x-axis, and the two vertical lines $x = 1$ and $x = n$. (A more detailed discussion of the logarithm will be found in Chapter VIII.) From an empirical study of prime number tables Gauss observed that the ratio A_n/n is approximately equal to $1/\log n$, and that this approximation appears to improve as n increases. The goodness of the approximation is given by the ratio $\frac{A_n/n}{1/\log n}$, whose values for n = 1000, 1,000,000, 1,000,000,000 are shown in the following table.

n	A_n/n	$1/\log n$	$\frac{A_n/n}{1/\log n}$
10^3	0.168	0.145	1.159
10^6	0.078498	0.072382	1.084
10^9	0.050847478	0.048254942	1.053
...	....	....	.....

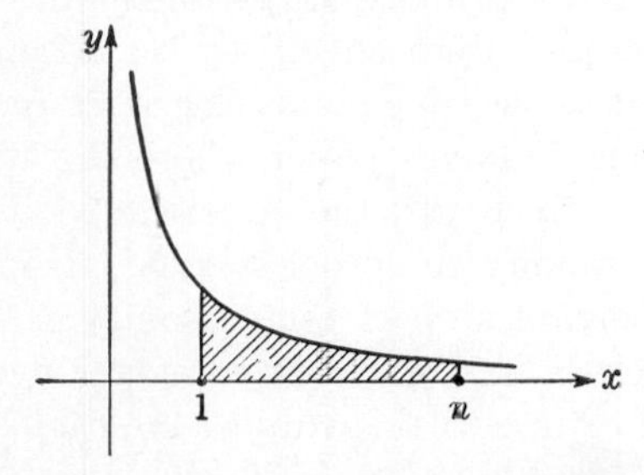

Fig. 5. The area of the shaded region under the hyperbola defines log n.

On the basis of such empirical evidence Gauss made the conjecture that the ratio A_n/n is "asymptotically equal" to $1/\log n$. By this is meant that if we take a sequence of larger and larger values of n, say n equal to

$$10, 10^2, 10^3, 10^4, \cdots$$

as before, then the ratio of A_n/n to $1/\log n$,

$$\frac{A_n/n}{1/\log n},$$

calculated for these successive values of n, will become more and more nearly equal to 1, and that the difference of this ratio from 1 can be made as small as we please by confining ourselves to sufficiently large values of n. This assertion is symbolically expressed by the sign $\sim$:

$$\frac{A_n}{n} \sim \frac{1}{\log n} \quad \text{means} \quad \frac{A_n/n}{1/\log n} \text{ tends to 1 as } n \text{ increases.}$$

That $\sim$ cannot be replaced by the ordinary sign $=$ of equality is clear from the fact that while A_n is always an integer, $n/\log n$ is not.

That the average behavior of the prime number distribution can be described by the logarithmic function is a very remarkable discovery, for it is surprising that two mathematical concepts which seem so unrelated should be in fact so intimately connected.

Although the statement of Gauss's conjecture is simple to understand, a rigorous mathematical proof was far beyond the powers of mathematical science in Gauss's time. To prove this theorem, concerned only with the most elementary concepts, it is necessary to employ the most powerful methods of modern mathematics. It took almost a hundred years before analysis was developed to the point where Hadamard (1896) in Paris and de la Vallée Poussin (1896) in Louvain could give a complete proof of the prime number theorem. Simplifications and important modifications were given by v. Mangoldt and Landau. Long before Hadamard, decisive pioneering work had been done by Riemann (1826–1866) in a famous paper where the strategic lines for the attack were drawn. Recently, the American mathematician Norbert Wiener was able to modify the proof so as to avoid the use of complex numbers at an important step of the reasoning. But the proof of the prime number theorem is still no easy matter even for an advanced student. We shall return to this subject on page 482 et seq.

d. Two Unsolved Problems Concerning Prime Numbers

While the problem of the average distribution of primes has been satisfactorily solved, there are many other conjectures which are supported by all the empirical evidence but which have not yet been proved to be true.

One of these is the famous *Goldbach conjecture.* Goldbach (1690–1764) has no significance in the history of mathematics except for this problem, which he proposed in 1742 in a letter to Euler. He observed that for every case he tried, any even number (except 2, which is itself a prime) could be represented as the sum of two primes. For example: $4 = 2 + 2$, $6 = 3 + 3$, $8 = 5 + 3$, $10 = 5 + 5$, $12 = 5 + 7$, $14 = 7 + 7$, $16 = 13 + 3$, $18 = 11 + 7$, $20 = 13 + 7$, $\cdots$, $48 = 29 + 19$, $\cdots$, $100 = 97 + 3$, etc.

Goldbach asked if Euler could prove this to be true for *all* even numbers, or if he could find an example disproving it. Euler never provided an answer, nor has one been given since. The empirical evidence in favor of the statement that every even number can be so represented is thoroughly convincing, as anyone can verify by trying a number of examples. The source of the difficulty is that primes are defined in terms of *multiplication*, while the problem involves *addition*. Generally speaking, it is difficult to establish connections between the multiplicative and the additive properties of integers.

Until recently, a proof of Goldbach's conjecture seemed completely inaccessible. Today a solution no longer seems out of reach. An important success, very unexpected and startling to all experts, was achieved in 1931 by a then unknown young Russian mathematician, Schnirelmann (1905–1938), who proved that *every positive integer can be represented as the sum of not more than 300,000 primes.* Though this result seems ludicrous in comparison with the original goal of proving Goldbach's conjecture, nevertheless it was a first step in that direction. The proof is a direct, constructive one, although it does not provide any practical method for finding the prime decomposition of an arbitrary integer. More recently, the Russian mathematician Vinogradoff, using methods due to Hardy, Littlewood and their great Indian collaborator Ramanujan, has succeeded in reducing the number from 300,000 to 4. This is much nearer to a solution of Goldbach's problem. But there is a striking difference between Schnirelmann's result and Vinogradoff's; more significant, perhaps, than the difference between 300,000 and 4. Vinogradoff's theorem was proved only for all "sufficiently large" integers; more precisely, Vinogradoff proved that there *exists* an integer N such that any integer $n > N$ can be represented as the sum of at most 4 primes. Vinogradoff's proof does not permit us to appraise N; in contrast to Schnirelmann's theorem it is essentially indirect and non-constructive. What Vinogradoff really proved is that the assumption that infinitely many integers cannot be decomposed into at most 4 prime summands leads to an absurdity. Here we have a good example of the profound difference between the two types of proof, direct and indirect. (See the general discussion on p. 86.)

The following even more striking problem than Goldbach's has come nowhere near a solution. It has been observed that primes frequently occur in pairs of the form p and $p + 2$. Such are 3 and 5, 11 and 13, 29 and 31, etc. The statement that there are infinitely many such pairs is believed to be correct, but as yet not the slightest definite step has been taken towards a proof.

§2. CONGRUENCES

1. General Concepts

Whenever the question of the divisibility of integers by a fixed integer d occurs, the concept and the notation of "congruence" (due to Gauss) serves to clarify and simplify the reasoning.

To introduce this concept let us examine the remainders left when integers are divided by the number 5. We have

$$\begin{array}{lll} 0 = 0\cdot 5 + 0 & 7 = 1\cdot 5 + 2 & -1 = -1\cdot 5 + 4 \\ 1 = 0\cdot 5 + 1 & 8 = 1\cdot 5 + 3 & -2 = -1\cdot 5 + 3 \\ 2 = 0\cdot 5 + 2 & 9 = 1\cdot 5 + 4 & -3 = -1\cdot 5 + 2 \\ 3 = 0\cdot 5 + 3 & 10 = 2\cdot 5 + 0 & -4 = -1\cdot 5 + 1 \\ 4 = 0\cdot 5 + 4 & 11 = 2\cdot 5 + 1 & -5 = -1\cdot 5 + 0 \\ 5 = 1\cdot 5 + 0 & 12 = 2\cdot 5 + 2 & -6 = -2\cdot 5 + 4 \\ 6 = 1\cdot 5 + 1 & \text{etc.} & \text{etc.} \end{array}$$

We observe that the remainder left when any integer is divided by 5 is one of the five integers 0, 1, 2, 3, 4. We say that two integers a and b are "congruent modulo 5" if they leave the *same remainder* on division by 5. Thus 2, 7, 12, 17, 22, $\cdots$, -3, -8, -13, -18, $\cdots$ are all congruent modulo 5, since they leave the remainder 2. In general, we say that two integers a and b are *congruent modulo* d, where d is a fixed integer, if a and b leave the same remainder on division by d, i.e., if there is an integer n such that $a - b = nd$. For example, 27 and 15 are congruent modulo 4, since

$$27 = 6\cdot 4 + 3, \qquad 15 = 3\cdot 4 + 3.$$

The concept of congruence is so useful that it is desirable to have a brief notation for it. We write

$$a \equiv b \pmod{d}$$

to express the fact that a and b are congruent modulo d. If there is no doubt concerning the modulus, the "mod d" of the formula may be omitted. (If a is not congruent to b modulo d, we shall write $a \not\equiv b \pmod{d}$.)

Congruences occur frequently in daily life. For example, the hands on a clock indicate the hour modulo 12, and the mileage indicator on a car gives the total miles traveled modulo 100,000.

Before proceeding with the detailed discussion of congruences the reader should observe that the following statements are all equivalent:

1. a is congruent to b modulo d.
2. $a = b + nd$ for some integer n.
3. d divides $a - b$.

The usefulness of Gauss's congruence notation lies in the fact that congruence with respect to a fixed modulus has many of the formal properties of ordinary equality. The most important formal properties of the relation $a = b$ are the following:

1) Always $a = a$.
2) If $a = b$, then $b = a$.
3) If $a = b$ and $b = c$, then $a = c$.

Moreover, if $a = a'$ and $b = b'$, then

4) $a + b = a' + b'$.
5) $a - b = a' - b'$.
6) $ab = a'b'$.

These properties remain true when the relation $a = b$ is replaced by the congruence relation $a \equiv b \pmod{d}$. Thus

1′) Always $a \equiv a \pmod{d}$.
2′) If $a \equiv b \pmod{d}$ then $b \equiv a \pmod{d}$.
3′) If $a \equiv b \pmod{d}$ and $b \equiv c \pmod{d}$, then $a \equiv c \pmod{d}$.

The trivial verification of these facts is left to the reader.

Moreover, if $a \equiv a' \pmod{d}$ and $b \equiv b' \pmod{d}$, then

4′) $a + b \equiv a' + b' \pmod{d}$.
5′) $a - b \equiv a' - b' \pmod{d}$.
6′) $ab \equiv a'b' \pmod{d}$.

Thus *congruences with respect to the same modulus may be added, subtracted, and multiplied.* To prove these three statements we need only observe that if

$$a = a' + rd, \qquad b = b' + sd,$$

then

$$a + b = a' + b' + (r + s)d,$$
$$a - b = a' - b' + (r - s)d,$$
$$ab = a'b' + (a's + b'r + rsd)d,$$

from which the desired conclusions follow.

The concept of congruence has an illuminating geometrical interpretation. Usually, if we wish to represent the integers geometrically, we choose a segment of unit length and extend it by multiples of its own length in both directions. In this way we can find a point on the line corresponding to each integer, as in Figure 6. But when we are dealing with the integers modulo d, any two congruent numbers are considered the same as far as their behavior on division by a is concerned, since they leave the same remainder. In order to show this geometrically, we use a circle divided into d equal parts. Any integer when divided by d leaves as remainder one of the d numbers $0, 1, \cdots, d-1$, which are placed at equal intervals on the circumference of the circle. Every integer is congruent modulo d to one of these numbers, and hence is represented geometrically by one of these points; two numbers are congruent if they are represented by the same point. Figure 7 is drawn for the case $d = 6$. The face of a clock is another illustration from daily life.

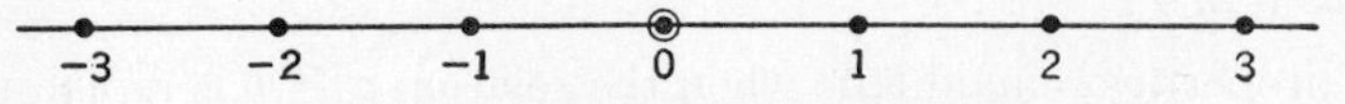

Fig. 6. Geometrical representation of the integers.

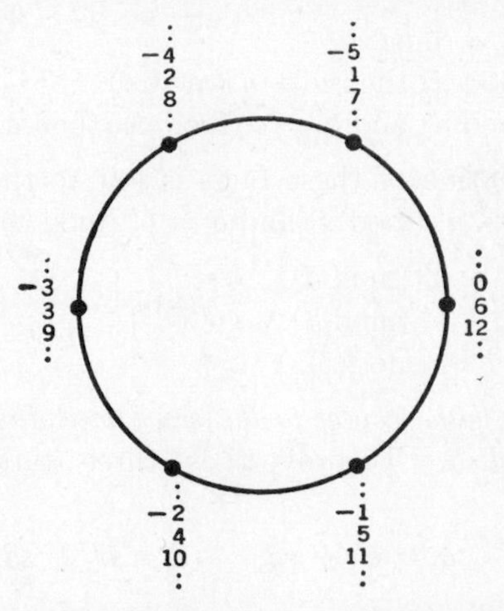

Fig. 7. Geometrical representation of the integers modulo 6.

As an example of the use of the multiplicative property 6′) of congruences we may determine the remainders left when successive powers of 10 are divided by a given number. For example,

$$10 \equiv -1 \pmod{11},$$

since $10 = -1 + 11$. Successively multiplying this congruence by itself, we obtain

$$10^2 \equiv (-1)(-1) = 1 \pmod{11},$$
$$10^3 \equiv -1 \quad \text{“} \quad ,$$
$$10^4 \equiv 1 \quad \text{“} \quad , \text{etc.}$$

From this we can show that any integer

$$z = a_0 + a_1 \cdot 10 + a_2 \cdot 10^2 + \cdots + a_n \cdot 10^n,$$

expressed in the decimal system, leaves the same remainder on division by 11 as does the sum of its digits, taken with alternating signs,

$$t = a_0 - a_1 + a_2 - a_3 + \cdots.$$

For we may write

$$z - t = a_1 \cdot 11 + a_2(10^2 - 1) + a_3(10^3 + 1) + a_4(10^4 - 1) + \cdots.$$

Since all the numbers $11, 10^2 - 1, 10^3 + 1, \cdots$ are congruent to 0 modulo 11, $z - t$ is also, and therefore z leaves the same remainder on division by 11 as does t. It follows in particular that a number is divisible by 11 (i.e. leaves the remainder 0) if and only if the alternating sum of its digits is divisible by 11. For example, since $3 - 1 + 6 - 2 + 8 - 1 + 9 = 22$, the number $z = 3162819$ is divisible by 11. To find a rule for divisibility by 3 or 9 is even simpler, since $10 \equiv 1$ (mod 3 or 9), and therefore $10^n \equiv 1$ (mod 3 or 9) for any n. It follows that a number z is divisible by 3 or 9 if and only if the sum of its digits

$$s = a_0 + a_1 + a_2 + \cdots + a_n$$

is likewise divisible by 3 or 9, respectively.

For congruences modulo 7 we have

$$10 \equiv 3, \quad 10^2 \equiv 2, \quad 10^3 \equiv -1, \quad 10^4 \equiv -3, \quad 10^5 \equiv -2, \quad 10^6 \equiv 1.$$

The successive remainders then repeat. Thus z is divisible by 7 if and only if the expression

$$r = a_0 + 3a_1 + 2a_2 - a_3 - 3a_4 - 2a_5 + a_6 + 3a_7 + \cdots$$

is divisible by 7.

Exercise: Find a similar rule for divisibility by 13.

In adding or multiplying congruences with respect to a fixed modulus, say $d = 5$, we may keep the numbers involved from getting too large by always replacing any number a by the number from the set

$$0, \ 1, \ 2, \ 3, \ 4$$

to which it is congruent. Thus, in order to calculate sums and products of integers modulo 5, we need only use the following addition and multiplication tables.

$a + b$

$b \equiv$	0	1	2	3	4
$a \equiv 0$	0	1	2	3	4
1	1	2	3	4	0
2	2	3	4	0	1
3	3	4	0	1	2
4	4	0	1	2	3

$a \cdot b$

$b \equiv$	0	1	2	3	4
$a \equiv 0$	0	0	0	0	0
1	0	1	2	3	4
2	0	2	4	1	3
3	0	3	1	4	2
4	0	4	3	2	1

From the second of these tables it appears that a product ab is congruent to 0 (mod 5) only if a or b is $\equiv 0$ (mod 5). This suggests the general law

7) $ab \equiv 0 \pmod{d}$ only if either $a \equiv 0$ or $b \equiv 0 \pmod{d}$,

which is an extension of the ordinary law for integers which states that $ab = 0$ only if $a = 0$ or $b = 0$. *The law* 7) *holds only when the modulus* d *is a prime.* For the congruence

$$ab \equiv 0 \pmod{d}$$

means that d divides ab, and we have seen that a prime d divides a product ab only if it divides a or b; that is, only if

$$a \equiv 0 \pmod{d} \quad \text{or} \quad b \equiv 0 \pmod{d}.$$

If d is not a prime the law need not hold; for we can write $d = r \cdot s$, where r and s are less than d, so that

$$r \not\equiv 0 \pmod{d}, \qquad s \not\equiv 0 \pmod{d},$$

but

$$rs = d \equiv 0 \pmod{d}.$$

For example, $2 \not\equiv 0 \pmod{6}$ and $3 \not\equiv 0 \pmod{6}$, but $2 \cdot 3 = 6 \equiv 0 \pmod{6}$.

Exercise: Show that the following *law of cancellation* holds for congruences with respect to a prime modulus:

If $ab \equiv ac$ and $a \not\equiv 0$, then $b \equiv c$.

Exercises: 1) To what number between 0 and 6 inclusive is the product $11 \cdot 18 \cdot 2322 \cdot 13 \cdot 19$ congruent modulo 7?

2) To what number between 0 and 12 inclusive is $3 \cdot 7 \cdot 11 \cdot 17 \cdot 19 \cdot 23 \cdot 29 \cdot 113$ congruent modulo 13?

3) To what number between 0 and 4 inclusive is the sum $1 + 2 + 2^2 + \cdots + 2^{19}$ congruent modulo 5?

2. Fermat's Theorem

In the seventeenth century, Fermat, the founder of modern number theory, discovered a most important theorem: *If* p *is any prime which does not divide the integer* a, *then*

$$a^{p-1} \equiv 1 \qquad (\text{mod } p).$$

This means that the $(p - 1)$st power of a leaves the remainder 1 upon division by p.

Some of our previous calculations confirm this theorem; for example, we found that $10^6 \equiv 1 \pmod 7$, $10^2 \equiv 1 \pmod 3$, and $10^{10} \equiv 1 \pmod{11}$. Likewise we may show that $2^{12} \equiv 1 \pmod{13}$ and $5^{10} \equiv 1 \pmod{11}$. To check the latter congruences we need not actually calculate such high powers, since we may take advantage of the multiplicative property of congruences:

$$\begin{aligned} 2^4 &= 16 \equiv 3 \quad (\text{mod } 13), & 5^2 &\equiv 3 \quad (\text{mod } 11),\\ 2^8 &\equiv 9 \equiv -4 \quad \text{"} \quad , & 5^4 &\equiv 9 \equiv -2 \quad \text{"} \quad ,\\ 2^{12} &\equiv -4 \cdot 3 = -12 \equiv 1 \quad \text{"} \quad . & 5^8 &\equiv 4 \quad \text{"} \quad ,\\ & & 5^{10} &\equiv 3 \cdot 4 = 12 \equiv 1 \quad \text{"} \quad . \end{aligned}$$

To prove Fermat's theorem, we consider the multiples of a

$$m_1 = a, \qquad m_2 = 2a, \qquad m_3 = 3a, \; \cdots, \; m_{p-1} = (p - 1)a.$$

No two of these integers can be congruent modulo p, for then p would be a factor of $m_r - m_s = (r - s)a$ for some pair of integers r, s with $1 \leq r < s \leq (p - 1)$. But the law 7) shows that this cannot occur; for since $s - r$ is less than p, p is not a factor of $s - r$, while by assumption p is not a factor of a. Likewise, none of these numbers can be congruent to 0. Therefore the numbers $m_1, m_2, \cdots, m_{p-1}$ must be respectively congruent to the numbers $1, 2, 3, \cdots, p - 1$, in some arrangement. It follows that

$$m_1 m_2 \cdots m_{p-1} = 1 \cdot 2 \cdot 3 \cdots (p - 1)a^{p-1} \equiv 1 \cdot 2 \cdot 3 \cdots (p - 1) \pmod p,$$

or, if for brevity we write K for $1 \cdot 2 \cdot 3 \cdots (p - 1)$,

$$K(a^{p-1} - 1) \equiv 0 \qquad (\text{mod } p).$$

But K is not divisible by p, since none of its factors is; hence by the law 7), $(a^{p-1} - 1)$ must be divisible by p, i.e.

$$a^{p-1} - 1 \equiv 0 \qquad (\text{mod } p).$$

This is Fermat's theorem.

To check the theorem once more, let us take $p = 23$ and $a = 5$. We then have, all modulo 23, $5^2 \equiv 2$, $5^4 \equiv 4$, $5^8 \equiv 16 \equiv -7$, $5^{16} \equiv 49 \equiv 3$, $5^{20} \equiv 12$, $5^{22} \equiv 24 \equiv 1$. With $a = 4$ instead of 5, we get, again modulo 23, $4^2 \equiv -7$, $4^3 \equiv -28 \equiv -5$, $4^4 \equiv -20 \equiv 3$, $4^8 \equiv 9$, $4^{11} \equiv -45 \equiv 1$, $4^{22} \equiv 1$.

In the example above with $a = 4$, $p = 23$, and in others, we observe that not only the $(p - 1)$st power of a, but also a smaller power may be congruent to 1. It is always true that the smallest such power, in this case 11, is a divisor of $p - 1$. (See the following Exercise 3.)

Exercises: 1) Show by similar computation that $2^8 \equiv 1$ (mod 17); $3^8 \equiv -1$ (mod 17); $3^{14} \equiv -1$ (mod 29); $2^{14} \equiv -1$ (mod 29); $4^{14} \equiv 1$ (mod 29); $5^{14} \equiv 1$ (mod 29).

2) Check Fermat's theorem for $p = 5, 7, 11, 17$, and 23 with different values of a.

3) Prove the general theorem: The smallest positive integer e for which $a^e \equiv 1$ (mod p) must be a divisor of $p - 1$. (Hint: Divide $p - 1$ by e, obtaining

$$p - 1 = ke + r,$$

where $0 \leq r < e$, and use the fact that $a^{p-1} \equiv a^e \equiv 1$ (mod p).)

3. Quadratic Residues

Referring to the examples for Fermat's theorem, we find that not only is $a^{p-1} \equiv 1$ (mod p) always, but (if p is a prime different from 2, therefore odd and of the form $p = 2p' + 1$) that for some values of a, $a^{p'} = a^{(p-1)/2} \equiv 1$ (mod p). This fact suggests a chain of interesting investigations. We may write the theorem in the following form:

$$a^{p-1} - 1 = a^{2p'} - 1 = (a^{p'} - 1)(a^{p'} + 1) \equiv 0 \qquad (\text{mod } p).$$

Since a product is divisible by p only if one of the factors is, it appears immediately that either $a^{p'} - 1$ or $a^{p'} + 1$ must be divisible by p, so that for any prime $p > 2$ and any number a not divisible by p, either

$$a^{(p-1)/2} \equiv 1 \quad \text{or} \quad a^{(p-1)/2} \equiv -1 \qquad (\text{mod } p).$$

From the beginning of modern number theory mathematicians have been interested in finding out for what numbers a we have the first case and for what numbers the second. Suppose a is congruent modulo p to the square of some number x,

$$a \equiv x^2 \qquad (\text{mod } p).$$

Then $a^{(p-1)/2} \equiv x^{p-1}$, which according to Fermat's theorem is congruent to 1 modulo p. A number a, not a multiple of p, which is congruent modulo p to the square of some number is called a *quadratic residue of* p, while a number b, not a multiple of p, which is not congruent to any square is called a *quadratic non-residue of* p. We have just seen that every quadratic residue a of p satisfies the congruence $a^{(p-1)/2} \equiv 1 \pmod{p}$. Without serious difficulty it can be proved that for every non-residue b we have the congruence $b^{(p-1)/2} \equiv -1 \pmod{p}$. Moreover, we shall presently show that among the numbers $1, 2, 3, \cdots, p-1$ there are exactly $(p-1)/2$ quadratic residues and $(p-1)/2$ non-residues.

Although much empirical data could be gathered by direct computation, it was not easy at first to discover general laws governing the distribution of quadratic residues and non-residues. The first deep-lying property of these residues was observed by Legendre (1752–1833), and later called by Gauss the *Law of Quadratic Reciprocity*. This law concerns the behavior of two different primes p and q, and states that q is a quadratic residue of p if and only if p is a quadratic residue of q, provided that the product $\left(\frac{p-1}{2}\right)\cdot\left(\frac{q-1}{2}\right)$ is *even*. In case this product is *odd*, the situation is reversed, so that p is a residue of q if and only if q is a *non-residue* of p. One of the achievements of the young Gauss was to give the first rigorous proof of this remarkable theorem, which had long been a challenge to mathematicians. Gauss's first proof was by no means simple, and the reciprocity law is not too easy to establish even today, although a great many different proofs have been published. Its true significance has come to light only recently in connection with modern developments in algebraic number theory.

As an example illustrating the distribution of quadratic residues, let us choose $p = 7$. Then, since

$$0^2 \equiv 0, \quad 1^2 \equiv 1, \quad 2^2 \equiv 4, \quad 3^2 \equiv 2, \quad 4^2 \equiv 2, \quad 5^2 \equiv 4, \quad 6^2 \equiv 1,$$

all modulo 7, and since the remaining squares repeat this sequence, the quadratic residues of 7 are the numbers congruent to 1, 2, or 4, while the non-residues are congruent to 3, 5, or 6. In the general case, the quadratic residues of p consist of the numbers congruent to 1^2, $2^2, \cdots, (p-1)^2$. But these are congruent in pairs, for

$$x^2 \equiv (p-x)^2 \pmod{p} \qquad (\text{e.g., } 2^2 \equiv 5^2 \pmod{7}),$$

since $(p - x)^2 = p^2 - 2px + x^2 \equiv x^2 \pmod{p}$. Hence half the numbers $1, 2, \cdots, p - 1$ are quadratic residues of p and half are quadratic non-residues.

To illustrate the quadratic reciprocity law, let us choose $p = 5$, $q = 11$. Since $11 \equiv 1^2 \pmod{5}$, 11 is a quadratic residue (mod 5); since the product $[(5 - 1)/2][(11 - 1)/2]$ is even, the reciprocity law tells us that 5 is a quadratic residue (mod 11). In confirmation of this, we observe that $5 \equiv 4^2 \pmod{11}$. On the other hand, if $p = 7$, $q = 11$, the product $[(7 - 1)/2][(11 - 1)/2]$ is odd, and indeed 11 is a residue (mod 7) (since $11 \equiv 2^2 \pmod{7}$), while 7 is a non-residue (mod 11).

Exercises: 1. $6^2 = 36 \equiv 13 \pmod{23}$. Is 23 a quadratic residue (mod 13)?

2. We have seen that $x^2 \equiv (p - x)^2 \pmod{p}$. Show that these are the *only* congruences among the numbers $1^2, 2^2, 3^2, \cdots, (p - 1)^2$.

§3. PYTHAGOREAN NUMBERS AND FERMAT'S LAST THEOREM

An interesting question in number theory is connected with the Pythagorean theorem. The Greeks knew that a triangle with sides 3, 4, 5 is a right triangle. This suggests the general question: What other right triangles have sides whose lengths are integral multiples of a unit length? The Pythagorean theorem is expressed algebraically by the equation

$$a^2 + b^2 = c^2, \tag{1}$$

where a and b are the lengths of the legs of a right triangle and c is the length of the hypotenuse. The problem of finding *all* right triangles with sides of integral length is thus equivalent to the problem of finding all integer solutions (a, b, c) of equation (1). Any such triple of numbers is called a *Pythagorean number triple.*

The problem of finding all Pythagorean number triples can be solved very simply. If a, b and c form a Pythagorean number triple, so that $a^2 + b^2 = c^2$, then we put, for abbreviation, $a/c = x$, $b/c = y$. x and y are rational numbers for which $x^2 + y^2 = 1$. We then have $y^2 = (1 - x)(1 + x)$, or $y/(1 + x) = (1 - x)/y$. The common value of the two sides of this equation is a number t which is expressible as the quotient of two integers, u/v. We can now write $y = t(1 + x)$ and $(1 - x) = ty$, or

$$tx - y = -t, \qquad x + ty = 1.$$

From these simultaneous equations we find immediately that

$$x = \frac{1 - t^2}{1 + t^2}, \qquad y = \frac{2t}{1 + t^2}.$$

Substituting for x, y and t, we have

$$\frac{a}{c} = \frac{v^2 - u^2}{u^2 + v^2}, \qquad \frac{b}{c} = \frac{2uv}{u^2 + v^2}.$$

Therefore

$$\begin{aligned} a &= (v^2 - u^2)r, \\ b &= (2uv)r, \\ c &= (u^2 + v^2)r, \end{aligned} \tag{2}$$

for some rational factor of proportionality r. This shows that if (a, b, c) is a Pythagorean number triple, then a, b, c are proportional to $v^2 - u^2$, $2uv$, $u^2 + v^2$, respectively. Conversely, it is easy to see that any triple (a, b, c) defined by (2) is a Pythagorean triple, for from (2) we obtain

$$\begin{aligned} a^2 &= (u^4 - 2u^2v^2 + v^4)r^2, \\ b^2 &= (4u^2v^2)r^2, \\ c^2 &= (u^4 + 2u^2v^2 + v^4)r^2, \end{aligned}$$

so that $a^2 + b^2 = c^2$.

This result may be simplified somewhat. From any Pythagorean number triple (a, b, c) we may derive infinitely many other Pythagorean triples (sa, sb, sc) for any positive integer s. Thus, from (3, 4, 5) we obtain (6, 8, 10), (9, 12, 15), etc. Such triples are not essentially distinct, since they correspond to similar right triangles. We shall therefore define a *primitive* Pythagorean number triple to be one where a, b, and c have no common factor. It can then be shown that *the formulas*

$$\begin{aligned} a &= v^2 - u^2, \\ b &= 2uv, \\ c &= u^2 + v^2, \end{aligned}$$

for any positive integers u *and* v *with* v > u, *where* u *and* v *have no common factor and are not both odd, yield all primitive Pythagorean number triples.*

**Exercise:* Prove the last statement.

As examples of primitive Pythagorean number triples we have $u = 2$, $v = 1$: (3, 4, 5), $u = 3$, $v = 2$: (5, 12, 13), $u = 4$, $v = 3$: (7, 24, 25), $\cdots$, $u = 10$, $v = 7$: (51, 140, 149), etc.

This result concerning Pythagorean numbers naturally raises the question as to whether integers a, b, c can be found for which $a^3 + b^3 = c^3$ or $a^4 + b^4 = c^4$, or, in general, whether, for a given positive integral exponent $n > 2$, the equation

$$a^n + b^n = c^n \tag{3}$$

can be solved with positive integers a, b, c. An answer was provided by Fermat in a spectacular way. Fermat had studied the work of Diophantus, the ancient contributor to number theory, and was accustomed to making comments in the margin of his copy. Although he stated many theorems there without bothering to give proofs, all of them have subsequently been proved, with but one significant exception. While commenting on Pythagorean numbers, Fermat stated that *the equation* (3) *is not solvable in integers for any* n > 2, but that the elegant proof which he had found was unfortunately too long for the margin in which he was writing.

Fermat's general statement has never been proved true or false, despite the efforts of some of the greatest mathematicians since his time. The theorem has indeed been proved for many values of n, in particular, for all $n < 619$, but not for all n, although no counter-example has ever been produced. Although the theorem itself is not so important mathematically, attempts to prove it have given rise to many important investigations in number theory. The problem has also aroused much interest in non-mathematical circles, due in part to a prize of 100,000 marks offered to the person who should first give a solution and held in trust at the Royal Academy at Göttingen. Until the post-war German inflation wiped out the monetary value of this prize, a great number of incorrect "solutions" was presented each year to the trustees. Even serious mathematicians sometimes deceived themselves into handing in or publishing proofs which collapsed after some superficial mistake was discovered. General interest in the question seems to have abated since the devaluation of the mark, though from time to time there is an announcement in the press that the problem has been solved by some hitherto unknown genius.

§4. THE EUCLIDEAN ALGORITHM

1. General Theory

The reader is familiar with the ordinary process of long division of one integer a by another integer b and knows that the process can be carried out until the remainder is smaller than the divisor. Thus if $a = 648$ and $b = 7$ we have a quotient $q = 92$ and a remainder $r = 4$.

$$\begin{array}{r|l} & \overline{92} \\ 7 & 648 \\ & \underline{63} \\ & 18 \\ & \underline{14} \\ & 4 \end{array} \qquad 648 = 7 \cdot 92 + 4.$$

We may state this as a general theorem: *If* a *is any integer and* b *is any integer greater than* 0, *then we can always find an integer* q *such that*

$$a = b \cdot q + r, \tag{1}$$

where r *is an integer satisfying the inequality* $0 \leq r < b$.

To prove this statement without making use of the process of long division we need only observe that any integer a is either itself a multiple of b,

$$a = bq,$$

or lies between two successive multiples of b,

$$bq < a < b(q + 1) = bq + b.$$

In the first case the equation (1) holds with $r = 0$. In the second case we have, from the first of the inequalities above,

$$a - bq = r > 0,$$

while from the second inequality we have

$$a - bq = r < b,$$

so that $0 < r < b$ as required by (1).

From this simple fact we shall deduce a variety of important consequences. The first of these is a method for finding the greatest common divisor of two integers.

Let a and b be any two integers, not both equal to 0, and consider the set of all positive integers which divide both a and b. This set is certainly finite, since if a, for example, is $\neq 0$, then no integer greater in magnitude than a can be a divisor of a, to say nothing of b. Hence there can be but a finite number of common divisors of a and b, and of these let d be the greatest. The integer d is called the *greatest common divisor* of a and b, and written $d = (a, b)$. Thus for $a = 8$ and $b = 12$ we find by direct trial that $(8, 12) = 4$, while for $a = 5$ and $b = 9$ we find that $(5, 9) = 1$. When a and b are large, say $a = 1804$ and $b = 328$, the attempt to find (a, b) by trial and error would be quite wearisome.

A short and certain method is provided by the *Euclidean algorithm.* (An algorithm is a systematic method for computation.) It is based on the fact that from any relation of the form

$$a = b \cdot q + r \tag{2}$$

it follows that

$$(a, b) = (b, r). \tag{3}$$

For any number u which divides both a and b,

$$a = su, \qquad b = tu,$$

also divides r, since $r = a - bq = su - qtu = (s - qt)u$; and conversely, every number v which divides b and r,

$$b = s'v, \qquad r = t'v,$$

also divides a, since $a = bq + r = s'vq + t'v = (s'q + t')v$. Hence *every* common divisor of a and b is at the same time a common divisor of b and r, and conversely. Since, therefore, the set of *all* common divisors of a and b is identical with the set of all common divisors of b and r, the *greatest* common divisor of a and b must be equal to the greatest common divisor of b and r, which establishes (3). The usefulness of this relation will be seen immediately.

Let us return to the question of finding the greatest common divisor of 1804 and 328. By ordinary long division

$$\begin{array}{r|l} & 5 \\ \hline 328 & 1804 \\ & \underline{1640} \\ & 164 \end{array}$$

we find that

$$1804 = 5 \cdot 328 + 164.$$

Hence from (3) we conclude that

$$(1804, 328) = (328, 164).$$

Observe that the problem of finding (1804, 328) has been replaced by a problem involving smaller numbers. We may continue the process. Since

$$\begin{array}{r|l} & 2 \\ \hline 164 & 328 \\ & \underline{328} \\ & 0, \end{array}$$

we have $328 = 2 \cdot 164 + 0$, so that $(328, 164) = (164, 0) = 164$. Hence $(1804, 328) = (328, 164) = (164, 0) = 164$, which is the desired result.

This process for finding the greatest common divisor of two numbers is given in a geometric form in Euclid's *Elements*. For arbitrary integers a and b, not both 0, it may be described arithmetically in the following terms.

We may suppose that $b \neq 0$, since $(a, 0) = a$. Then by successive division we can write

$$(4) \qquad \begin{aligned} a &= bq_1 + r_1 & (0 < r_1 < b) \\ b &= r_1 q_2 + r_2 & (0 < r_2 < r_1) \\ r_1 &= r_2 q_3 + r_3 & (0 < r_3 < r_2) \\ r_2 &= r_3 q_4 + r_4 & (0 < r_4 < r_3) \\ &\cdots\cdots\cdots & \cdots\cdots\cdots \end{aligned}$$

so long as the remainders $r_1, r_2, r_3, \cdots$ are not 0. From an inspection of the inequalities at the right, we see that the successive remainders form a steadily decreasing sequence of positive numbers:

$$(5) \qquad b > r_1 > r_2 > r_3 > r_4 > \cdots > 0.$$

Hence after at most b steps (often many fewer, since the difference between two successive r's is usually greater than 1) the remainder 0 must appear:

$$\begin{aligned} r_{n-2} &= r_{n-1} q_n + r_n \\ r_{n-1} &= r_n q_{n+1} + 0. \end{aligned}$$

When this occurs we know that

$$(a, b) = r_n;$$

in other words, (a, b) *is the last positive remainder in the sequence* (5). This follows from successive application of the equality (3) to the equations (4), since from successive lines of (4) we have

$$(a, b) = (b, r_1), \qquad (b, r_1) = (r_1, r_2), \qquad (r_1, r_2) = (r_2, r_3),$$
$$(r_2, r_3) = (r_3, r_4), \cdots, (r_{n-1}, r_n) = (r_n, 0) = r_n.$$

Exercise: Carry out the Euclidean algorithm for finding the greatest common divisor of (a) 187, 77. (b) 105, 385. (c) 245, 193.

An extremely important property of (a, b) can be derived from equations (4). *If* $d = (a, b)$, *then positive or negative integers* k *and* l *can be found such that*

$$d = ka + lb. \tag{6}$$

To show this, let us consider the sequence (5) of successive remainders. From the first equation in (4)

$$r_1 = a - q_1 b,$$

so that r_1 can be written in the form $k_1 a + l_1 b$ (in this case $k_1 = 1$, $l_1 = -q_1$). From the next equation,

$$\begin{aligned} r_2 = b - q_2 r_1 &= b - q_2(k_1 a + l_1 b) \\ &= (-q_2 k_1)a + (1 - q_2 l_1)b = k_2 a + l_2 b. \end{aligned}$$

Clearly this process can be repeated through the successive remainders $r_3, r_4, \cdots$ until we arrive at a representation

$$r_n = ka + lb,$$

as was to be proved.

As an example, consider the Euclidean algorithm for finding (61, 24); the greatest common divisor is 1 and the desired representation for 1 can be computed from the equations

$$61 = 2 \cdot 24 + 13, \qquad 24 = 1 \cdot 13 + 11, \qquad 13 = 1 \cdot 11 + 2,$$
$$11 = 5 \cdot 2 + 1, \qquad 2 = 2 \cdot 1 + 0.$$

We have from the first of these equations

$$13 = 61 - 2 \cdot 24,$$

from the second,

$$11 = 24 - 13 = 24 - (61 - 2 \cdot 24) = -61 + 3 \cdot 24,$$

from the third,

$$2 = 13 - 11 = (61 - 2 \cdot 24) - (-61 + 3 \cdot 24) = 2 \cdot 61 - 5 \cdot 24,$$

and from the fourth,

$$1 = 11 - 5 \cdot 2 = (-61 + 3 \cdot 24) - 5(2 \cdot 61 - 5 \cdot 24) = -11 \cdot 61 + 28 \cdot 24.$$

2. Application to the Fundamental Theorem of Arithmetic

The fact that $d = (a, b)$ can always be written in the form $d = ka + lb$ may be used to give a proof of the fundamental theorem of arithmetic that is independent of the proof given on page 23. First we shall prove, as a lemma, the corollary of page 24, and then from this lemma we shall deduce the fundamental theorem, thus reversing the previous order of proof.

Lemma: If a prime p *divides a product* ab, *then* p *must divide* a *or* b.

If a prime p does not divide the integer a, then $(a, p) = 1$, since the only divisors of p are p and 1. Hence we can find integers k and l such that

$$1 = ka + lp.$$

Multiplying both sides of this equation by b we obtain

$$b = kab + lpb.$$

Now if p divides ab we can write

$$ab = pr,$$

so that

$$b = kpr + lpb = p(kr + lb).$$

from which it is evident that p divides b. Thus we have shown that if p divides ab but does not divide a then it must divide b, so that in any event p must divide a or b if it divides ab.

The extension to products of more than two integers is immediate. For example, if p divides abc, then by twice applying the lemma we can show that p must divide at least one of the integers a, b, and c. For if p divides neither a, b, nor c, then it cannot divide ab and hence cannot divide $(ab)c = abc$.

Exercise: The extension of this argument to products of any number n of integers requires the explicit or implicit use of the principle of mathematical induction. Supply the details of this argument.

From this result the fundamental theorem of arithmetic follows at once. Let us suppose given any two decompositions of a positive integer N into primes:

$$N = p_1p_2 \cdots p_r = q_1q_2 \cdots q_s.$$

Since p_1 divides the left side of this equation, it must also divide the right, and hence, by the previous exercise, must divide one of the factors q_k. But q_k is a prime, therefore p_1 must be equal to this q_k. After these equal factors have been cancelled from the equation, it follows that p_2 must divide one of the remaining factors q_t, and hence must be equal to it. Striking out p_2 and q_t, we proceed similarly with $p_3, \cdots, p_r$. At the end of this process all the p's will be cancelled, leaving only 1 on the left side. No q can remain on the right side, since all the q's are larger than one. Hence the p's and q's will be paired off into equal couples, which proves that, except perhaps for the order of the factors, the two decompositions were identical.

3. Euler's φ Function. Fermat's Theorem Again

Two integers a and b are said to be *relatively prime* if their greatest common divisor is 1:

$$(a, b) = 1.$$

For example, 24 and 35 are relatively prime, while 12 and 18 are not. *If* a *and* b *are relatively prime, then for suitably chosen positive or negative integers* k *and* l *we can write*

$$ka + lb = 1.$$

This follows from the property of (a, b) stated on page 45.

Exercise: Prove the theorem: *If an integer* r *divides a product* ab *and is relatively prime to* a, *then* r *must divide* b. (Hint: if r is relatively prime to a then we can find integers k and l such that

$$kr + la = 1.$$

Multiply both sides of this equation by b.) This theorem includes the lemma of page 46 as a special case, since a prime p is relatively prime to an integer a if and only if p does not divide a.

For any positive integer n, let $\varphi(n)$ denote the *number of integers from* 1 *to* n *which are relatively prime to* n. This function $\varphi(n)$, first introduced by Euler, is a "number-theoretical function" of great importance. The values of $\varphi(n)$ for the first few values of n are easily computed:

$\varphi(1) = 1$ since 1 is relatively prime to 1,
$\varphi(2) = 1$ since 1 is relatively prime to 2,
$\varphi(3) = 2$ since 1 and 2 are relatively prime to 3,
$\varphi(4) = 2$ since 1 and 3 are relatively prime to 4,
$\varphi(5) = 4$ " 1, 2, 3, 4 are relatively prime to 5,
$\varphi(6) = 2$ " 1, 5 " " " " 6,
$\varphi(7) = 6$ " 1, 2, 3, 4, 5, 6 are relatively prime to 7,
$\varphi(8) = 4$ " 1, 3, 5, 7 " " " " 8,
$\varphi(9) = 6$ " 1, 2, 4, 5, 7, 8 " " " " 9,
$\varphi(10) = 4$ " 1, 3, 7, 9 " " " " 10,
etc.

We observe that $\varphi(p) = p - 1$ if p is a prime; for a prime p has no divisors other than itself and 1, and hence it is relatively prime to all of the integers 1, 2, 3, $\cdots$, $p - 1$. If n is composite, with prime decomposition

$$n = p_1^{\alpha_1} p_2^{\alpha_2} \cdots p_r^{\alpha_r},$$

where the p's represent distinct primes, each raised to a certain power, then

$$\varphi(n) = n\left(1 - \frac{1}{p_1}\right) \cdot \left(1 - \frac{1}{p_2}\right) \cdots \left(1 - \frac{1}{p_r}\right).$$

For example, since $12 = 2^2 \cdot 3$,

$$\varphi(12) = 12(1 - \tfrac{1}{2})(1 - \tfrac{1}{3}) = 12(\tfrac{1}{2})(\tfrac{2}{3}) = 4,$$

as it should be. The proof is quite elementary, but will be omitted here.

* *Exercise:* Using Euler's φ function, generalize Fermat's theorem of page 37. The general theorem states: *If* n *is any integer, and* a *is relatively prime to* n, *then*

$$a^{\varphi(n)} \equiv 1 \pmod{n}.$$

4. Continued Fractions. Diophantine Equations

The Euclidean algorithm for finding the greatest common divisor of two integers leads immediately to an important method for representing the quotient of two integers as a composite fraction.

Applied to the numbers 840 and 611, for example, the Euclidean algorithm yields the series of equations,

$$840 = 1 \cdot 611 + 229, \qquad 611 = 2 \cdot 229 + 153,$$

$$229 = 1 \cdot 153 + 76, \qquad 153 = 2 \cdot 76 + 1,$$

which show, incidentally, that $(840, 611) = 1$. From these equations we may derive the following expressions:

$$\frac{840}{611} = 1 + \frac{229}{611} = 1 + \frac{1}{611/229},$$

$$\frac{611}{229} = 2 + \frac{153}{229} = 2 + \frac{1}{229/153},$$

$$\frac{229}{153} = 1 + \frac{76}{153} = 1 + \frac{1}{153/76},$$

$$\frac{153}{76} = 2 + \frac{1}{76}.$$

On combining these equations we obtain the development of the rational number $\frac{840}{611}$ in the form

$$\frac{840}{611} = 1 + \cfrac{1}{2 + \cfrac{1}{1 + \cfrac{1}{2 + \cfrac{1}{76}}}}.$$

An expression of the form

$$(7) \qquad a = a_0 + \cfrac{1}{a_1 + \cfrac{1}{a_2 + \ddots + \cfrac{1}{a_n}}},$$

where the a's are positive integers, is called a *continued fraction.* The Euclidean algorithm gives us a method for expressing any rational number in this form.

Exercise: Find the continued fraction developments of

$$\frac{2}{5}, \frac{43}{30}, \frac{169}{70}.$$

* Continued fractions are of great importance in the branch of higher arithmetic known as Diophantine analysis. A *Diophantine equation* is an algebraic equation in one or more unknowns with *integer* coefficients, for which *integer* solutions are sought. Such an equation may have no solutions, a finite number, or an infinite number of solutions. The simplest case is the *linear* Diophantine equation in two unknowns,

$$(8) \qquad ax + by = c,$$

where a, b, and c are given integers, and integer solutions x, y are desired. The complete solution of an equation of this form may be found by the Euclidean algorithm.

To begin with, let us find $d = (a, b)$ by the Euclidean algorithm; then for proper choice of the integers k and l,

$$(9) \qquad ak + bl = d.$$

Hence the equation (8) has the particular solution $x = k$, $y = l$ for the case $c = d$. More generally, if c is any multiple of d:

$$c = d \cdot q,$$

then from (9) we obtain

$$a(kq) + b(lq) = dq = c,$$

so that (8) has the particular solution $x = x^* = kq$, $y = y^* = lq$. Conversely, if (8) has any solution x, y for a given c, then c must be a multiple of $d = (a, b)$; for d divides both a and b, and hence must divide c. We have therefore proved that the equation (8) has a solution if and only if c is a multiple of (a, b).

To determine the other solutions of (8) we observe that if $x = x'$, $y = y'$ is any solution other than the one, $x = x^*$, $y = y^*$, found above by the Euclidean algorithm, then $x = x' - x^*$, $y = y' - y^*$ is a solution of the "homogeneous" equation

$$ax + by = 0. \tag{10}$$

For if

$$ax' + by' = c \quad \text{and} \quad ax^* + by^* = c,$$

then on subtracting the second equation from the first we find that

$$a(x' - x^*) + b(y' - y^*) = 0.$$

Now the most general solution of the equation (10) is $x = rb/(a, b)$, $y = -ra/(a, b)$, where r is any integer. (We leave the proof as an exercise. Hint: Divide by (a, b) and use the Exercise on page 48.) It follows immediately that

$$x = x^* + rb/(a, b), \qquad y = y^* - ra/(a, b).$$

To summarize: The linear Diophantine equation $ax + by = c$, where a, b, and c are integers, has a solution in integers if and only if c is a multiple of (a, b). In the latter case, a particular solution $x = x^*$, $y = y^*$ may be found by the Euclidean algorithm, and the most general solution is of the form

$$x = x^* + rb/(a, b), \qquad y = y^* - ra/(a, b),$$

where r is any integer.

Examples: The equation $3x + 6y = 22$ has no integral solution, since $(3, 6) = 3$, which does not divide 22.

The equation $7x + 11y = 13$ has the particular solution $x = -39$, $y = 26$, found as follows:

$$11 = 1 \cdot 7 + 4, \qquad 7 = 1 \cdot 4 + 3, \qquad 4 = 1 \cdot 3 + 1, \qquad (7, 11) = 1.$$

$$1 = 4 - 3 = 4 - (7 - 4) = 2 \cdot 4 - 7 = 2(11 - 7) - 7 = 2 \cdot 11 - 3 \cdot 7.$$

Hence

$$7 \cdot (-3) + 11(2) = 1,$$
$$7 \cdot (-39) + 11(26) = 13.$$

The other solutions are given by

$$x = -39 + 11r, \qquad y = 26 - 7r,$$

where r is any integer.

Exercise: Solve the Diophantine equations (a) $3x - 4y = 29$. (b) $11x + 12y = 58$. (c) $153x - 34y = 51$.

CHAPTER II

THE NUMBER SYSTEM OF MATHEMATICS

INTRODUCTION

We must greatly extend the original concept of number as natural number in order to create an instrument powerful enough for the needs of practice and theory. In a long and hesitant evolution zero, negative integers, and fractions were gradually accepted on the same footing as the positive integers, and today the rules of operation with these numbers are mastered by the average school child. But to gain complete freedom in algebraic operations we must go further by including irrational and complex quantities in the number concept. Although these extensions of the concept of natural number have been in use for centuries and are at the basis of all modern mathematics it is only in recent times that they have been put on a logically sound basis. In the present chapter we shall give an account of this development.

§1. THE RATIONAL NUMBERS

1. Rational Numbers as a Device for Measuring

The integers are abstractions from the process of counting finite collections of objects. But in daily life we need not only to *count* individual *objects,* but also to *measure quantities* such as length, area, weight, and time. If we want to operate freely with the measures of these quantities, which are capable of arbitrarily fine subdivision, it is necessary to extend the realm of arithmetic beyond the integers. The first step is *to reduce the problem of measuring to the problem of counting.* First we select, quite arbitrarily, a *unit of measurement*—foot, yard, inch, pound, gram, or second as the case may be—to which we assign the measure 1. Then we count the number of these units which together make up the quantity to be measured. A given mass of lead may weigh exactly 54 pounds. In general, however, the process of counting units will not "come out even," and the given quantity will not be exactly measurable in terms of integral multiples of the chosen unit. The most we can say is that it lies between two successive multiples of this unit, say between 53 and 54 pounds. When this occurs, we take a further step by introducing new sub-units, obtained by subdividing the original unit into a number n of equal parts. In ordinary language, these new sub-units may have special names; for example, the foot is divided into 12 inches, the meter into 100 centimeters, the pound into 16 ounces, the hour into 60 minutes, the minute into 60 seconds, etc. In the symbolism of mathematics, however, a sub-unit obtained by dividing the original unit 1 into n equal parts is denoted by the symbol $1/n$; and if a given quantity contains exactly m of these sub-units, its measure is denoted by the symbol m/n. This symbol is called a *fraction* or *ratio* (sometimes written $m:n$). The next and decisive step was consciously taken only after centuries of groping effort: the symbol m/n was divested of its concrete reference to the process of measuring and the quantities measured, and instead considered as a pure *number*, an entity in itself, on the same footing with the natural numbers. When m and n are natural numbers, the symbol m/n is called a *rational number*.

The use of the word number (originally meaning natural number only) for these new symbols is justified by the fact that addition and multiplication of these symbols obey the same laws that govern the operations with natural numbers. To show this, addition, multiplication, and equality of rational numbers must first be defined. As everyone knows, these definitions are:

$$\frac{a}{b}+\frac{c}{d}=\frac{ad+bc}{bd}, \qquad \frac{a}{b}\cdot\frac{c}{d}=\frac{ac}{bd},$$

$$(1) \qquad \frac{a}{a}=1, \qquad \frac{a}{b}=\frac{c}{d} \text{ if } ad=bc,$$

for any integers a, b, c, d. For example:

$$\frac{2}{3}+\frac{4}{5}=\frac{2\cdot 5+3\cdot 4}{3\cdot 5}=\frac{10+12}{15}=\frac{22}{15}, \qquad \frac{2}{3}\cdot\frac{4}{5}=\frac{2\cdot 4}{3\cdot 5}=\frac{8}{15},$$

$$\frac{3}{3}=1, \qquad \frac{8}{12}=\frac{6}{9}=\frac{2}{3}.$$

Precisely these definitions are forced upon us if we wish to use the rational numbers as measures for lengths, areas, etc. But strictly speaking, these rules for the addition, multiplication, and equality of our symbols are established by our own definition and are not imposed upon us by any prior necessity other than that of consistency and usefulness for applications. On the basis of the definitions (1) we can show that *the fundamental laws of the arithmetic of natural numbers continue to hold in the domain of rational numbers*:

$$
\begin{aligned}
p + q &= q + p && \text{(commutative law of addition),} \\
p + (q + r) &= (p + q) + r && \text{(associative law of addition),} \\
pq &= qp && \text{(commutative law of multiplication),} \\
p(qr) &= (pq)r && \text{(associative law of multiplication),} \\
p(q + r) &= pq + pr && \text{(distributive law).}
\end{aligned} \tag{2}
$$

For example, the proof of the commutative law of addition for fractions is exhibited by the equations

$$\frac{a}{b} + \frac{c}{d} = \frac{ad + bc}{bd} = \frac{cb + da}{db} = \frac{c}{d} + \frac{a}{b},$$

of which the first and last equality signs correspond to the definition (1) of addition, while the middle one is a consequence of the commutative laws of addition and multiplication of natural numbers. The reader may verify the other four laws in the same way.

For a real understanding of these facts it must be emphasized once more that the rational numbers are our own creations, and that the rules (1) are imposed at our volition. We might whimsically decree some other rule for addition, such as $\frac{a}{b} + \frac{c}{d} = \frac{a + c}{b + d}$, which in particular would yield $\frac{1}{2} + \frac{1}{2} = 2/4$, an absurd result from the point of view of measuring. Rules of this type, though logically permissible, would make the arithmetic of our symbols a meaningless game. The free play of the intellect is guided here by the necessity of creating a suitable instrument for handling measurements.

2. Intrinsic Need for the Rational Numbers. Principle of Generalization

Aside from the "practical" reason for the introduction of rational numbers, there is a more intrinsic and in some ways an even more compelling one, which we shall now discuss quite independently of the preceding argument. It is of an entirely arithmetical character, and is typical of a dominant tendency in mathematical procedure.

In the ordinary arithmetic of natural numbers we can always carry out the two fundamental operations, addition and multiplication. But the "inverse operations" of *subtraction* and *division* are not always possible. The difference $b - a$ of two integers a, b is the integer c

such that $a + c = b$, i.e. it is the solution of the equation $a + x = b$. But in the domain of natural numbers the symbol $b - a$ has a meaning only under the restriction $b > a$, for only then does the equation $a + x = b$ have a natural number x as a solution. It was a very great step towards removing this restriction when the symbol 0 was introduced by setting $a - a = 0$. It was of even greater importance when, through the introduction of the symbols $-1, -2, -3, \cdots$, together with the definition

$$b - a = -(a - b)$$

for the case $b < a$, it was assured that subtraction could be performed without restriction *in the domain of positive and negative integers.* To include the new symbols $-1, -2, -3, \cdots$ in an enlarged arithmetic which embraces both positive and negative integers we must, of course, *define operations* with them in such a way that *the original rules of arithmetical operations are preserved.* For example, the rule

$$(-1)(-1) = 1, \tag{3}$$

which we set up to govern the multiplication of negative integers, is a consequence of our desire to preserve the distributive law $a(b + c) = ab + ac$. For if we had ruled that $(-1)(-1) = -1$, then, on setting $a = -1, b = 1, c = -1$, we should have had $-1(1 - 1) = -1 - 1 = -2$, while on the other hand we actually have $-1(1 - 1) = -1 \cdot 0 = 0$. It took a long time for mathematicians to realize that the "rule of signs" (3), together with all the other definitions governing negative integers and fractions cannot be "proved." They are *created* by us in order to attain freedom of operation while preserving the fundamental laws of arithmetic. What *can*—and must—be proved is only that on the basis of these definitions the commutative, associative, and distributive laws of arithmetic are preserved. Even the great Euler resorted to a thoroughly unconvincing argument to show that $(-1)(-1)$ "must" be equal to $+1$. For, as he reasoned, it must either be $+1$ or -1, and cannot be -1, since $-1 = (+1)(-1)$.

Just as the introduction of the negative integers and zero clears the way for unrestricted subtraction, so the introduction of fractional numbers removes the analogous arithmetical obstacle to division. The quotient $x = b/a$ of two integers a and b, defined by the equation

$$ax = b, \tag{4}$$

exists *as an integer* only if a is a factor of b. If this is not the case, as for example when $a = 2$, $b = 3$, we simply introduce a new symbol b/a, which we call a fraction, subject to the rule that $a(b/a) = a$, so that b/a is a solution of (4) "by definition." The invention of the fractions as new number symbols makes division possible without restriction—except for *division by zero*, which is *excluded once for all.*

Expressions like 1/0, 3/0, 0/0, etc. will be for us meaningless symbols. For if division by 0 were permitted, we could deduce from the true equation $0 \cdot 1 = 0 \cdot 2$ the absurd consequence $1 = 2$. It is, however, sometimes useful to denote such expressions by the symbol ∞ (read, "infinity"), *provided that one does not attempt to operate with the symbol* ∞ *as though it were subject to the ordinary rules of calculation with numbers.*

The purely arithmetical significance of the system of *all rational numbers*—integers and fractions, positive and negative—is now apparent. For in this extended number domain not only do the formal associative, commutative, and distributive laws hold, but the equations $a + x = b$ and $ax = b$ now have solutions, $x = b - a$ and $x = b/a$, without restriction, provided in the latter case that $a \neq 0$. In other words, in the domain of rational numbers the so-called *rational operations*—addition, subtraction, multiplication, and division—may be performed without restriction and will never lead out of this domain. Such a closed domain of numbers is called a *field.* We shall meet with other examples of fields later in this chapter and in Chapter III.

Extending a domain by introducing new symbols in such a way that the laws which hold in the original domain continue to hold in the larger domain is one aspect of the characteristic mathematical process of *generalization.* The generalization from the natural to the rational numbers satisfies both the theoretical need for removing the restrictions on subtraction and division, and the practical need for numbers to express the results of measurement. It is the fact that the rational numbers fill this two-fold need that gives them their true significance. As we have seen, this extension of the number concept was made possible by the creation of new numbers in the form of abstract symbols like 0, -2, and 3/4. Today, when we deal with such numbers as a matter of course, it is hard to believe that as late as the seventeenth century they were not generally credited with the same legitimacy as the positive integers, and that they were used, when necessary, with a certain amount of doubt and trepidation. The inherent human tendency to cling to the "concrete," as exemplified by the natural numbers, was responsible for this slowness in taking an inevitable step. Only in the realm of the abstract can a satisfactory system of arithmetic be created.

3. Geometrical Interpretation of Rational Numbers

An illuminating geometrical interpretation of the rational number system is given by the following construction.

On a straight line, the "number axis," we mark off a segment 0 to 1, as in Fig. 8. This establishes the length of the segment from 0 to 1 as the unit length, which we may choose at will. The positive and negative integers are then represented as a set of equidistant points on the number axis, the positive numbers to the right of the point 0 and the

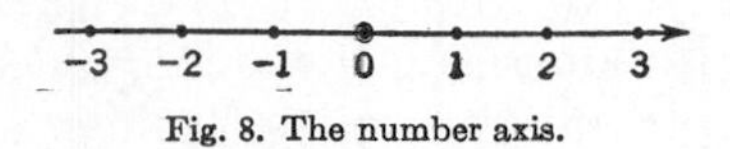

Fig. 8. The number axis.

negative numbers to the left. To represent fractions with the denominator n, we divide each of the segments of unit length into n equal parts; the points of subdivision then represent the fractions with denominator n. If we do this for every integer n, then all the rational numbers will be represented by points of the number axis. We shall call such points *rational points*, and we shall use the terms "rational number" and "rational point" interchangeably.

In Chapter I, §1, we defined the relation $A < B$ for natural numbers. This has its analog on the number axis in the fact that if natural number A is less than natural number B, then point A lies to the left of point B. Since the geometrical relation holds between *all* rational points, we are led to try to extend the arithmetical relation in such a way as to preserve the relative geometrical order of the corresponding points. This is achieved by the following definition: The rational number A is said to be *less than* the rational number B ($A < B$), and B is said to be *greater than* A ($B > A$), if $B - A$ is positive. It then follows that, if $A < B$, the points (numbers) *between* A and B are those which are both $>A$ and $<B$. Any such pair of distinct points, together with the points between them, is called a *segment*, or *interval*, $[A, B]$.

The distance of a point, A, from the origin, considered as positive, is called the *absolute value* of A and is indicated by the symbol

$$|A|.$$

In words, if $A \geq 0$, we have $|A| = A$; if $A \leq 0$, we have $|A| = -A$. It is clear that if A and B have the same sign, the equation $|A + B| = |A| + |B|$ holds, while if A and B have different signs, we have

$|A + B| < |A| + |B|$. Hence, combining these two statements, we have the general inequality

$$|A + B| \leq |A| + |B|,$$

which is valid irrespective of the signs of A and B.

A fact of fundamental importance is expressed in the statement: *The rational points are dense on the line.* By this we mean that within each interval, no matter how small, there are rational points. We need only take a denominator n large enough so that the interval $[0, 1/n]$ is smaller than the interval $[A, B]$ in question; then at least one of the fractions m/n must lie within the interval. Hence there is no interval on the line, however small, which is free from rational points. It follows, moreover, that there must be infinitely many rational points in any interval; for, if there were only a finite number, the interval between any two adjacent rational points would be devoid of rational points, which we have just seen to be impossible.

§2. INCOMMENSURABLE SEGMENTS, IRRATIONAL NUMBERS, AND THE CONCEPT OF LIMIT

1. Introduction

In comparing the magnitudes of two line segments a and b, it may happen that a is contained in b an exact integral number r of times. In this case we can express the measure of the segment b in terms of that of a by saying that the length of b is r times that of a. Or it may turn out that while no integral multiple of a equals b, we can divide a into, say, n equal segments, each of length a/n, such that some integral multiple m of the segment a/n is equal to b:

$$b = \frac{m}{n}a. \tag{1}$$

When an equation of the form (1) holds we say that the two segments a and b are *commensurable*, since they have as a common measure the segment a/n which goes n times into a and m times into b. The totality

Fig. 9. Rational points.

of all segments commensurable with a will be those whose length can be expressed in the form (1) for some choice of integers m and n ($n \neq 0$). If we choose a as the unit segment, $[0, 1]$, in Figure 9, then the segments commensurable with the unit segment will correspond to all the rational points m/n on the number axis. For all practical purposes of measuring, the rational numbers are entirely sufficient. Even from a theoretical viewpoint, since the set of rational points covers the line densely, it might seem that all points on the line are rational points. If this were true, then any segment would be commensurable with the unit. It was one of the most surprising discoveries of early Greek mathematics (the Pythagorean school) that the situation is by no means so simple. There exist *incommensurable segments* or, if we assume that to every segment corresponds a number giving its length in terms of the unit, *irrational numbers*. This revelation was a scientific event of the highest importance. Quite possibly it marked the origin of what we consider to be the specifically Greek contribution to rigorous procedure in mathematics. Certainly it has profoundly affected mathematics and philosophy from the time of the Greeks to the present day.

Eudoxus' theory of incommensurables, presented in geometrical form in Euclid's *Elements*, is a masterpiece of Greek mathematics, though it is usually omitted from the diluted high-school versions of this classical work. The theory became fully appreciated only in the late nineteenth century, after Dedekind, Cantor, and Weierstrass had constructed a rigorous theory of irrational numbers. We shall present the theory in the modern arithmetical way.

First we show: *The diagonal of a square is incommensurable with its side.* We may suppose that the side of the given square is chosen as the unit of length, and that the diagonal has the length x. Then, by the Pythagorean theorem, we have

$$x^2 = 1^2 + 1^2 = 2.$$

(We may denote x by the symbol $\sqrt{2}$.) Now if x were commensurable with 1, we could find two integers p and q such that $x = p/q$ and

$$p^2 = 2q^2. \tag{2}$$

We may suppose that p/q is already in lowest terms, since any common factor in numerator and denominator could be cancelled out at the beginning. Since 2 appears as a factor of the right side, p^2 is an even number, and hence p itself is even, because the square of an odd number is odd. We may therefore write $p = 2r$. Equation (2) then becomes

$$4r^2 = 2q^2, \text{ or } 2r^2 = q^2.$$

Since 2 is a factor of the left side, q^2, and hence q must also be even. Thus p and q are both divisible by 2, which contradicts the assumption that p and q had no common factor. Therefore, equation (2) cannot hold, and x cannot be a rational number.

Our result can be expressed by the statement that there is no rational number equal to $\sqrt{2}$.

The argument of the preceding paragraph shows that a very simple geometrical construction may result in a segment incommensurable with the unit. If such a segment is marked off on the number axis by means of a compass, the point so constructed cannot coincide with any of the

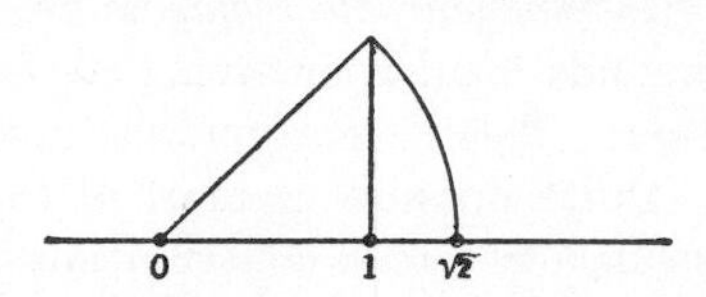

Fig. 10. Construction of $\sqrt{2}$.

rational points: *The system of rational points,* although it is everywhere dense, *does not cover all of the number axis.* To the naive mind it must certainly appear very strange and paradoxical that the dense set of rational points does not cover the whole line. Nothing in our "intuition" can help us to "see" the irrational points as distinct from the rational ones. No wonder that the discovery of the incommensurable stirred the Greek philosophers and mathematicians, and that it has retained even today its provocative effect on thoughtful minds.

It would be very easy to construct as many segments incommensurable with the unit as we want. The end-points of such segments, if marked off from the point 0 on the number axis, are called *irrational points.* Now, the guiding principle in introducing fractions was the *measuring of lengths by numbers,* and we should like to maintain this principle in dealing with segments incommensurable with the unit. If we demand that there should be a *mutual correspondence between numbers* on the one hand *and points of a straight line* on the other, it is necessary to introduce *irrational numbers.*

Summarizing the situation thus far we may say that an irrational number represents the length of a segment incommensurable with the unit. In the following sections we shall refine this somewhat vague and entirely geometrical definition, until we arrive at one more satisfactory from the point of view of logical rigor. Our first approach to the subject will be by way of the decimal fractions.

Exercises: 1) Prove that $\sqrt[3]{2}$, $\sqrt{3}$, $\sqrt{5}$, $\sqrt[3]{3}$ are not rational. (Hint: Use the lemma of p. 47).

2) Prove that $\sqrt{2} + \sqrt{3}$ and $\sqrt{2} + \sqrt[3]{2}$ are not rational. (Hint: If e.g. the

first of these numbers were equal to a rational number r then, writing $\sqrt{3} = r - \sqrt{2}$ and squaring, $\sqrt{2}$ would be rational.)

3) Prove that $\sqrt{2} + \sqrt{3} + \sqrt{5}$ is irrational. Try to make up similar and more general examples.

2. Decimal Fractions. Infinite Decimals

In order to cover the number axis with a set of points everywhere dense, we do not need the totality of *all* rational numbers; for example, it suffices to consider only those numbers which originate by subdivision of each unit interval into 10, then 100, 1000, etc. equal segments. The points so obtained correspond to the "decimal fractions." For example, the point $0.12 = 1/10 + 2/100$ corresponds to the point lying in the first unit interval, in the second subinterval of length 10^{-1}, and at the initial point of the third "sub-sub-" interval of length 10^{-2}. (a^{-n} means $1/a^n$.) Such a *decimal fraction*, if it contains n digits after the decimal point, has the form

$$f = z + a_1 10^{-1} + a_2 10^{-2} + a_3 10^{-3} + \cdots + a_n 10^{-n},$$

where z is an integer and the a's are digits—0, 1, 2, $\cdots$, 9—indicating the tenths, hundredths and so on. The number f is represented in the decimal system by the abbreviated symbol $z.a_1a_2a_3 \cdots a_n$. We see immediately that these decimal fractions can be written in the ordinary form of a fraction p/q where $q = 10^n$; for example, $f = 1.314 = 1 + 3/10 + 1/100 + 4/1000 = 1314/1000$. If p and q have a common divisor, the decimal fraction may then be reduced to a fraction with a denominator which is some divisor of 10^n. On the other hand, no fraction in lowest terms whose denominator is not a divisor of some power of 10 can be represented as a decimal fraction. For example, $\frac{1}{5} = \frac{2}{10} = 0.2$, and $\frac{1}{250} = \frac{4}{1000} = 0.004$; but $\frac{1}{3}$ cannot be written as a decimal fraction with a finite number n of decimal places, however great n be chosen, for an equation of the form

$$\tfrac{1}{3} = b/10^n$$

would imply

$$10^n = 3b,$$

which is absurd, since 3 is not a factor of any power of 10.

Now let us choose any point P on the number axis which does not correspond to a decimal fraction; e.g. the rational point $\frac{1}{3}$ or the irrational point $\sqrt{2}$. Then in the process of subdividing the unit interval into ten equal parts, and so on, P will never occur as the initial point of a subinterval. Still, P can be included within smaller and smaller intervals of the decimal division with any desired degree of approximation. This approximation process may be described as follows.

Suppose that P lies in the first unit interval. We subdivide this interval into 10 equal parts, each of length 10^{-1}, and find, say, that P lies in the third such interval. At this stage we can say that P lies *between* the decimal fractions 0.2 and 0.3. We subdivide the interval from 0.2 to 0.3 into 10 equal parts, each of length 10^{-2}, and find that P lies, say, in the fourth such interval. Subdividing this in turn, we find that P lies in the first interval of length 10^{-3}. We can now say that P lies between 0.230 and 0.231. This process can be continued indefinitely, and leads to an unending sequence of digits, $a_1, a_2, a_3, \cdots, a_n, \cdots$, with the following property: whatever number n we choose, the point P is included in the interval I_n whose left-hand end-point is the decimal fraction $0.a_1a_2a_3 \cdots a_{n-1}a_n$ and whose right-hand end-point is $0.a_1a_2a_3 \cdots a_{n-1}(a_n + 1)$, the length of I_n being 10^{-n}. If we choose in succession $n = 1, 2, 3, 4, \cdots$, we see that each of these intervals, $I_1, I_2, I_3, \cdots$, is contained in the preceding one, while their lengths, $10^{-1}, 10^{-2}, 10^{-3}, \cdots$, tend to zero. We say that the point P is contained in a *nested sequence of decimal intervals.* For example, if P is the rational point $\frac{1}{3}$, then all the digits $a_1, a_2, a_3, \cdots$ are equal to 3, and P is contained in every interval I_n which extends from $0.333 \cdots 33$ to $0.333 \cdots 34$; i.e., $\frac{1}{3}$ is greater than $0.333 \cdots 33$ but less than $0.333 \cdots 34$, where the number of digits may be taken arbitrarily large. We express this fact by saying that the n-digit decimal fraction $0.333 \cdots 33$ "tends to $\frac{1}{3}$" as n increases. We write

$$\tfrac{1}{3} = 0.333 \cdots,$$

the dots indicating that the decimal fraction is to be extended "indefinitely."

The irrational point $\sqrt{2}$ defined in Article 1 also leads to an indefinitely extended decimal fraction. Here, however, the law which determines the values of the digits in the sequence is by no means obvious. In fact, no explicit formula that determines the successive digits is known, although one may calculate as many digits as desired:

$$1^2 = 1 < 2 < 2^2 = 4$$
$$(1.4)^2 = 1.96 < 2 < (1.5)^2 = 2.25$$
$$(1.41)^2 = 1.9881 < 2 < (1.42)^2 = 2.0264$$
$$(1.414)^2 = 1.999396 < 2 < (1.415)^2 = 2.002225$$
$$(1.4142)^2 = 1.99996164 < 2 < (1.4143)^2 = 2.00024449, \text{ etc.}$$

As a general definition we say that a point P that is not represented by any decimal fraction with a finite number n of digits is represented by the *infinite decimal fraction*, $z.a_1a_2a_3 \cdots$, if for every value of n the point P lies in the interval of length 10^{-n} with $z.a_1a_2a_3 \cdots a_n$ as its initial point.

In this manner there is established a correspondence between all the points on the number axis and all the *finite and infinite* decimal fractions We offer the tentative definition: a "number" is a *finite or infinite* decimal. Those infinite decimals which do not represent rational numbers are called *irrational numbers.*

Until the middle of the nineteenth century these considerations were accepted as a satisfactory explanation of the system of rational and irrational numbers, the *continuum of numbers.* The enormous advance of mathematics since the seventeenth century, in particular the development of analytic geometry and of the differential and integral calculus, proceeded safely with this concept of the number system as a basis. But during the period of critical re-examination of principles and consolidation of results, it was felt more and more that the concept of irrational number required a more precise analysis. As a preliminary to our account of the modern theory of the number continuum we shall discuss in a more or less intuitive fashion the basic concept of *limit.*

Exercise: Calculate $\sqrt[3]{2}$ and $\sqrt[3]{5}$ with an accuracy of at least 10^{-2}.

3. Limits. Infinite Geometrical Series

As we saw in the preceding section, it sometimes happens that a certain rational number s is approximated by a sequence of other rational numbers s_n, where the index n assumes consecutively all the values $1, 2, 3, \cdots$. For example, if $s = 1/3$, then $s_1 = 0.3$, $s_2 = 0.33$, $s_3 = 0.333$, etc. As another example, let us divide the unit interval into two halves, the second half again into two equal parts, the second of these again into two equal parts, and so forth, until the smallest intervals thus obtained have the length 2^{-n}, where n is chosen arbitrarily large, e.g. $n = 100$, $n = 100{,}000$, or any number we please. Then by adding together all the intervals except the very last one we obtain a total length equal to

$$s_n = \frac{1}{2} + \frac{1}{4} + \frac{1}{8} + \frac{1}{16} \cdots + \frac{1}{2^n}. \tag{3}$$

We see that s_n differs from 1 by $(\frac{1}{2})^n$, and that this difference becomes arbitrarily small, or "tends to zero" as n increases indefinitely. It makes no sense to say that the difference *is* zero if n *is* infinite. The infinite enters only in the unending *procedure* and not as an actual *quantity*. We describe the behavior of s_n by saying that *the sum* s_n *approaches the limit* 1 *as* n *tends to infinity,* and by writing

$$1 = \frac{1}{2} + \frac{1}{2^2} + \frac{1}{2^3} + \frac{1}{2^4} + \cdots, \tag{4}$$

where on the right we have an *infinite series.* This "equation" does not mean that we actually have to add infinitely many terms; it is only an abbreviated expression for the fact that 1 is the limit of the finite sum s_n as n *tends to* infinity (by no means *is* infinity). Thus equation (4) with its incomplete symbol "$+ \cdots$" is merely mathematical shorthand for the precise statement

1 = the limit as n tends to infinity of the quantity

$$s_n = \frac{1}{2} + \frac{1}{2^2} + \frac{1}{2^3} + \cdots + \frac{1}{2^n}. \tag{5}$$

In an even more abbreviated but expressive form we write

$$s_n \to 1 \text{ as } n \to \infty. \tag{6}$$

As another example of limit, we consider the powers of a number q. If $-1 < q < 1$, e.g. $q = 1/3$ or $q = -4/5$, then the successive powers of q,

$$q, q^2, q^3, q^4, \cdots, q^n, \cdots,$$

will approach zero as n increases. If q is negative, the sign of q^n will alternate from $+$ to $-$, and q^n will tend to zero from alternate sides Thus if $q = 1/3$, then $q^2 = 1/9$, $q^3 = 1/27$, $q^4 = 1/81, \cdots$, while if $q = -1/2$, then $q^2 = 1/4$, $q^3 = -1/8$, $q^4 = 1/16, \cdots$. We say that the *limit of* q^n, *as* n *tends to infinity, is zero,* or, in symbols,

$$q^n \to 0 \text{ as } n \to \infty, \text{ for } -1 < q < 1. \tag{7}$$

(Incidentally, if $q > 1$ or $q < -1$ then q^n does not tend to zero, but increases in magnitude without limit.)

To give a rigorous proof of the assertion (7) we start with the inequality proved on page 15, which states that $(1 + p)^n \geq 1 + np$ for any positive integer n and $p > -1$. If q is any fixed number between 0 and 1, e.g. $q = 9/10$, we have $q = 1/(1 + p)$, where $p > 0$. Hence

$$\frac{1}{q^n} = (1 + p)^n \geq 1 + np > np,$$

or (see rule 4, p. 322)

$$0 < q^n < \frac{1}{p} \cdot \frac{1}{n}.$$

q^n is therefore included between the fixed bound 0 and the bound $(1/p)(1/n)$ which approaches zero as n increases, since p is fixed. This makes it evident that $q^n \to 0$. If q is negative, we have $q = -1/(1+p)$ and the bounds become $(-1/p)(1/n)$ and $(1/p)(1/n)$ instead of 0 and $(1/p)(1/n)$. Otherwise the reasoning remains unchanged.

We now consider the *geometrical series*

$$s_n = 1 + q + q^2 + q^3 + \cdots + q^n. \tag{8}$$

(The case $q = 1/2$ was discussed above.) As shown on page 13, we can express the sum s_n in a simple and concise form. If we multiply s_n by q, we find

$$qs_n = q + q^2 + q^3 + q^4 + \cdots + q^{n+1}, \tag{8a}$$

and by subtraction of (8a) from (8) we see that all terms except 1 and q^{n+1} cancel out. We obtain by this device

$$(1 - q)s_n = 1 - q^{n+1},$$

or, by division,

$$s_n = \frac{1 - q^{n+1}}{1 - q} = \frac{1}{1-q} - \frac{q^{n+1}}{1-q}.$$

The concept of limit comes into play if we let n increase. As we have seen, $q^{n+1} = q \cdot q^n$ tends to zero if $-1 < q < 1$, and we obtain the limiting relation

$$s_n \to \frac{1}{1-q} \text{ as } n \to \infty, \text{ for } -1 < q < 1. \tag{9}$$

Written as an *infinite geometrical series* this becomes

$$1 + q + q^2 + q^3 + \cdots = \frac{1}{1-q}, \text{ for } -1 < q < 1. \tag{10}$$

For example,

$$1 + \frac{1}{2} + \frac{1}{2^2} + \frac{1}{2^3} + \cdots = \frac{1}{1 - \frac{1}{2}} = 2,$$

in agreement with equation (4), and similarly

$$\frac{9}{10} + \frac{9}{10^2} + \frac{9}{10^3} + \frac{9}{10^4} + \cdots = \frac{9}{10}\,\frac{1}{1 - 1/10} = 1,$$

so that $0.99999 \cdots = 1$. Similarly, the finite decimal 0.2374 and the infinite decimal $0.23739999999 \cdots$ represent the same number.

In Chapter VI we shall resume the general discussion of the limit concept in the modern spirit of rigor.

Exercises: 1) Prove that $1 - q + q^2 - q^3 + q^4 - \cdots = \frac{1}{1+q}$, if $|q| < 1$.

2) What is the limit of the sequence $a_1, a_2, a_3, \cdots$, where $a_n = n/(n+1)$? (Hint: Write the expression in the form $n/(n+1) = 1 - 1/(n+1)$ and observe that the second term tends to zero.)

3) What is the limit of $\frac{n^2+n+1}{n^2-n+1}$ for $n \to \infty$? (Hint: Write the expression in the form

$$\frac{1+\frac{1}{n}+\frac{1}{n^2}}{1-\frac{1}{n}+\frac{1}{n^2}}.)$$

4) Prove, for $|q| < 1$, that $1 + 2q + 3q^2 + 4q^3 + \cdots = \frac{1}{(1-q)^2}$. (Hint: Use the result of exercise 3 on p. 18.)

5) What is the limit of the infinite series

$$1 - 2q + 3q^2 - 4q^3 + \cdots ?$$

6) What is the limit of $\frac{1+2+3+\cdots+n}{n^2}$, of $\frac{1^2+2^2+\cdots+n^2}{n^3}$, and of $\frac{1^3+2^3+\cdots+n^3}{n^4}$? (Hint: Use the results of pp. 12, 14, 15.)

4. Rational Numbers and Periodic Decimals

Those rational numbers p/q which are not finite decimal fractions can be expanded into infinite decimal fractions by performing the elementary process of long division. At each stage in this process there must be a non-zero remainder, for otherwise the decimal fraction would be finite. All the different remainders that arise in the process of division will be integers between 1 and $q-1$, so that there are at most $q-1$ different possibilities for the values of the remainders. This means that within at most q divisions some remainder k will turn up for a second time. But then all subsequent remainders will repeat in the same order in which they appeared after the remainder k first appeared. This shows that *the decimal expression for any rational number is periodic*; after some finite set of digits has appeared initially, the same digit or group of digits will repeat itself infinitely often. For example, $1/6 = 0.166666666 \cdots$; $1/7 = 0.142857142857142857 \cdots$; $1/11 = 0.09090909 \cdots$; $122/1100 = 0.1109090909 \cdots$; $11/90 = 0.122222222 \cdots$; etc. (Those rational numbers which can be represented as finite decimal fractions may be thought of as having periodic decimal expansions with the figure 0 repeating itself infinitely often after a finite number of digits.) We see, incidentally, that some of these periodic decimals have a non-periodic head before the periodic tail begins.

Conversely, it may be shown that *all periodic decimals are rational numbers.* As an example, let us take the infinite periodic decimal

$$p = 0.3322222 \cdots .$$

We have $p = 33/100 + 10^{-3}2(1 + 10^{-1} + 10^{-2} + \cdots)$. The expression in parentheses is the infinite geometrical series

$$1 + 10^{-1} + 10^{-2} + 10^{-3} + \cdots = \frac{1}{1 - 1/10} = \frac{10}{9}.$$

Hence

$$p = \frac{33}{100} + 2 \cdot 10^{-3} \cdot \frac{10}{9} = \frac{2970 + 20}{9 \cdot 10^3} = \frac{2990}{9000} = \frac{299}{900}.$$

The proof in the general case is essentially the same, but requires a more general notation. In the general periodic decimal

$$p = 0.a_1a_2a_3 \cdots a_mb_1b_2 \cdots b_nb_1b_2 \cdots b_nb_1b_2 \cdots b_n \cdots$$

we set $0.b_1b_2 \cdots b_n = B$, so that B represents the periodic part of the decimal. Then p becomes

$$p = 0.a_1a_2 \cdots a_m + 10^{-m}B(1 + 10^{-n} + 10^{-2n} + 10^{-3n} \cdots).$$

The expression in parentheses is an infinite geometrical series with $q = 10^{-n}$. Its sum, according to equation (10) of the previous article, is $1/(1 - 10^{-n})$, and therefore

$$p = 0.a_1a_2 \cdots a_m + \frac{10^{-m}B}{1 - 10^{-n}}.$$

Exercises: 1) Expand the fractions $\frac{1}{11}, \frac{1}{13}, \frac{2}{13}, \frac{3}{13}, \frac{1}{17}, \frac{2}{17}$ into decimal fractions and determine the period.

*2) The number 142,857 has the property that multiplication with any one of the numbers 2, 3, 4, 5, or 6 produces only a cyclic permutation of its digits. Explain this property, using the expansion of $\frac{1}{7}$ into a decimal fraction.

3) Expand the rational numbers of exercise 1 as "decimals" with bases 5, 7, and 12.

4) Expand one-third as a dyadic number.

5) Write .11212121 $\cdots$ as a fraction. Find the value of this symbol if it is meant in the systems with the bases 3 or 5.

5. General Definition of Irrational Numbers by Nested Intervals

On page 63 we adopted the tentative definition: a "number" is a finite or infinite decimal. We agreed that those infinite decimals which do not represent rational numbers should be called irrational numbers. On the basis of the results of the preceding section we may now formulate this definition as follows: *the continuum of numbers,* or *real number system* ("real" in contrast to the "imaginary" or "complex" numbers to be introduced in §5) *is the totality of infinite decimals.* (Finite decimals may be considered as a special case where all digits from a certain point on are zero, or one might just as well prescribe that, instead of taking a finite decimal the last digit of which is a, we write down an infinite decimal with $a-1$ in place of a, followed by an infinite number of digits all equal to 9. This expresses the fact that $.999 \cdots = 1$, according to Article 3.) The *rational* numbers are the *periodic* decimals; the *irrational* numbers are the *non-periodic* decimals. Even this definition does not seem entirely satisfactory; for, as we have seen in Chapter I, the decimal system is in no way singled out by the nature of things. We might just as well have gone through the reasoning with the dyadic or any other system. For this reason it is desirable to give a more general definition of the number continuum, detached from special reference to the base ten. Perhaps the simplest way to do this is the following:

Let us consider any sequence $I_1, I_2, \cdots, I_n, \cdots$ of intervals on the number axis with rational end-points, each of which is contained in the preceding one, and such that the length of the n-th interval I_n tends to zero as n increases. Such a sequence is called a *sequence of nested intervals.* In the case of decimal intervals the length of I_n is 10^{-n} but it may just as well be 2^{-n} or merely restricted to the milder requirement that it be less than $1/n$. Now we formulate as a basic postulate of geometry: *corresponding to each such sequence of nested intervals there is precisely one point on the number-axis which is contained in all of them.* (It is seen directly that there cannot be more than *one* point common to all the intervals, for the lengths of the intervals tend to zero, and two different points could not both be contained in any interval smaller than the distance between them.) This point is called by definition a *real number*; if it is not a rational point it is called an *irrational number.* By this definition we establish a perfect correspondence between points and numbers. It is nothing but a more general formulation of what was expressed by the definition using infinite decimals.

Here the reader may be troubled by an entirely legitimate doubt. What *is* this "point" on the number axis, which we assumed to belong to all the intervals of a nested sequence, in case it is not a rational point? Our answer is: the existence on the number axis (regarded as a line) of a point contained in every nested sequence of intervals with rational end-points is a fundamental *postulate of geometry*. No logical reduction of this postulate to other mathematical facts is required. We accept it, just as we accept other axioms or postulates in mathematics, because of its intuitive plausibility and its usefulness in building a consistent system of mathematical thought. From a purely formal point of view, we may start with a line made up only of rational points and then *define*

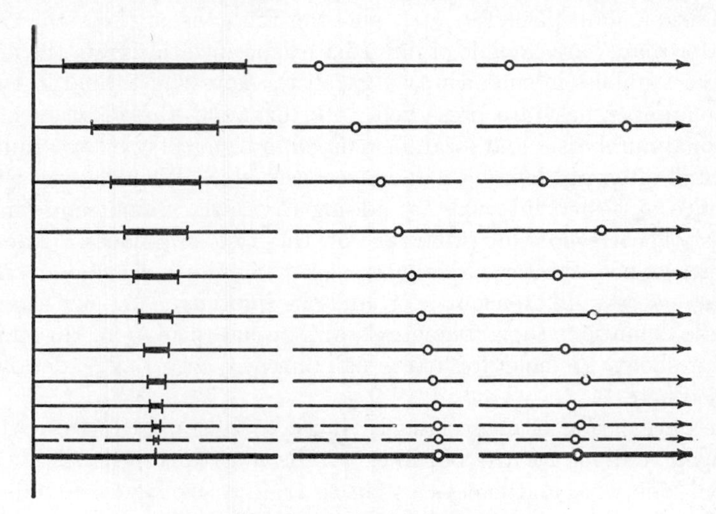

Fig. 11. Nested intervals. Limits of sequences.

an irrational point as just a *symbol for a certain sequence of nested rational intervals*. An irrational point is completely described by a sequence of nested rational intervals with lengths tending to zero. Hence our fundamental postulate really amounts to a definition. To make this definition after having been led to a sequence of nested rational intervals by an intuitive feeling that the irrational point "exists," is to throw away the intuitive crutch with which our reasoning proceeded and to realize that all the *mathematical properties* of irrational points may be expressed as properties of nested sequences of rational intervals.

We have here a typical instance of the philosophical position described in the introduction to this book; to discard the naive "realistic" approach that regards a mathematical object as a "thing in itself" of which we humbly investigate the properties, and instead to realize that the only relevant existence of mathematical objects lies in their mathematical properties and in the relations by which they are interconnected. These relations and properties exhaust the possible aspects under which an object can enter the realm of mathematical activity. We give up the mathematical "thing in itself" as physics gave up the unobservable ether. This is the meaning of the "intrinsic" definition of an irrational number as a nested sequence of rational intervals.

The mathematically important point here is that for these irrational numbers, defined as nested sequences of rational intervals, the operations of addition, multiplication, etc., and the relations of "less than" and "greater than," are capable of immediate generalization from the field of rational numbers in such a way that all the laws which hold in the rational number field are preserved. For example, the addition of two irrational numbers α and β can be defined in terms of the two sequences of nested intervals defining α and β respectively. We construct a third sequence of nested intervals by adding the initial values and the end values of corresponding intervals of the two sequences. The new sequence of nested intervals defines $\alpha + \beta$. Similarly, we may define the product $\alpha\beta$, the difference $\alpha - \beta$, and the quotient α/β. On the basis of these definitions the arithmetical laws discussed in §1 of this chapter can be shown to hold for irrational numbers also. The details are omitted here.

The verification of these laws is simple and straightforward, though somewhat tedious for the beginner who is more anxious to learn what can be done with mathematics than to analyze its logical foundations. Some modern textbooks on mathematics repel many students by starting with a pedantically complete analysis of the real number system. The reader who simply disregards these introductions may find comfort in the thought that until late in the nineteenth century all the great mathematicians made their discoveries on the basis of the "naive" concept of the number system supplied by their intuition.

From a physical point of view, the definition of an irrational number by a sequence of nested intervals corresponds to the determination of the value of some observable quantity by a sequence of measurements of greater and greater accuracy. Any given operation for determining, say, a length, will have a practical meaning only within the limits of a certain possible error which measures the precision of the operation. Since the rational numbers are dense on the line, it is impossible to determine by any physical operation, however precise, whether a given length is rational or irrational. Thus it might seem that the irrational numbers are unnecessary for the adequate description of physical phenomena. But as we shall see more clearly in Chapter VI, the real advantage which the introduction of irrational numbers brings to the mathematical description of physical phenomena is that this description is enormously simplified by the free use of the limit concept, for which the number continuum is the basis.

*6. Alternative Methods of Defining Irrational Numbers. Dedekind Cuts

A somewhat different way of defining irrational numbers was chosen by Richard Dedekind (1831–1916), one of the great pioneers in the logical and philosophical analysis of the foundations of mathematics. His essays, *Stetigkeit und irrationale Zahlen* (1872) and *Was sind und was sollen die Zahlen?* (1887), exercised a profound influence on studies in the foundations of mathematics. Dedekind preferred to operate with general abstract ideas rather than with specific sequences of nested intervals. His procedure is based on the definition of a "cut," which we shall describe briefly.

Suppose there is given some method for dividing the set of *all rational numbers* into two classes, A and B, such that every element b of class B is greater than every element a of class A. Any classification of this sort is called a *cut* in the set of rational numbers. For a cut there are just three possibilities, one and only one of which must hold:

1) *There is a largest element* a^* *of* A. This is the case, for example, if A consists of all rational numbers ≤ 1 and B of all rational numbers > 1.
2) *There is a smallest element* b^* *of* B. This is the case, for example, if A consists of all rational numbers < 1 and B of all rational numbers ≥ 1.
3) *There is neither a largest element in* A *nor a smallest element in* B. This is the case, for example, if A consists of all negative rational numbers, 0, and all positive rational numbers with square less than 2 and B of all rational numbers with square greater than 2. A and B together include all rational numbers, for we have proved that there is no rational number whose square is equal to 2.

The case in which A has a largest element a^* *and* B a smallest element b^* is impossible, for then the rational number $(a^* + b^*)/2$, which lies halfway between a^* and b^*, would be larger than the largest element of

A and smaller than the smallest element of B, and hence could belong to neither.

In the third case, where there is neither a largest rational number in A nor a smallest rational number in B, the cut is said by Dedekind to define or simply to *be* an irrational number. It is easily seen that this definition is in agreement with the definition by nested intervals; any sequence $I_1, I_2, I_3, \cdots$ of nested intervals defines a cut if we place in the class A all those rational numbers which are exceeded by the left-hand end-point of at least one of the intervals I_n, and in B all other rational numbers.

Philosophically, Dedekind's definition of irrational numbers involves a rather high degree of abstraction, since it places no restrictions on the nature of the mathematical law which defines the two classes A and B. A more concrete method of defining the real number continuum is due to Georg Cantor (1845–1918). Although at first sight quite different from the method of nested intervals or of cuts, it is equivalent to either of them, in the sense that the number systems defined in these three ways have the same properties. Cantor's idea was suggested by the facts that 1) real numbers may be regarded as infinite decimals, and 2) infinite decimals are limits of finite decimal fractions. Freeing ourselves from dependence on the decimal system, we may state with Cantor that any sequence $a_1, a_2, a_3, \cdots$ of rational numbers defines a *real number* if it "converges." Convergence is understood to mean that the difference $(a_m - a_n)$ between any two members of the sequence tends to zero when a_m and a_n are sufficiently far out in the sequence, i.e. as m and n tend to infinity. (The successive decimal approximations to any number have this property, since any two after the nth can differ by at most 10^{-n}.) Since there are many ways of approaching the same real number by a sequence of rational numbers, we say that two convergent sequences of rationals $a_1, a_2, a_3, \cdots$ and $b_1, b_2, b_3, \cdots$ define the same real number if $a_n - b_n$ tends to zero as n increases indefinitely. The operations of addition, etc., for such sequences are quite easy to define.

§3. REMARKS ON ANALYTIC GEOMETRY†

1. The Basic Principle

The number continuum, whether it is accepted as a matter of course or only after a critical examination, has been the basis of mathematics—and in particular of analytic geometry and the calculus—since the seventeenth century.

Introducing the continuum of numbers makes it possible to associate with each line segment a definite real number as its length. But we may

† For readers who are not familiar with the subject, a series of exercises on the elements of analytic geometry will be found in the appendix at the end of the book, pp. 489–494.

go much farther. Not only length, but *every geometrical object and every geometrical operation can be referred to the realm of numbers.* The decisive steps in this arithmetization of geometry were taken as early as 1629 by Fermat (1601–1655) and 1637 by Descartes (1596–1650). The fundamental idea of analytic geometry is the introduction of "coördinates," that is, *numbers* attached to or coördinated with a *geometrical object* and characterizing this object completely. Known to most readers are the so-called rectangular or Cartesian coördinates which serve to characterize the position of an arbitrary point P in a plane. We start with two fixed perpendicular lines in the plane, the "x-axis" and the "y-axis," to which we refer every point. These lines are regarded as directed number axes, and measured with the same unit. To each point P, as in Figure 12, two coördinates, x and y, are assigned. These are

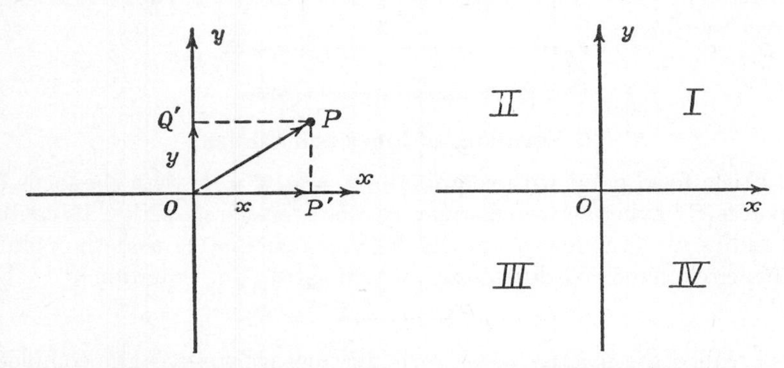

Fig. 12. Rectangular coördinates of a point.

Fig. 13. The four quadrants.

obtained as follows: we consider the directed segment from the "origin" O to the point P, and project this directed segment, sometimes called the "position vector" of the point P, perpendicularly on the two axes, obtaining the directed segment OP' on the x-axis, with the number x measuring its directed length from O, and likewise the directed segment OQ' on the y-axis, with the number y measuring its directed length from O. The two numbers x and y are called the *coördinates* of P. Conversely, if x and y are two arbitrarily prescribed numbers, then the corresponding point P is uniquely determined. If x and y are both positive, P is in the *first quadrant* of the coördinate system (see Fig. 13); if both are negative, P is in the third quadrant; if x is positive and y negative, it is in the fourth, and if x is negative and y positive, in the second.

The distance between the point P_1 with coördinates x_1, y_1 and the point P_2 with coördinates x_2, y_2 is given by the formula

$$d^2 = (x_1 - x_2)^2 + (y_1 - y_2)^2. \tag{1}$$

This follows immediately from the Pythagorean theorem, as may be seen from Figure 14.

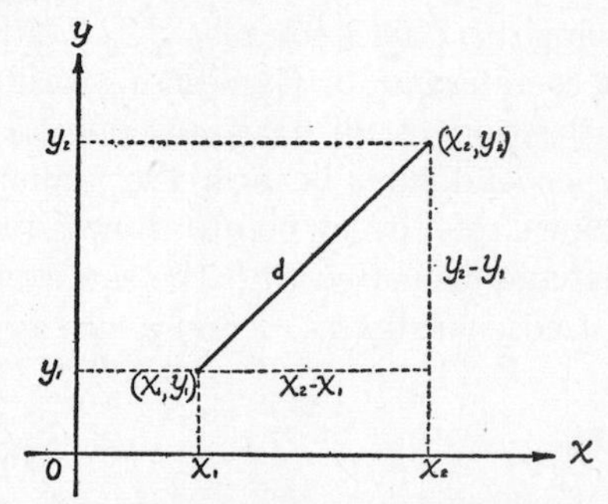

Fig. 14. The distance between two points.

*2. Equations of Lines and Curves

If C is a fixed point with coördinates $x = a$, $y = b$, then the locus of all points P having a given distance r from C is a circle with C as center and radius r. It follows from the distance formula (1) that the points of this circle have coördinates x, y which satisfy the equation

$$(x - a)^2 + (y - b)^2 = r^2. \tag{2}$$

This is called the *equation of the circle*, because it expresses the complete (necessary and sufficient) condition on the coördinates x, y of a point P

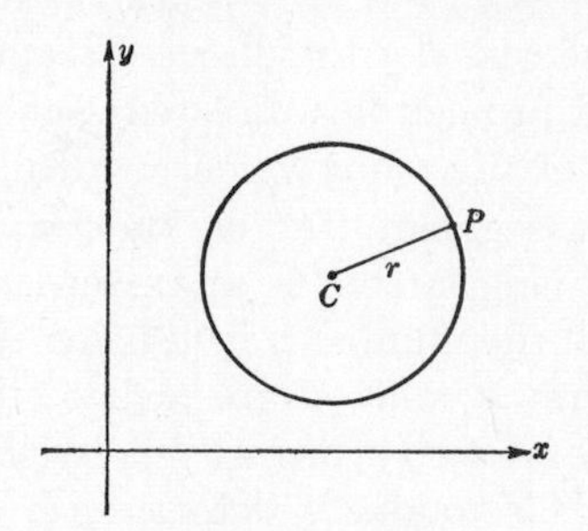

Fig. 15. The circle.

that lies on the circle around C with radius r. If the parentheses are expanded, equation (2) takes the form

$$x^2 + y^2 - 2ax - 2by = k, \tag{3}$$

where $k = r^2 - a^2 - b^2$. Conversely, if an equation of the form (3) is given, where a, b, and k are arbitrary constants such that $k + a^2 + b^2$ is positive, then by the algebraic process of "completing the square" we can write the equation in the form

$$(x - a)^2 + (y - b)^2 = r^2,$$

where $r^2 = k + a^2 + b^2$. It follows that the equation (3) defines a circle of radius r around the point C with coördinates a and b.

The equations of straight lines are even simpler in form. For example, the x-axis has the equation $y = 0$, since $y = 0$ for all points on the x-axis and for no other points. The y-axis has the equation $x = 0$. The lines through the origin bisecting the angles between the axes have the equations $x = y$ and $x = -y$. It is easily shown that any straight line has an equation of the form

$$ax + by = c, \tag{4}$$

where a, b, c are fixed constants characterizing the line. The meaning of equation (4) is again that all pairs of real numbers x, y which satisfy this equation are the coördinates of a point of the line, and conversely.

The reader may have learned that the equation

$$\frac{x^2}{p^2} + \frac{y^2}{q^2} = 1 \tag{5}$$

represents an ellipse (Fig. 16). This curve cuts the x-axis at the points $A(p, 0)$ and $A'(-p, 0)$, and the y-axis at $B(0, q)$ and $B'(0, -q)$. (The notation $P(x, y)$ or simply (x, y) is used as a shorter way of writing "the point P with coördinates x and y.") If $p > q$, the segment AA', of length $2p$, is called the major axis of the ellipse, while the segment BB', of length $2q$, is called the minor axis. This ellipse is the locus of all points P the sum of whose distances from the points $F(\sqrt{p^2 - q^2}, 0)$ and $F'(-\sqrt{p^2 - q^2}, 0)$ is $2p$. As an exercise the reader may verify this by using formula (1). The points F and F' are called the *foci* (singular, *focus*) of the ellipse, and the ratio $e = \dfrac{\sqrt{p^2 - q^2}}{p}$ is called the *eccentricity* of the ellipse.

An equation of the form

$$\frac{x^2}{p^2} - \frac{y^2}{q^2} = 1 \tag{6}$$

represents a hyperbola. This curve consists of two branches which cut the x-axis at $A(p, 0)$ and $A'(-p, 0)$ (Fig. 17) respectively. The segment

AA', of length $2p$, is called the transverse axis of the hyperbola. The hyperbola approaches more and more nearly the two straight lines $qx \pm py = 0$ as we go out farther and farther from the origin, but it never actually reaches these lines. They are called the *asymptotes* of the hyperbola. The hyperbola is the locus of all points P the *difference* of whose distances to the two points $F(\sqrt{p^2 + q^2}, 0)$ and $F'(-\sqrt{p^2 + q^2}, 0)$ is $2p$. These points are again called the foci of the hyperbola; by its eccentricity we mean the ratio $e = \dfrac{\sqrt{p^2 + q^2}}{p}$.

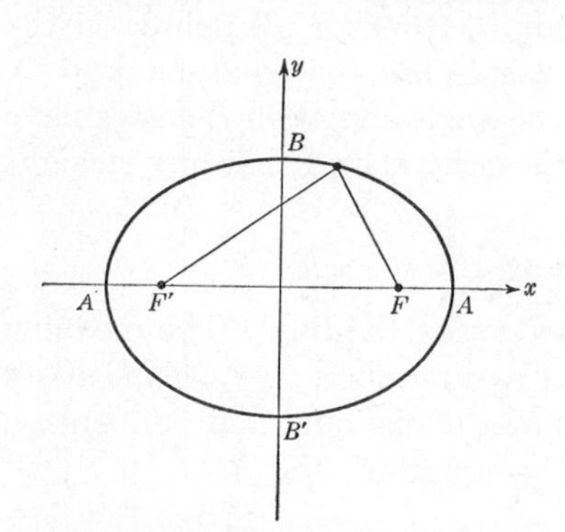

Fig. 16. The ellipse; F and F' are the foci.

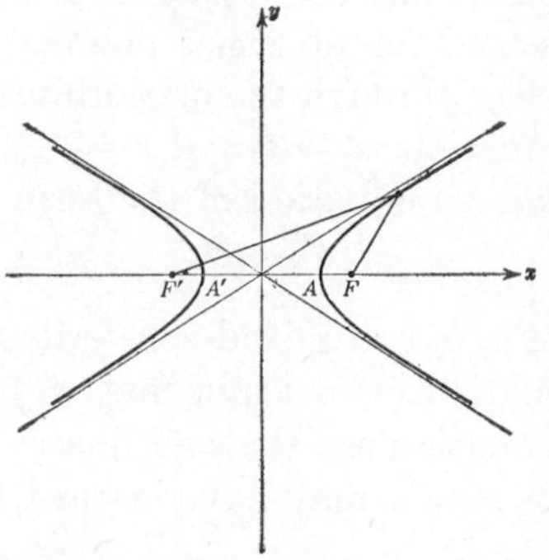

Fig. 17. The hyperbola; F and F' are the foci.

The equation

$$xy = 1 \tag{7}$$

also defines a hyperbola, whose asymptotes now are the two axes (Fig. 18). The equation of this "equilateral" hyperbola indicates that the area of the rectangle determined by P is equal to 1 for every point P on the curve. An equilateral hyperbola whose equation is

$$xy = c, \tag{7a}$$

c being a constant, is only a special case of the general hyperbola, just as the circle is a special case of the ellipse. The special character of the equilateral hyperbola lies in the fact that its two asymptotes (in this case the two coördinate axes) are perpendicular to each other.

For us the main point here is the fundamental idea that geometrical objects may be completely represented in numerical and algebraic terms, and that the same is true of geometrical operations. For example, if we want to find the point of intersection of two lines, we consider their two equations

$$(8)\qquad \begin{aligned} ax + by &= c \\ a'x + b'y &= c'. \end{aligned}$$

The point common to the two lines is then found simply by determining its coördinates as the solution x, y of the two simultaneous equations (8). Similarly, the points of intersection of any two curves, such as the circle $x^2 + y^2 - 2ax - 2by = k$ and the straight line $ax + by = c$, are found by solving the two corresponding equations simultaneously.

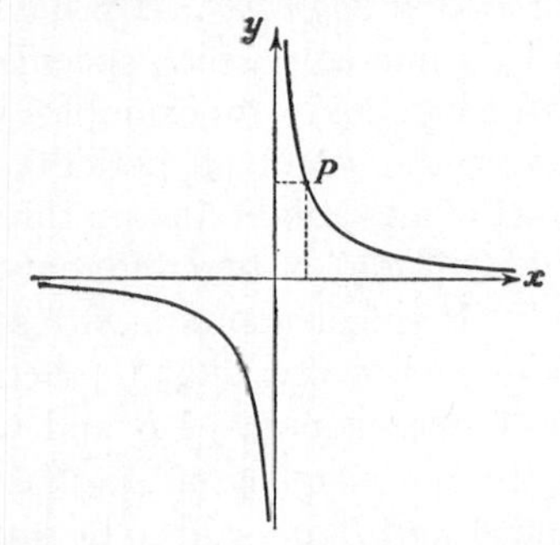

Fig. 18. The equilateral hyperbola $xy = 1$. The area xy of the rectangle determined by the point P (x, y) is equal to 1.

§4. THE MATHEMATICAL ANALYSIS OF INFINITY

1. Fundamental Concepts

The sequence of positive integers

$$1, 2, 3, \cdots$$

is the first and most important example of an infinite set. There is no mystery about the fact that this sequence has no end, no "finis"; for, however large be the integer n, the next integer, $n + 1$, can always be formed. But in the passage from the *adjective* "infinite," meaning simply "without end," to the *noun* "infinity" we must not make the assumption that "infinity," usually expressed by the special symbol ∞, can be considered as though it were an ordinary *number*. We cannot include the symbol ∞ in the real number system and at the same time preserve the fundamental rules of arithmetic. Nevertheless, the concept of the infinite pervades all of mathematics, since mathematical objects are usually studied, not as individuals, but as members of classes or aggregates containing infinitely many objects of the same type, such as the totality of integers, or of real numbers, or of triangles in a plane. For this reason it is necessary to analyze the mathematical infinite in a precise way. The modern theory of sets, created by Georg Cantor and his school at the end of the nineteenth century, has met this challenge with striking success. Cantor's theory of sets has penetrated and strongly influenced many fields of mathematics, and has become of basic importance in the study of the logical and philosophical foundations of mathematics. The point of departure is the general concept of a *set* or *aggregate*. By this is meant any collection of objects defined by some rule which specifies exactly which objects belong to the given collection. As examples we may consider the set of all positive integers, the set of all periodic decimals, the set of all real numbers, or the set of all straight lines in three-dimensional space.

For comparing the "magnitude" of two different sets the basic notion is that of "equivalence." If the elements in two sets A and B may be paired with each other in such a way that to each element of A there corresponds one and only one element of B and to each element of B corresponds one and only one element of A, then the correspondence is said to be *biunique* and A and B are said to be *equivalent*. The notion of equivalence for *finite* sets coincides with the ordinary notion of *equality of number*, since two finite sets have the same number of elements if and only if the elements of the two sets can be put into biunique correspondence. This is in fact the very idea of counting, for when we count a finite set of objects, we simply establish a biunique correspondence between these objects and a set of number symbols $1, 2, 3, \cdots, n$.

It is not always necessary to count the objects in two finite sets to establish their equivalence. For example, we can assert without counting that any finite set of circles of radius 1 is equivalent to the set of their centers.

Cantor's idea was to extend the concept of equivalence to infinite sets in order to define an "arithmetic" of infinities. The set of all real numbers and the set of all points on a straight line are equivalent, since the choice of an origin and a unit allows us to associate in a biunique manner with every point P of the line a definite real number x as its coördinate:

$$P \leftrightarrow x.$$

The *even integers* form a proper subset of the set of *all integers*, and the *integers* form a proper subset of the set of all *rational numbers*. (By the phrase *proper subset* of a set S, we mean a set S' consisting of some, but not all, of the objects in S.) Clearly, *if a set is finite*, i.e. if it contains some number n of elements and no more, *then it cannot be equivalent to any one of its proper subsets,* since any proper subset could contain at most $n - 1$ elements. But, *if a set contains infinitely many objects,* then, paradoxically enough, *it may be equivalent to a proper subset of itself.* For example, the coördination

$$\begin{array}{cccccccc} 1 & 2 & 3 & 4 & 5 & \cdots & n & \cdots \\ \updownarrow & \updownarrow & \updownarrow & \updownarrow & \updownarrow & & \updownarrow & \\ 2 & 4 & 6 & 8 & 10 & \cdots & 2n & \cdots \end{array}$$

establishes a biunique correspondence between the set of *positive integers* and the proper subset of *even integers,* which are thereby shown to be equivalent. This contradiction to the familiar truth, "the whole is greater than any of its parts," shows what surprises are to be expected in the domain of the infinite.

2. The Denumerability of the Rational Numbers and the Non-Denumerability of the Continuum

One of Cantor's first discoveries in his analysis of the infinite was that the set of *rational numbers* (which contains the infinite set of integers as a subset and is therefore itself infinite) is equivalent to the *set of integers.* At first sight it seems very strange that the dense set of rational numbers should be on the same footing as its sparsely sown subset of integers. True, one cannot arrange the positive rational numbers *in order of size* (as one can the integers) by saying that a is the first rational number, b the next larger, and so forth, because there are infinitely many rational numbers between any two given ones, and hence there is no "next larger." But, as Cantor observed, by disregarding the relation of magnitude between successive elements, it is possible to arrange all the rational numbers in a single row, $r_1, r_2, r_3, r_4, \cdots$, like that of the integers. In this sequence there will be a first rational number, a second, a third, and so forth, and every rational number will appear exactly once. Such an arrangement of a set of objects in a sequence like that of the integers is called a *denumeration* of the set. By exhibiting such a denumeration Cantor showed the set of rational numbers to be equivalent with the set of integers, since the correspondence

$$\begin{array}{ccccccc} 1 & 2 & 3 & 4 & \cdots & n & \cdots \\ \updownarrow & \updownarrow & \updownarrow & \updownarrow & & \updownarrow & \\ r_1 & r_2 & r_3 & r_4 & \cdots & r_n & \cdots \end{array}$$

is biunique. One way of denumerating the rational numbers will now be described.

Every rational number can be written in the form a/b, where a and b are integers, and all these numbers can be put in an array, with a/b in the ath column and bth row. For example, 3/4 is found in the third column and fourth row of the table below. All the positive rational numbers may now be arranged according to the following scheme: in the array just defined we draw a continuous, broken line that goes through all the numbers in the array. Starting at 1, we go horizontally to the next place on the right, obtaining 2 as the second member of the sequence, then diagonally down to the left until the first column is reached at the position occupied by 1/2, then vertically down one place to 1/3, diagonally up until the first row is reached again at 3, across to 4, diagonally down to 1/4, and so on, as shown in the figure. Travelling along this broken line we arrive at a sequence 1, 2, 1/2, 1/3, 2/2, 3, 4, 3/2, 2/3, 1/4, 1/5, 2/4, 3/3, 4/2, 5, $\cdots$ containing the rational numbers in the order in which they occur along the broken line. In this sequence we now cancel all those numbers a/b for which a and b have a common factor, so that each rational number r will appear exactly once and in its simplest form. Thus we obtain a sequence

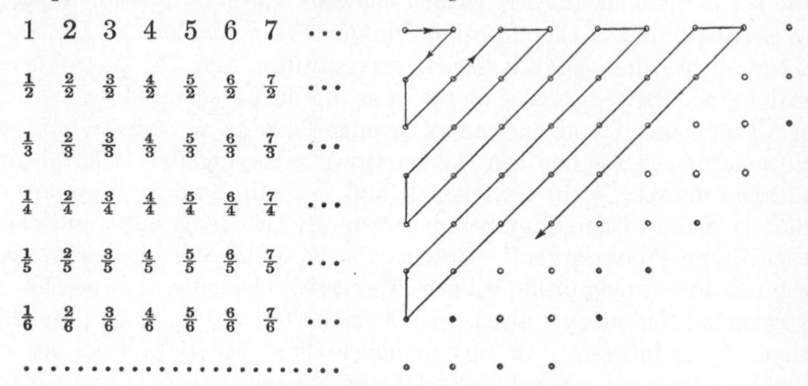

Fig. 19. Denumeration of the rational numbers.

1, 2, 1/2, 1/3, 3, 4, 3/2, 2/3, 1/4, 1/5, 5, $\cdots$ which contains each positive rational number once and only once. This shows that the set of all positive rational numbers is denumerable. In view of the fact that the rational numbers correspond in a biunique manner with the rational points on a line, we have proved at the same time that the set of positive rational points on a line is denumerable.

Exercises: 1) Show that the set of all positive and negative integers is denumerable. Show that the set of all positive and negative rational numbers is denumerable.

2) Show that the set $S + T$ (see p. 110) is denumerable if S and T are denumerable sets. Show the same for the sum of three, four, or any number, n, of sets, and finally for a set composed of denumerably many denumerable sets.

Since the rational numbers have been shown to be denumerable, one might suspect that *any* infinite set is denumerable, and that this is the ultimate result of the analysis of the infinite. This is far from being the case. Cantor made the very significant discovery that *the set of all real numbers*, rational and irrational, *is not denumerable.* In other words, the totality of real numbers presents a radically different and, so to speak, higher type of infinity than that of the integers or of the rational numbers alone. Cantor's ingenious indirect proof of this fact has become a model for many mathematical demonstrations. The outline of the proof is as follows. We start with the tentative assumption that all the real numbers have actually been denumerated in a sequence, and then we exhibit a number which does not occur in the assumed denumeration. This provides a contradiction, since the assumption was that *all* the real numbers were included in the denumeration, and this assumption must be false if even one number has been left out. Thus the assumption that a denumeration of the real numbers is possible is shown to be untenable, and hence the opposite, i.e. Cantor's statement that the set of real numbers is not denumerable, is shown to be true.

To carry out this program, let us suppose that we have denumerated all the real numbers by arranging them in a table of infinite decimals,

1st number	$N_1 . a_1a_2a_3a_4a_5 \cdots$
2nd number	$N_2 . b_1b_2b_3b_4b_5 \cdots$
3rd number	$N_3 . c_1c_2c_3c_4c_5 \cdots$
....	....

where the N's denote the integral parts and the small letters denote the digits after the decimal point. We assume that this sequence of decimal fractions contains *all* the real numbers. The essential point in the proof is now to construct by a "diagonal process" a new number which we can show to be not included in this sequence. To do this we first choose a digit a which differs from a_1 and is neither 0 nor 9 (to avoid possible ambiguities which may arise from equalities like $0.999 \cdots = 1.000 \cdots$), then a digit b different from b_2 and again unequal to 0 or 9, similarly c different from c_3, and so on. (For example, we might simply choose $a = 1$ unless $a_1 = 1$, in which case we choose $a = 2$, and similarly down the table for all the digits b, c, d, e, $\cdots$.) Now consider the infinite decimal

$$z = 0.abcde \cdots .$$

This new number z is certainly different from any one of the numbers in the table above; it cannot be equal to the first because it differs from it in the first digit after the decimal point; it cannot be equal to the second since it differs from it in the second digit; and, in general, it cannot be identical with the nth number in the table since it differs from it in the nth digit. This shows that our table of consecutively arranged decimals does *not* contain all the real numbers. Hence this set is not denumerable.

The reader may perhaps imagine that the reason for the non-denumerability of the number continuum lies in the fact that the straight line is infinite in extent, and that a finite segment of the line would contain only a denumerable infinity of points. This is not the case, for

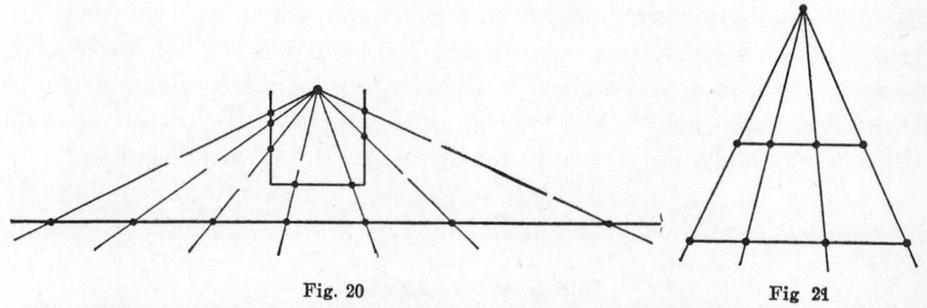

Fig. 20 Fig 21

Fig. 20. Biunique correspondence between the points of a bent segment and a whole straight line.
Fig. 21. Biunique correspondence between the points of two segments of different length.

it is easy to show that the entire number continuum is equivalent to any finite segment, say the segment from 0 to 1 with the endpoints excluded. The desired biunique correspondence may be obtained by bending the segment at $\frac{1}{3}$ and $\frac{2}{3}$ and projecting from a point, as shown in Figure 20. It follows that even a finite segment of the number axis contains a non-denumerable infinity of points.

Exercise: Show that any interval $[A, B]$ of the number axis is equivalent to any other interval $[C, D]$.

It is worthwhile to indicate another and perhaps more intuitive proof of the non-denumerability of the number continuum. In view of what we have just proved it will be sufficient to confine our attention to the set of points between 0 and 1. Again the proof is indirect. Let us suppose that the set of all points on the line between 0 and 1 can be arranged in a sequence

$$a_1, a_2, a_3, \cdots. \tag{1}$$

Let us enclose the point with coördinate a_1 in an interval of length $1/10$, the point with coördinate a_2 in an interval of length $1/10^2$, and so on. If all points between 0 and 1 were included in the sequence (1), the unit interval would be entirely covered by an infinite sequence of possibly overlapping subintervals of lengths $1/10, 1/10^2, \cdots$. (The fact that some of these extend beyond the unit interval does not influence our proof.) The sum of these lengths is given by the geometric series

$$1/10 + 1/10^2 + 1/10^3 + \cdots = \frac{1}{10}\left[\frac{1}{1-\frac{1}{10}}\right] = \frac{1}{9}.$$

Thus the assumption that the sequence (1) contains all real numbers from 0 to 1 leads to the possibility of covering the whole of an interval of length 1 by a set of intervals of total length $1/9$, which is intuitively absurd. We might accept this contradiction as a proof, although from a logical point of view it would require fuller analysis.

The reasoning of the preceding paragraph serves to establish a theorem of great importance in the modern theory of "measure". Replacing the intervals above by smaller intervals of length $\epsilon/10^n$, where ϵ is an arbitrary small positive number, we see that any denumerable set of points on the line can be included in a set of intervals of total length $\epsilon/9$. Since ϵ was arbitrary, the latter number can be made as small as we please. In the terminology of measure theory we say that a denumerable set of points has the *measure zero*.

Exercise: Prove that the same result holds for a denumerable set of points in the plane, replacing lengths of intervals by areas of squares.

3. Cantor's "Cardinal Numbers"

In summary of the results thus far: The number of elements in a *finite* set A cannot equal the number of elements in a finite set B if A contains *more* elements than B. If we replace the concept of "sets with the same (finite) number of elements" by the more general concept of *equivalent sets,* then with infinite sets the previous statement does not hold; the set of all integers contains more elements than the set of even integers, and the set of rational numbers more than the set of integers, but we have seen that these sets are equivalent. One might suspect that *all* infinite sets are equivalent and that distinctions other than that between finite numbers and infinity could not be made, but

Cantor's result disproves this; there is a set, the real number continuum, which is not equivalent to any denumerable set.

Thus there are at least two different types of "infinity," the denumerable infinity of the integers and the non-denumerable infinity of the continuum. If two sets A and B, finite or infinite, are equivalent, we shall say that they have the *same cardinal number*. This reduces to the ordinary notion of *same natural number* if A and B are finite, and may be regarded as a valid generalization of this concept. Moreover, if a set A is equivalent with some subset of B, while B is not equivalent to A or to any of its subsets, we shall say, following Cantor, that the set B has a *greater cardinal number* than the set A. This use of the word "number" also agrees with the ordinary notion of greater number for finite sets. The set of integers is a subset of the set of real numbers, while the set of real numbers is neither equivalent to the set of integers nor to any subset of it (i.e. the set of real numbers is neither denumerable nor finite). Hence, according to our definition, the continuum of real numbers has a greater cardinal number than the set of integers.

* As a matter of fact, Cantor actually showed how to construct a whole sequence of infinite sets with greater and greater cardinal numbers. Since we may start with the set of positive integers, it clearly suffices to show that *given any set* A *it is possible to construct another set* B *with a greater cardinal number*. Because of the great generality of this theorem, the proof is necessarily somewhat abstract. We define the set B to be the set whose elements are all the different subsets of the set A. By the word "subset" we shall include not only the proper subsets of A but also the set A itself, and the empty "subset" 0, containing no elements at all. (Thus, if A consists of the three integers 1, 2, 3, then B contains the 8 different elements $\{1, 2, 3\}$, $\{1, 2\}$, $\{1, 3\}$, $\{2, 3\}$, $\{1\}$, $\{2\}$, $\{3\}$, and 0.) Each element of the set B is itself a *set*, consisting of certain elements of A. Now suppose that B is equivalent to A or to some subset of it, i.e. that there is some rule which correlates in a biunique manner the elements of A or of a subset of A with all the elements of B, i.e. with the subsets of A:

$$a \longleftrightarrow S_a, \tag{2}$$

where we denote by S_a the subset of A corresponding to the element a of A. We shall arrive at a contradiction by exhibiting an element of B (i.e. a subset T of A) which cannot have any element a correlated with it. In order to construct this subset we observe that for any element x of A two possibilities exist: either the set S_x assigned to x in the given correspondence (2) contains the element x, or S_x does not contain x. *We define* T *as the subset of* A *consisting of all those elements* x *such that* S_x *does not contain* x. This subset differs from every S_a by at least the element a, since if S_a contains a, T does not, while if S_a does not contain a, T does. Hence T is not included in the correspondence (2). This shows that it is impossible to set up a biunique correspondence between the elements of A or of any subset of A and those of B. But the correlation

$$a \longleftrightarrow \{a\}$$

defines a biunique correspondence between the elements of A and the subset of B consisting of all one-element subsets of A. Hence, by the definition of the last paragraph, B has a greater cardinal number than A.

* *Exercise:* If A contains n elements, where n is a positive integer, show that B, defined as above, contains 2^n elements. If A consists of the set of all positive integers, show that B is equivalent to the continuum of real numbers from 0 to 1. (Hint: Symbolize a subset of A in the first case by a finite and in the second case by an infinite sequence of the symbols 0 and 1,

$$a_1a_2a_3 \cdots,$$

where $a_n = 1$ or 0, according as the nth element of A does or does not belong to the given subset.)

One might think it a simple matter to find a set of *points* with a greater cardinal number than the set of real numbers from 0 to 1. Certainly a square, being "two-dimensional," would appear to contain "more" points than a "one-dimensional" segment. Surprisingly enough, this is not so; *the cardinal number of the set of points in a square is the same as the cardinal number of the set of points on a segment.* To prove this we set up the following correspondence.

If (x, y) is a point of the unit square, x and y may be written in decimal form as

$$x = 0.a_1a_2a_3a_4 \cdots,$$

$$y = 0.b_1b_2b_3b_4 \cdots,$$

where to avoid ambiguity we choose, for example, $0.250000 \cdots$ instead of $0.249999 \cdots$ for the rational number $\frac{1}{4}$. To the point (x, y) of the square we then assign the point

$$z = 0.a_1b_1a_2b_2a_3b_3a_4b_4 \cdots$$

of the segment from 0 to 1. Clearly, different points (x, y) and (x', y') of the square will correspond to different points z and z' of the segment, so that the cardinal number of the square cannot exceed that of the segment.

(As a matter of fact, the correspondence just defined is biunique between the set of all points of the square and a proper subset of the unit segment; no point of the square could correspond to the point $0.2140909090 \cdots$, for example, since the form $0.25000 \cdots$ rather than $0.24999 \cdots$ was chosen for the number $\frac{1}{4}$. But it is possible to modify the correspondence slightly so that it will be biunique between the whole square and the whole segment, which are thus seen to have the same cardinal number.)

A similar argument shows that the cardinal number of the points in a cube is no greater than the cardinal number of the segment.

Although these results seem to contradict the intuitive notion of dimensionality, we must remember that the correspondence we have defined is not "continuous"; if we travel along the segment from 0 to 1 continuously, the corresponding points in the square will not form a continuous curve but will appear in a completely chaotic order. The dimension of a set of points depends not only on the cardinal number of the set, but also on the manner in which the points are distributed in space. In Chapter V we shall return to this subject.

4. The Indirect Method of Proof

The theory of cardinal numbers is but one aspect of the general theory of sets, created by Cantor in the face of severe criticism by some of the most distinguished mathematicians of the time. Many of these critics, such as Kronecker and Poincaré, objected to the vagueness of the general concept of "set," and to the non-constructive character of the reasoning used to define certain sets.

The objections to non-constructive reasoning refer to what may be called *essentially indirect* proofs. Indirect proofs themselves are a familiar sort of mathematical reasoning: to establish the truth of a statement A, one makes the tentative assumption that A', the contrary of A, is true. Then by some chain of reasoning one produces a contradiction to A', thus demonstrating the absurdity of A'. Hence, on the basis of the fundamental logical principle of the "excluded middle," the absurdity of A' establishes the truth of A.

Throughout this book we shall meet with examples where an indirect proof can easily be converted into a direct proof, though the indirect form of proof often has the advantages of brevity and freedom from details not necessary for the immediate objective. But there are some theorems for which it has not yet been possible to give other than indirect proofs. There are even theorems, provable by the indirect method, for which direct constructive proofs could not possibly be given even in principle, because of the very nature of the theorems themselves. Such, for example, is the theorem on page 81. On different occasions in the history of mathematics, when the efforts of mathematicians were directed towards *constructing* solutions for certain problems in order to show their solvability, someone else came along and sidestepped the task of construction by giving an indirect and non-constructive proof.

There is an essential difference between proving the existence of an object of a certain type by constructing a tangible example of such an object, and showing that if none existed one could deduce contradictory results. In the first case one has a tangible object, while in the second case one has only the contradiction. Some distinguished mathematicians have recently advocated the more or less complete banishment from mathematics of all non-constructive proofs. Even if such a program were desirable, it would at present involve tremendous complication and even the partial destruction of the body of living mathematics. For this reason it is no wonder that the school of "intuitionism," which has adopted this program, has met with strong resistance, and that even the most thoroughgoing intuitionists cannot always live up to their convictions.

5. The Paradoxes of the Infinite

Although the uncompromising position of the intuitionists is far too extreme for most mathematicians, a serious threat to the beautiful theory of infinite aggregates arose when outright logical paradoxes in the theory became apparent. It was soon observed that unrestricted freedom in using the concept of "set" must lead to contradiction. One of the paradoxes, exhibited by Bertrand Russell, may be formulated as follows. Most sets do not contain themselves as elements. For example, the set A of all integers contains as elements only integers; A, being itself not an integer but a *set of integers*, does not contain itself as element. Such a set we may call "ordinary." There may possibly be sets which do contain themselves as elements; for example, the set S defined as follows: "S contains as elements all sets definable by an English phrase of less than twenty words" could be considered to contain itself as an element. Such sets we might call "extraordinary" sets. In any case, however, most sets will be ordinary, and we may exclude the erratic behavior of "extraordinary" sets by confining our attention to the *set of all ordinary sets*. Call this set C. Each element of the set C is itself a set; in fact an ordinary set. The question now arises, is C itself an ordinary set or an extraordinary set? It must be one or the other. If C is ordinary, it contains itself as an element, since C is defined as containing *all* ordinary sets. This being so, C must be extraordinary, since the extraordinary sets are those containing themselves as members. This is a contradiction. Hence C must be extraordinary. But then C contains as a member an extraordinary set (namely C itself), which contradicts the definition whereby C was to contain ordinary sets only. Thus in either case we see that the assumption of the mere existence of the set C has led us to a contradiction.

6. The Foundations of Mathematics

Paradoxes like this have led Russell and others to a systematic study of the foundations of mathematics and logic. The ultimate aim of their efforts is to provide for mathematical reasoning a logical basis which can be shown to be free from possible contradiction, and which still covers everything that is considered important by all (or some) mathematicians. While this ambitious goal has not been attained and perhaps cannot ever be attained, the subject of mathematical logic has attracted the attention of increasing numbers of students. Many problems in this field which can be stated in very simple terms are very difficult to solve. As an example, we mention the *Hypothesis of the Continuum*, which states that there is no set whose cardinal number is greater than that of the set of the integers but less than that of the set of real numbers. Many interesting consequences can be deduced from this hypothesis, but up to now it has neither been proved nor disproved, though it has recently been shown by Kurt Gödel that if the usual postulates at the basis of set theory are consistent, then the enlarged set of postulates obtained by adding the Hypothesis of the Continuum is also consistent. Questions such as this ultimately reduce to the question of what is meant by the concept of *mathematical existence.* Luckily, the existence of mathematics does not depend on a satisfactory answer. The school of "formalists," led by the great mathematician Hilbert, asserts that in mathematics "existence" simply means "freedom from contradiction." It then becomes necessary to construct a set of postulates from which all of mathematics can be deduced by purely formal reasoning, and to show that this set of postulates will never lead to a contradiction. Recent results by Gödel and others seem to show that this program, at least as originally conceived by Hilbert, cannot be carried out. Significantly, Hilbert's theory of the formalized structure of mathematics is essentially based on intuitive procedure. In some way or other, openly or hidden, even under the most uncompromising formalistic, logical, or postulational aspect, constructive intuition always remains the vital element in mathematics.

§5. COMPLEX NUMBERS

1. The Origin of Complex Numbers

For many reasons the concept of number has had to be extended even beyond the real number continuum by the introduction of the so-called complex numbers. One must realize that in the historical and psychological development of mathematics, all these extensions and new inventions were by no means the products of some one individual's efforts. They appear rather as the outcome of a gradual and hesitant evolution for which no single person can receive major credit. It was the need for more freedom in formal calculations that brought about the use of negative and rational numbers. Only at the end of the middle ages did mathematicians begin to lose their feeling of uneasiness in using these concepts, which did not appear to have the same intuitive and concrete character as do the natural numbers. It was not until the middle of the nineteenth century that mathematicians fully realized that the essential logical and philosophical basis for operating in an extended number domain is formalistic; that extensions have to be created by definitions which, as such, are free, but which are useless if not made in such a way that the prevailing rules and properties of the original domain are preserved in the larger domain. That these extensions may sometimes be linked with "real" objects and in this way provide tools for new applications is of the highest importance, but this can provide only a motivation and not a logical proof of the validity of the extension.

The process which first requires the use of complex numbers is that of *solving quadratic equations*. We recall the concept of the linear equation, $ax = b$, where the unknown quantity x is to be determined. The solution is simply $x = b/a$, and the requirement that every linear equation with integral coefficients $a \neq 0$ and b shall have a solution necessitated the introduction of the rational numbers. Equations such as

$$x^2 = 2, \tag{1}$$

which has no solution x in the field of rational numbers, led us to construct the wider field of real numbers in which a solution does exist. But even the field of real numbers is not wide enough to provide a complete theory of quadratic equations. A simple equation like

$$x^2 = -1 \tag{2}$$

has no real solution, since the square of any real number is never negative.

We must either be content with the statement that this simple equation is not solvable, or follow the familiar path of extending our concept of number by introducing numbers that will make the equation solvable. This is exactly what is done when we introduce the new symbol i by defining $i^2 = -1$. Of course this object i, the "imaginary unit," has nothing to do with the concept of a number as a means of *counting*. It is purely a *symbol*, subject to the fundamental rule $i^2 = -1$, and its value will depend entirely on whether by this introduction a really useful and workable extension of the number system can be effected.

Since we wish to add and multiply with the symbol i as with an ordinary real number, we should be able to form symbols like $2i$, $3i$, $-i$,

$2 + 5i$, or more generally, $a + bi$, where a and b are any two real numbers. If these symbols are to obey the familiar commutative, associative, and distributive laws of addition and multiplication, then, for example,

$$(2 + 3i) + (1 + 4i) = (2 + 1) + (3 + 4)i = 3 + 7i,$$
$$(2 + 3i)(1 + 4i) = 2 + 8i + 3i + 12i^2$$
$$= (2 - 12) + (8 + 3)i = -10 + 11i.$$

Guided by these considerations we begin our systematic exposition by making the following *definition*: A symbol of the form $a + bi$, where a and b are any two real numbers, shall be called a *complex number* with *real part* a and *imaginary part* b. The operations of addition and multiplication shall be performed with these symbols just as though i were an ordinary real number, except that i^2 shall always be replaced by -1. More precisely, we define addition and multiplication of complex numbers by the rules

$$(3) \qquad \begin{aligned} (a + bi) + (c + di) &= (a + c) + (b + d)i, \\ (a + bi)(c + di) &= (ac - bd) + (ad + bc)i. \end{aligned}$$

In particular, we have

$$(4) \qquad (a + bi)(a - bi) = a^2 - abi + abi - b^2i^2 = a^2 + b^2.$$

On the basis of these definitions it is easily verified that the commutative, associative, and distributive laws hold for complex numbers. Moreover, not only addition and multiplication, but also subtraction and division of two complex numbers lead again to numbers of the form $a + bi$, so that the complex numbers form a *field* (see p. 56):

$$(5) \qquad \begin{aligned} (a + bi) - (c + di) &= (a - c) + (b - d)i, \\ \frac{a + bi}{c + di} = \frac{(a + bi)}{(c + di)}\frac{(c - di)}{(c - di)} &= \left(\frac{ac + bd}{c^2 + d^2}\right) + \left(\frac{bc - ad}{c^2 + d^2}\right)i. \end{aligned}$$

(The second equation is meaningless when $c + di = 0 + 0i$, for then $c^2 + d^2 = 0$. So again *we must exclude division by zero*, i.e. by $0 + 0i$.) For example,

$$(2 + 3i) - (1 + 4i) = 1 - i,$$

$$\frac{2 + 3i}{1 + 4i} = \frac{2 + 3i}{1 + 4i}\cdot\frac{1 - 4i}{1 - 4i} = \frac{2 - 8i + 3i + 12}{1 + 16} = \frac{14}{17} - \frac{5}{17}i.$$

The field of complex numbers includes the field of real numbers as a subfield, for the complex number $a + 0i$ is regarded as the same as the real number a. On the other hand, a complex number of the form $0 + bi = bi$ is called a pure imaginary number.

Exercises: 1) Express $\dfrac{(1+i)(2+i)(3+i)}{(1-i)}$ in the form $a + bi$.

2) Express

$$\left(-\frac{1}{2} + i\frac{\sqrt{3}}{2}\right)^3$$

in the form $a + bi$.

3) Express in the form $a + bi$:

$$\frac{1+i}{1-i},\ \frac{1+i}{2-i},\ \frac{1}{i^5},\ \frac{1}{(-2+i)(1-3i)},\ \frac{(4-5i)^2}{(2-3i)^2}.$$

4) Calculate $\sqrt{5 + 12i}$. (Hint: Write $\sqrt{5+12i} = x + yi$, square, and equate real and imaginary parts.)

By the introduction of the symbol i we have extended the field of real numbers to a field of symbols $a + bi$ in which the special quadratic equation

$$x^2 = -1$$

has the two solutions $x = i$ and $x = -i$. For by definition, $i \cdot i = (-i)(-i) = i^2 = -1$. In reality we have gained much more: we can easily verify that now *every quadratic equation*, which we may write in the form

$$ax^2 + bx + c = 0, \tag{6}$$

has a solution. For from (6) we have

$$x^2 + \frac{b}{a}x = -\frac{c}{a},$$

$$x^2 + \frac{b}{a}x + \frac{b^2}{4a^2} = \frac{b^2}{4a^2} - \frac{c}{a},$$

$$\left(x + \frac{b}{2a}\right)^2 = \frac{b^2 - 4ac}{4a^2}, \tag{7}$$

$$x + \frac{b}{2a} = \frac{\pm\sqrt{b^2 - 4ac}}{2a},$$

$$x = \frac{-b \pm \sqrt{b^2 - 4ac}}{2a}.$$

Now if $b^2 - 4ac \geq 0$, then $\sqrt{b^2 - 4ac}$ is an ordinary real number, and the solutions (7) are real, while if $b^2 - 4ac < 0$, then $4ac - b^2 > 0$ and $\sqrt{b^2 - 4ac} = \sqrt{-(4ac - b^2)} = \sqrt{4ac - b^2} \cdot i$, so that the solutions (7) are complex numbers. For example, the solutions of the equation

$$x^2 - 5x + 6 = 0$$

are $x = (5 \pm \sqrt{25 - 24})/2 = (5 \pm 1)/2 = 2$ or 3, while the solutions of the equation

$$x^2 - 2x + 2 = 0,$$

are $x = (2 \pm \sqrt{4 - 8})/2 = (2 \pm 2i)/2 = 1 + i$ or $1 - i$.

2. The Geometrical Interpretation of Complex Numbers

As early as the sixteenth century mathematicians were compelled to introduce expressions for square roots of negative numbers in order to solve all quadratic and cubic equations. But they were at a loss to explain the exact meaning of these expressions, which they regarded with superstitious awe. The name "imaginary" is a reminder of the fact that these expressions were considered to be somehow fictitious and unreal. Finally, early in the nineteenth century, when the importance of these numbers in many branches of mathematics had become manifest, a simple geometric interpretation of the operations with complex numbers was provided which set to rest the lingering doubts about their validity. Of course, such an interpretation is unnecessary from the modern point of view in which the justification of formal calculations with complex numbers is given directly on the basis of the formal definitions of addition and multiplication. But the geometric interpretation, given at about the same time by Wessel (1745–1818), Argand (1768–1822) and Gauss, made these operations seem more natural from an intuitive standpoint, and has ever since been of the utmost importance in applications of complex numbers in mathematics and the physical sciences.

This geometrical interpretation consists simply in representing the complex number $z = x + yi$ by the point in the plane with rectangular coördinates x, y. Thus the real part of z is its x-coördinate, and the imaginary part is its y-coördinate. A correspondence is thereby established between the complex numbers and the points in a "number plane," just as a correspondence was established in §2 between the real numbers and the points on a line, the number axis. The points on the x-axis of the number plane correspond to the real numbers

$z = x + 0i$, while the points on the y-axis correspond to the pure imaginary numbers $z = 0 + yi$.

If

$$z = x + yi$$

is any complex number, we call the complex number

$$\bar{z} = x - yi$$

the *conjugate* of z. The point $\bar{z}$ is represented in the number plane by the reflection of the point z in the x-axis as in a mirror. If we denote

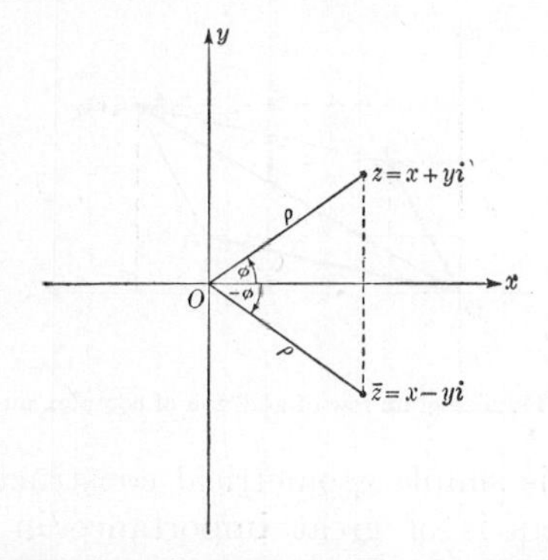

Fig. 22. Geometrical representation of complex numbers. The point z has the rectangular coördinates x, y.

the distance of the point z from the origin by ρ, then by the Pythagorean theorem

$$\rho^2 = x^2 + y^2 = (x + yi)(x - yi) = z \cdot \bar{z}.$$

The real number $\rho = \sqrt{x^2 + y^2}$ is called the *modulus* of z, and written

$$\rho = | z |.$$

If z lies on the real axis, its modulus is its ordinary absolute value. The complex numbers with modulus 1 lie on the "unit circle" with center at the origin and radius 1.

If $| z | = 0$ *then* $z = 0$. This follows from the definition of $| z |$ as the distance of z from the origin. Moreover *the modulus of the product of two complex numbers is equal to the product of their moduli*:

$$| z_1 \cdot z_2 | = | z_1 | \cdot | z_2 |.$$

This will follow from a more general theorem to be proved on page 95.

Exercises: 1. Prove this theorem directly from the definition of multiplication of two complex numbers, $z_1 = x_1 + y_1 i$ and $z_2 = x_2 + y_2 i$.

2. From the fact that the product of two *real* numbers is 0 only if one of the factors is 0, prove the corresponding theorem for *complex* numbers. (Hint: Use the two theorems just stated.)

From the definition of addition of two complex numbers, $z_1 = x_1 + y_1 i$ and $z_2 = x_2 + y_2 i$, we have

$$z_1 + z_2 = (x_1 + x_2) + (y_1 + y_2)i.$$

Hence the point $z_1 + z_2$ is represented in the number plane by the fourth vertex of a parallelogram, three of whose vertices are the

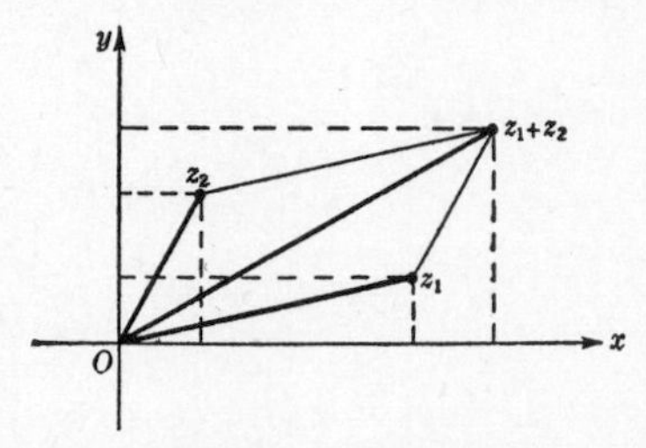

Fig. 23. Parallelogram law of addition of complex numbers.

points O, z_1, z_2. This simple geometrical construction for the sum of two complex numbers is of great importance in many applications. From it we can deduce the important consequence that *the modulus of the sum of two complex numbers does not exceed the sum of the moduli* (compare p. 58):

$$|z_1 + z_2| \leq |z_1| + |z_2|.$$

This follows from the fact that the length of any side of a triangle cannot exceed the sum of the lengths of the other two sides.

Exercise: When does the equality $|z_1 + z_2| = |z_1| + |z_2|$ hold?

The angle between the positive direction of the x-axis and the line Oz is called the *angle* of z, and is denoted by ϕ (Fig. 22). The modulus of $\bar{z}$ is the same as the modulus of z,

$$|\bar{z}| = |z|,$$

but the angle of $\bar{z}$ is the negative of the angle of z,

$$\bar{\phi} = -\phi.$$

Of course, the angle of z is not uniquely determined, since any integral multiple of 360° can be added to or subtracted from an angle without affecting the position of its terminal side. Thus

$$\phi, \phi + 360°, \phi + 720°, \phi + 1080°, \cdots,$$
$$\phi - 360°, \phi - 720°, \phi - 1080°, \cdots$$

all represent graphically the same angle. By means of the modulus ρ and the angle ϕ, the complex number z can be written in the form

$$z = x + yi = \rho(\cos\phi + i\sin\phi); \tag{8}$$

for, by the definition of sine and cosine (see p. 277),

$$x = \rho\cos\phi, \qquad y = \rho\sin\phi.$$

E.g. for $z = i$, $\rho = 1$, $\phi = 90°$, so that $i = 1\,(\cos 90° + i\sin 90°)$;

for $\quad z = 1 + i, \qquad \rho = \sqrt{2}, \phi = 45°, \quad$ so that

$$1 + i = \sqrt{2}\,(\cos 45° + i\sin 45°);$$

for $\quad z = 1 - i, \qquad \rho = \sqrt{2}, \phi = -45°, \quad$ so that

$$1 - i = \sqrt{2}\,[\cos(-45°) + i\sin(-45°)];$$

for $\quad z = -1 + \sqrt{3}\,i, \qquad \rho = 2, \phi = 120°, \quad$ so that

$$-1 + \sqrt{3}\,i = 2\,(\cos 120° + i\sin 120°).$$

The reader should confirm these statements by substituting the values of the trigonometrical functions.

The trigonometrical representation (8) is of great value when two complex numbers are to be multiplied. If

$$z = \rho(\cos\phi + i\sin\phi),$$

and

$$z' = \rho'(\cos\phi' + i\sin\phi'),$$

then

$$zz' = \rho\rho'\{(\cos\phi\cos\phi' - \sin\phi\sin\phi') + i(\cos\phi\sin\phi' + \sin\phi\cos\phi')\}$$

Now, by the fundamental addition theorems for the sine and cosine,

$$\cos\phi\cos\phi' - \sin\phi\sin\phi' = \cos(\phi + \phi'),$$
$$\cos\phi\sin\phi' + \sin\phi\cos\phi' = \sin(\phi + \phi').$$

Hence

$$zz' = \rho\rho'\{\cos(\phi + \phi') + i\sin(\phi + \phi')\}. \tag{9}$$

This is the trigonometrical form of the complex number with modulus $\rho\rho'$ and angle $\phi + \phi'$. In other words, *to multiply two complex numbers, we multiply their moduli and add their angles* (Fig. 24). Thus we see that multiplication of complex numbers has something to do with *rotation.* To be more precise, let us call the directed line segment pointing from the origin to the point z the *vector* z; then $\rho = |z|$ will be its length. Let z' be a number on the unit circle, so that $\rho' = 1$; then multiplying z by z' simply rotates the vector z through the angle ϕ'. If $\rho' \neq 1$, the length of the vector has to be multiplied by ρ' after the rotation. The reader may illustrate these facts by multiplying various numbers by $z_1 = i$ (rotating by 90°); $z_2 = -i$ (rotating by 90° in the opposite sense); $z_3 = 1 + i$; and $z_4 = 1 - i$.

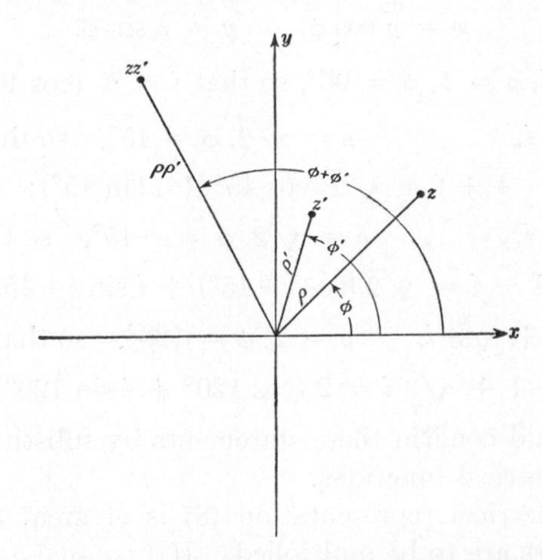

Fig. 24. Multiplication of two complex numbers; the angles are added and the moduli multiplied.

Formula (9) has a particularly important consequence when $z = z'$, for then we have

$$z^2 = \rho^2(\cos 2\phi + i \sin 2\phi).$$

Multiplying this result again by z we obtain

$$z^3 = \rho^3(\cos 3\phi + i \sin 3\phi),$$

and continuing indefinitely in this way,

(10) $$z^n = \rho^n (\cos n\phi + i \sin n\phi) \qquad \text{for any integer } n.$$

In particular, if z is a point on the *unit circle,* with $\rho = 1$, we obtain the formula discovered by the English mathematician A. De Moivre (1667–1754):

(11) $$(\cos \phi + i \sin \phi)^n = \cos n\phi + i \sin n\phi.$$

This formula is one of the most remarkable and useful relations in elementary mathematics. An example will illustrate this. We may apply the formula for $n = 3$ and expand the left hand side according to the binomial formula,

$$(u + v)^3 = u^3 + 3u^2v + 3uv^2 + v^3,$$

obtaining the relation

$$\cos 3\phi + i \sin 3\phi = \cos^3 \phi - 3 \cos \phi \sin^2 \phi + i(3 \cos^2 \phi \sin \phi - \sin^3 \phi).$$

A single equation such as this *between two complex numbers amounts to a pair of equations between real numbers.* For when two complex numbers are equal, both real and imaginary parts must be equal. Hence we may write

$$\cos 3\phi = \cos^3 \phi - 3 \cos \phi \sin^2 \phi, \qquad \sin 3\phi = 3 \cos^2 \phi \sin \phi - \sin^3 \phi.$$

Using the relation

$$\cos^2 \phi + \sin^2 \phi = 1,$$

we have finally

$$\cos 3\phi = \cos^3 \phi - 3 \cos \phi(1 - \cos^2 \phi) = 4 \cos^3 \phi - 3 \cos \phi,$$
$$\sin 3\phi = -4 \sin^3 \phi + 3 \sin \phi.$$

Similar formulas, expressing $\sin n\phi$ and $\cos n\phi$ in terms of powers of $\sin \phi$ and $\cos \phi$ respectively, can easily be obtained for any value of n.

Exercises: 1) Find the corresponding formulas for $\sin 4\phi$ and $\cos 4\phi$.

2) Prove that for a point, $z = \cos \phi + i \sin \phi$, on the unit circle, $1/z = \cos \phi - i \sin \phi$.

3) Prove without calculation that $(a + bi)/(a - bi)$ always has the absolute value 1.

4) If z_1 and z_2 are two complex numbers prove that the angle of $z_1 - z_2$ is equal to the angle between the real axis and the vector pointing from z_2 to z_1.

5) Interpret the angle of the complex number $(z_1 - z_2)/(z_1 - z_3)$ in the triangle formed by the points z_1, z_2, and z_3.

6) Prove that the quotient of two complex numbers with the same angle is real.

7) Prove that if for four complex numbers z_1, z_2, z_3, z_4 the angles of $\dfrac{z_3 - z_1}{z_3 - z_2}$ and $\dfrac{z_4 - z_1}{z_4 - z_2}$ are the same, then the four numbers lie on a circle or on a straight line, and conversely.

8) Prove that four points z_1, z_2, z_3, z_4 lie on a circle or on a straight line if and only if

$$\frac{z_3 - z_1}{z_3 - z_2} \bigg/ \frac{z_4 - z_1}{z_4 - z_2}$$

is real.

3. De Moivre's Formula and the Roots of Unity

By an nth root of a number a we mean a number b such that $b^n = a$. In particular, the number 1 has two square roots, 1 and -1, since $1^2 = (-1)^2 = 1$. The number 1 has only one real cube root, 1, while it has four fourth roots: the real numbers 1 and -1, and the imaginary numbers i and $-i$. These facts suggest that there may be two more cube roots of 1 in the complex domain, making a total of three in all. That this is the case may be shown at once from De Moivre's formula.

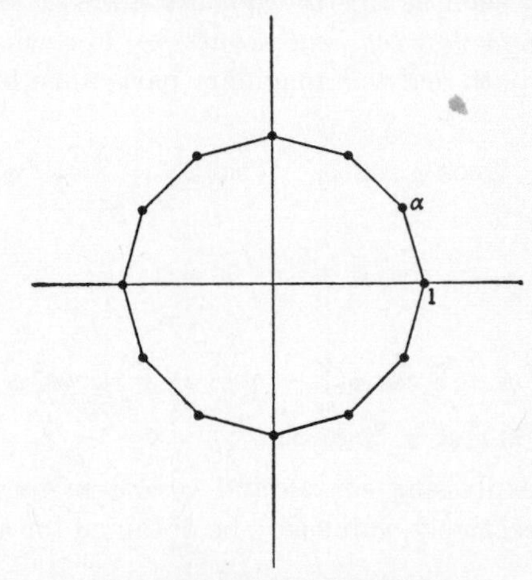

Fig. 25. The twelve twelfth roots of 1.

We shall see that *in the field of complex numbers there are exactly* n *different* n*th roots of* 1. *They are represented by the vertices of the regular* n*-sided polygon inscribed in the unit circle and having the point* z = 1 *as one of its vertices.* This is almost immediately clear from Figure 25 (drawn for the case $n = 12$). The first vertex of the polygon is 1. The next is

$$\alpha = \cos\frac{360°}{n} + i\sin\frac{360°}{n}, \tag{12}$$

since its angle must be the nth part of the total angle 360°. The next vertex is $\alpha \cdot \alpha = \alpha^2$, since we obtain it by rotating the vector α through the angle $\frac{360°}{n}$. The next vertex is α^3, etc., and finally, after n steps, we are back at the vertex 1, i.e., we have

$$\alpha^n = 1,$$

which also follows from formula (11), since

$$\left[\cos\frac{360^\circ}{n} + i\sin\frac{360^\circ}{n}\right]^n = \cos 360^\circ + i\sin 360^\circ = 1 + 0i.$$

It follows that $\alpha^1 = \alpha$ is a root of the equation $x^n = 1$. The same is true for the next vertex $\alpha^2 = \cos\left(\frac{720^\circ}{n}\right) + i\sin\left(\frac{720^\circ}{n}\right)$. We can see this by writing

$$(\alpha^2)^n = \alpha^{2n} = (\alpha^n)^2 = (1)^2 = 1,$$

or from De Moivre's formula:

$$(\alpha^2)^n = \cos\left(n\frac{720^\circ}{n}\right) + i\sin\left(n\frac{720^\circ}{n}\right)$$
$$= \cos 720^\circ + i\sin 720^\circ = 1 + 0i = 1.$$

In the same way we see that all the n numbers

$$1, \alpha, \alpha^2, \alpha^3, \cdots, \alpha^{n-1}$$

are nth roots of 1. To go farther in the sequence of exponents or to use negative exponents would yield no new roots. For $\alpha^{-1} = 1/\alpha = \alpha^n/\alpha = \alpha^{n-1}$ and $\alpha^n = 1$, $\alpha^{n+1} = (\alpha)^n\alpha = 1\cdot\alpha = \alpha$, etc., so that the previous values would simply be repeated. It is left as an exercise to show that there are no other nth roots.

If n is even, then one of the vertices of the n-sided polygon will lie at the point -1, in accordance with the algebraic fact that in this case -1 is an nth root of 1.

The equation satisfied by the nth roots of 1

$$x^n - 1 = 0 \tag{13}$$

is of the nth degree, but it can easily be reduced to an equation of the $(n - 1)$st degree. We use the algebraic formula

$$x^n - 1 = (x - 1)(x^{n-1} + x^{n-2} + x^{n-3} + \cdots + 1). \tag{14}$$

Since the product of two numbers is 0 if *and only if* at least one of the two numbers is 0, the left hand side of (14) vanishes only if one of the two factors on the right hand side is zero, i.e. only if either $x = 1$, or the equation

$$x^{n-1} + x^{n-2} + x^{n-3} + \cdots + x + 1 = 0 \tag{15}$$

is satisfied. This, then, is the equation which must be satisfied by the roots $\alpha, \alpha^2, \cdots \alpha^{n-1}$; it is called the *cyclotomic* (circle-dividing) *equation.* For example, the complex cube roots of 1,

$$\alpha = \cos 120° + i \sin 120° = \tfrac{1}{2}(-1 + i\sqrt{3}),$$

$$\alpha^2 = \cos 240° + i \sin 240° = \tfrac{1}{2}(-1 - i\sqrt{3}),$$

are the roots of the equation

$$x^2 + x + 1 = 0,$$

as the reader will readily see by direct substitution. Likewise the fifth roots of 1, other than 1 itself, satisfy the equation

$$x^4 + x^3 + x^2 + x + 1 = 0. \tag{16}$$

To construct a regular pentagon, we have to solve this fourth degree equation. By a simple algebraic device it can be reduced to a quadratic equation in the quantity $w = x + 1/x$. We divide (16) by x^2 and rearrange the terms:

$$x^2 + \frac{1}{x^2} + x + \frac{1}{x} + 1 = 0,$$

or, since $(x + 1/x)^2 = x^2 + 1/x^2 + 2$, we obtain the equation

$$w^2 + w - 1 = 0.$$

By formula (7) of Article 1 this equation has the roots

$$w_1 = \frac{-1 + \sqrt{5}}{2}, \qquad w_2 = \frac{-1 - \sqrt{5}}{2}.$$

Hence the complex fifth roots of 1 are the roots of the two quadratic equations

$$x + \frac{1}{x} = w_1, \quad \text{or} \quad x^2 + \tfrac{1}{2}(\sqrt{5} - 1)x + 1 = 0,$$

and

$$x + \frac{1}{x} = w_2, \quad \text{or} \quad x^2 - \tfrac{1}{2}(\sqrt{5} + 1)x + 1 = 0,$$

which the reader may solve by the formula already used.

Exercise: 1) Find the 6th roots of 1. 2) Find $(1 + i)^{11}$. 3) Find all the different values of $\sqrt{1 + i}$, $\sqrt[3]{7 - 4i}$, $\sqrt[3]{i}$, $\sqrt[5]{-i}$. 4) Calculate $\frac{1}{2i}(i^7 - i^{-7})$.

*4. The Fundamental Theorem of Algebra

Not only is every equation of the form $ax^2 + bx + c = 0$ or of the form $x^n - 1 = 0$ solvable in the field of complex numbers, but far more is true: *Every algebraic equation of any degree* n *with real or complex coefficients,*

$$(17) \quad f(x) = x^n + a_{n-1}x^{n-1} + a_{n-2}x^{n-2} + \cdots + a_1x + a_0 = 0,$$

has solutions in the field of complex numbers. For equations of the 3rd and 4th degrees this was established in the sixteenth century by Tartaglia, Cardan, and others, who solved such equations by formulas essentially similar to that for the quadratic equation, although much more complicated. For almost two hundred years the general equations of 5th and higher degree were intensively studied, but all efforts to solve them by similar methods failed. It was a great achievement when the young Gauss in his doctoral thesis (1799) succeeded in giving the first complete proof that solutions *exist*, although the question of generalizing the classical formulas, which express the solutions of equations of degree less than 5 in terms of the rational operations plus root extraction, remained unanswered at the time. (See p. 118.)

Gauss's theorem states that *for any algebraic equation of the form* (17), *where* n *is a positive integer and the* a*'s are any real or even complex numbers, there exists at least one complex number* $\alpha = \mathrm{c} + \mathrm{di}$ *such that*

$$f(\alpha) = 0.$$

The number α is called a *root* of the equation (17). A proof of this theorem will be given on page 269. Assuming its truth for the moment, we can prove what is known as the *fundamental theorem of algebra* (it should more fittingly be called the fundamental theorem of the complex number system): *Every polynomial of degree* n,

$$(18) \qquad f(x) = x^n + a_{n-1}x^{n-1} + \cdots + a_1x + a_0,$$

can be factored into the product of exactly n *factors,*

$$(19) \qquad f(x) = (x - \alpha_1)(x - \alpha_2) \cdots (x - \alpha_n),$$

where $\alpha_1, \alpha_2, \alpha_3, \cdots, \alpha_n$ *are complex numbers, the roots of the equation* $\mathrm{f(x)} = 0$. As an example illustrating this theorem, the polynomial

$$f(x) = x^4 - 1$$

may be factored into the form

$$f(x) = (x - 1)(x - i)(x + i)(x + 1).$$

That the α's are roots of the equation $f(x) = 0$ is evident from the factorization (19), since for $x = \alpha_r$ one factor of $f(x)$, and hence $f(x)$ itself, is equal to zero.

In some cases the factors $(x - \alpha_1)$, $(x - \alpha_2)$, $\cdots$ of a polynomial $f(x)$ of degree n will not all be distinct, as in the example

$$f(x) = x^2 - 2x + 1 = (x - 1)(x - 1),$$

which has but one root, $x = 1$, "counted twice" or "of multiplicity 2." In any case, a polynomial of degree n can have no more than n distinct factors $(x - \alpha)$ and the corresponding equation n roots.

To prove the factorization theorem we again make use of the algebraic identity

$$(20) \quad x^k - \alpha^k = (x - \alpha)(x^{k-1} + \alpha x^{k-2} + \alpha^2 x^{k-3} + \cdots + \alpha^{k-2} x + \alpha^{k-1}),$$

which for $\alpha = 1$ is merely the formula for the geometrical series. Since we are assuming the truth of Gauss's theorem, we may suppose that $\alpha = \alpha_1$ is a root of equation (17), so that

$$f(\alpha_1) = \alpha_1^n + a_{n-1}\alpha_1^{n-1} + a_{n-2}\alpha_1^{n-2} + \cdots + a_1\alpha_1 + a_0 = 0.$$

Subtracting this from $f(x)$ and rearranging the terms, we obtain the identity

$$(21) \qquad f(x) = f(x) - f(\alpha_1) = (x^n - \alpha_1^n) + a_{n-1}(x^{n-1} - \alpha_1^{n-1}) + \cdots + a_1(x - \alpha_1).$$

Now, because of (20), we may factor out $(x - \alpha_1)$ from every term of (21), so that the degree of the other factor of each term is reduced by 1. Hence, on rearranging the terms again, we find that

$$f(x) = (x - \alpha_1)g(x),$$

where $g(x)$ is a polynomial of degree $n - 1$:

$$g(x) = x^{n-1} + b_{n-2}x^{n-2} + \cdots + b_1 x + b_0 .$$

(For our purposes it is quite unnecessary to calculate the coefficients b_k.) Now we may apply the same procedure to $g(x)$. By Gauss's theorem there exists a root α_2 of the equation $g(x) = 0$, so that

$$g(x) = (x - \alpha_2)h(x),$$

where $h(x)$ is a polynomial of degree $n - 2$. Proceeding a total of $(n - 1)$ times in the same way (of course, this phrase is merely a substitute for an argument by mathematical induction) we finally obtain the complete factorization

$$(22) \qquad f(x) = (x - \alpha_1)(x - \alpha_2)(x - \alpha_3) \cdots (x - \alpha_n).$$

From (22) it follows not only that the complex numbers $\alpha_1, \alpha_2, \cdots, \alpha_n$ are roots of the equation (17), but also that they are the *only* roots. For if y were a root of equation (17), then by (22)

$$f(y) = (y - \alpha_1)(y - \alpha_2) \cdots (y - \alpha_n) = 0.$$

We have seen on page 94 that a product of complex numbers is equal to 0 if *and only if* one of the factors is equal to 0. Hence one of the factors $(y - \alpha_r)$ must be 0, and y must be equal to α_r, as was to be shown.

*§6. ALGEBRAIC AND TRANSCENDENTAL NUMBERS

1. Definition and Existence

An *algebraic number* is any number x, real or complex, that satisfies some algebraic equation of the form

$$(1) \quad a_n x^n + a_{n-1} x^{n-1} + \cdots + a_1 x + a_0 = 0 \qquad (n \geq 1,\ a_n \neq 0)$$

where the a_k are *integers*. For example, $\sqrt{2}$ is an algebraic number, since it satisfies the equation

$$x^2 - 2 = 0.$$

Similarly, any root of an equation with integer coefficients of third, fourth, fifth, or any higher degree, is an algebraic number, whether or not the roots can be expressed in terms of radicals. The concept of algebraic number is a natural generalization of rational number, which constitutes the special case when $n = 1$.

Not every real number is algebraic. This may be shown by a proof, due to Cantor, that the totality of all algebraic numbers is *denumerable*. Since the set of all real numbers is non-denumerable, there must exist real numbers which are not algebraic.

A method for denumerating the set of algebraic numbers is as follows: To each equation of the form (1) the positive integer

$$h = |a_n| + |a_{n-1}| + \cdots + |a_1| + |a_0| + n$$

is assigned as its "height." For any *fixed* value of h there are only a *finite* number of equations (1) with height h. Each of these equations can have at most n different roots. Therefore there can be but a finite number of algebraic numbers whose equations are of height h, and we can arrange all the algebraic numbers in a sequence by starting with those of height 1, then taking those of height 2, and so on.

This proof that the set of algebraic numbers is denumerable assures the existence of real numbers which are not algebraic; such numbers are called *transcendental*, for, as Euler said, they "transcend the power of algebraic methods."

Cantor's proof of the existence of transcendental numbers can hardly be called constructive. Theoretically, one could construct a transcendental number by applying Cantor's diagonal process to a denumerated table of decimal expressions for the roots of algebraic equations, but this procedure would be quite impractical and would not lead to any number whose expression in the decimal or any other system could actually be written down. Moreover, the most interesting problems concerning transcendental numbers lie in proving that certain definite numbers such as π and e (these numbers will be defined on pages 297 and 299) are actually transcendental.

**2. Liouville's Theorem and the Construction of Transcendental Numbers

A proof for the existence of transcendental numbers which antedates Cantor's was given by J. Liouville (1809–1882). Liouville's proof actually permits the *construction* of examples of such numbers. It is somewhat more difficult than Cantor's proof, as are most constructions when compared with mere existence proofs. The proof is included here for the more advanced reader only, though it requires no more than high school mathematics.

Liouville showed that irrational algebraic numbers are those which cannot be approximated by rational numbers with a very high degree of accuracy unless the denominators of the approximating fractions are quite large.

Suppose the number z satisfies the algebraic equation with integer coefficients

$$(2) \qquad f(x) = a_0 + a_1x + a_2x^2 + \cdots + a_nx^n = 0 \qquad (a_n \neq 0),$$

but no such equation of lower degree. Then z is said to be an algebraic number *of degree n*. For example, $z = \sqrt{2}$ is an algebraic number of degree 2, since it satisfies the equation $x^2 - 2 = 0$ but no equation of the first degree; $z = \sqrt[3]{2}$ is of the third degree because it satisfies the equation $x^3 - 2 = 0$ and, as we shall see in Chapter III, no equation of lower degree. An algebraic number of degree $n > 1$ cannot be rational, since a rational number $z = p/q$ satisfies the equation $qx - p = 0$ of degree 1. Now each irrational number z can be approximated to any desired degree of accuracy by a rational number; this means that we can find a sequence

$$\frac{p_1}{q_1}, \frac{p_2}{q_2}, \cdots$$

of rational numbers with larger and larger denominators such that

$$\frac{p_r}{q_r} \to z.$$

Liouville's theorem asserts: For any algebraic number z of degree $n > 1$ such an approximation must be less accurate than $1/q^{n+1}$; i.e., the inequality

$$\left| z - \frac{p}{q} \right| > \frac{1}{q^{n+1}} \tag{3}$$

must hold for sufficiently large denominators q.

We shall prove this theorem presently, but first we shall show how it permits the construction of transcendental numbers. Let us take the number (see p. 17 for the definition of the symbol $n!$)

$$z = a_1 \cdot 10^{-1!} + a_2 \cdot 10^{-2!} + a_3 \cdot 10^{-3!} + \cdots + a_m \cdot 10^{-m!} + a_{m+1} \cdot 10^{-(m+1)!} + \cdots$$

$$= 0.a_1a_2000a_3000000000000000000a_4000000 \quad \cdot,$$

where the a_i are arbitrary digits from 1 to 9 (we could, for example, choose all the a_i equal to 1). Such a number is characterized by rapidly increasing stretches of 0's, interrupted by single non-zero digits. Let us denote by z_m the finite decimal fraction formed by taking only the terms of z up to and including $a_m \cdot 10^{-m!}$. Then

$$| z - z_m | < 10 \cdot 10^{-(m+1)!}. \tag{4}$$

Suppose that z were algebraic of degree n. Then in (3) let us set $p/q = z_m = p/10^{m!}$, obtaining

$$| z - z_m | > \frac{1}{10^{(n+1)m!}}$$

for sufficiently large m. Combining this with (4), we should have

$$\frac{1}{10^{(n+1)m!}} < \frac{10}{10^{(m+1)!}} = \frac{1}{10^{(m+1)!-1}},$$

so that $(n + 1)m! > (m + 1)! - 1$ for all sufficiently large m. But this is false for any value of m greater than n (the reader should give a detailed proof of this statement), which gives a contradiction. Hence z is transcendental.

It remains to prove Liouville's theorem. Suppose z is an algebraic number of degree $n > 1$ which satisfies (1), so that

$$f(z) = 0. \tag{5}$$

Let $z_m = p_m/q_m$ be a sequence of rational numbers with $z_m \to z$. Then $f(z_m) = f(z_m) - f(z) = a_1(z_m - z) + a_2(z_m^2 - z^2) + \cdots + a_n(z_m^n - z^n)$. Dividing both sides of this equation by $z_m - z$, and using the algebraic formula

$$\frac{u^n - v^n}{u - v} = u^{n-1} + u^{n-2}v + u^{n-3}v^2 + \cdots + uv^{n-2} + v^{n-1},$$

we obtain

$$\frac{f(z_m)}{z_m - z} = a_1 + a_2(z_m + z) + a_3(z_m^2 + z_m z + z^2) + \cdots + a_n(z_m^{n-1} + \cdots + z^{n-1}). \tag{6}$$

Since z_m tends to z as a limit, it will differ from z by less than 1 for sufficiently large m. We can therefore write the following rough estimate for sufficiently large m:

$$\left| \frac{f(z_m)}{z_m - z} \right| < |a_1| + 2|a_2|(|z| + 1) + 3|a_3|(|z| + 1)^2 + \cdots + n|a_n|(|z| + 1)^{n-1} = M, \tag{7}$$

which is a fixed number, since z is fixed in our reasoning. If now we choose m so large that in $z_m = \frac{p_m}{q_m}$ the denominator q_m is larger than M, then

$$|z - z_m| > \frac{|f(z_m)|}{M} > \frac{|f(z_m)|}{q_m}. \tag{8}$$

For brevity let us denote p_m by p and q_m by q. Then

$$|f(z_m)| = \left| \frac{a_0 q^n + a_1 q^{n-1} p + \cdots + a_n p^n}{q^n} \right|. \tag{9}$$

Now the rational number $z_m = p/q$ cannot be a root of $f(x) = 0$, for if it were we could factor out $(x - z_m)$ from $f(x)$, and z would satisfy an equation of degree less than n. Hence $f(z_m) \neq 0$. But the numerator of the right hand side of (9) is an integer, so it must be at least equal to 1. Hence from (8) and (9) we have

$$(10) \qquad | z - z_m | > \frac{1}{q}\frac{1}{q^n} = \frac{1}{q^{n+1}},$$

which proves the theorem.

During the last few decades, investigations into the possibility of approximating algebraic numbers by rational numbers have been carried much farther. For example, the Norwegian mathematician A. Thue (1863–1922) proved that in Liouville's inequality (3) the exponent $n + 1$ may be replaced by $(n/2) + 1$. C. L. Siegel later showed that the even sharper statement (sharper for large n) with the exponent $2\sqrt{n}$ holds.

The subject of transcendental numbers has always fascinated mathematicians. But until recently, very few examples of numbers interesting in themselves were known which could be shown to be transcendental. (In Chapter III we shall discuss the transcendental character of π, from which follows the impossibility of squaring the circle with ruler and compass.) In a famous address to the international congress of mathematicians at Paris in 1900, David Hilbert proposed thirty mathematical problems which were easy to formulate, some of them in elementary and popular language, but none of which had been solved nor seemed immediately accessible to the mathematical technique then existing. These "Hilbert problems" stood as a challenge to the subsequent period of mathematical development. Almost all have been solved in the meantime, and often the solution meant definite progress in mathematical insight and general methods One of the problems that seemed most hopeless was to prove that

$$2^{\sqrt{2}}$$

is a transcendental, or even that it is an irrational number. For almost three decades there was not the slightest suggestion of a promising line of attack on this problem. Finally Siegel and, independently, the young Russian, A. Gelfond, discovered new methods for proving the transcendental character of many numbers significant in mathematics, including the Hilbert number $2^{\sqrt{2}}$ and, more generally, any number a^b where a is an algebraic number $\neq 0$ or 1 and b is any irrational algebraic number.

SUPPLEMENT TO CHAPTER II

THE ALGEBRA OF SETS

1. General Theory

The concept of a *class* or *set* of objects is one of the most fundamental in mathematics. A set is defined by any property or attribute $\mathfrak{A}$ which each object considered must either possess or not possess; those objects which possess the property form a corresponding set A. Thus, if we consider the integers, and the property $\mathfrak{A}$ is that of being a prime, the corresponding set A is the set of all primes 2, 3, 5, 7, $\cdots$.

The mathematical study of sets is based on the fact that sets may be combined by certain operations to form other sets, just as numbers may be combined by addition and multiplication to form other numbers. The study of operations on sets comprises the "algebra of sets," which has many formal similarities with, as well as differences from, the algebra of numbers. The fact that algebraic methods can be applied to the study of non-numerical objects like sets illustrates the great generality of the concepts of modern mathematics. In recent years it has become apparent that the algebra of sets illuminates many branches of mathematics such as measure theory and the theory of probability; it is also helpful in the systematic reduction of mathematical concepts to their logical basis.

In what follows, I will denote a fixed set of objects of any nature, called the universal set or universe of discourse, and $A, B, C, \cdots$ will denote arbitrary subsets of I. If I denotes the set of all integers, A may denote the set of all even integers, B the set of all odd integers, C the set of all primes, etc. Or I might denote the set of all points of a fixed plane, A the set of all points within some circle in the plane, B the set of all points within some other circle in the plane, etc. For convenience we include as "subsets" of I the set I itself and the "empty set" O which contains no elements. The aim of this artificial extension is to preserve the rule that to each property $\mathfrak{A}$ corresponds the subset A of all elements of I possessing this property. In case $\mathfrak{A}$ is some universally valid property such as the one specified by the trivial equation $x = x$, the corresponding subset of I will be I itself, since every object satisfies this equation, while if $\mathfrak{A}$ is some self-contradictory property like $x \neq x$, the corresponding subset will contain no objects, and may be denoted by the symbol O.

The set A is said to be a *subset* of the set B if there is no object in A that is not also in B. When this is the case we write

$$A \subset B \qquad \text{or} \qquad B \supset A.$$

For example, the set A of all integers that are multiples of 10 is a subset of the set B of all integers that are multiples of 5, since every multiple of 10 is also a multiple of 5. The statement $A \subset B$ does not exclude the possibility that $B \subset A$. If both relations hold, we say that the sets A and B are equal, and write

$$A = B.$$

For this to be true every element of A must be an element of B, and conversely, so that the sets A and B contain exactly the same elements.

The relation $A \subset B$ has many similarities with the order relation $a \leq b$ between real numbers. In particular, it is true that

1) $A \subset A$.
2) If $A \subset B$ and $B \subset A$, then $A = B$.
3) If $A \subset B$ and $B \subset C$, then $A \subset C$.

For this reason we also call the relation $A \subset B$ an "order relation." Its chief difference from the relation $a \leq b$ for numbers is that, while for *every* pair of numbers a and b at least one of the relations $a \leq b$ or $b \leq a$ always holds, this is not true for sets. For example, if A denotes the set consisting of the integers 1, 2, 3,

$$A = \{1, 2, 3\},$$

and B the set consisting of the integers 2, 3, 4,

$$B = \{2, 3, 4\},$$

then neither $A \subset B$ nor $B \subset A$. For this reason, the relation $A \subset B$ is said to determine a partial ordering among sets, whereas the relation $a \leq b$ determines a complete ordering among numbers.

In passing, we may remark that from the definition of the relation $A \subset B$ it follows that

4) $O \subset A$ for any set A, and,

5) $A \subset I$,

where A is any subset of the universe of discourse I. The relation 4) may seem somewhat paradoxical, but it is in agreement with a strict interpretation of the definition of the sign $\subset$. For the statement $O \subset A$ could be false only if the empty set O contained an object not in A, and since the empty set contains no objects at all, this is impossible no matter what the set A.

We shall now define two operations on sets which have many of the algebraic properties of ordinary addition and multiplication of numbers, though they are conceptually quite distinct from those operations. To this end, let A and B be any two sets. By the "union" or "logical sum" of A and B we mean the set which consists of all the objects which are in *either* A or B (including any that may be in both). This set we denote by the symbol $A + B$. By the "intersection" or "logical product" of A and B we mean the set consisting only of those elements which are in *both* A and B. This set we denote by the symbol $A \cdot B$ or simply AB. To illustrate these operations, we may again choose as A and B the sets

$$A = \{1, 2, 3\}, \qquad B = \{2, 3, 4\}.$$

Then

$$A + B = \{1, 2, 3, 4\}, \qquad AB = \{2, 3\}.$$

Among the important algebraic properties of the operations $A + B$ and AB we list the following. They should be verified by the reader on the basis of the definition of these operations:

6) $A + B = B + A$ 7) $AB = BA$

8) $A + (B + C) = (A + B) + C$ 9) $A(BC) = (AB)C$

10) $A + A = A$ 11) $AA = A$

12) $A(B + C) = (AB + AC)$ 13) $A + (BC) = (A + B)(A + C)$

14) $A + O = A$ 15) $AI = A$

16) $A + I = I$ 17) $AO = O$

18) the relation $A \subset B$ is equivalent to either of the two relations $A + B = B$, $AB = A$.

The verification of these laws is a matter of elementary logic. For example, 10) states that the set consisting of those objects which are either in A or in A is precisely the set A, while 12) states that the set consisting of those objects which are in A and also in either B or C is the same as the set consisting of those objects which are either in both

A and B or in both A and C. The logical reasoning involved in this and other arguments may be illustrated by representing the sets A, B, C as areas in a plane, provided that one is careful to provide for all the possibilities of the sets involved having elements distinct from and in common with each other.

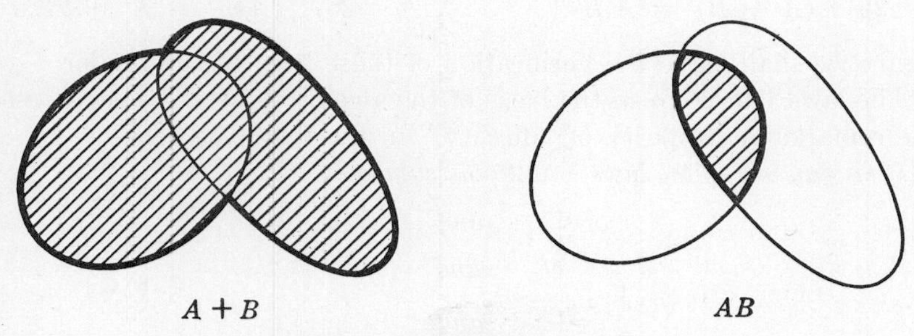

Fig. 26. Union and intersection of sets.

The reader will have observed that the laws 6, 7, 8, 9, and 12 are identical with the familiar commutative, associative, and distributive laws of algebra. It follows that all rules of the ordinary algebra of numbers which are consequences of the commutative, associative, and distributive laws are also valid in the algebra of sets. The laws 10, 11, and 13, however, have no numerical analogs, and give the algebra of sets a simpler structure than the algebra of numbers. For example, the binomial theorem of ordinary algebra is replaced in the algebra of sets by the equality

$$(A + B)^n = (A + B) \cdot (A + B) \cdot \cdots \cdot (A + B) = A + B$$

which is a consequence of 11. Laws 14, 15, and 17 indicate that the properties of O and I with respect to union and intersection of sets are largely similar to the properties of the numbers 0 and 1 with respect to ordinary addition and multiplication. Law 16 has no analog in the algebra of numbers.

It remains to define one further operation in the algebra of sets. Let A be any subset of the universal set I. Then by the *complement* of A in I we mean the set which consists of all the objects in I which are *not* in A. This set we denote by the symbol A'. Thus if I is the set of all natural numbers and A the set of primes, A' consists of 1 and the composites. The operation A', which has no exact analog in the algebra of numbers, possesses the following properties:

19) $A + A' = I$ 20) $AA' = O$

21) $O' = I$ 22) $I' = O$

23) $A'' = A$

24) The relation $A \subset B$ is equivalent to the relation $B' \subset A'$.

25) $(A + B)' = A'B'$ 26) $(AB)' = A' + B'$.

Again we shall leave the verification of these laws to the reader.

The laws 1 to 26 form the basis of the algebra of sets. They possess the remarkable property of "duality," in the following sense:

If in any one of the laws 1 *to* 26 *the symbols*

$$\subset \quad \textit{and} \quad \supset$$

$$O \quad \textit{and} \quad I$$

$$+ \quad \textit{and} \quad \cdot$$

are everywhere interchanged (insofar as they appear), then the result is again one of these laws.

For example, the law 6 becomes 7, 12 becomes 13, 17 becomes 16, etc. It follows that *to any theorem which can be proved on the basis of the laws* 1 *to* 26 *there corresponds another, "dual," theorem, obtained by making the interchanges above.* For, since the proof of any theorem will consist of the successive application at each step of certain of the laws 1 to 26, the application at each step of the dual law will provide a proof of the dual theorem. (For a similar duality in geometry, see Chapt. IV.)

2. Application to Mathematical Logic

The verification of the laws of the algebra of sets rested on the analysis of the logical meaning of the relation $A \subset B$ and the operations $A + B$, AB, and A'. We can now reverse this process and use the laws 1 to 26 as the basis for an "algebra of logic." More precisely, that part of logic which concerns sets, or what is equivalent, properties or *attributes* of objects, may be reduced to a formal algebraic system based on the laws 1 to 26. The logical "universe of discourse" defines the set I; *each property or attribute* $\mathfrak{A}$ *of objects defines the set A consisting of all objects in* I *which possess this attribute.* The rules for translating the usual logical terminology into the language of sets may be illustrated by the following examples:

"Either A or B" $A + B$

"Both A and B" AB

"Not A"	A'
"Neither A nor B"	$(A + B)'$, or equivalently, $A'B'$
"Not both A and B"	$(AB)'$, or equivalently, $A' + B'$
"All A are B" or "If A then B" or "A implies B"	$A \subset B$
"Some A are B"	$AB \neq O$
"No A are B"	$AB = O$
"Some A are not B"	$AB' \neq O$
"There are no A"	$A = O$

In terms of the algebra of sets, the syllogism "Barbara," which states: "If all A are B, and all B are C, then all A are C," becomes simply

3) If $A \subset B$ and $B \subset C$ then $A \subset C$.

Likewise, the "law of contradiction," which states: "An object cannot both possess an attribute and not possess it," becomes

20) $AA' = O$,

while the "law of excluded middle" which states: "An object must either possess a given attribute or not possess it" becomes

19) $A + A' = I$.

Thus the part of logic which is expressible in terms of the symbols $\subset$, $+$, $\cdot$, and $'$ can be treated as a formal algebraic system, subject to the laws 1 to 26. This fusion of the logical analysis of mathematics with the mathematical analysis of logic has resulted in the creation of a new discipline, *mathematical logic*, which is now in the process of vigorous development.

From the point of view of axiomatics, it is a remarkable fact that the statements 1 to 26, together with all other theorems of the algebra of sets, can be deduced from the following three equations:

27)
$$A + B = B + A$$
$$(A + B) + C = A + (B + C)$$
$$(A' + B')' + (A' + B)' = A.$$

It follows that the algebra of sets can be constructed as a purely deductive theory like Euclidean geometry on the basis of these three statements taken as axioms. When this is done, the operation AB and the order relation $A \subset B$ are *defined* in terms of $A + B$ and A':

$$AB \text{ means the set } (A' + B')'$$

$$A \subset B \text{ means that } A + B = B.$$

A quite different example of a mathematical system satisfying all the formal laws of the algebra of sets is provided by the eight numbers 1, 2, 3, 5, 6, 10, 15, 30, where $a + b$ is defined to mean the least common multiple of a and b, ab the greatest common divisor of a and b, $a \subset b$ the statement "a is a factor of b," and a' the number $30/a$. The existence of such examples has led to the study of general algebraic systems satisfying the laws 27). These systems are called "Boolean algebras" in honor of George Boole (1815–1864), an English mathematician and logician whose book, *An Investigation of the Laws of Thought*, appeared in 1854.

3. An Application to the Theory of Probability

The algebra of sets greatly illuminates the theory of probability. To consider only the simplest case, let us imagine an experiment with a finite number of possible outcomes, all of which are assumed to be "equally likely." The experiment may, for example, consist of drawing a card at random from a well-shuffled deck of 52 cards. If the set of possible outcomes of the experiment is denoted by I, and if A denotes any subset of I, then the probability that the outcome of the experiment will belong to the subset A is defined to be the ratio

$$p(A) = \frac{\text{number of elements in } A}{\text{number of elements in } I}.$$

If we denote the number of elements in any set A by the symbol $n(A)$, then this definition may be written in the form

$$p(A) = \frac{n(A)}{n(I)}. \tag{1}$$

In our example, if A denotes the subset of hearts, then $n(A) = 13$, $n(I) = 52$, and $p(A) = \frac{13}{52} = \frac{1}{4}$.

The concepts of the algebra of sets enter into the calculation of probabilities when the probabilities of certain sets are known and the probability of others are required. For example, from a knowledge of $p(A)$, $p(B)$, and $p(AB)$ we may compute the probability of $p(A + B)$:

$$p(A + B) = p(A) + p(B) - p(AB). \tag{2}$$

The proof is simple. We have

$$n(A + B) = n(A) + n(B) - n(AB),$$

since the elements common to A and B, i.e. the elements in AB, will be counted twice in the sum $n(A) + n(B)$, and hence we must subtract $n(AB)$ from this sum in order to obtain the correct count for $n(A + B)$. Dividing each term of this equation by $n(I)$, we obtain equation (2).

A more interesting formula arises when we consider three subsets, A, B, C, of I. From (2) we have

$$p(A + B + C) = p[(A + B) + C] = p(A + B) + p(C) - p[(A + B)C].$$

From (12) of the preceding section we know that $(A + B)C = AC + BC$. Hence

$$p[(A + B)C] = p(AC + BC) = p(AC) + p(BC) - p(ABC).$$

Substituting in the previous equation this value for $p[(A + B)C]$ and the value of $p(A + B)$ given by (2), we obtain the desired formula:

$$p(A + B + C) = p(A) + p(B) + p(C) - p(AB) - p(AC) - p(BC) + p(ABC). \tag{3}$$

As an example, let us consider the following experiment. The three digits 1, 2, 3 are written down in random order. What is the probability that at least one digit will occupy its proper place? Let A denote the set of all arrangements in which the digit 1 comes first, B the set of all arrangements in which the digit 2 comes second, and C the set of all arrangements in which the digit 3 comes third. Then we wish to calculate $p(A + B + C)$. It is clear that

$$p(A) = p(B) = p(C) = \tfrac{2}{6} = \tfrac{1}{3};$$

for when one digit occupies its proper place there are two possible orders for the remaining digits, out of a total of $3 \cdot 2 \cdot 1 = 6$ possible arrangements of the three digits. Moreover,

$$p(AB) = p(AC) = p(BC) = \tfrac{1}{6}$$

and

$$p(ABC) = \tfrac{1}{6},$$

since there is only one way in which each of these cases may occur. It follows from (3) that

$$\begin{aligned} p(A + B + C) &= 3 \cdot \tfrac{1}{3} - 3(\tfrac{1}{6}) + \tfrac{1}{6} \\ &= 1 - \tfrac{1}{2} + \tfrac{1}{6} = \tfrac{2}{3} = 0.6666 \cdots \end{aligned}$$

Exercise: Find a corresponding formula for $p(A + B + C + D)$ and apply it to the case of four digits. The corresponding probability is $\tfrac{5}{8} = 0.6250$.

The general formula for the union of n subsets is

$$p(A_1 + A_2 + \cdots + A_n) = \sum_1 p(A_i) - \sum_2 p(A_iA_j) + \sum_3 p(A_iA_jA_k) - \cdots \pm p(A_1A_2 \cdots A_n), \tag{4}$$

where the symbols $\sum_1, \sum_2, \sum_3, \cdots, \sum_{n-1}$ stand for summation of the possible combinations of the sets $A_1, A_2, \cdots, A_n$ taken one, two, three, $\cdots$, $(n - 1)$ at a time. This formula may be established by mathematical induction in precisely the same way that we derived (3) from (2). From (4) it is easy to show that if the n digits 1, 2, 3, $\cdots$, n are written down in random order, the probability that at least one digit will occupy its proper place is

$$p_n = 1 - \frac{1}{2!} + \frac{1}{3!} - \frac{1}{4!} + \cdots \pm \frac{1}{n!}, \tag{5}$$

where the last term is taken with a plus or minus sign according as n is odd or even. In particular, for $n = 5$ the probability is

$$p_5 = 1 - \frac{1}{2!} + \frac{1}{3!} - \frac{1}{4!} + \frac{1}{5!} = \frac{19}{30} = 0.63333 \cdots.$$

We shall see in Chapter VIII that as n tends to infinity the expression

$$S_n = \frac{1}{2!} - \frac{1}{3!} + \frac{1}{4!} - \cdots \pm \frac{1}{n!}$$

tends to a limit, $1/e$, whose value to five places of decimals is .36788. Since from (5) $p_n = 1 - S_n$, this shows that as n tends to infinity

$$p_n \to 1 - 1/e = .63212.$$

CHAPTER III

GEOMETRICAL CONSTRUCTIONS. THE ALGEBRA OF NUMBER FIELDS

INTRODUCTION

Construction problems have always been a favorite subject in geometry. With ruler and compass alone a great variety of constructions may be performed, as the reader will remember from school: a line segment or an angle may be bisected, a line may be drawn from a point perpendicular to a given line, a regular hexagon may be inscribed in a circle, etc. In all these problems the ruler is used merely as a straightedge, an instrument for drawing a straight line but not for measuring or marking off distances. The traditional restriction to ruler and compass alone goes back to antiquity, although the Greeks themselves did not hesitate to use other instruments.

One of the most famous of the classical construction problems is the so-called contact problem of Apollonius (circa 200 B.C.) in which three arbitrary circles in the plane are given and a fourth circle tangent to all three is required. In particular, it is permitted that one or more of the given circles have degenerated into a point or a straight line (a "circle" with radius zero or "infinity," respectively). For example, it may be required to construct a circle tangent to two given straight lines and passing through a given point. While such special cases are rather easily dealt with, the general problem is considerably more difficult.

Of all construction problems, that of constructing with ruler and compass a regular polygon of n sides has perhaps the greatest interest. For certain values of n—e.g. $n = 3, 4, 5, 6$—the solution has been known since antiquity, and forms an important part of school geometry. But for the regular heptagon ($n = 7$) the construction has been proved impossible. There are three other classical Greek problems for which a solution has been sought in vain: to trisect an arbitrary given angle, to double a given cube (i.e. to find the edge of a cube whose volume shall be twice that of a cube with a given segment as its edge) and to square the circle (i.e. to construct a square having the same area as a given circle). In all these problems, ruler and compass are the only instruments permitted.

Unsolved problems of this sort gave rise to one of the most remarkable and novel developments in mathematics, when, after centuries of futile search for solutions, the suspicion grew that these problems might be definitely unsolvable. Thus mathematicians were challenged to investigate the question: *How is it possible to prove that certain problems cannot be solved?*

In algebra, it was the problem of solving equations of degree 5 and higher which led to this new way of thinking. During the sixteenth century mathematicians had learned that algebraic equations of degree 3 or 4 could be solved by a process similar to the elementary method for solving quadratic equations. All these methods have the following characteristic in common: the solutions or "roots" of the equation can be written as algebraic expressions obtained from the coefficients of the equation by a sequence of operations, each of which is either a rational operation—addition, subtraction, multiplication, or division—or the extraction of a square root, cube root, or fourth root. One says that algebraic equations up to the fourth degree can be solved "by radicals" (radix is the Latin word for root). Nothing seemed more natural than to extend this procedure to equations of degree 5 and higher, by using roots of higher order. All such attempts failed. Even distinguished mathematicians of the eighteenth century deceived themselves into thinking that they had found the solution. It was not until early in the nineteenth century that the Italian Ruffini (1765–1822) and the Norwegian genius N. H. Abel (1802–1829) conceived the then revolutionary idea of proving the *impossibility of the solution of the general algebraic equation of degree* n *by means of radicals.* One must clearly understand that the question is not whether any algebraic equation of degree n *possesses* solutions. This fact was first proved by Gauss in his doctoral thesis in 1799. So there is no doubt about the *existence* of the roots of an equation, especially since these roots can be found by suitable procedures to any degree of accuracy. The art of the numerical solution of equations is, of course, very important and highly developed. But the problem of Abel and Ruffini was quite different: can the solution be effected *by means of rational operations and radicals alone*? It was the desire to attain full clarity about this question that inspired the magnificent development of modern algebra and group theory started by Ruffini, Abel, and Galois (1811–1832).

The question of proving the impossibility of certain geometrical constructions provides one of the simplest examples of this trend in algebra. By the use of algebraic concepts we shall be able in this chapter to prove the impossibility of trisecting the angle, constructing the regular heptagon, or doubling the cube, by ruler and compass alone. (The problem of squaring the circle is much more difficult to dispose of; see p. 140.) Our point of departure will be not so much the negative question of the impossibility of certain constructions, but rather the positive question: How can all constructible problems be completely characterized? After we have answered this question, it will be an easy matter to show that the problems mentioned above do not fall into this category.

At the age of seventeen Gauss investigated the constructibility of regular "p-gons" (polygons with p sides), where p is a prime number. The construction was then known only for $p = 3$ and $p = 5$. Gauss discovered that the regular p-gon is constructible if and only if p is a prime "Fermat number,"

$$p = 2^{2^n} + 1.$$

The first Fermat numbers are 3, 5, 17, 257, 65537 (see p. 26). So overwhelmed was young Gauss by his discovery that he at once gave up his intention of becoming a philologist and resolved to devote his life to mathematics and its applications. He always looked back on this first of his great feats with particular pride. After his death, a bronze statue of him was erected in Goettingen, and no more fitting honor could be devised than to shape the pedestal in the form of a regular 17-gon.

When dealing with a geometrical construction, one must never forget that the problem is not that of drawing figures in practice with a certain degree of accuracy, but of whether, by the use of straightedge and compass alone, the solution can be found theoretically, supposing our instruments to have perfect precision. What Gauss proved is that his constructions could be performed in principle. His theory does not concern the simplest way actually to perform them or the devices which could be used to simplify and to cut down the number of necessary steps. This is a question of much less theoretical importance. From a practical point of view, no such construction would give as satisfactory a result as could be obtained by the use of a good protractor. Failure properly to understand the theoretical character of the question of geometrical construction and stubbornness in refusing to take cognizance of well-established scientific facts are responsible for the persistence of an unending line of angle-trisectors and circle-squarers. Those among them who are able to understand elementary mathematics might profit by studying this chapter.

Once more it should be emphasized that in some ways our concept of geometrical construction seems artificial. Ruler and compass are certainly the simplest instruments for drawing, but the restriction to these instruments is by no means inherent in geometry. As the Greek mathematicians recognized long ago, certain problems—for example that of doubling the cube—can be solved if, e.g., the use of a ruler in the form of a right angle is permitted; it is just as easy to invent instruments other than the compass by means of which one can draw ellipses, hyperbolas, and more complicated curves, and whose use enlarges considerably the domain of constructible figures. In the next sections, however, we shall adhere to the standard concept of geometrical constructions using only ruler and compass.

PART I

IMPOSSIBILITY PROOFS AND ALGEBRA

§1. FUNDAMENTAL GEOMETRICAL CONSTRUCTIONS

1. Construction of Fields and Square Root Extraction

To shape our general ideas we shall begin by examining a few of the classical constructions. The key to a more profound understanding lies in translating the geometrical problems into the language of algebra. Any geometrical construction problem is of the following type: a certain set of line segments, say $a, b, c, \cdots$, is given, and one or more other segments $x, y, \cdots$, are sought. It is always possible to formulate problems in this way, even when at first glance they have a quite different aspect. The required segments may appear as sides of a triangle to be constructed, as radii of circles, or as the rectangular coördinates of certain points (see e.g. p. 137). For simplicity we shall suppose that only one segment x is required. The geometrical construction then amounts to solving an algebraic problem: first we must find a relationship (equation) between the required quantity x and the given quantities $a, b, c, \cdots$; next we must find the unknown quantity x by solving this equation, and finally we must determine whether this solution can be obtained by algebraic processes that correspond to ruler and compass constructions. It is the principle of analytic geometry, the quantitative characterization of geometrical objects by real numbers, based on the introduction of the real number continuum, that provides the foundation for the whole theory.

First we observe that some of the simplest algebraic operations correspond to elementary geometrical constructions. If two segments are given with lengths a and b (as measured by a given "unit" segment), then it is very easy to construct $a + b$, $a - b$, ra (where r is any rational number), a/b, and ab.

To construct $a + b$ (Fig. 27) we draw a straight line and on it mark off with the compass the distances $OA = a$ and $AB = b$. Then $OB = a + b$. Similarly, for $a - b$ we mark off $OA = a$ and $AB = b$, but this time with AB in the opposite direction from OA. Then $OB = a - b$. To construct $3a$ we simply add $a + a + a$; similarly we can

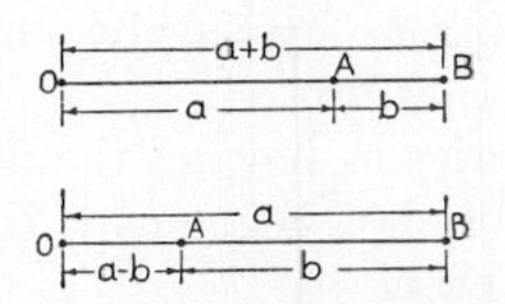

Fig. 27. Construction of $a + b$ and $a - b$.

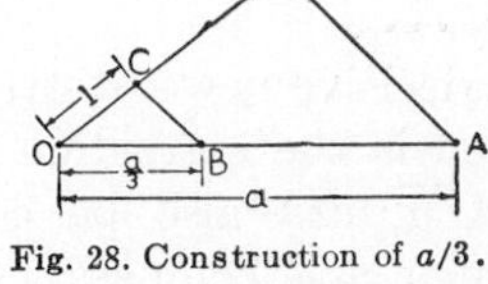

Fig. 28. Construction of $a/3$.

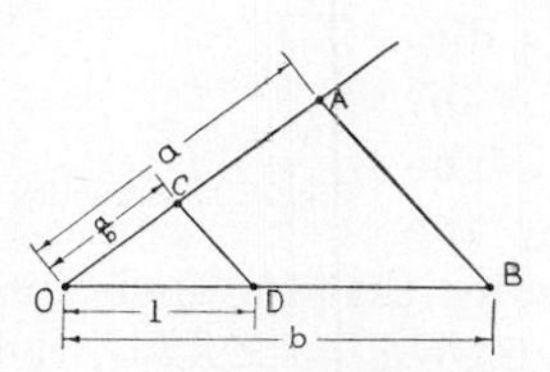

Fig. 29. Construction of a/b.

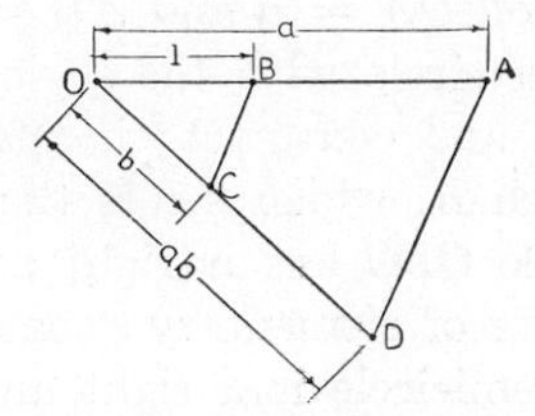

Fig. 30. Construction of ab.

construct pa, where p is any integer. We construct $a/3$ by the following device (Fig. 28): we mark off $OA = a$ on one line, and draw any second line through O. On this line we mark off an arbitrary segment $OC = c$, and construct $OD = 3c$. We connect A and D, and draw a line through C parallel to AD, intersecting OA at B. The triangles OBC and OAD are similar; hence $OB/a = OB/OA = OC/OD = 1/3$, and $OB = a/3$. In the same way we can construct a/q, where q is any integer. By performing this operation on the segment pa, we can thus construct ra, where $r = p/q$ is any rational number.

To construct a/b (Fig. 29) we mark off $OB = b$ and $OA = a$ on the sides of any angle O, and on OB we mark off $OD = 1$. Through D we draw a line parallel to AB meeting OA in C. Then OC will have the length a/b. The construction of ab is shown in Figure 30, where AD is a line parallel to BC through A.

From these considerations it follows that *the "rational" algebraic processes,*—addition, subtraction, multiplication, and division of known quantities—*can be performed by geometrical constructions.* From any given segments, measured by real numbers $a, b, c, \cdots$, we can, by successive application of these simple constructions, construct any quantity that is expressible in terms of $a, b, c, \cdots$ in a rational way, i.e. by repeated application of addition, subtraction, multiplication and division. The totality of quantities that can be obtained in this way from $a, b, c, \cdots$ constitute what is called a *number field,* a set of numbers such that any rational operations applied to two or more members of the set again yield a number of the set. We recall that the rational numbers, the real numbers, and the complex numbers form such fields. In the present case, the field is said to be *generated* by the given numbers $a, b, c, \cdots$.

The decisive new construction which carries us beyond the field just obtained is the extraction of a square root: if a segment a is given, then $\sqrt{a}$ can also be constructed by using only ruler and compass. On a straight line we mark off $OA = a$ and $AB = 1$ (Fig. 31). We draw a circle with the segment OB as its diameter and construct the perpendicular to OB through A, which meets the circle in C. The triangle OBC has a right angle at C, by the theorem of elementary geometry which states that an angle inscribed in a semicircle is a right angle. Hence, $\angle OCA = \angle ABC$, the right triangles OAC and CAB are similar, and we have for $x = AC$,

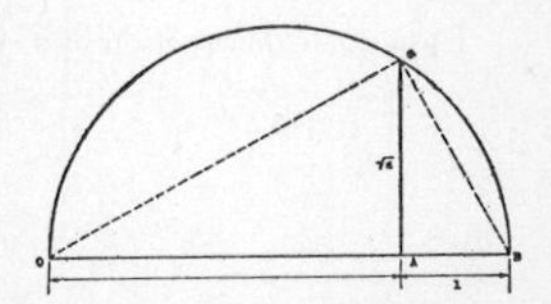
Fig. 31. Construction of $\sqrt{a}$.

$$\frac{a}{x} = \frac{x}{1}, \qquad x^2 = a, \qquad x = \sqrt{a}.$$

2. Regular Polygons

Let us now consider a few somewhat more elaborate construction problems. We begin with the *regular decagon.* Suppose that a regular decagon is inscribed in a circle with radius 1 (Fig. 32), and call its side x. Since x will subtend an angle of 36° at the center of the circle, the other two angles of the large triangle will each be 72°, and hence the dotted line which bisects angle A divides triangle OAB into two isosceles triangles, each with equal sides of length x. The radius of the circle is thus divided into two segments, x and $1 - x$. Since OAB is similar to the smaller isosceles triangle, we have $1/x = x/(1-x)$. From this proportion we get the quadratic equation $x^2 + x - 1 = 0$, the solution of which is $x = (\sqrt{5} - 1)/2$. (The other solution of the equation is irrelevant, since it yields a negative x.) From this it is clear that x can be constructed geometrically. Having the length x, we may now construct the regular decagon by marking off this length ten times as a chord of the circle. The regular pentagon may now be constructed by joining alternate vertices of the regular decagon.

Instead of constructing $\sqrt{5}$ by the method of Figure 31 we can also obtain it as the hypotenuse of a right triangle whose other sides have lengths 1 and 2. We then obtain x by subtracting the unit length from $\sqrt{5}$ and bisecting the result.

The ratio $OB:AB$ of the preceding problem has been called the golden ratio, because the Greek mathematicians considered a rectangle

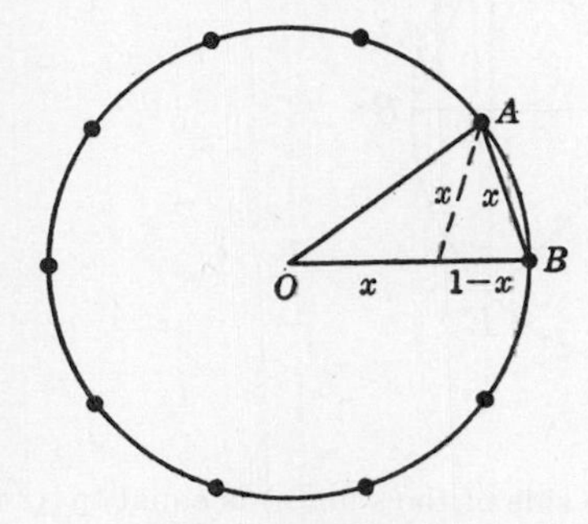

Fig. 32 Regular decagon.

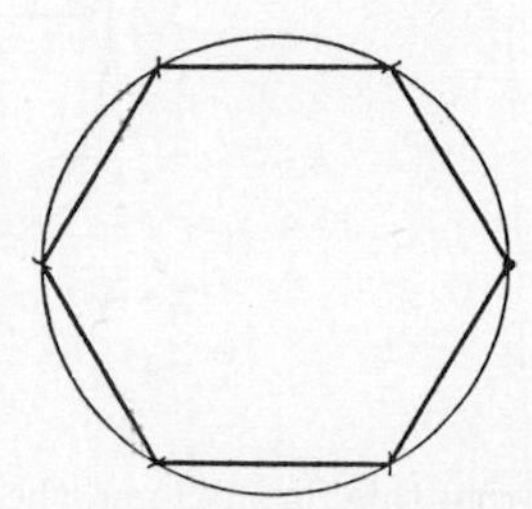
Fig. 33. Regular hexagon.

whose two sides are in this ratio to be aesthetically the most pleasing. Its value, incidentally, is about 1.62.

Of all the regular polygons the hexagon is simplest to construct. We start with a circle of radius r; the length of the side of a regular hexagon inscribed in this circle will then be equal to r. The hexagon itself can be constructed by successively marking off from any point of the circle chords of length r until all six vertices are obtained.

From the regular n-gon we can obtain the regular $2n$-gon by bisecting the arc subtended on the circumscribed circle by each edge of the n-gon, using the additional points thus found as well as the original vertices for the required $2n$-gon. Starting with the diameter of a circle (a "2-gon"), we can therefore construct the 4, 8, 16, $\cdots$, 2^n-gon. Similarly, we can obtain the 12-, 24-, 48-gon, etc. from the hexagon, and the 20-, 40-gon, etc. from the decagon.

If s_n denotes the length of the side of the regular n-gon inscribed in the unit circle (circle with radius 1), then the side of the $2n$-gon is of length

$$s_{2n} = \sqrt{2 - \sqrt{4 - s_n^2}}.$$

This may be proved as follows: In Figure 34 s_n is equal to $DE = 2DC$, s_{2n} equal to DB, and AB equal to 2. The area of the right triangle ABD is given by $\frac{1}{2}BD \cdot AD$ and by $\frac{1}{2}AB \cdot CD$. Since $AD = \sqrt{AB^2 - DB^2}$, we find, by substituting $AB = 2$, $BD = s_{2n}$, $CD = \frac{1}{2}s_n$, and by equating the two expressions for the area,

$$s_n = s_{2n}\sqrt{4 - s_{2n}^2} \qquad \text{or} \qquad s_n^2 = s_{2n}^2\,(4 - s_{2n}^2).$$

Solving this quadratic equation for $x = s_{2n}^2$ and observing that x must be less than 2, one easily finds the formula given above.

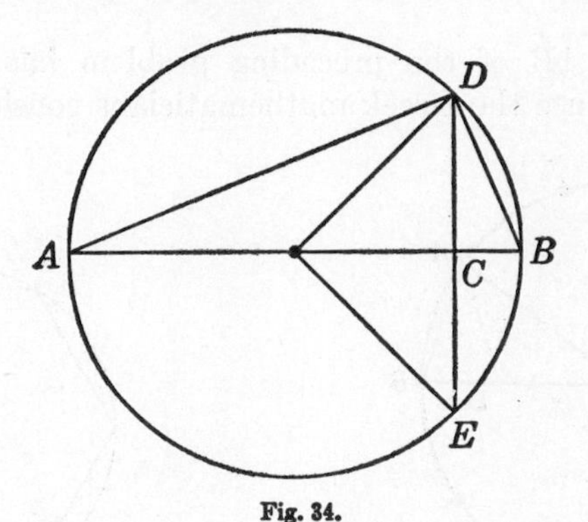

Fig. 34.

From this formula and the fact that s_4 (the side of the square) is equal to $\sqrt{2}$ it follows that

$$s_8 = \sqrt{2 - \sqrt{2}}, \qquad s_{16} = \sqrt{2 - \sqrt{2 + \sqrt{2}}},$$

$$s_{32} = \sqrt{2 - \sqrt{2 + \sqrt{2 + \sqrt{2}}}}, \text{ etc.}$$

As a general formula we obtain, for $n > 2$,

$$s_{2^n} = \sqrt{2 - \sqrt{2 + \sqrt{2 + \cdots + \sqrt{2}}}}$$

with $n - 1$ nested square roots. The circumference of the 2^n-gon in the circle is $2^n s_{2^n}$. As n tends to infinity, the 2^n-gon tends to the circle. Hence $2^n s_{2^n}$ approaches the length of the circumference of the unit circle, which is by definition 2π. Thus we obtain, by substituting m for $n - 1$ and cancelling a factor 2, the limiting formula for π:

$$\underbrace{2^m \sqrt{2 - \sqrt{2 + \sqrt{2 + \cdots + \sqrt{2}}}}}_{m \text{ square roots}} \to \pi \quad \text{as} \quad m \to \infty,$$

Exercise: Since $2^m \longrightarrow \infty$, prove as a consequence that

$$\underbrace{\sqrt{2+\sqrt{2+\cdots+\sqrt{2}}}}_{n \text{ square roots.}} \rightarrow 2 \quad \text{as} \quad n \rightarrow \infty.$$

The results obtained thus far exhibit the following characteristic feature: *The sides of the 2^n-gon, the $5 \cdot 2^n$-gon, and the $3 \cdot 2^n$-gon, can all be found entirely by the processes of addition, subtraction, multiplication, division, and the extraction of square roots.*

*3. Apollonius' Problem

Another construction problem that becomes quite simple from the algebraic standpoint is the famous contact problem of Apollonius already mentioned. In the present context it is unnecessary for us to find a particularly elegant construction. What matters here is that in principle the problem can be solved by straightedge and compass alone. We shall give a brief indication of the proof, leaving the question of a more elegant method of construction to page 161.

Let the centers of the three given circles have coördinates (x_1, y_1), (x_2, y_2) and (x_3, y_3), respectively, with radii r_1, r_2, and r_3. Denote the center and radius of the required circle by (x, y) and r. Then the condition that the required circle be tangent to the three given circles is obtained by observing that the distance between the centers of two tangent circles is equal to the sum or difference of the radii, according as the circles are tangent externally or internally. This yields the equations

$$(1) \qquad (x - x_1)^2 + (y - y_1)^2 - (r \pm r_1)^2 = 0,$$

$$(2) \qquad (x - x_2)^2 + (y - y_2)^2 - (r \pm r_2)^2 = 0,$$

$$(3) \qquad (x - x_3)^2 + (y - y_3)^2 - (r \pm r_3)^2 = 0,$$

or

$$(1a) \quad x^2 + y^2 - r^2 - 2xx_1 - 2yy_1 \pm 2rr_1 + x_1^2 + y_1^2 - r_1^2 = 0,$$

etc. The plus or minus sign is to be chosen in each of these equations according as the circles are to be externally or internally tangent. (See Fig. 35.) Equations (1), (2), (3) are three quadratic equations in three unknowns x, y, r with the property that the second degree terms are the same in each equation, as is seen from the expanded form (1a). Hence, by subtracting (2) from (1), we get a linear equation in x, y, r:

$$(4) \qquad ax + by + cr = d,$$

where $a = 2(x_2 - x_1)$, etc. Similarly, by subtracting (3) from (1), we get another linear equation,

$$(5) \qquad a'x + b'y + c'r = d'.$$

Solving (4) and (5) for x and y in terms of r and then substituting in (1) we get a quadratic equation in r, which can be solved by rational operations and the extraction of a square root (see p. 91). There will in general be two solutions of this equation, of which only one will be positive. After finding r from this equation we obtain x and y from the two linear equations (4) and (5). The circle with center (x, y) and radius r will be tangent to the three given circles. In the whole process we have used only rational operations and square root extractions. It follows that r, x, and y can be constructed by ruler and compass alone.

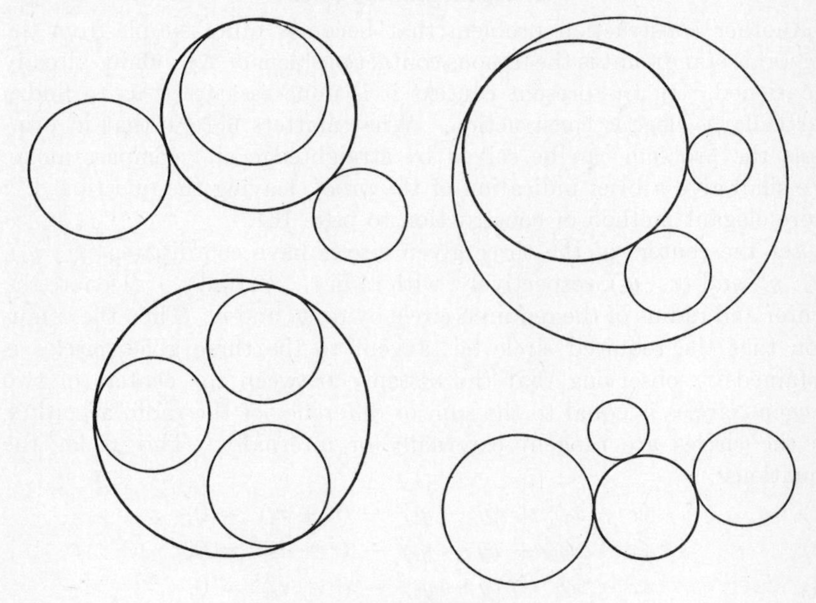

Fig. 35. Apollonius' circles.

There will in general be eight solutions of the problem of Apollonius, corresponding to the $2 \cdot 2 \cdot 2 = 8$ possible combinations of $+$ and $-$ signs in equations (1), (2), and (3). These choices correspond to the conditions that the desired circles be externally or internally tangent to each of the three given circles. It may happen that our algebraic procedure does not actually yield real values for x, y, and r. This will be the case, for example, if the three given circles are concentric, so that no solution to the geometrical problem exists. Likewise, we must expect possible "degenerations" of the solution, as in the case when the three given circles degenerate into three points on a line. Then the Apollonius circle degenerates into this line. We shall not discuss these possibilities in detail; a reader with some algebraic experience will be able to complete the analysis.

*§2. CONSTRUCTIBLE NUMBERS AND NUMBER FIELDS

1. General Theory

Our previous discussion indicates the general algebraic background of geometrical constructions. Every ruler and compass construction consists of a sequence of steps, each of which is one of the following: 1) connecting two points by a straight line, 2) finding the point of intersection of two lines, 3) drawing a circle with a given radius about a point, 4) finding the points of intersection of a circle with another circle or with a line. An element (point, line, or circle) is considered to be known if it was given at the outset or if it has been constructed in some previous step. For a theoretical analysis we may refer the whole construction to a coördinate system x, y (see p. 73). The given elements will then be represented by points or segments in the x, y plane. If only one segment is given at the outset, we may take this as the unit length, which fixes the point $x = 1$, $y = 0$. Sometimes there appear "arbitrary" elements: arbitrary lines are drawn, arbitrary points or radii are chosen. (An example of such an arbitrary element appears in constructing the midpoint of a segment; we draw two circles of equal but arbitrary radius from each endpoint of the segment, and join their intersections.) In such cases we may choose the element to be rational; i.e. arbitrary points may be chosen with rational coordinates x, y, arbitrary lines $ax + by + c = 0$ with rational coefficients a, b, c, arbitrary circles with centers having rational coördinates and with rational radii. We shall make such a choice of rational arbitrary elements throughout; if the elements are indeed arbitrary this restriction cannot affect the result of a construction.

For the sake of simplicity, we shall assume in the following discussion that only one element, the unit length 1, is given at the outset. Then according to §1 we can construct by ruler and compass all numbers that can be obtained from unity by the rational processes of addition, subtraction, multiplication and division, i.e. all the rational numbers r/s, where r and s are integers. The system of rational numbers is "closed" with respect to the rational operations; that is, the sum, difference, product, or quotient of any two rational numbers—excluding division by 0, as always—is again a rational number. Any set of numbers possessing this property of closure with respect to the four rational operations is called a *number field.*

Exercise: Show that every field contains all the rational numbers at least. (Hint: If $a \neq 0$ is a number in the field F, then $a/a = 1$ belongs to F, and from 1 we can obtain any rational number by rational operations.)

Starting from the unit, we can thus construct the whole rational number field and hence all the rational points (i.e. points with both coördinates rational) in the x, y plane. We can reach new, irrational, numbers by using the compass to construct e.g. the number $\sqrt{2}$ which, as we know from Chapter II, §2, is not in the rational field. Having constructed $\sqrt{2}$ we may then, by the "rational" constructions of §1, find all numbers of the form

$$a + b\sqrt{2}, \tag{1}$$

where a, b are rational, and therefore are themselves constructible. We may likewise construct all numbers of the form

$$\frac{a + b\sqrt{2}}{c + d\sqrt{2}} \quad \text{or} \quad (a + b\sqrt{2})(c + d\sqrt{2}),$$

where a, b, c, d are rational. These numbers, however, may always be written in the form (1). For we have

$$\frac{a + b\sqrt{2}}{c + d\sqrt{2}} = \frac{a + b\sqrt{2}}{c + d\sqrt{2}} \cdot \frac{c - d\sqrt{2}}{c - d\sqrt{2}}$$

$$= \frac{ac - 2bd}{c^2 - 2d^2} + \frac{bc - ad}{c^2 - 2d^2}\sqrt{2} = p + q\sqrt{2},$$

where p, q are rational. (The denominator $c^2 - 2d^2$ cannot be zero, for if $c^2 - 2d^2 = 0$, then $\sqrt{2} = c/d$, contrary to the fact that $\sqrt{2}$ is irrational.) Likewise

$$(a + b\sqrt{2})(c + d\sqrt{2}) = (ac + 2bd) + (bc + ad)\sqrt{2} = r + s\sqrt{2},$$

where r, s are rational. Hence all that we reach by the construction of $\sqrt{2}$ is the set of numbers of the form (1), with arbitrary rational a, b.

Exercises: From $p = 1 + \sqrt{2}$, $q = 2 - \sqrt{2}$, $r = -3 + \sqrt{2}$ obtain the numbers

$$\frac{p}{q},\ p + p^2,\ (p - p^2)\frac{q}{r},\ \frac{pqr}{1 + r^2},\ \frac{p + qr}{q + pr^2},$$

in the form (1).

These numbers (1) again form a field, as the preceding discussion shows. (That the sum and difference of two numbers of the form (1) are also of the form (1) is obvious.) This field is larger than the rational field, which is a part or *subfield* of it. But, of course, it is smaller than the field of *all* real numbers. Let us call the rational field F_0 and the new field of numbers of the form (1), F_1. The constructibility of every number in the "extension field" F_1 has been established. We may now extend the scope of our constructions, e.g. by taking a number of F_1, say $k = 1 + \sqrt{2}$, and extracting the square root, thus obtaining the constructible number

$$\sqrt{1+\sqrt{2}} = \sqrt{k},$$

and with it, according to §1, the field consisting of all the numbers

$$p + q\sqrt{k}, \tag{2}$$

where now p and q may be arbitrary numbers of F_1, i.e. of the form $a + b\sqrt{2}$, with a, b in F_0, i.e. rational.

Exercises: Represent

$$(\sqrt{k})^3,\ \frac{1+(\sqrt{k})^2}{1+\sqrt{k}},\ \frac{\sqrt{2}\sqrt{k}+\frac{1}{\sqrt{2}}}{(\sqrt{k})^3-3},\ \frac{(1+\sqrt{k})(2-\sqrt{k})\left(\sqrt{2}+\frac{1}{\sqrt{k}}\right)}{1+\sqrt{2}\,k},$$

in the form (2).

All these numbers have been constructed on the assumption that only one segment was given at the outset. If two segments are given we may select one of them as the unit length. In terms of this unit suppose that the length of the other segment is α. Then we can construct the field G consisting of all numbers of the form

$$\frac{a_m\alpha^m + a_{m-1}\alpha^{m-1} + \cdots + a_1\alpha + a_0}{b_n\alpha^n + b_{n-1}\alpha^{n-1} + \cdots + b_1\alpha + b_0}$$

where the numbers $a_0, \cdots, a_m$ and $b_0, \cdots, b_n$ are rational, and m and n are arbitrary positive integers.

Exercise: If two segments of lengths 1 and α are given, give actual constructions for $1 + \alpha + \alpha^2$, $(1 + \alpha)/(1 - \alpha)$, α^3.

Now let us assume more generally that we are able to construct all the numbers of some number field F. We shall show that *the use of the ruler alone will never lead us out of the field* F. The equation of the straight line through two points whose coördinates a_1, b_1 and a_2, b_2 are in F is $(b_1 - b_2)x + (a_2 - a_1)y + (a_1b_2 - a_2b_1) = 0$ (see p. 491); its coefficients are rational expressions formed from numbers in F, and therefore, by definition of a field, are themselves in F. Moreover, if we have two lines, $\alpha x + \beta y - \gamma = 0$ and $\alpha' x + \beta' y - \gamma' = 0$, with coefficients in F, then the coördinates of their point of intersection, found by solving these two simultaneous equations, are $x = \dfrac{\gamma\beta' - \beta\gamma'}{\alpha\beta' - \beta\alpha'}$, $y = \dfrac{\alpha\gamma' - \gamma\alpha'}{\alpha\beta' - \beta\alpha'}$. Since these are likewise numbers of F, it is clear that the use of the ruler alone cannot take us beyond the confines of the field F.

Exercises: The lines $x + \sqrt{2}y - 1 = 0$, $2x - y + \sqrt{2} = 0$, have coefficients in the field (1). Calculate the coördinates of their point of intersection, and verify that these have the form (1).—Join the points $(1, \sqrt{2})$ and $(\sqrt{2}, 1 - \sqrt{2})$ by a line $ax + by + c = 0$, and verify that the coefficients are of the form (1).—Do the same with respect to the field (2) for the lines $\sqrt{1 + \sqrt{2}}x + \sqrt{2}y = 1$, $(1 + \sqrt{2})x - y = 1 - \sqrt{1 + \sqrt{2}}$, and the points $(\sqrt{2}, -1)$, $(1 + \sqrt{2}, \sqrt{1 + \sqrt{2}})$, respectively.

We can only break through the walls of F by using the compass. For this purpose we select an element k of F which is such that $\sqrt{k}$ is not in F. Then we can construct $\sqrt{k}$ and therefore all the numbers

$$a + b\sqrt{k}, \tag{3}$$

where a and b are rational, or even arbitrary elements of F. The sum and the difference of two numbers $a + b\sqrt{k}$ and $c + d\sqrt{k}$, their product, $(a + b\sqrt{k})(c + d\sqrt{k}) = (ac + kbd) + (ad + bc)\sqrt{k}$, and their quotient,

$$\frac{a + b\sqrt{k}}{c + d\sqrt{k}} = \frac{(a + b\sqrt{k})(c - d\sqrt{k})}{c^2 - kd^2} = \frac{ac - kbd}{c^2 - kd^2} + \frac{bc - ad}{c^2 - kd^2}\sqrt{k},$$

are again of the form $p + q\sqrt{k}$ with p and q in F. (The denominator $c^2 - kd^2$ cannot vanish unless c and d are both zero; for otherwise we would have $\sqrt{k} = c/d$, a number in F, contrary to the assumption that $\sqrt{k}$ is not in F.) Hence the set of numbers of the form $a + b\sqrt{k}$ forms a field F'. The field F' contains the original field F, for we may, in particular, choose $b = 0$. F' is called an *extension field* of F, and F a *subfield* of F'.

As an example, let F be the field $a + b\sqrt{2}$ with rational a, b, and take $k = \sqrt{2}$. Then the numbers of the extension field F' are represented by $p + q\sqrt[4]{2}$, where p and q are in F, $p = a + b\sqrt{2}$, $q = a' + b'\sqrt{2}$, with rational a, b, a', b'. Any number in F' can be reduced to that form; for example

$$\begin{aligned}
\frac{1}{\sqrt{2} + \sqrt[4]{2}} &= \frac{\sqrt{2} - \sqrt[4]{2}}{(\sqrt{2} + \sqrt[4]{2})(\sqrt{2} - \sqrt[4]{2})} = \frac{\sqrt{2} - \sqrt[4]{2}}{2 - \sqrt{2}} \\
&= \frac{\sqrt{2}}{2 - \sqrt{2}} - \frac{\sqrt[4]{2}}{2 - \sqrt{2}} = \frac{\sqrt{2}\,(2 + \sqrt{2})}{4 - 2} - \frac{(2 + \sqrt{2})}{4 - 2}\sqrt[4]{2} \\
&= (1 + \sqrt{2}) - (1 + \tfrac{1}{2}\sqrt{2})\sqrt[4]{2}
\end{aligned}$$

Exercise: Let F be the field $p + q\sqrt{2 + \sqrt{2}}$, where p and q are of the form $a + b\sqrt{2}$, a, b rational. Represent $\dfrac{1 + \sqrt{2 + \sqrt{2}}}{2 - 3\sqrt{2 + \sqrt{2}}}$ in this form.

We have seen that if we start with any field F of constructible numbers containing the number k, then by use of the ruler and a single application of the compass we can construct $\sqrt{k}$ and hence any number of the form $a + b\sqrt{k}$, where a, b, are in F.

We now show, conversely, that by a single application of the compass we can obtain *only* numbers of this form. For what the compass does in a construction is to define points (or their coördinates) as points of intersection of a circle with a straight line, or of two circles. A circle with center ξ, η and radius r has the equation $(x - \xi)^2 + (y - \eta)^2 = r^2$; hence, if ξ, η, r are in F, the equation of the circle can be written in the form

$$x^2 + y^2 + 2\alpha x + 2\beta y + \gamma = 0,$$

with the coefficients α, β, γ in F. A straight line,

$$ax + by + c = 0,$$

joining any two points whose coördinates are in F, has coefficients a, b, c in F, as we have seen on page 129. By eliminating y from these simultaneous equations, we obtain for the x-coördinate of a point of intersection of the circle and line a quadratic equation of the form

$$Ax^2 + Bx + C = 0,$$

with coefficients A, B, C in F (explicitly: $A = a^2 + b^2$, $B = 2(ac + b^2\alpha - ab\beta)$, $C = c^2 - 2bc\beta + b^2\gamma$). The solution is given by the formula

$$x = \frac{-B \pm \sqrt{B^2 - 4AC}}{2A},$$

which is of the form $p + q\sqrt{k}$, with p, q, k in F. A similar formula holds for the y-coördinate of a point of intersection.

Again, if we have two circles,

$$x^2 + y^2 + 2\alpha x + 2\beta y + \gamma = 0,$$
$$x^2 + y^2 + 2\alpha' x + 2\beta' y + \gamma' = 0,$$

then by subtracting the second equation from the first we obtain the linear equation

$$2(\alpha - \alpha')x + 2(\beta - \beta')y + (\gamma - \gamma') = 0,$$

which may be solved with the equation of the first circle as before. In either case, the construction yields the x- and y-coördinates of either one or two new points, and these new quantities are of the form $p + q\sqrt{k}$, with p, q, k in F. In particular, of course, $\sqrt{k}$ may itself belong to F, e.g., when $k = 4$. Then the construction does not yield anything essentially new, and we remain in F. But in general this will not be the case.

Exercises: Consider the circle with radius $2\sqrt{2}$ about the origin, and the line joining the points $(1/2, 0)$, $(4\sqrt{2}, \sqrt{2})$. Find the field F' determined by the coördinates of the points of intersection of the circle and the line. Do the same with respect to the intersection of the given circle with the circle with radius $\sqrt{2}/2$ and center $(0, 2\sqrt{2})$.

Summarizing again: If certain quantities are given at the outset, then we can construct with a straightedge alone all the quantities in the field F generated by rational processes from the given quantities. Using the compass we can then extend the field F of constructible quantities to a wider extension field by selecting any number k of F, extracting the square root of k, and constructing the field F' consisting of the numbers $a + b\sqrt{k}$, where a and b are in F. F is called a subfield of F'; all quantities in F are also contained in F', since in the expression $a + b\sqrt{k}$ we may choose $b = 0$. (It is assumed that $\sqrt{k}$ is a new number not lying in F, since otherwise the process of adjunction of $\sqrt{k}$ would not lead to anything new, and F' would be identical with F.) We have shown that any step in a geometrical construction (drawing a line through two known points, drawing a circle with known center and radius, or marking the intersection of two known lines or circles) will either produce new quantities lying in the field already known to consist of constructible numbers, or, by the construction of a square root, will open up a new extension field of constructible numbers.

The totality of all constructible numbers can now be described with precision. We start with a given field F_0, defined by whatever quantities are given at the outset, e.g. the field of rational numbers if only a single segment, chosen as the unit, is given. Next, by the adjunction of $\sqrt{k_0}$, where k_0 is in F_0, but $\sqrt{k_0}$ is not, we construct an extension field F_1 of constructible numbers, consisting of all numbers of the form $a_0 + b_0\sqrt{k_0}$, where a_0 and b_0 may be any numbers of F_0. Then F_2, a new extension field of F_1, is defined by the numbers $a_1 + b_1\sqrt{k_1}$, where a_1 and b_1 are any numbers of F_1, and k_1 is some number of F_1 whose square root does not lie in F_1. Repeating this procedure, we shall reach a field F_n after n adjunctions of square roots. *Constructible numbers are those and only those which can be reached by such a sequence of extension fields; that is, which lie in a field* F_n *of the type described.* The size of the number n of necessary extensions does not matter; in a way it measures the degree of complexity of the problem.

The following example may illustrate the process. We want to reach the number

$$\sqrt{6} + \sqrt{\sqrt{\sqrt{1 + \sqrt{2}} + \sqrt{3}} + 5}.$$

Let F_0 denote the rational field. Putting $k_0 = 2$, we obtain the field F_1, which contains the number $1 + \sqrt{2}$. We now take $k_1 = 1 + \sqrt{2}$ and $k_2 = 3$. As a matter of fact, 3 is in the original field F_0, and *a fortiori* in the field F_2, so that it is perfectly permissible to take $k_2 = 3$. We then take $k_3 = \sqrt{1 + \sqrt{2}} + \sqrt{3}$, and finally $k_4 = \sqrt{\sqrt{1 + \sqrt{2}} + \sqrt{3}} + 5$. The field F_5 thus constructed contains the desired number, for $\sqrt{6}$ is also in F_5, since $\sqrt{2}$ and $\sqrt{3}$, and therefore their product, are in F_3 and therefore also in F_5.

Exercises: Verify that, starting with the rational field, the side of the regular 2^m-gon (see p. 124) is a constructible number, with $n = m - 1$. Determine the sequence of extension fields. Do the same for the numbers

$$\sqrt{1 + \sqrt{2} + \sqrt{3} + \sqrt{5}}, \qquad (\sqrt{5} + \sqrt{11})/(1 + \sqrt{7 - \sqrt{3}}),$$
$$(\sqrt{2 + \sqrt{3}})(\sqrt[3]{2} + \sqrt{1 + \sqrt{2 + \sqrt{5}} + \sqrt{3 - \sqrt{7}}}).$$

2. All Constructible Numbers are Algebraic

If the initial field F_0 is the rational field generated by a single segment, then all constructible numbers will be algebraic. (For the definition of algebraic numbers see p. 103). The numbers of the field F_1 are roots of quadratic equations, those of F_2 are roots of fourth degree equations, and, in general, the numbers of F_k are roots of equations of degree 2^k, with rational coefficients. To show this for a field F_2 we may first consider as an example $x = \sqrt{2} + \sqrt{3 + \sqrt{2}}$. We have $(x - \sqrt{2})^2 = 3 + \sqrt{2}$, $x^2 + 2 - 2\sqrt{2}x = 3 + \sqrt{2}$, or $x^2 - 1 = \sqrt{2}(2x + 1)$, a quadratic equation with coefficients in a field F_1. By squaring, we finally obtain

$$(x^2 - 1)^2 = 2(2x + 1)^2,$$

which is an equation of the fourth degree with rational coefficients.

In general, any number in a field F_2 has the form

$$x = p + q\sqrt{w}, \tag{4}$$

where p, q, w are in a field F_1, and hence have the form $p = a + b\sqrt{s}$, $q = c + d\sqrt{s}$, $w = e + f\sqrt{s}$, where a, b, c, d, e, f, s are rational. From (4) we have

$$x^2 - 2px + p^2 = q^2w,$$

where all the coefficients are in a field F_1, generated by $\sqrt{s}$. Hence this equation may be rewritten in the form

$$x^2 + ux + v = \sqrt{s}(rx + t),$$

where r, s, t, u, v are rational. By squaring both sides we obtain an equation of the fourth degree

$$(5) \qquad (x^2 + ux + v)^2 = s(rx + t)^2$$

with rational coefficients, as stated.

Exercises: 1) Find the equations with rational coefficients for a) $x = \sqrt{2 + \sqrt{3}}$; b) $x = \sqrt{2} + \sqrt{3}$; c) $x = 1/\sqrt{5 + \sqrt{3}}$.

2) Find by a similar method equations of the eighth degree for a) $x = \sqrt{2 + \sqrt{2 + \sqrt{2}}}$; b) $x = \sqrt{2} + \sqrt{1 + \sqrt{3}}$; c) $x = 1 + \sqrt{5 + \sqrt{3 + \sqrt{2}}}$.

To prove the theorem in general for x in a field F_k with arbitrary k, we show by the procedure used above that x satisfies a quadratic equation with coefficients in a field F_{k-1}. Repeating the procedure, we find that x satisfies an equation of degree $2^2 = 4$ with coefficients in a field F_{k-2}, etc.

Exercise: Complete the general proof by using mathematical induction to show that x satisfies an equation of degree 2^l with coefficients in a field F_{k-l}, $0 < l \leq k$. This statement for $l = k$ is the desired theorem.

*§3. THE UNSOLVABILITY OF THE THREE GREEK PROBLEMS

1. Doubling the Cube

Now we are well prepared to investigate the old problems of trisecting the angle, doubling the cube, and constructing the regular heptagon. We consider first the problem of doubling the cube. If the given cube has an edge of unit length, its volume will be the cubic unit; it is required that we find the edge x of a cube with twice this volume. The required edge x will therefore satisfy the simple cubic equation

$$(1) \qquad x^3 - 2 = 0.$$

Our proof that this number x cannot be constructed by ruler and compass alone is indirect. We assume tentatively that a construction is possible. According to the preceding discussion this means that x lies in some field F_k obtained, as above, from the rational field by successive extensions through adjunction of square roots. As we shall show, this assumption leads to an absurd consequence.

We already know that x cannot lie in the rational field F_0, for $\sqrt[3]{2}$ is an irrational number (see Exercise 1, p. 60). Hence x can only lie in some extension field F_k, where k is a positive integer. We may as well assume that k is the *least* positive integer such that x lies in some F_k. It follows that x can be written in the form

$$x = p + q\sqrt{w},$$

where p, q, and w belong to some F_{k-1}, but $\sqrt{w}$ does not. Now, by a simple but important type of algebraic reasoning, we shall show that if $x = p + q\sqrt{w}$ is a solution of the cubic equation (1), then $y = p - q\sqrt{w}$ is also a solution. Since x is in the field F_k, x^3 and $x^3 - 2$ are also in F_k, and we have

$$x^3 - 2 = a + b\sqrt{w}, \tag{2}$$

where a and b are in F_{k-1}. By an easy calculation we can show that $a = p^3 + 3pq^2w - 2$, $b = 3p^2q + q^3w$. If we put

$$y = p - q\sqrt{w},$$

then a substitution of $-q$ for q in these expressions for a and b shows that

$$y^3 - 2 = a - b\sqrt{w}. \tag{2'}$$

Now x was supposed to be a root of $x^3 - 2 = 0$, hence

$$a + b\sqrt{w} = 0. \tag{3}$$

This implies—and here is the key to the argument—that a and b must both be zero. If b were not zero, we would infer from (3) that $\sqrt{w} = -a/b$. But then $\sqrt{w}$ would be a number of the field F_{k-1} in which a and b lie, contrary to our assumption. Hence $b = 0$, and it follows immediately from (3) that $a = 0$ also.

Now that we have shown that $a = b = 0$, we immediately infer from (2′) that $y = p - q\sqrt{w}$ is also a solution of the cubic equation (1), since $y^3 - 2$ is equal to zero. Furthermore, $y \neq x$, i.e. $x - y \neq 0$; for, $x - y = 2q\sqrt{w}$ can only vanish if $q = 0$; and if this were so then $x = p$ would lie in F_{k-1}, contrary to our assumption.

We have therefore shown that, if $x = p + q\sqrt{w}$ is a root of the cubic equation (1), then $y = p - q\sqrt{w}$ is a different root of this equation. This leads immediately to a contradiction. For there is only one real number x which is a cube root of 2, the other cube roots of 2 being imaginary (see p. 98); $y = p - q\sqrt{w}$ is obviously real, since p, q, and $\sqrt{w}$ were real.

Thus our basic assumption has led to an absurdity, and hence is proved to be wrong; a solution of (1) cannot lie in a field F_k, so that doubling the cube by ruler and compass is impossible.

2. A Theorem on Cubic Equations

Our concluding algebraic argument was especially adapted to the particular problem at hand. If we want to dispose of the two other Greek problems, it is desirable to proceed on a more general basis. All three problems depend algebraically on cubic equations. It is a fundamental fact concerning the cubic equation

$$z^3 + az^2 + bz + c = 0 \tag{4}$$

that, if x_1, x_2, x_3 are the three roots of this equation, then

$$x_1 + x_2 + x_3 = -a.\dagger \tag{5}$$

Let us consider any cubic equation (4) where the coefficients a, b, c are rational numbers. It may be that one of the roots of the equation is rational; for example, the equation $x^3 - 1 = 0$ has the rational root 1, while the two other roots, given by the quadratic equation $x^2 + x + 1 = 0$, are necessarily imaginary. But we can easily prove the general theorem: *If a cubic equation with rational coefficients has no rational root, then none of its roots is constructible starting from the rational field* F_0.

Again we give the proof by an indirect method. Suppose x were a constructible root of (4). Then x would lie in the last field F_k of some chain of extension fields, $F_0, F_1, \cdots, F_k$, as above. We may assume that k is the *smallest* integer such that a root of the cubic equation (4) lies in an extension field F_k. Certainly k must be greater than zero, since in the statement of the theorem it is assumed that no root x lies in the rational field F_0. Hence x can be written in the form

$$x = p + q\sqrt{w},$$

where p, q, w are in the preceding field, F_{k-1}, but $\sqrt{w}$ is not. It follows, exactly as for the special equation, $z^3 - 2 = 0$, of the preceding article, that another number of F_k,

$$y = p - q\sqrt{w},$$

will also be a root of the equation (4). As before, we see that $q \neq 0$ and hence $x \neq y$.

From (5) we know that the third root u of the equation (4) is given by $u = -a - x - y$. But since $x + y = 2p$, this means that

$$u = -a - 2p,$$

† The polynomial $z^3 + az^2 + bz + c$ may be factored into the product $(z - x_1)(z - x_2)(z - x_3)$, where x_1, x_2, x_3, are the three roots of the equation (4) (see p. 101). Hence

$$z^3 + az^2 + bz + c = z^3 - (x_1 + x_2 + x_3)z^2 + (x_1x_2 + x_1x_3 + x_2x_3)z - x_1x_2x_3,$$

so that, since the coefficient of each power of z must be the same on both sides,

$$-a = x_1 + x_2 + x_3, \qquad b = x_1x_2 + x_1x_3 + x_2x_3, \qquad -c = x_1x_2x_3.$$

where $\sqrt{w}$ has disappeared, so that u is a number in the field F_{k-1}. This contradicts the hypothesis that k is the *smallest* number such that some F_k contains a root of (4). Hence the hypothesis is absurd, and no root of (4) can lie in such a field F_k. The general theorem is proved. On the basis of this theorem, a construction by ruler and compass alone is proved to be impossible if the algebraic equivalent of the problem is the solution of a cubic equation with no rational roots. This equivalence was at once obvious for the problem of doubling the cube, and will now be established for the other two Greek problems.

3. Trisecting the Angle

We shall now prove that the trisection of the angle by ruler and compass alone is *in general* impossible. Of course, there are angles, such as 90° and 180°, for which the trisection can be performed. What we have to show is that the trisection cannot be effected by a procedure valid for *every* angle. For the proof, it is quite sufficient to exhibit only one angle that cannot be trisected, since a valid *general method* would have to cover every single example. Hence the non-existence of a general method will be proved if we can demonstrate, for example, that the angle 60° cannot be trisected by ruler and compass alone.

We can obtain an algebraic equivalent of this problem in different ways; the simplest is to consider an angle θ as given by its cosine: $\cos\theta = g$. Then the problem is equivalent to that of finding the quantity $x = \cos(\theta/3)$. By a simple trigonometrical formula (see p. 97), the cosine of $\theta/3$ is connected with that of θ by the equation

$$\cos\theta = g = 4\cos^3(\theta/3) - 3\cos(\theta/3).$$

In other words, the problem of trisecting the angle θ with $\cos\theta = g$ amounts to constructing a solution of the cubic equation

$$4z^3 - 3z - g = 0. \tag{6}$$

To show that this cannot in general be done, we take $\theta = 60°$, so that $g = \cos 60° = \frac{1}{2}$. Equation (6) then becomes

$$8z^3 - 6z = 1. \tag{7}$$

By virtue of the theorem proved in the preceding article, we need only show that this equation has no rational root. Let $v = 2z$. Then the equation becomes

$$v^3 - 3v = 1. \tag{8}$$

If there were a rational number $v = r/s$ satisfying this equation, where r and s are integers without a common factor > 1, we should have $r^3 - 3s^2r = s^3$. From this it follows that $s^3 = r(r^2 - 3s^2)$ is divisible by r, which means that r and s have a common factor unless $r = \pm 1$. Likewise, s^2 is a factor of $r^3 = s^2(s + 3r)$, which means that r and s have a common factor unless $s = \pm 1$. Since we assumed that r and s had no common factor, we have shown that the only rational numbers which could possibly satisfy equation (8) are $+1$ or -1. By substituting $+1$ and -1 for v in equation (8) we see that neither value satisfies it. Hence (8), and consequently (7), has no rational root, and the impossibility of trisecting the angle is proved.

The theorem that the general angle cannot be trisected with ruler and compass alone is true only when the ruler is regarded as an instrument for drawing a straight line through any two given points and *nothing else*. In our general

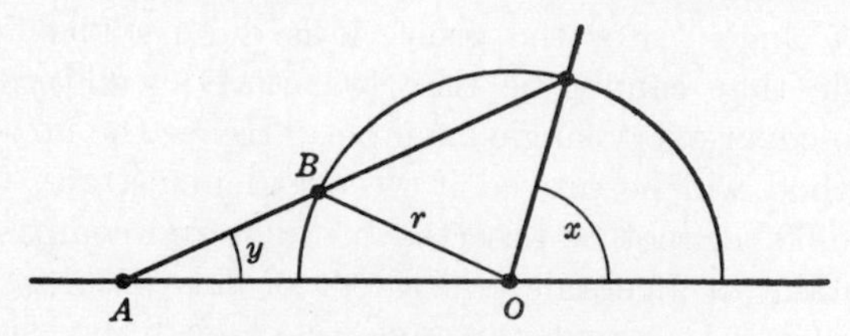

Fig. 36. Archimedes' trisection of an angle.

characterization of constructible numbers the use of the ruler was always limited to this operation only. By permitting other uses of the ruler the totality of possible constructions may be greatly extended. The following method for trisecting the angle, found in the works of Archimedes, is a good example.

Let an arbitrary angle x be given, as in Fig. 36. Extend the base of the angle to the left, and swing a semicircle with O as center and arbitrary radius r. Mark two points A and B on the edge of the ruler such that $AB = r$. Keeping the point B on the semicircle, slide the ruler into the position where A lies on the extended base of the angle x, while the edge of the ruler passes through the intersection of the terminal side of the angle x with the semicircle about O. With the ruler in this position draw a straight line, making an angle y with the extended base of the original angle x.

Exercise: Show that this construction actually yields $y = x/3$.

4. The Regular Heptagon

We shall now consider the problem of finding the side x of a regular heptagon inscribed in the unit circle. The simplest way to dispose of this problem is by means of complex numbers (see Ch. II, §5). We know that the vertices of the heptagon are given by the roots of the equation

$$z^7 - 1 = 0, \tag{9}$$

the coördinates x, y of the vertices being considered as the real and imaginary parts of complex numbers $z = x + yi$. One root of this equation is $z = 1$, and the others are the roots of the equation

$$\frac{z^7 - 1}{z - 1} = z^6 + z^5 + z^4 + z^3 + z^2 + z + 1 = 0, \tag{10}$$

obtained from (9) by factoring out $z - 1$ (see p. 99). Dividing (10) by z^3, we obtain the equation

$$z^3 + 1/z^3 + z^2 + 1/z^2 + z + 1/z + 1 = 0. \tag{11}$$

By a simple algebraic transformation this may be written in the form

$$(z + 1/z)^3 - 3(z + 1/z) + (z + 1/z)^2 - 2 + (z + 1/z) + 1 = 0. \tag{12}$$

Denoting the quantity $z + 1/z$ by y, we find from (12) that

$$y^3 + y^2 - 2y - 1 = 0. \tag{13}$$

We know that z, the seventh root of unity, is given by

$$z = \cos\phi + i\sin\phi, \tag{14}$$

where $\phi = 360°/7$ is the angle subtended at the center of the circle by the edge of the regular heptagon; likewise we know from Exercise 2, page 97, that $1/z = \cos\phi - i\sin\phi$, so that $y = z + 1/z = 2\cos\phi$. If we can construct y, we can also construct $\cos\phi$, and conversely. Hence, if we can prove that y is not constructible, we shall at the same time show that z, and therefore the heptagon, is not constructible. Thus, considering the theorem of Article 2, it remains merely to show that the equation (13) has no rational roots. This, too, is proved indirectly. Assume that (13) has a rational root r/s, where r and s are integers having no common factor. Then we have

$$r^3 + r^2s - 2rs^2 - s^3 = 0; \tag{15}$$

whence it is seen as above that r^3 has the factor s, and s^3 the factor r. Since r and s have no common factor, each must be ± 1; therefore y can have only the possible values $+1$ and -1, if it is to be rational. On substituting these numbers in the equation, we see that neither of them satisfies it. Hence y, and therefore the edge of the regular heptagon, is not constructible.

5. Remarks on the Problem of Squaring the Circle

We have been able to dispose of the problems of doubling the cube, trisecting the angle, and constructing the regular heptagon, by comparatively elementary methods. The problem of squaring the circle is much more difficult and requires the technique of advanced mathematical analysis. Since a circle with radius r has the area πr^2, the problem of constructing a square with area equal to that of a given circle whose radius is the unit length 1 amounts to the construction of a segment of length $\sqrt{\pi}$ as the edge of the required square. This segment will be constructible if and only if the number π is constructible. In the light of our general characterization of constructible numbers, we could show the impossibility of squaring the circle by showing that the number π cannot be contained in any field F_k that can be reached by the successive adjunction of square roots to the rational field F_0. Since all the members of any such field are algebraic numbers, i.e. numbers that satisfy algebraic equations with integer coefficients, it will be sufficient if the number π can be shown to be not algebraic, i.e. to be transcendental (see p. 104).

The technique necessary for proving that π is a transcendental number was created by Charles Hermite (1822–1905), who proved the number e to be transcendental. By a slight extension of Hermite's method F. Lindemann succeeded (1882) in proving the transcendence of π, and thus definitely settled the age-old question of squaring the circle. The proof is within the reach of the student of advanced analysis, but is beyond the scope of this book.

PART II

VARIOUS METHODS FOR PERFORMING CONSTRUCTIONS

§4. GEOMETRICAL TRANSFORMATIONS. INVERSION

1. General Remarks

In the second part of this chapter we shall discuss in a systematic way some general principles that may be applied to construction problems. Many of these problems can be more clearly viewed from the general standpoint of "geometrical transformations"; instead of studying an individual construction, we shall consider simultaneously a whole class of problems connected by certain processes of transformation. The clarifying power of the concept of a class of geometrical transformations is by no means restricted to construction problems, but affects almost everything in geometry. In Chapters IV and V we shall deal with this general aspect of geometrical transformations. Here we shall study a particular type of transformation, the inversion of the plane in a circle, which is a generalization of ordinary reflection in a straight line.

By a *transformation*, or *mapping*, of the plane onto itself we mean a rule which assigns to every point P of the plane another point P', called the *image* of P under the transformation; the point P is called the *antecedent* of P'. A simple example of such a transformation is given by the *reflection* of the plane in a given straight line L as in a mirror: a point P on one side of L has as its image the point P', on the other side of L, and such that L is the perpendicular bisector of the segment PP'. A transformation may leave certain points of the plane fixed; in the case of a reflection this is true of the points on L.

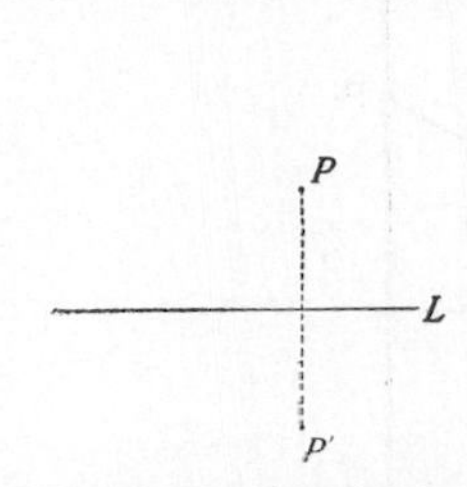

Fig. 37. Reflection of a point in a line.

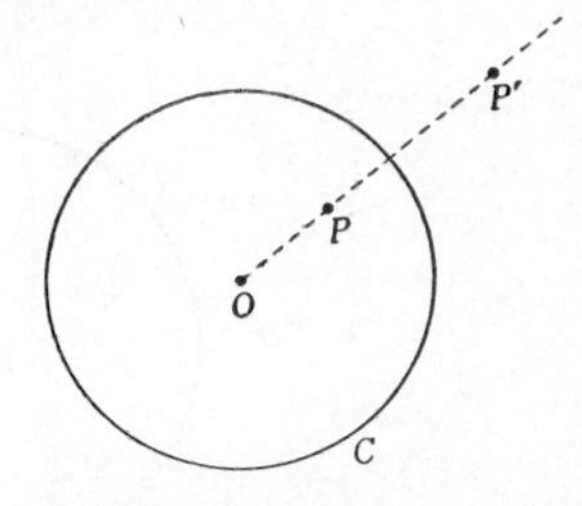

Fig. 38. Inversion of a point in a circle.

Other examples of transformations are the *rotations* of the plane about a fixed point O, the parallel *translations*, which move every point a distance d in a given direction (such a transformation has no fixed points), and, more generally, the *rigid motions* of the plane, which may be thought of as compounded of rotations and parallel translations.

The particular class of transformations of interest to us now are the *inversions* with respect to circles. (These are sometimes known as circular reflections, because to a certain approximation they represent the relation between original and image in reflection by a circular mirror.) In a fixed plane let C be a given circle with center O (called the center of inversion) and radius r. The image of a point P is defined to be the point P' lying on the line OP on the same side of O as P and such that

$$OP \cdot OP' = r^2. \tag{1}$$

The points P and P' are said to be *inverse points* with respect to C. From this definition it follows that, if P' is the inverse point of P, then P is the inverse of P'. An inversion interchanges the inside and outside of the circle C, since for $OP < r$ we have $OP' > r$, and for $OP > r$, we have $OP' < r$. The only points of the plane that remain fixed under the inversion are the points on the circle C itself.

Rule (1) does not define an image for the center O. It is clear that if a moving point P approaches O, the image P' will recede farther and farther out in the plane. For this reason we sometimes say that O itself corresponds to the *point at infinity* under the inversion. The usefulness of this terminology lies in the fact that it enables us to state that an inversion sets up a correspondence between the points of the plane and their images which is biunique without exception: each point of the plane has one and only one image and is itself the image of one and only one point. This property is shared by all the transformations previously considered.

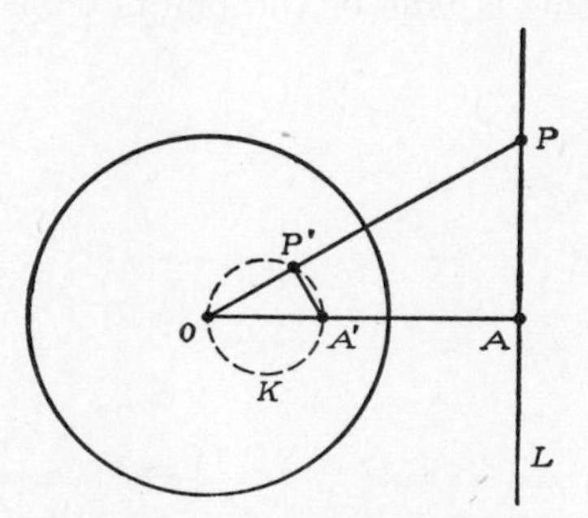

Fig. 39. Inversion of a line L in a circle.

2. Properties of Inversion

The most important property of an inversion is that it transforms straight lines and circles into straight lines and circles. More precisely, we shall show that after an inversion

(a)	a line through O becomes	a line through O,
(b)	a line not through O becomes	a circle through O,
(c)	a circle through O becomes	a line not through O,
(d)	a circle not through O becomes	a circle not through O.

Statement (a) is obvious, since from the definition of inversion any point on the straight line has as image another point on the same line, so that although the points on the line are interchanged, the line as a whole is transformed into itself.

To prove statement (b), drop a perpendicular from O to the straight line L (Fig. 39). Let A be the point where this perpendicular meets L, and let A' be the inverse point to A. Mark any point P on L, and let P' be its inverse point. Since $OA' \cdot OA = OP' \cdot OP = r^2$, it follows that

$$\frac{OA'}{OP'} = \frac{OP}{OA}.$$

Hence the triangles $OP'A'$ and OAP are similar and angle $OP'A'$ is a right angle. From elementary geometry it follows that P' lies on the circle K with diameter OA', so that the inverse of L is this circle. This proves (b). Statement (c) now follows from the fact that since the inverse of L is K, the inverse of K is L.

It remains to prove statement (d). Let K be any circle not passing through O, with center M and radius k. To obtain its image, we draw a line through O intersecting K at A and B, and then determine how the

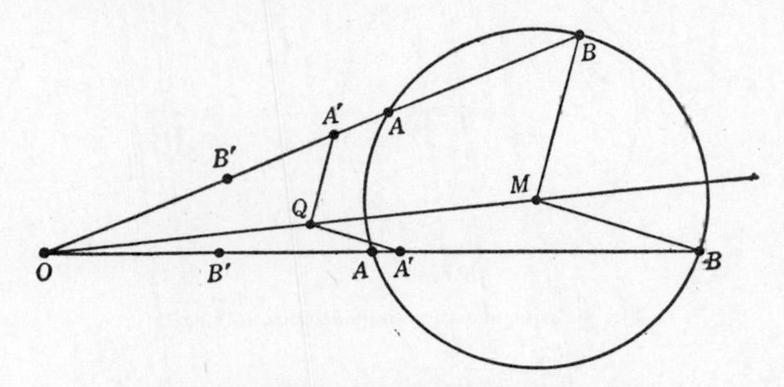

Fig. 40. Inversion of a circle.

images A', B' vary when the line through O intersects K in all possible ways. Denote the distances OA, OB, OA', OB', OM by a, b, a', b', m, and let t be the length of a tangent to K from O. We have $aa' = bb' = r^2$, by definition of inversion, and $ab = t^2$, by an elementary geometrical property of the circle. If we divide the first relations by the second, we get

$$a'/b = b'/a = r^2/t^2 = c^2,$$

where c^2 is a constant that depends only upon r and t, and is the same for all positions of A and B. Through A' we draw a line parallel to BM meeting OM at Q. Let $OQ = q$ and $A'Q = \rho$. Then $q/m = a'/b = \rho/k$, or

$$q = ma'/b = mc^2, \qquad \rho = ka'/b = kc^2.$$

This means that for all positions of A and B, Q will always be the same point on OM, and the distance $A'Q$ will always have the same value.

Likewise $B'Q = \rho$, since $a'/b = b'/a$. Thus the images of all points A, B on K are points whose distance from Q is always ρ, i.e. the image of K is a circle. This proves (d).

3. Geometrical Construction of Inverse Points

The following theorem will be useful in Article 4 of this section: *The point* P′ *inverse to a given point* P *with respect to a circle* C *may be constructed geometrically by the use of the compass alone.* We consider first the case where the given point P is exterior to C. With OP as radius and P as center we describe an arc intersecting C at the points R and S. With these two points as centers we describe arcs with radius r which

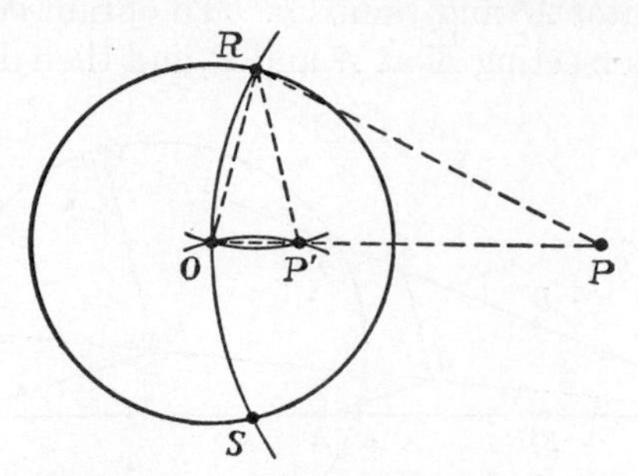

Fig. 41. Inversion of an outside point in a circle.

intersect at O and at a point P' on the line OP. In the isosceles triangles ORP and ORP',

$$\angle ORP = \angle POR = \angle OP'R,$$

so that these triangles are similar, and therefore

$$\frac{OP}{OR} = \frac{OR}{OP'}, \text{ i.e. } OP \cdot OP' = r^2.$$

Hence P' is the required inverse of P, which was to be constructed.

If the given point P lies inside C the same construction and proof hold, provided that the circle of radius OP about P intersects C in two points. If not, we can reduce the construction of the inverse point P' to the previous case by the following simple artifice.

First we observe that with the compass alone we can find a point C on the line joining two given points A, O and such that $AO = OC$. To do this, we draw a circle about O with radius $r = AO$, and mark off on this circle, starting from A, the points P, Q, C such that $AP = PQ = QC = r$. Then C is the desired point, as is seen from the fact that the triangles AOP, OPQ, OQC are equilateral, so that OA and OC form an angle of 180°, and $OC = OQ = AO$. By repeating this procedure, we can easily extend AO any desired number of times. Incidentally, since the length of the segment AQ is $r\sqrt{3}$, as the reader can easily verify, we have at the same time constructed $\sqrt{3}$ from the unit without using the straightedge.

Now we can find the inverse of any point P inside the circle C. First we find a point R on the line OP whose distance from O is an integral multiple of OP and which lies outside C,

$$OR = n \cdot OP.$$

We can do this by successively measuring off the distance OP with the compass until we land outside C. Now we find the point R' inverse to R by the construction previously given. Then

$$r^2 = OR' \cdot OR = OR' \cdot (n \cdot OP) = (n \cdot OR') \cdot OP.$$

Therefore the point P' for which $OP' = n \cdot OR'$ is the desired inverse.

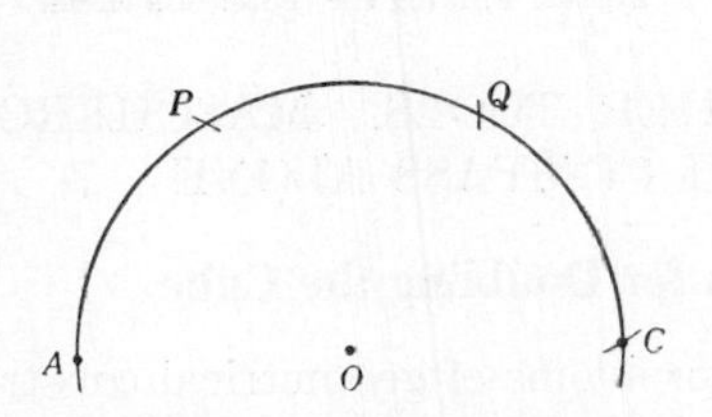

Fig. 42. Doubling of a segment.

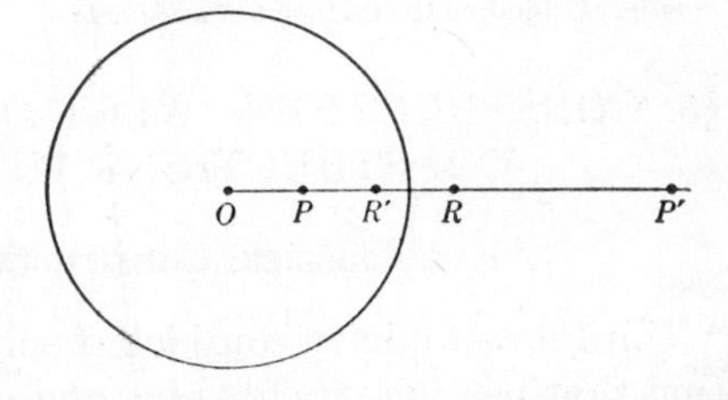

Fig. 43. Inversion of an inside point in a circle.

4. How to Bisect a Segment and Find the Center of a Circle with the Compass Alone

Now that we have learned how to find the inverse of a given point by using the compass alone, we can perform some interesting constructions. For example, we consider the problem of finding the point midway between two given points A and B by using the compass alone (no straight lines may be drawn!). Here is the solution: Draw the circle with radius AB about B as center, and mark off three arcs with radius AB, starting from A. The final point C will be on the line AB, with $AB = BC$. Now draw the circle with radius AB and center A, and let C' be the point inverse to C with respect to this circle. Then

$$AC' \cdot AC = AB^2$$
$$AC' \cdot 2AB = AB^2$$
$$2AC' = AB.$$

Hence C' is the desired midpoint.

Another compass construction using inverse points is that of finding the center of a circle whose circumference only is given, the center being unknown. We choose any point P on the circumference and about it draw a circle intersecting the given circle in the points R and S. With these as centers we draw arcs with the radii $RP = SP$, intersecting at the point Q. A comparison with Figure 41 shows that the unknown center, Q', is inverse to Q with respect to the circle about P, so that Q' can be constructed by compass alone.

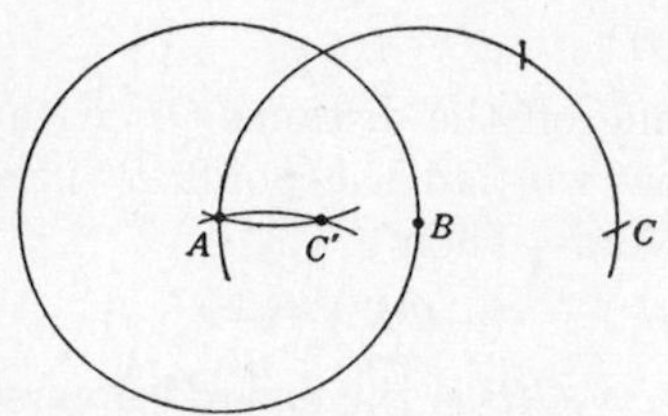

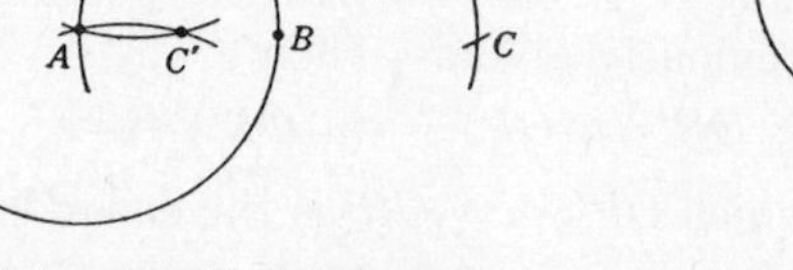

Fig. 44. Finding the midpoint of a segment.

Fig. 45. Finding the center of a circle.

§5. CONSTRUCTIONS WITH OTHER TOOLS. MASCHERONI CONSTRUCTIONS WITH COMPASS ALONE

*1. A Classical Construction for Doubling the Cube

Until now we have considered only problems of geometrical construction that use the straightedge and compass alone. When other instruments are allowed the variety of possible constructions naturally becomes more extensive. For example, the Greeks solved the problem of doubling the cube in the following way. Consider (as in Fig. 46) a rigid right angle MZN and a movable right-angled cross B, VW, PQ. Two additional edges RS and TU are allowed to slide perpendicularly to the arms of the right angle. On the cross let two fixed points E and G be chosen such that $GB = a$ and $BE = f$ have prescribed lengths. By placing the cross so that the points E and G lie on NZ and MZ respectively, and sliding the edges TU and RS, we can bring the entire apparatus into a position where we have a rectangle $ADEZ$ through whose vertices A, D, E pass the arms BW, BQ, BV of the cross. Such an arrangement is always possible if $f > a$. We see at once that $a:x = x:y = y:f$, whence, if f is set equal to $2a$ in the apparatus, $x^3 = 2a^3$. Hence x will be the edge of a cube whose volume is double that of the cube with edge a. This is what is required for doubling the cube.

2. Restriction to the Use of the Compass Alone

While it is only natural that by permitting a greater variety of instruments we can solve a large collection of construction problems, one might expect that more restrictions on the tools allowed would narrow the class of possible constructions. Hence it was a very surprising discovery, made by the Italian Mascheroni (1750–1800), that *all geometrical constructions possible by straightedge and compass can be made by the compass alone.* Of course, one cannot draw the straight line joining two points with-

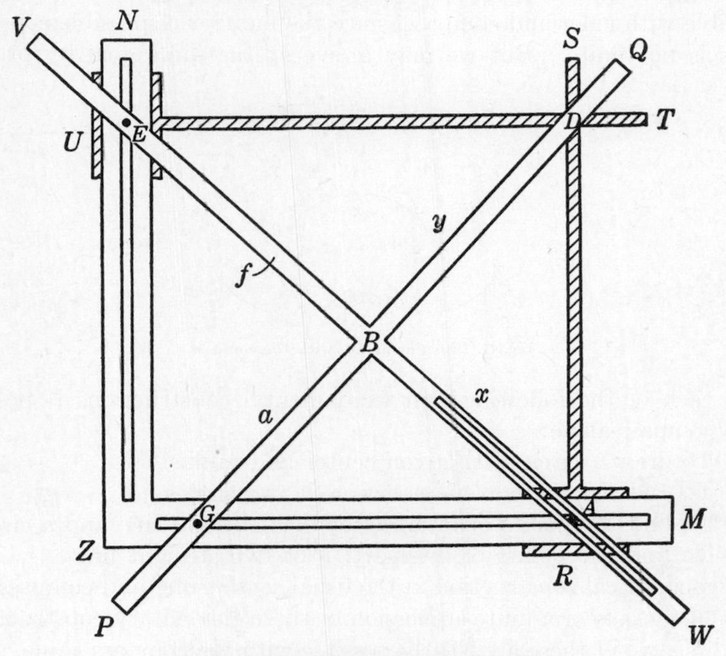

Fig. 46. An instrument for doubling the cube.

out a straightedge, so that this fundamental construction is not really covered by the Mascheroni theory. Instead, one must think of a straight line as given by any two points on it. By using the compass alone, one can find the point of intersection of two lines given in this way, and likewise the intersections of a given circle with a straight line.

Perhaps the simplest example of a Mascheroni construction is the doubling of a given segment AB. The solution was given on page 144. On page 145 we bisected a straight segment. Now we shall solve the problem of bisecting a given arc AB of a circle with given center O. The construction is as follows: from A and B as centers, swing two arcs with radius AO. From O lay off arcs OP and OQ equal to AB. Then swing two arcs with PB and QA as radii and with P and Q as centers, intersecting at R. Finally, with OR as radius, describe an arc with either P or Q as center until it intersects AB; this point of intersection is the required midpoint of the arc AB. The proof is left as an exercise for the reader.

It would be impossible to prove Mascheroni's general theorem by actually giving a construction by compass alone for every construction possible with ruler and compass, since the number of possible constructions is not finite. But we may arrive at the same goal by proving

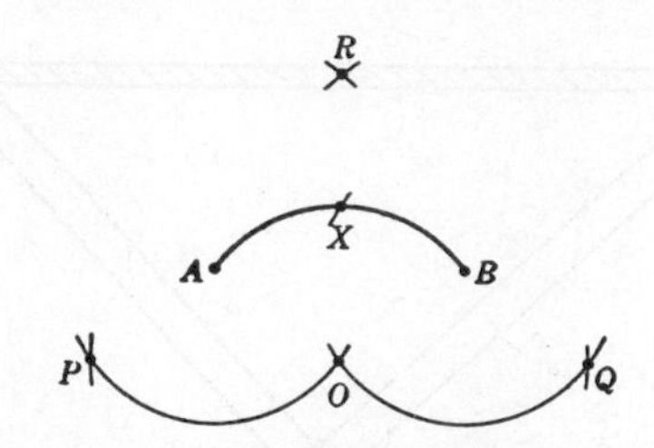

Fig. 47. Bisecting an arc with the compass.

that each of the following four fundamental constructions is possible with compass alone:

1. To draw a circle with given center and radius.
2. To find the points of intersection of two circles.
3. To find the points of intersection of a straight line and a circle.
4. To find the points of intersection of two straight lines.

Any geometrical construction in the usual sense, ruler and compass permitted, consists of a finite succession of these elementary constructions. The first two of these are clearly possible with the compass alone. The solutions of the more difficult problems 3 and 4 depend on the properties of inversion developed in the preceding section.

Let us solve problem 3, that of finding the points of intersection of a circle C and a straight line given by the two points A and B. With centers A and B and radii AO and BO, respectively, draw two arcs, intersecting again at P. Now determine the point Q inverse to P with respect to C, by the construction with compass alone given on p. 144. Draw the circle with center Q and radius QO (this circle must intersect C); the points of intersection X and X' of this circle with the given circle C are the required points. To prove this we need only show that

X and X' are equidistant from O and P, since A and B are so by construction. This follows from the fact that the inverse of Q is a point whose distance from X and X' is equal to the radius of C (p. 144). Note that the circle through X, X', and O is the inverse of the line AB, since this circle and the line AB intersect C at the same points. (Points on the circumference of a circle are their own inverses.)

The construction is invalid only if the line AB goes through the center of C. But then the points of intersection can be found, by the construction given on page 148, as the midpoints of arcs on C obtained by swinging around B an arbitrary circle which intersects C in B_1 and B_2.

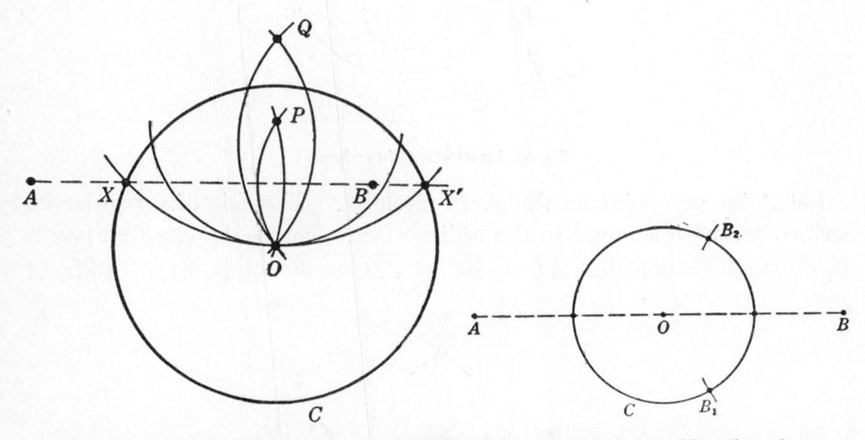

Fig. 48. Intersection of circle and line not through center.

Fig. 49. Intersection of circle and line through center.

The method of determining the circle inverse to the line joining two given points permits an immediate solution of problem 4. Let the lines be given by AB and $A'B'$ (Fig. 50). Draw any circle C in the plane, and by the preceding method find the circles inverse to AB and $A'B'$. These circles intersect at O and at a point Y. The point X inverse to Y is the required point of intersection, and can be constructed by the process already used. That X is the required point is evident from the fact that Y is the only point that is inverse to a point of both AB and $A'B'$; hence the point X inverse to Y must lie on both AB and $A'B'$.

With these two constructions we have completed the proof of the equivalence between Mascheroni constructions using only the compass and the conventional geometrical constructions with ruler and compass.

We have taken no pains to provide elegant solutions for individual problems, since our aim was rather to give some insight into the general scope of the Mascheroni constructions. We shall, however, give as an example the construction of the regular pentagon. More precisely, we shall find five points on a circle which will be the vertices of a regular inscribed pentagon.

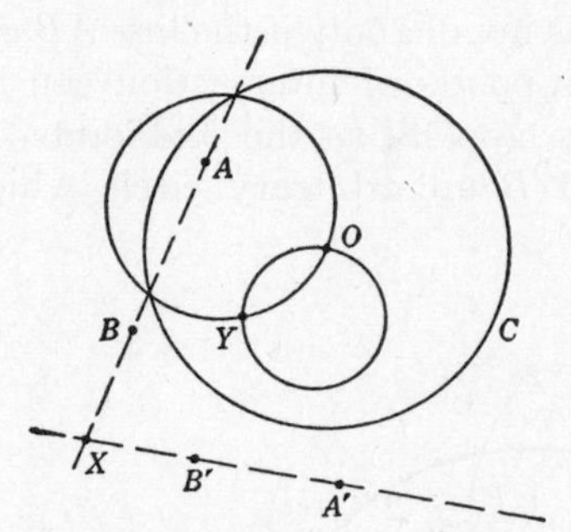

Fig. 50. Intersection of two lines.

Let A be any point on the given circle K. The side of a regular inscribed hexagon is equal to the radius of K. Hence we can find points B, C, D on K such that $\widehat{AB} = \widehat{BC} = \widehat{CD} = 60°$ (Fig. 51). With A

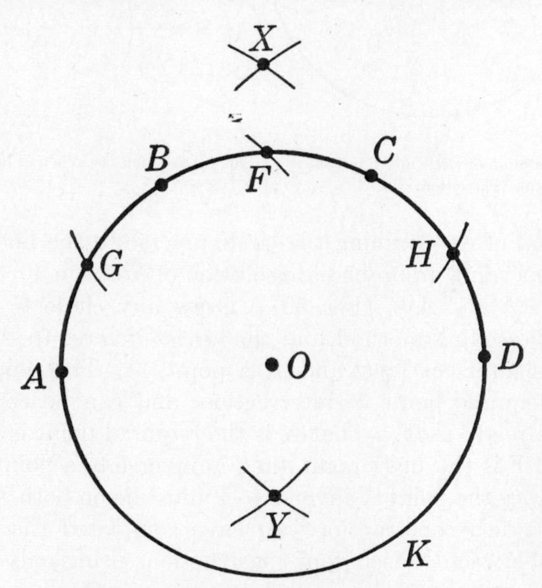

Fig. 51. Construction of the regular pentagon.

and D as centers and AC as radius we draw arcs meeting at X. Then if O is the center of K, an arc about A of radius OX will meet K at the midpoint F of $\widehat{BC}$ (see p. 148). Now with the radius of K we draw arcs about F meeting K at G and H. Let Y be a point whose distance from G and H is OX, and which is separated from X by O. Then AY will be equal to a side of the required pentagon. The proof is left as an exercise for the reader. Note that only three different radii were used in the construction.

In 1928 the Danish mathematician Hjelmslev found in a Copenhagen bookstore a copy of a book, *Euclides Danicus*, published in 1672 by an obscure author G. Mohr. From the title one might infer that this work was simply a version of, or a commentary on Euclid's *Elements*. But when Hjelmslev examined the book, he found to his surprise that it contained essentially the Mascheroni problem and its complete solution, found long before Mascheroni.

Exercises: The following is a description of Mohr's constructions. Check their validity. Why do they solve the Mascheroni problem?

1) On a segment AB of length p erect a perpendicular segment BC. (Hint: Extend AB by a point D such that $AB = BD$. Draw arbitrary circles around A and D and thus determine C.)

2) Two segments of length p and q with $p > q$ are given somewhere in the plane. Find a segment of the length $x = \sqrt{p^2 - q^2}$ by making use of 1).

3) From a given segment a construct the segment $a\sqrt{2}$. (Hint: Observe that $(a\sqrt{2})^2 = (a\sqrt{3})^2 - a^2$.)

4) With given segments p and q find a segment $x = \sqrt{p^2 + q^2}$. (Hint: Use the relation $x^2 = 2p^2 - (p^2 - q^2)$.) Find other similar constructions.

5) Using the previous results, find segments of length $p + q$ and $p - q$ if segments of length p and q are given somewhere in the plane.

6) Check and prove the following construction for the midpoint M of a given segment AB of length a. On the extension of AB find C and D such that $CA = AB = BD$. Construct the isosceles triangle ECD with $EC = ED = 2a$, and find M as the intersection of the circles with diameters EC and ED.

7) Find the orthogonal projection of a point A on a line BC.

8) Find x such that $x:a = p:q$, if a, p, and q are given segments.

9) Find $x = ab$, if a and b are given segments.

Inspired by Mascheroni, Jacob Steiner (1796–1863) tried to single out as a tool the straightedge instead of the compass. Of course, the straightedge alone does not lead out of a given number field, and hence cannot suffice for all geometrical constructions in the classical sense. It is all the more remarkable that Steiner was able to restrict the use of the compass to a single application. He proved that all constructions in the plane which are possible with straightedge and compass are possible with the straightedge alone, provided that a single fixed circle and its center are given. These constructions require projective methods and will be indicated later (see page 197).

* This circle and its center cannot be dispensed with. For example, if a circle, but not its center, is given, it is impossible to construct the latter by the use of the straightedge alone. To prove this we shall make use of a fact that will be discussed later (p. 220): There exists a transformation of the plane into itself which has the following properties: (a) the given circle is fixed under the transformation. (b) Any straight line is carried into a straight line. (c) The center of the circle is carried into some other point. The mere existence of such a transformation shows the impossibility of constructing with the straightedge alone the center of the given circle. For, whatever the construction might be, it would consist in drawing a certain number of straight lines and finding their intersections with one another and with the given circle. Now if the whole figure, consisting of the given circle together with all points and lines of the construction, is subjected to the transformation whose existence we have assumed, the transformed figure will satisfy all the requirements of the construction, but will yield as result a point other than the center of the given circle. Hence such a construction is impossible.

3. Drawing with Mechanical Instruments. Mechanical Curves. Cycloids

By devising mechanisms to draw curves other than the circle and the straight line we may greatly enlarge the domain of constructible figures. For example, if we have an instrument for drawing the hyperbolas $xy = k$, and another for drawing parabolas $y = ax^2 + bx + c$, then any problem leading to a cubic equation,

$$ax^3 + bx^2 + cx = k, \tag{1}$$

may be solved by construction, using only these instruments. For if we set

$$xy = k, \qquad y = ax^2 + bx + c, \tag{2}$$

then solving equation (1) amounts to solving the simultaneous equations (2) by eliminating y; i.e. the roots of (1) are the x-coördinates of the points of intersection of the hyperbola and parabola in (2). Thus the solutions of (1) can be constructed if we have instruments with which to draw the hyperbola and parabola of equations (2).

Since antiquity mathematicians have known that many interesting curves can be defined and drawn by simple mechanical instruments. Of these "mechanical curves" the *cycloids* are among the most remarkable. Ptolemy (circa 200 A.D.) used them in a very ingenious way to describe the movements of the planets in the heavens.

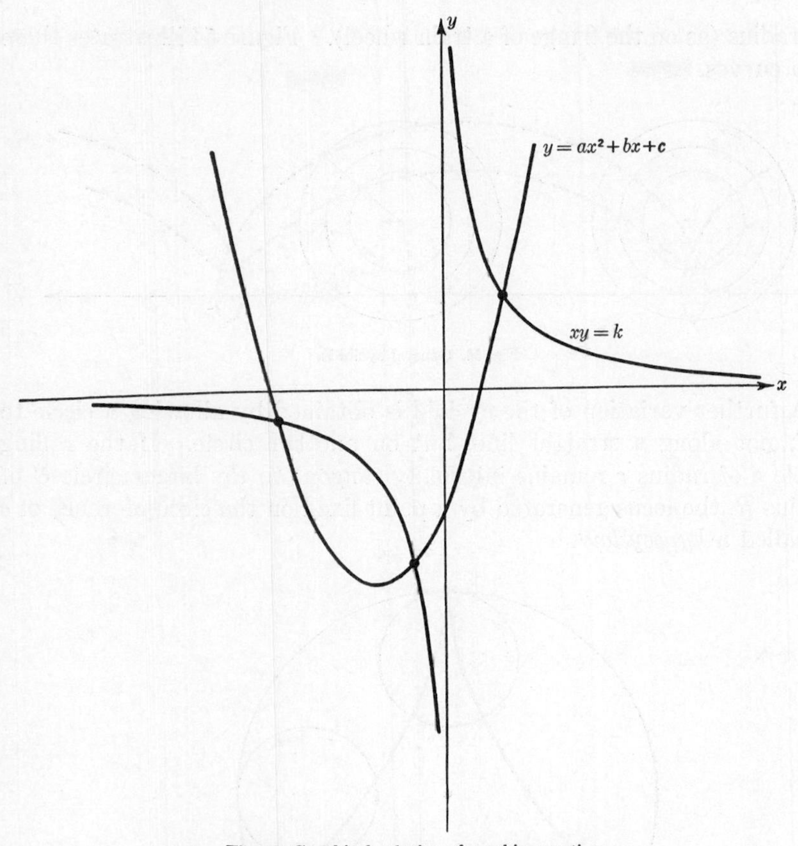

Fig. 52. Graphical solution of a cubic equation.

The simplest cycloid is the curve described by a fixed point on the circumference of a circle which rolls without slipping along a straight line. Figure 53 shows four positions of the point P on the rolling circle. The general appearance of the cycloid is that of a series of arches resting on the line.

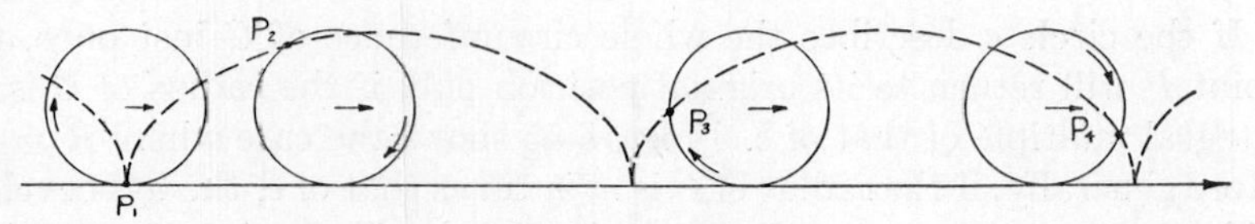

Fig. 53. The cycloid.

Variations of this curve may be obtained by choosing the point P either inside the circle (as on a spoke of a wheel) or on an extension of its radius (as on the flange of a train wheel). Figure 54 illustrates these two curves.

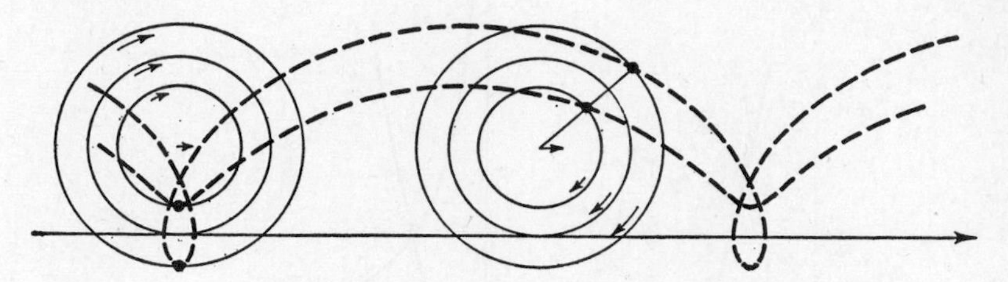

Fig. 54. General cycloids.

A further variation of the cycloid is obtained by allowing a circle to roll, not along a straight line, but on another circle. If the rolling circle c of radius r remains internally tangent to the larger circle C of radius R, the locus generated by a point fixed on the circumference of c is called a *hypocycloid*.

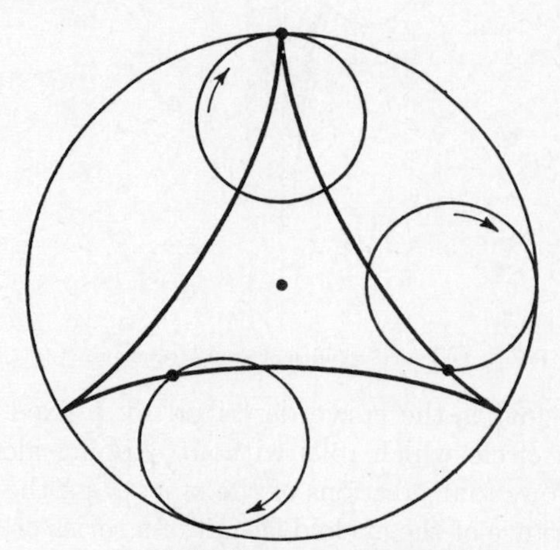

Fig. 55. Three-cusped hypocycloid.

If the circle c describes the whole circumference of C just once, the point P will return to its original position only if the radius of C is an integral multiple of that of c. Figure 55 shows the case where $R = 3r$. More generally, if the radius of C is m/n times that of c, the hypocycloid will close up after n circuits around C, and will consist of m arches. An interesting special case occurs if $R = 2r$. Any point P of the inner circle will then describe a diameter of the larger circle (Fig. 56). We propose the proof of this fact as a problem for the reader.

Still another type of cycloid can be generated by means of a rolling circle remaining externally tangent to a fixed circle. Such a curve is called an *epicycloid.*

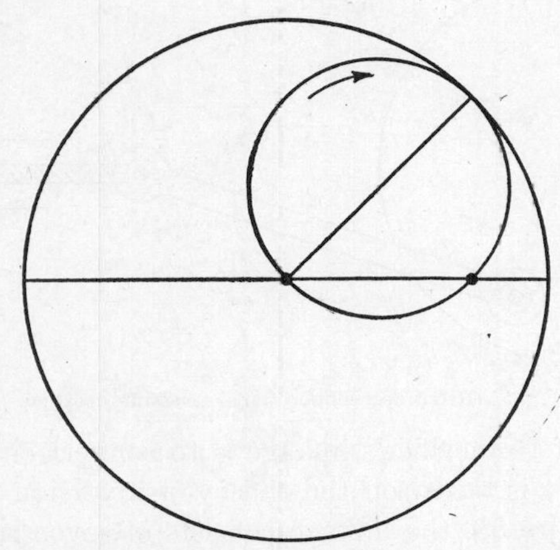

Fig. 56. Straight motion by points on a circle rolling in a circle of double radius.

*4. Linkages. Peaucellier's and Hart's Inversors

We leave for the present the subject of cycloids (they will appear again in an unexpected place) to consider other methods of generating curves. The simplest mechanical instruments for tracing curves are the *linkages.* A linkage consists of a set of rigid rods, connected in some manner at movable joints, in such a way that the whole system has just enough freedom to allow a point on it to describe a certain curve. The compass is really a simple linkage, consisting in principle of a single rod which is fastened at one point.

Linkages have long been used in machine construction. One of the historically famous examples, the "Watt parallelogram," was invented by James Watt to solve the problem of linking the piston of his steam engine to a point on the flywheel in such a way that the rotation of the flywheel would move the piston along a straight line. Watt's solution was only approximate, and despite the efforts of many distinguished mathematicians, the problem of constructing a linkage to move a point *precisely* on a straight line remained unsolved. At one time, when proofs for the impossibility of solutions to certain problems were attracting wide attention, the conjecture was made that the construction of such a linkage was impossible. It was a great surprise when, in 1864, a French naval officer, Peaucellier, invented a simple linkage that solved the problem. With the introduction of efficient lubricants the technical problem for steam engines had by then lost its significance.

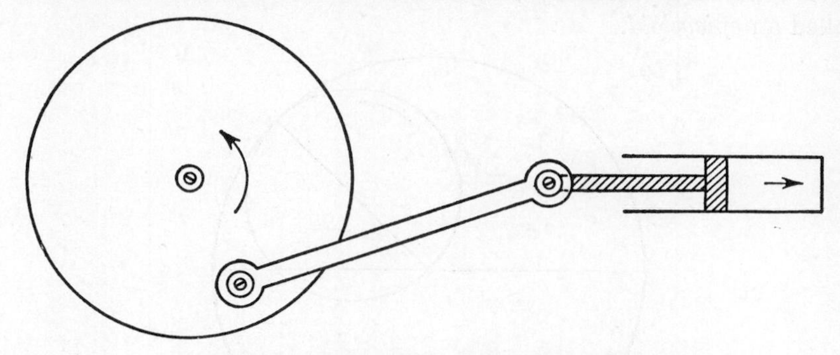

Fig. 57. Rectilinear motion transformed into rotation.

The purpose of Peaucellier's linkage is to convert circular into rectilinear motion. It is based on the theory of inversion discussed in §4. As shown in Figure 58, the linkage consists of seven rigid rods; two of length t, four of length s, and a seventh of arbitrary length. O and R are two fixed points, placed so that $OR = PR$. The entire apparatus is free to move, subject to the given conditions. We shall prove that,

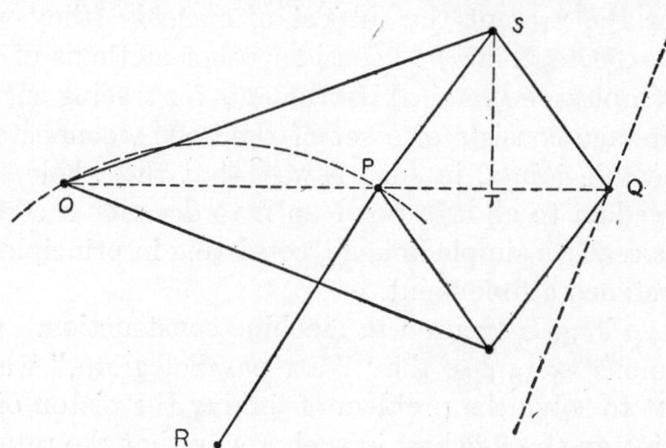

Fig. 58. Peaucellier's transformation of rotation into true rectilinear motion.

as P *describes an arc about* R *with radius* PR, Q *describes a segment of a straight line.* Denoting the foot of the perpendicular from S to OQ by T, we observe that

$$\begin{aligned} OP \cdot OQ &= (OT - PT)(OT + PT) = OT^2 - PT^2 \\ &= (OT^2 + ST^2) - (PT^2 + ST^2) \\ &= t^2 - s^2. \end{aligned}$$

The quantity $t^2 - s^2$ is a constant which we call r^2. Since $OP \cdot OQ = r^2$, P and Q are inverse points with respect to a circle with radius r and center O. As P describes its circular path (which passes through O), Q describes the curve inverse to the circle. This curve must be a straight line, for we have proved that the inverse of a circle passing through O is a straight line. Thus the path of Q is a straight line, drawn without using a straightedge.

Another linkage that solves the same problem is Hart's inversor. This consists of five rods connected as in Figure 59. Here $AB = CD$,

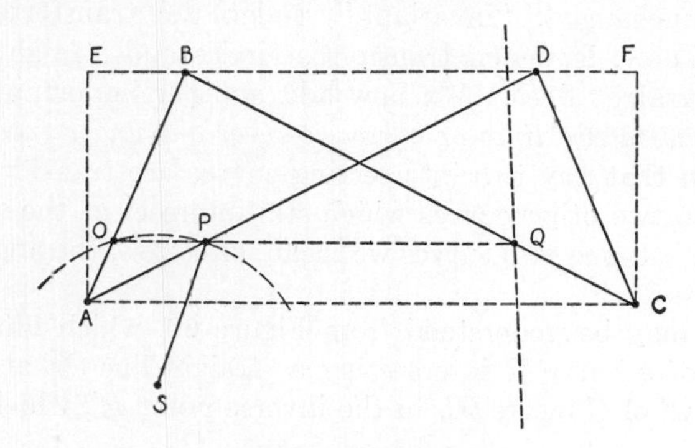

Fig. 59. Hart's inversor.

$BC = AD$. O, P and Q are points fixed on the rods AB, AD, CB, respectively, such that $AO/OB = AP/PD = CQ/QB = m/n$. Points O and S are fixed in the plane so that $OS = PS$, while the rest of the linkage is free to move. Evidently, AC is always parallel to BD. Hence, O, P and Q are collinear, and OP is parallel to AC. Draw AE and CF perpendicular to BD. We have

$$AC \cdot BD = EF \cdot BD = (ED + EB)(ED - EB) = ED^2 - EB^2.$$

But $ED^2 + AE^2 = AD^2$, and $EB^2 + AE^2 = AB^2$. Hence $ED^2 - EB^2 = AD^2 - AB^2$. Now

$$OP/BD = AO/AB = m/(m+n) \quad \text{and} \quad OQ/AC = OB/AB = n/(m+n).$$

Thus

$$OP \cdot OQ = [mn/(m+n)^2]BD \cdot AC = [mn/(m+n)^2](AD^2 - AB^2).$$

This quantity is the same for all possible positions of the linkage. Therefore P and Q are inverse points with respect to some circle about O. When the linkage is moved, P describes a circle about S which passes through O, while its inverse Q describes a straight line.

Other linkages can be constructed (at least in principle) which will draw ellipses, hyperbolas, and indeed any curve given by an algebraic equation $f(x, y) = 0$ of any degree.

§6. MORE ABOUT INVERSION AND ITS APPLICATIONS

1. Invariance of Angles. Families of Circles

Although inversion in a circle greatly changes the appearance of geometrical figures, it is a remarkable fact that the new figures continue to possess many of the properties of the old. These are the properties which are unchanged, "invariant," under the transformation. As we already know, inversion transforms circles and straight lines into circles and straight lines. We now add another important property: *The angle between two lines or curves is invariant under inversion.* By this we mean that any two intersecting curves are transformed by an inversion into two other curves which still intersect at the same angle. By the angle between two curves we mean, of course, the angle between their tangents.

The proof may be understood from Figure 60, which illustrates the special case of a curve C intersecting a straight line OL at a point P. The inverse C' of C meets OL in the inverse point P', which, since OL

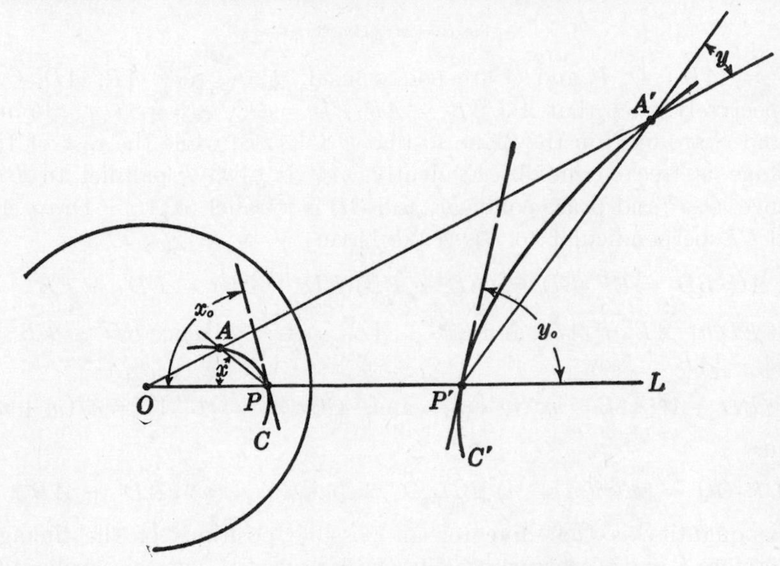

Fig. 60. Invariance of angles under inversion.

is its own inverse, lies on OL. We shall show that the angle x_0 between OL and the tangent to C at P is equal in magnitude to the corresponding angle y_0. To do this we choose a point A on the curve C near P, and draw the secant AP. The inverse of A is a point A' which, being on both the line OA and the curve C', must be at their intersection. We draw the secant $A'P'$. By the definition of inversion,

$$r^2 = OP \cdot OP' = OA \cdot OA',$$

or

$$\frac{OP}{OA} = \frac{OA'}{OP'},$$

i.e. the triangles OAP and $OA'P'$ are similar. Hence angle x is equal to angle $OA'P'$, which we call y. Our final step consists in letting the point A move along C and approach the point P. This causes the secant line AP to revolve into the position of the tangent line to C at P, while the angle x tends to x_0. At the same time A' will approach P', and $A'P'$ will revolve into the tangent at P'. The angle y approaches y_0. Since x is equal to y at every position of A, we must have in the limit, $x_0 = y_0$.

Our proof is only partially completed, however, since we have considered only the case of a curve intersecting a line through O. The general case of two curves C, C^* forming an angle z at P is now easily disposed of. For it is evident that the line OPP' divides z into two angles, each of which we know to be preserved by the inversion.

It should be noted that although inversion preserves the *magnitude* of angles, it reverses their *sense*; i.e. if a ray through P sweeps out the angle x_0 in a counterclockwise direction, its image will sweep out angle y_0 in a clockwise direction.

A particular consequence of the invariance of angle under inversion is that two circles or lines that are orthogonal, i.e. that intersect at right angles, remain orthogonal after an inversion, while two circles which are tangent, i.e. intersect at the angle zero, remain tangent.

Let us consider the family of all circles that pass through the center of inversion O and through another fixed point A of the plane. From §4, Article 2, we know that this family of circles is transformed into a family of straight lines that radiate from A', the image of A. The family of circles orthogonal to the original family goes over into circles orthogonal to the lines through A', as shown in Figure 61. (The orthogonal circles are shown by broken lines.) The simple picture of the radiating straight lines appears to be quite different from that of the circles, yet we see that they are closely related—indeed from the standpoint of the theory of inversion they are entirely equivalent.

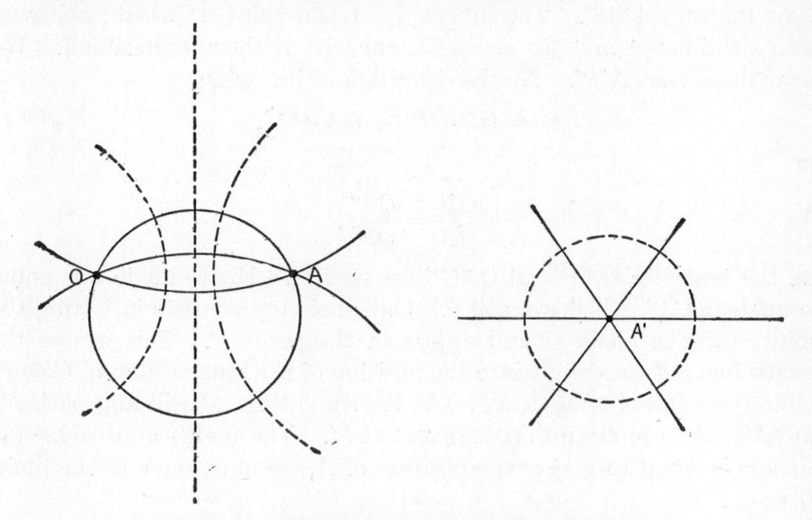

Fig. 61. Two systems of orthogonal circles related by inversion.

Another example of the effect of inversion is given by a family of circles tangent to each other at the center of inversion. After the transformation they become a system of parallel lines. For the images of the circles are straight lines, and no two of these lines intersect, since the original circles meet only at O.

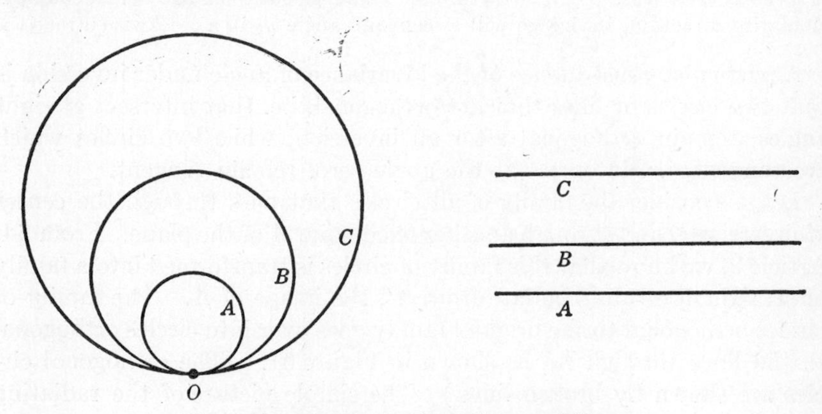

Fig. 62. Tangent circles transformed into parallel lines.

2. Application to the Problem of Apollonius

A good illustration of the usefulness of the theory of inversion is the following simple geometrical solution of the problem of Apollonius. By inversion with respect to any center, the Apollonius problem for three given circles can be transformed into the corresponding problem for three other circles (why is this?). Hence, if we can solve the problem for any one triple of circles, then it is solved for any other triple of circles obtained from the first by inversion. We shall exploit this fact by selecting among all these equivalent triples of circles one for which the problem is almost trivially simple.

We start with three circles having centers A, B, C, and we shall suppose the required circle U with center O and radius ρ to be externally tangent to the three given circles. If we increase the radii of the three given circles by the same quantity d, then the circle with the same center O and the radius $\rho - d$ will obviously solve the new problem.

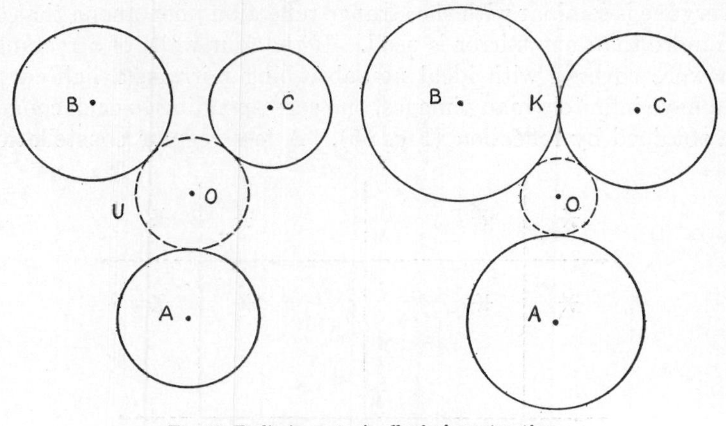

Fig. 63. Preliminary to Apollonius' construction.

By way of preparation we make use of this fact in order to replace the three given circles by three others such that two of them are tangent to each other at a point K (Fig. 63). Next we invert the whole figure in some circle with center K. The circles around B and C become parallel lines b and c, while the third circle becomes another circle a (Fig. 64). We know that a, b, c can all be constructed by ruler and compass. The unknown circle is transformed into a circle u which touches a, b, c. Its radius r is evidently half the distance between b and c. Its center O' is one of the two intersections of the line midway between b and c with the circle about A' (the center of a) having the radius $r + s$ (s being the radius of a). Finally, by constructing the circle inverse to u we find the center of the desired Apollonius circle U. (Its center, O, will be the inverse in the circle of inversion of the point inverse to K in u.)

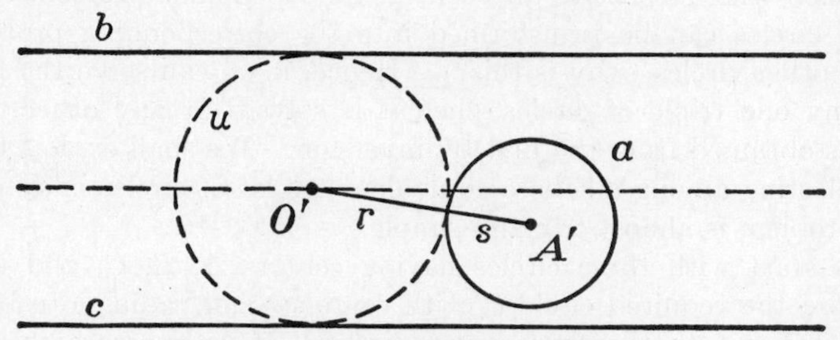

Fig. 64. Solution of Apollonius' problem.

*3. Repeated Reflections

Everyone is familiar with the strange reflection phenomena that occur when more than one mirror is used. If the four walls of a rectangular room were covered with ideal non-absorbing mirrors, a lighted point would have infinitely many images, one corresponding to each congruent room obtained by reflection (Fig. 65). A less regular constellation of

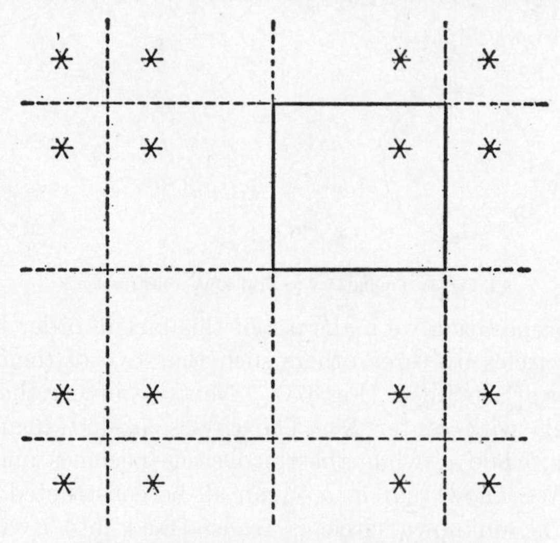

Fig. 65. Repeated reflection in rectangular walls.

mirrors, e.g. three mirrors, gives a much more complicated series of images. The resulting configuration can be described easily only when the reflected triangles form a non-overlapping covering of the plane. This occurs only for the case of the rectangular isosceles triangle, the equilateral triangle, and the rectangular half of the latter; see Figure 66.

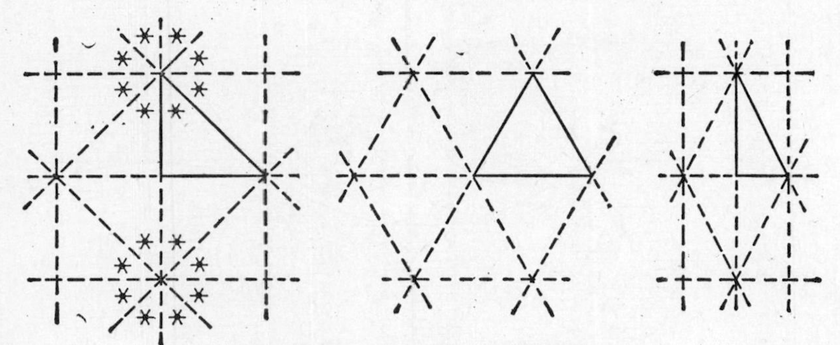

Fig. 66. Regular constellations of triangular mirrors.

The situation becomes much more interesting if we consider repeated inversion in a pair of circles. Standing between two concentric circular mirrors one would see an infinite number of other circles concentric with them. One sequence of these circles tends to infinity, while the other concentrates around the center. The case of two external circles is a

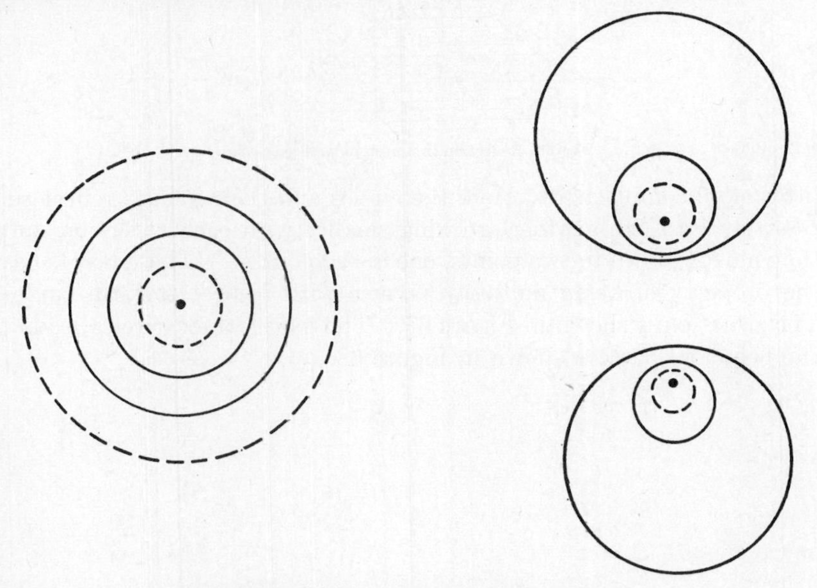

Fig. 67. Repeated reflection in systems of two circles.

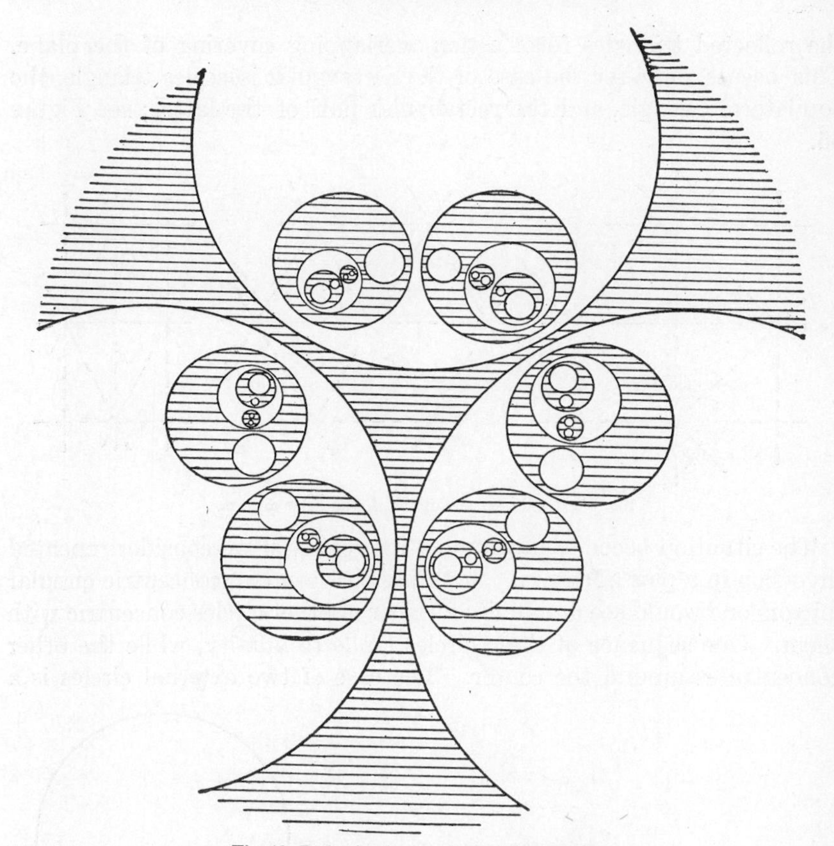

Fig. 68. Reflection in a system of three circles.

little more complicated. Here the circles and their images reflect successively into one another, growing smaller with each reflection, until they narrow down to two points, one in each circle. (These points have the property of being mutually inverse with respect to both circles.) The situation is shown in Figure 67. The use of three circles leads to the beautiful pattern shown in Figure 68.

CHAPTER IV

PROJECTIVE GEOMETRY. AXIOMATICS. NON-EUCLIDEAN GEOMETRIES

§1. INTRODUCTION

1. Classification of Geometrical Properties. Invariance under Transformations

Geometry deals with the properties of figures in the plane or in space. These properties are so numerous and so varied that some principle of classification is necessary to bring order into this wealth of knowledge. One might, for example, introduce a classification based on the method used in deriving the theorems. From this point of view a distinction is usually made between the "synthetic" and the "analytic" procedures. The first of these is the classical axiomatic method of Euclid, in which the subject is built upon purely geometrical foundations independent of algebra and the concept of the number continuum, and in which the theorems are deduced by logical reasoning from an initial body of statements called axioms or postulates. The second method is based on the introduction of numerical coördinates, and uses the technique of algebra. This method has brought about a profound change in mathematical science, resulting in a unification of geometry, analysis and algebra into one organic system.

In this chapter a classification according to method will be less important than a classification according to *content*, based on the character of the theorems themselves, irrespective of the methods used to prove them. In elementary plane geometry one distinguishes between theorems dealing with the congruence of figures, using the concepts of length and angle, and theorems dealing with the similarity of figures, using the concept of angle only. This particular distinction is not very important, since length and angle are so closely connected that it is rather artificial to separate them. (It is the study of this connection which makes up most of the subject of trigonometry.) Instead, we may say that the theorems of elementary geometry concern *magnitudes*—lengths, measures of angles, and areas. Two figures are equivalent from this point of view if they are *congruent*, that is, if one can be obtained from the other by a *rigid motion*, in which merely position but no magnitude is changed. The question now arises whether the concept of magnitude and the related concepts of congruence and similarity are essential to geometry, or whether geometrical figures may have even deeper properties that are not destroyed by transformations more drastic than the rigid motions. We shall see that this is indeed the case.

Suppose we draw a circle and a pair of its perpendicular diameters on a rectangular block of soft wood, as in Figure 69. If we place this

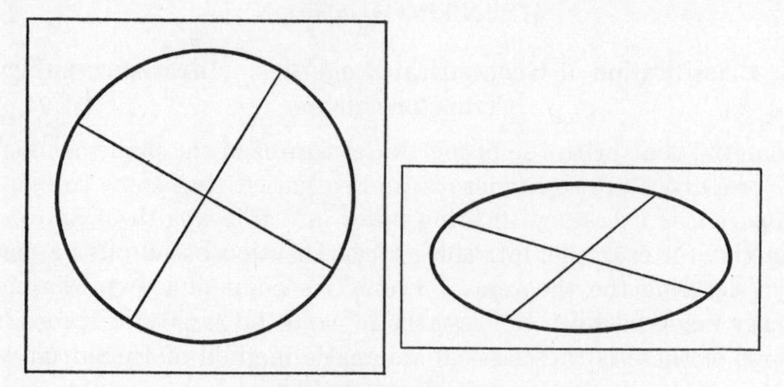

Fig. 69. Compression of a circle.

block between the jaws of a powerful vise and compress it to half its original width, the circle will become an ellipse and the angles between the diameters of the ellipse will no longer be right angles. The circle has the property that its points are equidistant from the center, while this does not hold true of the ellipse. Thus it might seem that all the geometrical properties of the original configuration are destroyed by the compression. But this is far from being the case; for example, the statement that the center bisects each diameter is true of both the circle and the ellipse. Here we have a property which persists even after a rather drastic change in the magnitudes of the original figure. This observation suggests the possibility of classifying theorems about a geometrical figure according to whether they remain true or become false when the figure is subjected to a uniform compression. More generally, given any definite class of transformations of a figure (such as the class of all rigid motions, compressions, inversion in circles, etc.), we may ask what properties of the figure will be unchanged under this class of transformations. The body of theorems dealing with these properties will be the *geometry associated with this class of transformations*. The idea of classifying the different branches of geometry according to the classes of transformations considered was proposed by Felix Klein (1849–1925) in a famous address (the "Erlanger program") given in 1872. Since that time it has greatly influenced geometrical thinking.

In Chapter V we shall discover the very surprising fact that certain properties of geometrical figures are so deeply inherent that they persist even after the figures are subjected to quite arbitrary deformations; figures drawn on a piece of rubber which is stretched or compressed in any manner still preserve some of their original characteristics. In this chapter, however, we shall be concerned with those properties which remain unchanged, or "invariant," under a special class of transformations which lies between the very restricted class of rigid motions on the one hand, and the most general class of arbitrary deformations on the other. This is the class of "projective transformations."

2. Projective Transformations

The study of these geometrical properties was forced upon mathematicians long ago by the problems of *perspective*, which were studied by artists such as Leonardo da Vinci and Albrecht Dürer. The image made by a painter can be regarded as a projection of the original onto the canvas, with the center of projection at the eye of the painter. In this process lengths and angles are necessarily distorted, in a way that depends on the relative positions of the various objects depicted. Still, the geometrical structure of the original can usually be recognized on the canvas. How is this possible? It must be because there exist geometrical properties "invariant under projection"—properties which appear unchanged in the image and make the identification possible. To find and analyze these properties is the object of projective geometry.

It is clear that the theorems in this branch of geometry cannot be statements about lengths and angles or about congruence. Some isolated facts of a projective nature have been known since the seventeenth century and even, as in the case of the "theorem of Menelaus," since antiquity. But a systematic study of projective geometry was first made at the end of the eighteenth century, when the École Polytechnique in Paris initiated a new period in mathematical progress, particularly in geometry. This school, a product of the French Revolution, produced many officers for the military services of the Republic. One of its graduates was J. V. Poncelet (1788–1867), who wrote his famous *Traité des propriétés projectives des figures* in 1813, while a prisoner of war in Russia. In the nineteenth century, under the influence of

Steiner, von Staudt, Chasles, and others, projective geometry became one of the chief subjects of mathematical research. Its popularity was due partly to its great aesthetic charm and partly to its clarifying effect on geometry as a whole and its intimate connection with non-Euclidean geometry and algebra.

§2. FUNDAMENTAL CONCEPTS

1. The Group of Projective Transformations

We first define the class, or "group,"† of projective transformations. Suppose we have two planes π and π' in space, not necessarily parallel to each other. We may then perform a *central projection* of π onto π' from a given center O not lying in π or π' by defining the image of each point P of π to be that point P' of π', such that P and P' lie on the same straight line through O. We may also perform a *parallel projection*, where the projecting lines are all parallel. In the same way, we can define the projection of a line l in a plane π onto another line l' in π from a point O in π or by a parallel projection.

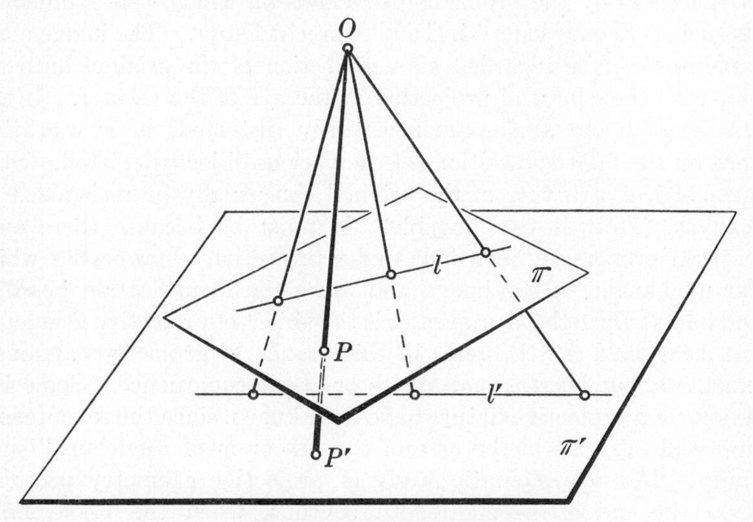

Fig. 70. Projection from a point.

† The term "group," when applied to a class of transformations, implies that the successive application of two transformations of the class results again in a transformation of the same class, and that the "inverse" of a transformation of the class again belongs to the class. Group properties of mathematical operations have played and are playing a very great rôle in many fields, although in geometry, perhaps, the importance of the group concept has been a little exaggerated.

Any mapping of one figure onto another by a central or parallel projection, or by a finite succession of such projections, is called a *projective transformation.*† The *projective geometry* of the plane or of the line consists of the body of those geometrical propositions which are unaffected by arbitrary projective transformations of the figures to which they refer. In contrast, we shall call *metric geometry* the body of those propositions dealing with the magnitudes of figures, invariant only under the class of rigid motions.

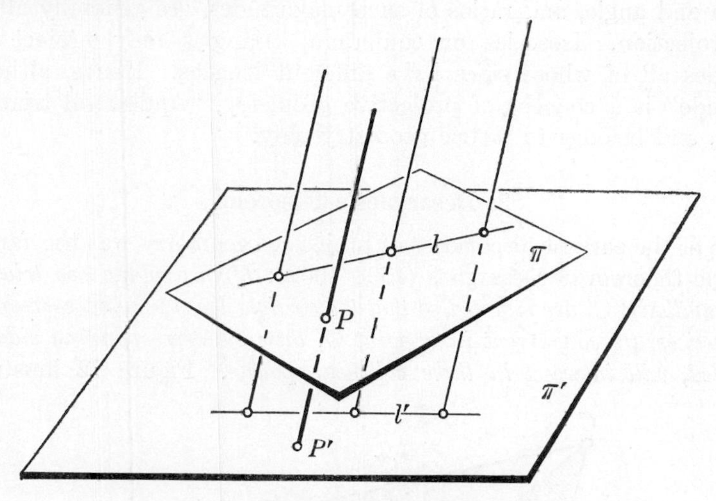

Fig. 71. Parallel projection.

Some projective properties can be recognized immediately. A point, of course, projects into a point. Moreover, *a straight line is projected into a straight line*; for, if the line l in π is projected onto the plane π', the intersection of π' with the plane through O and l will be the straight line l'.‡ If a point A and a straight line l are incident,†† then after any projection the corresponding point A' and line l' will again be incident.

† Two figures related by a *single* projection are commonly said to be in *perspective*. Thus a figure F is related by a projective transformation to a figure F' if F and F' are in perspective, or if we can find a succession of figures, $F, F_1, F_2, \cdots, F_n, F'$, such that each figure is in perspective with the following one.

‡ There are exceptions if the line OP (or if the plane through O and l) is parallel to the plane π'. These exceptions will be removed in §4.

†† A point and a line are called *incident* if the line goes through the point, or the point is on the line. The neutral word leaves it open whether the line or the point is considered more important.

Thus the *incidence of a point and a line is invariant under the projective group*. From this fact many simple but important consequences follow. If three or more points are *collinear*, i.e. incident with some straight line, then their images will also be collinear. Likewise, if in the plane π three or more straight lines are *concurrent*, i.e. incident with some point, then their images will also be concurrent straight lines. While these simple properties—incidence, collinearity, and concurrence—are *projective properties* (i.e. properties invariant under projections), measures of length and angle, and ratios of such magnitudes, are generally altered by projection. Isosceles or equilateral triangles may project into triangles all of whose sides have different lengths. Hence, although "triangle" is a concept of projective geometry, "equilateral triangle" is not, and belongs to metric geometry only.

2. Desargues's Theorem

One of the earliest discoveries of projective geometry was the famous triangle theorem of Desargues (1593–1662): *If in a plane two triangles* ABC *and* A'B'C' *are situated so that the straight lines joining corresponding vertices are concurrent in a point* O, *then the corresponding sides, if extended, will intersect in three collinear points.* Figure 72 illustrates

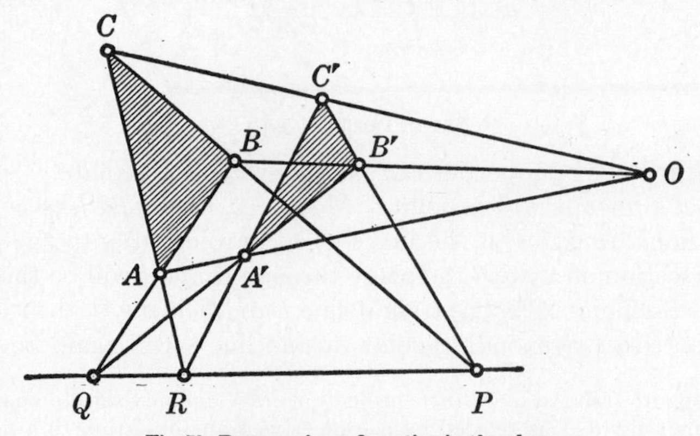

Fig. 72. Desargues's configuration in the plane.

the theorem, and the reader should draw other figures to test it by experiment. The proof is not trivial, in spite of the simplicity of the figure, which involves only straight lines. The theorem clearly belongs to projective geometry, for if we project the whole figure onto another plane, it will retain all the properties involved in the theorem.

We shall return to this theorem on page 187. At the moment we wish to call attention to the remarkable fact that Desargues's theorem is also true if the two triangles lie in two *different* (non-parallel) planes, and that this Desargues's theorem of three-dimensional geometry is very easily proved. Suppose that the lines AA', BB', and CC' intersect at O (Fig. 73), according to hypothesis. Then AB lies in the same plane

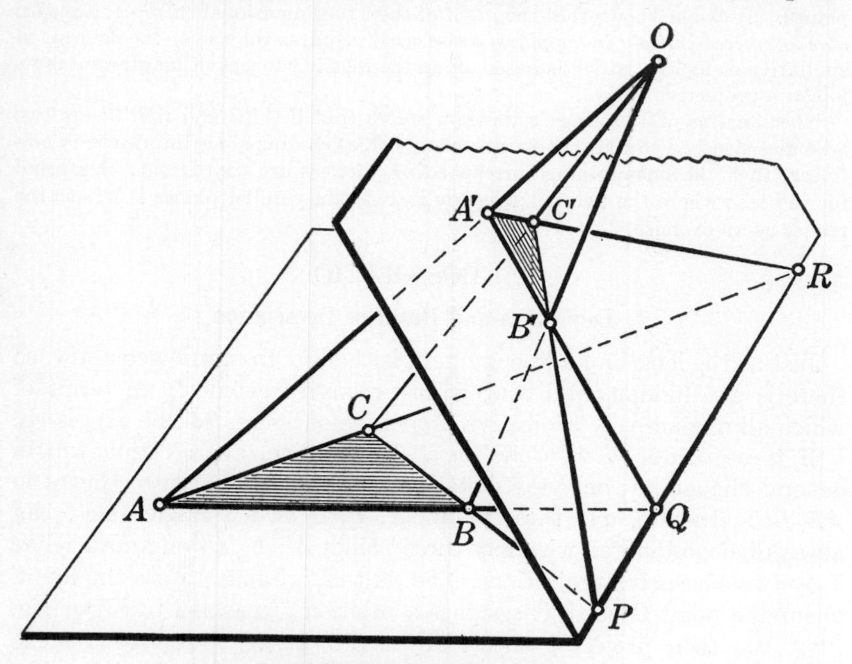

Fig. 73. Desargues's configuration in space.

as $A'B'$, so that these two lines intersect at some point Q; likewise AC and $A'C'$ intersect in R, and BC and $B'C'$ intersect in P. Since P, Q, and R are on extensions of the sides of ABC and $A'B'C'$, they lie in the same plane with each of these two triangles, and must consequently lie on the line of intersection of these two planes. Therefore P, Q, and R are collinear, as was to be proved.

This simple proof suggests that we might prove the theorem for two dimensions by, so to speak, a passage to the limit, letting the whole figure flatten out so that the two planes coincide in the limit and the point O, together with all the others, falls into this plane. There is, however, a certain difficulty in carrying out such a limiting process, because the line of intersection PQR is not uniquely determined when the planes coincide. However, the configuration of Figure 72 may be regarded as a perspective drawing of the space configuration of Figure 73, and this fact can be used to prove the theorem in the plane case.

There is actually a fundamental difference between Desargues's theorem in the plane and in space. Our proof in three dimensions used geometrical reasoning based solely on the concepts of incidence and intersection of points, lines, and planes. It can be shown that the proof of the two-dimensional theorem, *provided it is to proceed entirely in the plane,* necessarily requires the use of the concept of similarity of figures, which is based upon the metric concept of length and is no longer a projective notion.

The *converse* of Desargues's theorem states that if ABC and $A'B'C'$ are two triangles situated so that the points where corresponding sides intersect are collinear, then the lines joining corresponding vertices are concurrent. Its proof for the case where the two triangles are in two non-parallel planes is left to the reader as an exercise.

§3. CROSS-RATIO

1. Definition and Proof of Invariance

Just as the length of a line segment is the key to metric geometry, so there is one fundamental concept of projective geometry in terms of which all distinctively projective properties of figures can be expressed.

If three points A, B, C lie on a straight line, a projection will in general change not only the distances AB and BC but also the ratio AB/BC. In fact, *any* three points A, B, C on a straight line l can always be coördinated with any three points A', B', C' on another line l' by two successive projections. To do this, we may rotate the line l' about the point C' until it assumes a position l'' parallel to l (see Fig. 74). We then project l onto l'' by a projection parallel to the line joining C and C', defining three points, A'', B'', and C'' $(= C')$. The

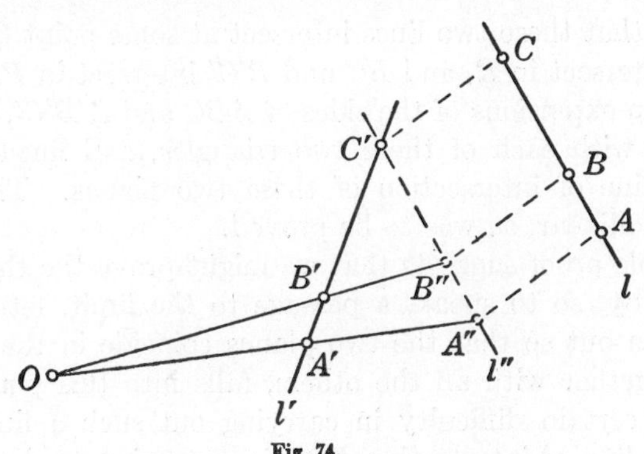

Fig. 74.

lines joining A', A'' and B', B'' will intersect in a point O, which we choose as the center of a second projection. These two projections accomplish the desired result.†

As we have just seen, no quantity that involves only three points on a line can be invariant under projection. But—and this is the decisive discovery of projective geometry—if we have *four* points A, B, C, D on a straight line, and project these into A', B', C', D' on another line, then there is a certain quantity, called the *cross-ratio* of the four points, that retains its value under the projection. Here is a mathematical property of a set of four points on a line that is not destroyed by projection and that can be recognized in any image of the line. The cross-ratio is neither a length, nor the ratio of two lengths, but the *ratio of two such ratios*: if we consider the ratios CA/CB and DA/DB, then their ratio,

A B C D

$$x = \frac{CA}{CB} \bigg/ \frac{DA}{DB},$$

is by definition the cross-ratio of the four points A, B, C, D, taken in that order.

We now show that *the cross-ratio of four points is invariant under projection*, i.e. that if A, B, C, D and A', B', C', D' are corresponding points on two lines related by a projection, then

$$\frac{CA}{CB} \bigg/ \frac{DA}{DB} = \frac{C'A'}{C'B'} \bigg/ \frac{D'A'}{D'B'}.$$

The proof follows by elementary means. We recall that the area of a triangle is equal to $\frac{1}{2}$(base $\times$ altitude) and is also given by half the product of any two sides by the sine of the included angle. We then have, in Figure 75,

$$\begin{aligned}
\text{area } OCA &= \tfrac{1}{2}h \cdot CA = \tfrac{1}{2}OA \cdot OC \sin \angle\, COA \\
\text{area } OCB &= \tfrac{1}{2}h \cdot CB = \tfrac{1}{2}OB \cdot OC \sin \angle\, COB \\
\text{area } ODA &= \tfrac{1}{2}h \cdot DA = \tfrac{1}{2}OA \cdot OD \sin \angle\, DOA \\
\text{area } ODB &= \tfrac{1}{2}h \cdot DB = \tfrac{1}{2}OB \cdot OD \sin \angle\, DOB.
\end{aligned}$$

† What if the lines joining A', A'' and B', B'' are parallel?

It follows that

$$\frac{CA}{CB} \Big/ \frac{DA}{DB} = \frac{CA}{CB} \cdot \frac{DB}{DA} = \frac{OA \cdot OC \cdot \sin \angle COA}{OB \cdot OC \cdot \sin \angle COB} \cdot \frac{OB \cdot OD \cdot \sin \angle DOB}{OA \cdot OD \cdot \sin \angle DOA}$$

$$= \frac{\sin \angle COA}{\sin \angle COB} \cdot \frac{\sin \angle DOB}{\sin \angle DOA}.$$

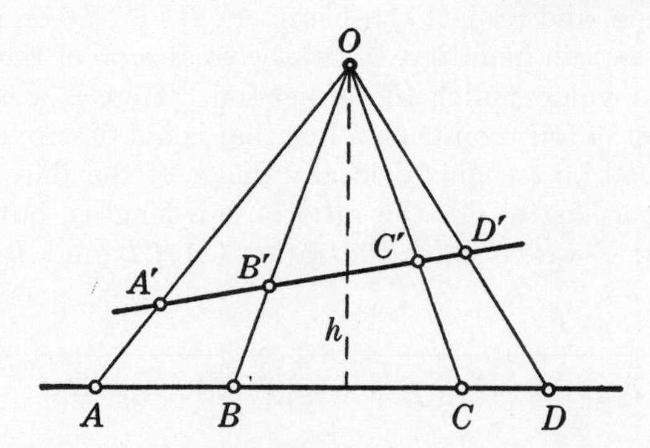

Fig. 75. Invariance of cross-ratio under central projection.

Hence the cross-ratio of A, B, C, D depends only on the angles subtended at O by the segments joining A, B, C, D. Since these angles are the same for any four points A', B', C', D' into which A, B, C, D may be projected from O, it follows that the cross-ratio remains unchanged by projection.

That the cross-ratio of four points remains unchanged by a *parallel* projection follows from elementary properties of similar triangles. The proof is left to the reader as an exercise.

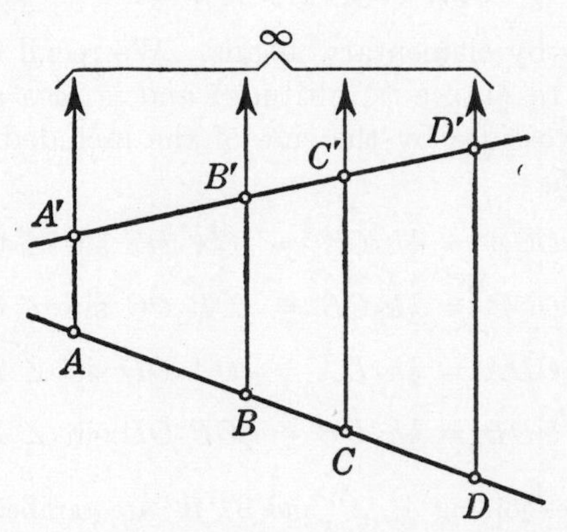

Fig. 76. Invariance of cross-ratio under parallel projection.

So far we have understood the cross-ratio of four points A, B, C, D on a line l to be a ratio involving positive lengths. It is more convenient to modify this definition as follows. We choose one direction on l as positive, and agree that lengths measured in this direction shall be positive, while lengths measured in the opposite direction shall be negative. We then define the cross-ratio of A, B, C, D in that order as the quantity

$$(1) \qquad (ABCD) = \frac{CA}{CB} \Big/ \frac{DA}{DB},$$

where the numbers CA, CB, DA, DB are understood to be taken with the proper sign. Since a reversal of the chosen positive direction on l will merely change the sign of every term of this ratio, the value of $(ABCD)$ will not depend on the direction chosen. It is easily seen that $(ABCD)$ will be negative or positive according as the pair of points A, B is or is not separated (i.e. interlocked) by the pair C, D. Since this separation property is invariant under projection, the signed cross-ratio $(ABCD)$ is invariant also. If we select a fixed point O on l as

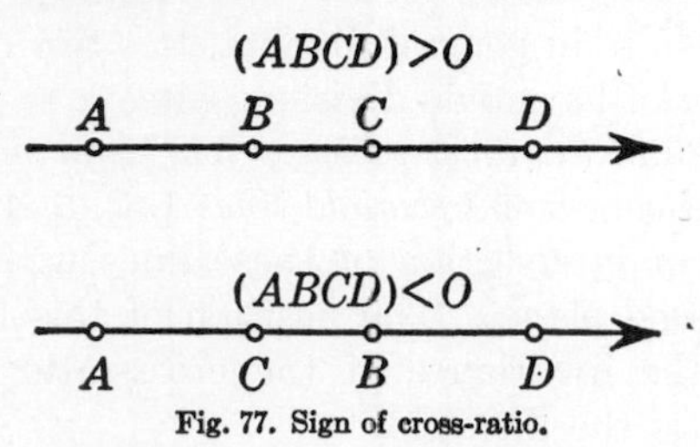

Fig. 77. Sign of cross-ratio.

origin and choose as the coördinate x of each point on l its directed distance from O, so that the coördinates of A, B, C, D are x_1, x_2, x_3, x_4, respectively, then

$$(ABCD) = \frac{CA}{CB} \Big/ \frac{DA}{DB} = \frac{x_3 - x_1}{x_3 - x_2} \Big/ \frac{x_4 - x_1}{x_4 - x_2} = \frac{x_3 - x_1}{x_3 - x_2} \cdot \frac{x_4 - x_2}{x_4 - x_1}.$$

When $(ABCD) = -1$, so that $CA/CB = -DA/DB$, then C and D

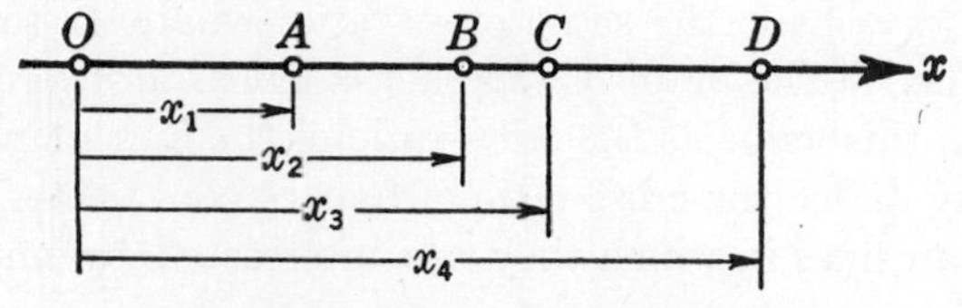

Fig. 78. Cross-ratio in terms of coördinates.

divide the segment AB internally and externally in the same ratio. In this case, C and D are said to divide the segment AB *harmonically*, and each of the points C, D is called the *harmonic conjugate* of the other with respect to the pair A, B. If $(ABCD) = 1$, then the points C and D (or A and B) coincide.

It should be kept in mind that the *order* in which A, B, C, D are taken is an essential part of the definition of the cross-ratio $(ABCD)$. For example, if $(ABCD) = \lambda$, then the cross-ratio $(BACD)$ is $1/\lambda$, while $(ACBD) = 1 - \lambda$, as the reader may easily verify. Four points A, B, C, D can be ordered in $4 \cdot 3 \cdot 2 \cdot 1 = 24$ different ways, each of which gives a certain value to their cross-ratio. Some of these permutations will yield the same value for the cross-ratio as the original arrangement A, B, C, D; e.g. $(ABCD) = (BADC)$. It is left as an exercise for the reader to show that there are only six different values of the cross-ratio for these 24 different permutations of the points, namely

$$\lambda, \quad 1 - \lambda, \quad 1/\lambda, \quad \frac{\lambda - 1}{\lambda}, \quad \frac{1}{1 - \lambda}, \quad \frac{\lambda}{\lambda - 1}.$$

These six quantities are in general distinct, but two of them may coincide—as in the case of harmonic division, when $\lambda = -1$.

We may also define *the cross-ratio of four coplanar* (i.e. lying in a common plane) *and concurrent straight lines* 1, 2, 3, 4 as the cross-ratio of the four points of intersection of these lines with another straight line lying in the same plane. The position of this fifth line is immaterial because of the invariance of the cross-ratio under projection. Equivalent to this is the definition

$$(1\,2\,3\,4) = \frac{\sin(1, 3)}{\sin(2, 3)} \Big/ \frac{\sin(1, 4)}{\sin(2, 4)},$$

taken with a plus or minus sign according as one pair of lines does not or does separate the other. (In this formula, (1, 3), for example, means the angle between the lines 1 and 3.) Finally, we may define *the cross-ratio of four coaxial planes* (four planes in space intersecting in a line l, their axis). If a straight line intersects the planes in four points, these points will always have the same cross-ratio, whatever the position of the line may be. (The proof of this fact is left as an exercise.) Hence we may assign this value as the cross-ratio of the four planes. Equivalently, we may define the cross-ratio of four coaxial planes as the cross-ratio of the four lines in which they are intersected by any fifth plane (see Fig. 79).

The concept of the cross-ratio of four planes leads naturally to the question of whether a projective transformation of *three-dimensional* space into itself can be defined. The definition by central projection

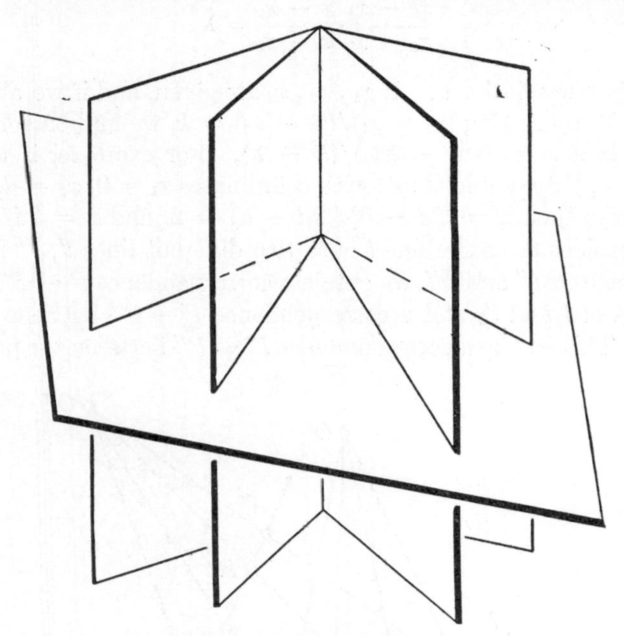

Fig. 79. Cross-ratio of coaxial planes.

cannot immediately be generalized from two to three dimensions. But it can be proved that every continuous transformation of a plane into itself that correlates in a biunique manner points with points and lines with lines is a projective transformation. This theorem suggests the following definition for three dimensions: A projective transformation of space is a continuous biunique transformation that preserves straight lines. It can be shown that these transformations leave the cross-ratio invariant.

The preceding statements may be supplemented by a few remarks. Suppose we have three distinct points, A, B, C, on a line, with coördinates x_1, x_2, x_3. Required, to find a fourth point D so that the cross-ratio $(ABCD) = \lambda$, where λ is prescribed. (The special case $\lambda = -1$, for which the problem amounts to the construction of the fourth harmonic point, will be taken up in more detail in the next article.) In general, the problem has one and only one solution; for, if x is the coördinate of the desired point D, then the equation

$$\frac{x_3 - x_1}{x_3 - x_2} \cdot \frac{x - x_2}{x - x_1} = \lambda \tag{2}$$

has exactly one solution x. If x_1, x_2, x_3 are given, and if we abbreviate equation (2) by setting $(x_3 - x_1)/(x_3 - x_2) = k$, we find on solving this equation that $x = (kx_2 - \lambda x_1)/(k - \lambda)$. For example, if the three points A, B, C are equidistant, with coördinates $x_1 = 0$, $x_2 = d$, $x_3 = 2d$ respectively; then $k = (2d - 0)/(2d - d) = 2$, and $x = 2d/(2 - \lambda)$.

If we project the same line l onto two different lines l', l'' from two different centers O' and O'', we obtain a correspondence $P \leftrightarrow P'$ between the points of l and l', and a correspondence $P \leftrightarrow P''$ between those of l and l''. This sets up a correspondence $P' \leftrightarrow P''$ between the points of l'

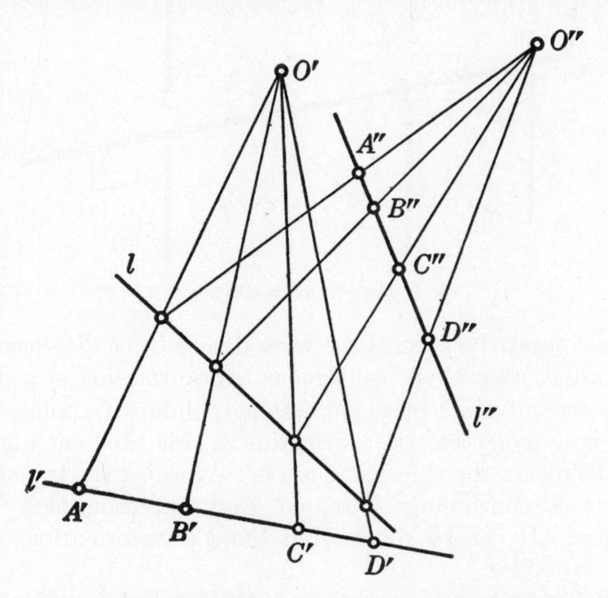

Fig. 80. Projective correspondence between the points on two lines.

and those of l'' which has the property that every set of four points A', B', C', D' on l' has the same cross-ratio as the corresponding set A'', B'', C'', D'' on l''. Any biunique correspondence between the points on two lines which has this property is called a *projective correspondence*, irrespective of how the correspondence is defined.

Exercises: 1) Prove that, given two lines together with a projective correspondence between their points, one can shift one line by a parallel displacement into such a position that the given correspondence is obtained by a simple projection. (Hint: Bring a pair of corresponding points of the two lines into coincidence.)

2) On the basis of the preceding result, show that if the points of two lines l and l' are coördinated by any finite succession of projections onto various intermediate lines, using arbitrary centers of projection, the same result can be obtained by only *two* projections.

2. Application to the Complete Quadrilateral

As an interesting application of the invariance of the cross-ratio we shall establish a simple but important theorem of projective geometry. It concerns the *complete quadrilateral*, a figure consisting of any four straight lines, no three of which are concurrent, and of the six points where they intersect. In Figure 81 the four lines are AE, BE, BI, AF. The lines through AB, EG, and IF are the *diagonals* of the quadrilateral. Take any diagonal, say AB, and mark on it the points C and D where it meets the other two diagonals. We then have the theorem: $(ABCD) = -1$; in words, *the points of intersection of one diagonal with the other two separate the vertices on that diagonal harmonically.* To prove this we simply observe that

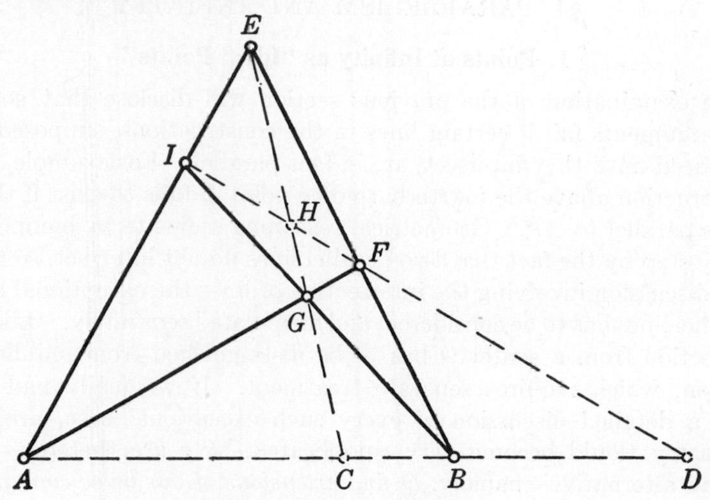

Fig. 81. Complete quadrilateral.

$$x = (ABCD) = (IFHD) \qquad \text{by projection from } E,$$
$$(IFHD) = (BACD) \qquad \text{by projection from } G.$$

But we know that $(BACD) = 1/(ABCD)$; so that $x = 1/x$, $x^2 = 1$, $x = \pm 1$. Since C, D separate A, B, the cross-ratio x is negative and must therefore be -1, which was to be proved.

This remarkable property of the complete quadrilateral enables us to find with the straightedge alone the harmonic conjugate with respect to A, B of any third collinear point C. We need only choose a point E off the line, draw EA, EB, EC, mark a point G on EC, draw AG and BG intersecting EB and EA at F and I respectively, and draw IF, which intersects the line of A, B, C in the required fourth harmonic point D.

Problem: Given a segment AB in the plane and a region R, as shown in Figure 82. It is desired to continue the line AB to the right of R. How may this be done with straightedge alone so that the straightedge never crosses R during the construction? (Hint: Choose two arbitrary points C, C' on the segment AB, then locate their harmonic conjugates D, D' respectively by means of four quadrilaterals having A, B as vertices.)

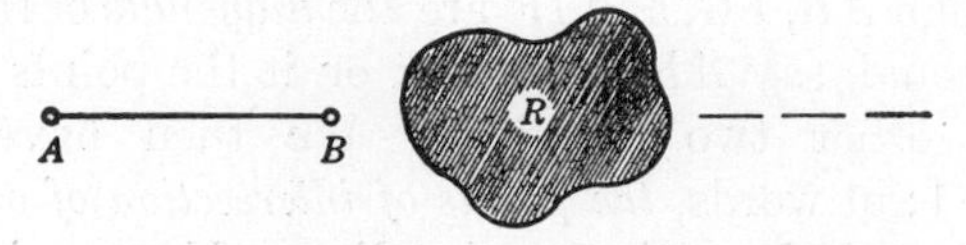

Fig. 82. Producing a line beyond an obstacle.

§4. PARALLELISM AND INFINITY

1. Points at Infinity as "Ideal Points"

An examination of the previous section will disclose that some of our arguments fail if certain lines in the constructions, supposed to be produced until they intersect, are in fact parallel. For example, in the construction above the fourth harmonic point D fails to exist if the line IF is parallel to AB. Geometrical reasoning seems to be hampered at every step by the fact that two parallel lines do not intersect, so that in any discussion involving the intersection of lines the exceptional case of parallel lines has to be considered, and formulated separately. Likewise, projection from a center O has to be distinguished from parallel projection, which requires separate treatment. If we really had to go into a detailed discussion of every such exceptional case, projective geometry would become very complicated. We are therefore led to try an alternative—namely, *to find extensions of our basic concepts that will eliminate the exceptions.*

Here geometrical intuition points the way: if a straight line that intersects another is rotated slowly towards a parallel position, then the point of intersection of the two lines will recede to infinity. We might naively say that the two lines intersect at a "point at infinity." The essential thing is then to give this vague statement a precise meaning, so that points at infinity, or, as they are sometimes called, ideal points, can be dealt with exactly as though they were ordinary points in the plane or in space. In other words, we want all rules concerning the behavior of points, lines, planes, etc. to persist, even when these geometric elements are ideal. To achieve this goal we can proceed either intuitively or formally, just as we did in extending the number system, where one approach was from the intuitive idea of measuring, and another from the formal rules of arithmetical operations.

First, let us realize that in synthetic geometry even the basic concepts of "ordinary" point and line are not mathematically defined. The so-called definitions of these concepts which are frequently found in textbooks on elementary geometry are only suggestive descriptions. In the case of ordinary geometrical elements our intuition makes us feel at ease as far as their "existence" is concerned. But all we really need in geometry, considered as a mathematical system, is the validity of certain rules by means of which we can operate with these concepts, as in joining points, finding the intersection of lines, etc. Logically considered, a "point" is not a "thing in itself," but is completely described by the totality of statements by which it is related to other objects. The mathematical existence of "points at infinity" will be assured as soon as we have stated in a clear and consistent manner the mathematical *properties* of these new entities, i.e. their relations to "ordinary" points and to each other. The ordinary axioms of geometry (e.g. Euclid's) are abstractions from the physical world of pencil and chalk marks, stretched strings, light rays, rigid rods, etc. The properties which these axioms attribute to mathematical points and lines are highly simplified and idealized descriptions of the behavior of their physical counterparts. Through any two actual pencil dots not one but many pencil lines can be drawn. If the dots become smaller and smaller in diameter then all these lines will have approximately the same appearance. This is what we have in mind when we state as an axiom of geometry that "through any two points *one and only one* straight line may be drawn"; we are referring not to physical points and lines but to the abstract and conceptual points and lines of geometry. Geometrical points and lines have essentially simpler properties than do any physical objects, and this simplification provides the essential condition for the development of geometry as a deductive science.

As we have noticed, the ordinary geometry of points and lines is greatly complicated by the fact that a pair of parallel lines do not intersect in a point. We are therefore led to make a further simplification in the structure of geometry by enlarging the concept of geometrical point in order to remove this exception, just as we enlarged the concept of number in order to remove the restrictions on subtraction and division. Here also we shall be guided throughout by the desire to preserve in the extended domain the laws which governed the original domain.

We shall therefore agree to add to the ordinary points on each line a single "ideal" point. This point shall be considered to belong to all the lines parallel to the given line and to no other lines. As a consequence of this convention *every* pair of lines in the plane will now intersect in a single point; if the lines are not parallel they will intersect in an ordinary point, while if the lines are parallel they will intersect in the ideal point common to the two lines. For intuitive reasons the ideal point on a line is called the *point at infinity* on the line.

The intuitive concept of a point on a line receding to infinity might suggest that we add *two* ideal points to each line, one for each direction along the line. The reason for adding only one, as we have done, is that we wish to preserve the law that through any two points *one and only one* line may be drawn. If a line contained two points at infinity in common with every parallel line then through these two "points" infinitely many parallel lines would pass.

Wc shall also agree to add to the ordinary lines in a plane a single "ideal" line (also called the line at infinity in the plane), *containing all the ideal points in the plane and no other points.* Precisely this convention is forced upon us if we wish to preserve the original law that through every two points one line may be drawn, and the newly gained law that every two lines intersect in a point. To see this, let us choose any two ideal points. Then the unique line which is required to pass through these points cannot be an ordinary line, since by our agreement any ordinary line contains but one ideal point. Moreover, this line cannot contain any ordinary points, since an ordinary point and one ideal point determine an ordinary line. Finally, this line must contain *all* the ideal points, since we wish it to have a point in common with every ordinary line. Hence this line must have precisely the properties which we have assigned to the ideal line in the plane.

According to our conventions, a point at infinity is determined or is represented by any family of parallel lines, just as an irrational number is determined by a sequence of nested rational intervals. The statement that the intersection of two parallel lines is a point at infinity has no mysterious connotation, but is only a convenient way of stating that the lines are parallel. This way of expressing parallelism, in the language originally reserved for intuitively different objects, has the sole purpose of making the enumeration of exceptional cases superfluous; they are now automatically covered by the same kind of linguistic expressions or other symbols that are used for the "ordinary" cases.

To sum up: our conventions regarding points at infinity have been so chosen that the laws governing the incidence relation between ordinary points and lines continue to hold in the extended domain of points, while the operation of finding the point of intersection of two lines, previously possible only if the lines are not parallel, may now be performed without restriction. The considerations which led to this formal simplification in the properties of the incidence relation may seem somewhat abstract. But they are amply justified by the result, as the reader will see in the following pages.

2. Ideal Elements and Projection

The introduction of the points at infinity and the line at infinity in a plane enables us to treat the projection of one plane onto another in a much more satisfactory way. Let us consider the projection of a plane π onto a plane π' from a center O (Fig. 83). This projection estab-

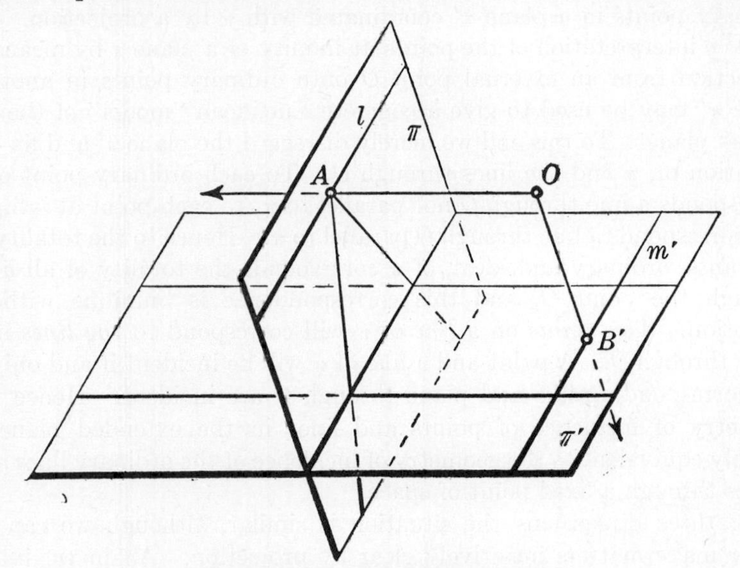

Fig. 83. Projection into elements at infinity.

lishes a correspondence between the points and lines of π and those of π'. To every point A of π corresponds a unique point A' of π', with the following exceptions: if the projecting ray through O is *parallel* to the plane π', then it intersects π in a point A to which no ordinary point of π' corresponds. These exceptional points of π lie on a line l to which no ordinary line of π' corresponds. But these exceptions are eliminated if we make the agreement that to A corresponds the point at infinity in π' in the direction of the line OA, and that to l corresponds the line at infinity in π'. In the same way, we assign a point at infinity in π to any point B' on the line m' in π' through which pass all the rays from O parallel to the plane π. To m' itself will correspond the line at infinity in π. Thus, by the introduction of the points and line at infinity in a plane, *a projection of one plane onto another establishes a correspondence between the points and lines of the two planes which is biunique without exception.* (This disposes of the exceptions mentioned in the footnote on p. 169.) Moreover, it is easily seen to be a consequence of our agreement that *a point lies on a line if and only if the projection of the point lies on the projection of the line.* Hence all statements about collinear points, concurrent lines, etc. that involve only points, lines, and the incidence relation, are seen to be invariant under projection in the extended sense. This enables us to operate with the points at infinity in a plane π simply by operating with the corresponding ordinary points in a plane π' coördinated with π by a projection.

*The interpretation of the points at infinity of a plane π by means of projection from an external point O onto ordinary points in another plane π' may be used to give a concrete Euclidean "model" of the extended plane. To this end we merely disregard the plane π' and fix our attention on π and the lines through O. To each ordinary point of π corresponds a line through O not parallel to π; to each point at infinity of π corresponds a line through O parallel to π. Hence to the totality of all points, ordinary and ideal, of π corresponds the totality of all lines through the point O, and this correspondence is biunique without exception. The *points* on a *line* of π will correspond to the *lines* in a *plane* through O. A point and a line of π will be incident if and only if the corresponding line and plane through O are incident. Hence the geometry of incidence of points and lines in the extended plane is entirely equivalent to the geometry of incidence of the ordinary lines and planes through a fixed point of space.

*In three dimensions the situation is similar, although we can no longer make matters intuitively clear by projection. Again we introduce a point at infinity associated with every family of parallel lines. In each plane we have a line at infinity. Next we have to introduce a new element, the *plane at infinity*, consisting of all points at infinity of the space and containing all lines at infinity. Each ordinary plane intersects the plane at infinity in its line at infinity.

3. Cross-Ratio with Elements at Infinity

A remark must be made about cross-ratios involving elements at infinity. Let us denote the point at infinity on a straight line l by the symbol ∞. If A, B, C are three ordinary points on l, then we may assign a value to the symbol $(ABC\infty)$ in the following way: choose a point P on l; then $(ABC\infty)$ should be the limit approached by $(ABCP)$ as P recedes to infinity along l. But

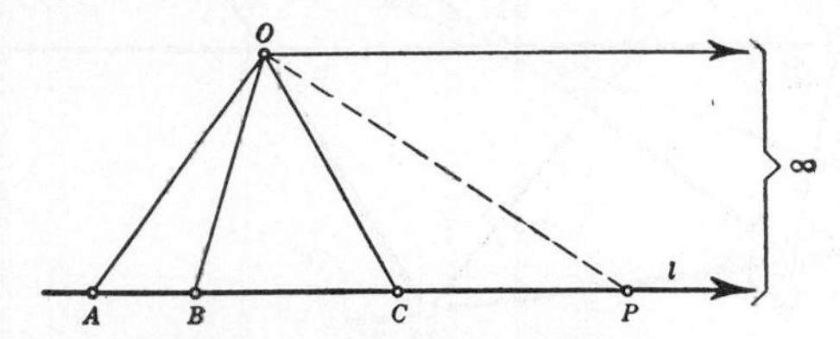

Fig. 84. Cross-ratio with a point at infinity.

$$(ABCP) = \frac{CA}{CB} \Big/ \frac{PA}{PB},$$

and as P recedes to infinity, PA/PB approaches 1. Hence we define

$$(ABC\infty) = CA/CB.$$

In particular, if $(ABC\infty) = -1$, then C is the midpoint of the segment AB: *the midpoint and the point at infinity in the direction of a segment divide the segment harmonically.*

Exercises: What is the cross-ratio of four lines l_1, l_2, l_3, l_4 if they are parallel? What is the cross-ratio if l_4 is the line at infinity?

§5. APPLICATIONS

1. Preliminary Remarks

With the introduction of elements at infinity it is no longer necessary to state explicitly the exceptional cases that arise in constructions and theorems when two or more lines are parallel. We need merely remember that when a point is at infinity all the lines through that point are parallel. The distinction between central and parallel projection need no longer be made, since the latter simply means projection from a point at infinity. In Figure 72 the point O or the line PQR may be at infinity (Fig. 85 shows the former case); it is left as an exercise for the reader to formulate in "finite" language the corresponding statements of Desargues's theorem.

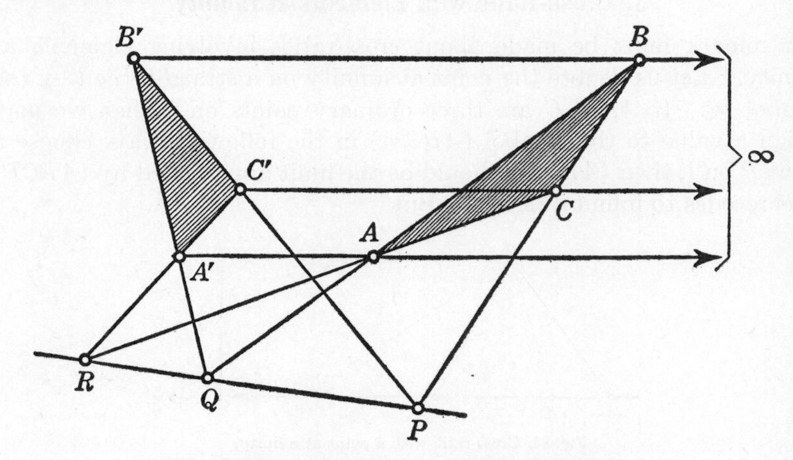

Fig. 85. Desargues's configuration with center at infinity.

Not only the *statement* but even the *proof* of a projective theorem is often made simpler by the use of elements at infinity. The general principle is the following. By the "projective class" of a geometrical figure F we mean the class of all figures into which F may be carried by projective transformations. The projective properties of F will be identical with those of any member of its projective class, since projective properties are by definition invariant under projection. Thus, any projective theorem (one involving only projective properties) that is true of F will be true of any member of the projective class of F, and conversely. Hence, in order to prove any such theorem for F, it suffices to prove it for any other member of the projective class of F. We may often take advantage of this by finding a special member of the projective class of F for which the theorem is simpler to prove than for F itself. For example, any two points A, B of a plane π can be projected to infinity by projecting from a center O onto a plane π' parallel to the plane of O, A, B; the straight lines through A and those through B will be transformed into two families of parallel lines. In the projective theorems to be proved in this section we shall make such a preliminary transformation.

The following elementary fact about parallel lines will be useful. Let two straight lines, intersecting at a point O, be cut by a pair of lines l_1

and l_2 at points A, B, C, D, as shown in Figure 86. If l_1 and l_2 are parallel then

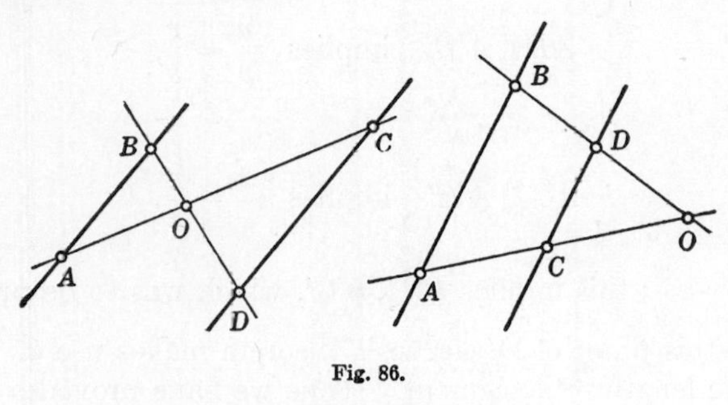

Fig. 86.

$$\frac{OA}{OC} = \frac{OB}{OD};$$

and conversely, if $\dfrac{OA}{OC} = \dfrac{OB}{OD}$ then l_1 and l_2 are parallel. The proof follows from elementary properties of similar triangles, and will be left to the reader.

2. Proof of Desargues's Theorem in the Plane

We now give the proof that for two triangles ABC and $A'B'C'$ in a plane situated as shown in Figure 72, where the lines through corresponding vertices meet in a point, the intersections P, Q, R of the corresponding sides lie on a straight line. To do this we first project the figure so that Q and R go to infinity. After the projection, AB will be parallel to $A'B'$, AC to $A'C'$, and the figure will appear as shown in Figure 87. As we have pointed out in Article 1 of this section, to

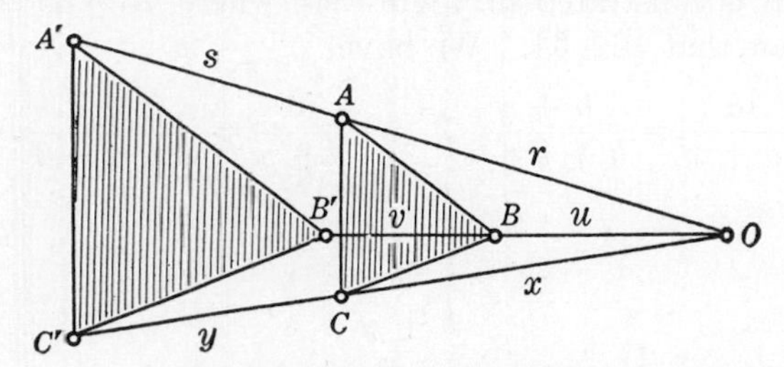

Fig. 87. Proof of Desargues's theorem.

prove Desargues's theorem in general it suffices to prove it for this special type of figure For this purpose we need only show that the intersection of BC and $B'C'$ also goes to infinity, so that BC is parallel to

$B'C'$; then P, Q, R will indeed be collinear (since they will lie on the line at infinity). Now

$$AB \parallel A'B' \quad \text{implies} \quad \frac{u}{v} = \frac{r}{s},$$

and

$$AC \parallel A'C' \quad \text{implies} \quad \frac{x}{y} = \frac{r}{s}.$$

Therefore $\frac{u}{v} = \frac{x}{y}$; this implies $BC \parallel B'C'$, which was to be proved.

Note that this proof of Desargues's theorem makes use of the metric notion of the length of a segment. Thus we have proved a projective theorem by metric means. Moreover, if projective transformations are defined "intrinsically" as plane transformations that preserve cross-ratio (see p. 177), then this proof remains entirely in the plane.

Exercise: Prove, in a similar manner, the converse of Desargues's theorem: If triangles ABC and $A'B'C'$ have the property that P, Q, R are collinear, then the lines AA', BB', CC', are concurrent.

3. Pascal's Theorem†

This theorem states: *If the vertices of a hexagon lie alternately on a pair of intersecting lines, then the three intersections* P, Q, R *of the opposite sides of the hexagon are collinear* (Fig. 88). (The hexagon may intersect itself. The "opposite" sides can be recognized from the schematic diagram of Fig. 89.)

By performing a preliminary projection we may assume that P and Q are at infinity. Then we need only show that R also is at infinity. The situation is illustrated in Figure 90, where $23 \parallel 56$ and $12 \parallel 45$. We must show that $16 \parallel 34$. We have

$$\frac{a}{a+x} = \frac{b+y}{b+y+s}, \qquad \frac{b}{b+y} = \frac{a+x}{a+x+r}.$$

Therefore

$$\frac{a}{b} = \frac{a+x+r}{b+y+s},$$

so that $16 \parallel 34$, as was to be proved.

† On p. 209 we shall discuss a more general theorem of the same type. The present special case is also known by the name of its discoverer, Pappus of Alexandria (third century A.D.).

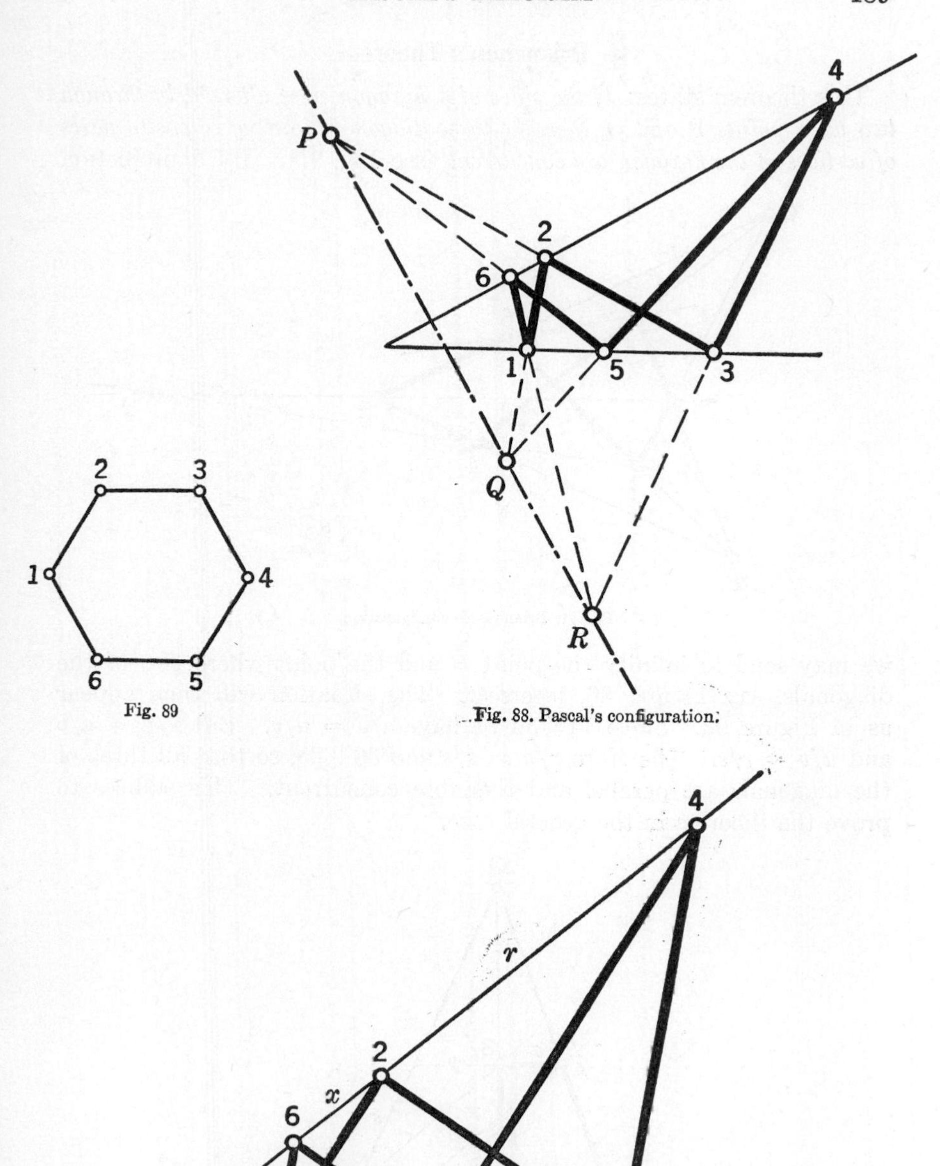

Fig. 89

Fig. 88. Pascal's configuration.

Fig. 90. Proof of Pascal's theorem.

4. Brianchon's Theorem

This theorem states: *If the sides of a hexagon pass alternately through two fixed points* P *and* Q, *then the three diagonals joining opposite pairs of vertices of the hexagon are concurrent* (see Fig. 91). By a projection

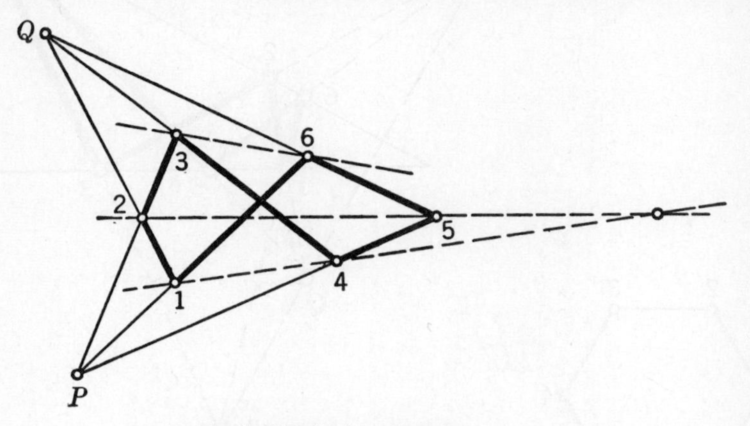

Fig. 91. Brianchon's configuration.

we may send to infinity the point P and the point where two of the diagonals, say 14 and 36, intersect. The situation will then appear as in Figure 92. Since $14 \parallel 36$ we have $a/b = u/v$. But $x/y = a/b$ and $u/v = r/s$. Therefore $x/y = r/s$ and $36 \parallel 25$, so that all three of the diagonals are parallel and therefore concurrent. This suffices to prove the theorem in the general case.

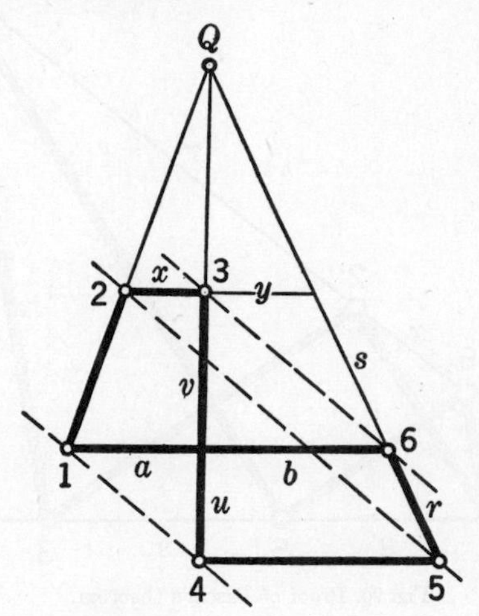

Fig. 92. Proof of Brianchon's theorem.

5. Remark on Duality

The reader may have noticed the remarkable similarity between the theorems of Pascal (1623–1662) and Brianchon (1785–1864). This similarity becomes particularly striking if we write the theorems side by side:

Pascal's Theorem	*Brianchon's Theorem*
If the *vertices* of a hexagon *lie alternately on* two *straight lines*, the *points where opposite sides meet* are *collinear.*	If the *sides* of a hexagon *pass alternately through* two *points*, the *lines joining opposite vertices* are *concurrent.*

Not only the theorems of Pascal and Brianchon, but all the theorems of projective geometry occur in pairs, each similar to the other, and, so to speak, identical in structure. This relationship is called *duality*. In plane geometry point and line are called *dual elements*. Drawing a line through a point, and marking a point on a line are *dual operations*. Two figures are dual if one may be obtained from the other by replacing each element and operation by its dual element or operation. Two theorems are dual if one becomes the other when all elements and operations are replaced by their duals. For example, Pascal's and Brianchon's theorems are dual, and the dual of Desargues's theorem is precisely its converse. This phenomenon of duality gives projective geometry a character quite distinct from that of elementary (metric) geometry, in which no such duality exists. (For example, it would be meaningless to speak of the dual of an angle of 37° or of a segment of length 2.) In many textbooks on projective geometry the *principle of duality*, which states that the *dual of any true theorem of projective geometry is likewise a true theorem of projective geometry*, is exhibited by placing the dual theorems together with their dual proofs in parallel columns on the page, as we have done above. The basic reason for this duality will be considered in the following section (see also p. 217).

§6. ANALYTIC REPRESENTATION

1. Introductory Remarks

In the early development of projective geometry there was a strong tendency to build everything on a synthetic and "purely geometric" basis, avoiding the use of numbers and of algebraic methods. This program met with great difficulties, since there always remained places where some algebraic formulation seemed unavoidable. Complete success in building up a purely synthetic projective geometry was only attained toward the end of the nineteenth century, at a rather high cost in complication. In this respect the methods of analytic geometry have been much more successful. The general tendency in modern mathematics is to base everything on the number concept, and in geometry this tendency, which started with Fermat and Descartes, has had decisive triumphs. Analytic geometry has developed from the status of a mere tool in geometrical reasoning to a subject where the intuitive geometrical interpretation of the operations and results is no longer the ultimate and exclusive goal, but has rather the function of a guiding principle that aids in suggesting and understanding the analytical results. This change in the meaning of geometry is the product of a gradual historical growth that has greatly enlarged the scope of the classical geometry, and at the same time has brought about an almost organic union of geometry and analysis.

In analytic geometry the "coördinates" of a geometrical object are any set of numbers which characterize that object uniquely. Thus a point is defined by giving its rectangular coördinates x, y or its polar coördinates ρ, θ, while a triangle can be defined by giving the coördinates of its three vertices, which requires six coördinates in all. We know that a straight line in the x, y-plane is the geometrical locus of all points $P\,(x, y)$ (see p. 75 for this notation) whose coördinates satisfy some linear equation

$$ax + by + c = 0. \tag{1}$$

We may therefore call the three numbers a, b, c the "coördinates" of this line. For example, $a = 0$, $b = 1$, $c = 0$ define the line $y = 0$, which is the x-axis; $a = 1$, $b = -1$, $c = 0$ define the line $x = y$, which bisects the angle between the positive x-axis and the positive y-axis. In the same way, quadratic equations define "conic sections":

$x^2 + y^2 = r^2$ a circle, center at origin, radius r,

$(x - a)^2 + (y - b)^2 = r^2$ a circle, center at (a, b), radius r,

$\dfrac{x^2}{a^2} + \dfrac{y^2}{b^2} = 1$ an ellipse,

etc.

The naive approach to analytic geometry is to start with purely "geometric" concepts—point, line, etc.—and then to translate these into the language of numbers. The modern viewpoint is the reverse. We start with the *set of all pairs of numbers* x, y and *call* each such pair a point, since we can, if we choose, *interpret* or *visualize* such a pair of numbers by the familiar notion of a geometrical point. Similarly, a linear equation between x and y is said to define a line. Such a shift of emphasis from the intuitive to the analytical aspect of geometry opens the way for a simple, yet rigorous, treatment of the points at infinity in projective geometry, and is indispensable for a deeper understanding of the whole subject. For those readers who possess a certain amount of preliminary training we shall give an account of this approach.

*2. Homogeneous Coördinates. The Algebraic Basis of Duality

In ordinary analytic geometry, the rectangular coördinates of a point in the plane are the signed distances of the point from a pair of perpendicular axes. This system breaks down for the points at infinity in the extended plane of projective geometry. Hence if we wish to apply analytic methods to projective geometry it is necessary to find a coördinate system which shall embrace the ideal as well as the ordinary points. The introduction of such a coördinate system is best described by supposing the given X, Y-plane π imbedded in three-dimensional space, where rectangular coördinates x, y, z (the signed distances of a point from the three coördinate planes determined by the x, y, and z axes) have been introduced. We place π parallel to the x, y coördinate plane and at a distance 1 above it, so that any point P of π will have the three-dimensional coördinates $(X, Y, 1)$. Taking the origin O of the coördinate system as the center of projection, we note that *each point* P *determines a unique line through* O *and conversely*. (See p. 184. The lines through O and parallel to π correspond to the points at infinity of π.)

We shall now describe a system of "homogeneous coördinates" for the points of π. To find the homogeneous coördinates of any ordinary point P of π, we take the line through O and P and choose *any* point Q other than O on this line (see Fig. 93). Then the ordinary three-dimensional coördinates x, y, z of Q are said to be *homogeneous coördinates* of P. In particular, the coördinates $(X, Y, 1)$ of P itself are a set of homogeneous coördinates for P. Moreover, any other set of numbers (tX, tY, t) with $t \neq 0$ will also be a set of homogeneous coördinates for P, since the coördinates of all points on the line OP other than O will be of this form. (We have excluded the point $(0, 0, 0)$ since it lies on all lines through O and does not serve to distinguish one from another.)

This method of introducing coördinates in the plane requires three numbers instead of two to specify the position of a point, and has the further disadvantage that the coördinates of a point are not determined uniquely but only up to an arbitrary factor t. But it has the great advantage that the points at infinity in π are now included in the coördinate representation. A point P at infinity in π is determined by a line through O parallel to π. Any point Q on this line will have coördinates of the form $(x, y, 0)$. Hence *the homogeneous coördinates of a point at infinity in π are of the form* (x, y, 0).

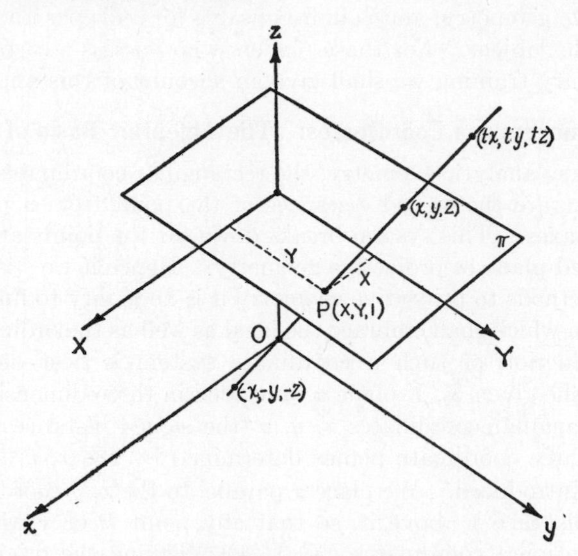

Fig. 93. Homogeneous coördinates.

The equation in homogeneous coördinates of a straight line in π is readily found by observing that the lines joining O to the points of this line lie in a plane through O. It is proved in analytic geometry that the equation of such a plane is of the form

$$ax + by + cz = 0.$$

Hence this is the equation in homogeneous coördinates of a straight line in π.

Now that the geometrical model of the points of π as lines through O has served its purpose, we may lay it aside and give the following purely analytic definition of the extended plane:

A *point* is an ordered triple of real numbers (x, y, z), not all of which are zero. Two such triples, (x_1, y_1, z_1) and (x_2, y_2, z_2), define the *same* point if for some $t \neq 0$,

$$x_2 = tx_1,$$

$$y_2 = ty_1,$$

$$z_2 = tz_1$$

In other words, the coördinates of any point may be multiplied by any non-zero factor without changing the point. (It is for this reason that they are called *homogeneous* coördinates.) A point (x, y, z) is an *ordinary* point if $z \neq 0$; if $z = 0$, it is a *point at infinity.*

A *straight line* in π consists of all points (x, y, z) which satisfy a linear equation of the form

$$ax + by + cz = 0, \tag{1'}$$

where a, b, c are any three constants, not all zero. In particular, the points at infinity in π all satisfy the linear equation

$$z = 0. \tag{2}$$

This is by definition a line, and is called the *line at infinity* in π. Since a line is defined by an equation of the form (1′), we call the triple of numbers (a, b, c) the *homogeneous coördinates of the line* (1′). It follows that (ta, tb, tc), for any $t \neq 0$, are also coördinates of the line (1′), since the equation

$$(ta)x + (tb)y + (tc)z = 0 \tag{3}$$

is satisfied by the same coördinate-triples (x, y, z) as (1′).

In these definitions we observe the perfect symmetry between point and line: each is specified by three homogeneous coördinates (u, v, w). The condition that the point (x, y, z) lie on the line (a, b, c) is that

$$ax + by + cz = 0,$$

and this is likewise the condition that the point whose coördinates are (a, b, c) lie on the line whose coördinates are (x, y, z). For example, the arithmetical identity

$$2 \cdot 3 + 1 \cdot 4 - 5 \cdot 2 = 0$$

may be interpreted equally well as meaning that the point $(3, 4, 2)$ lies on the line $(2, 1, -5)$ or that the point $(2, 1, -5)$ lies on the line $(3, 4, 2)$. This symmetry is the basis of the duality in projective geometry between point and line, for any relationship between points and lines becomes a relationship between lines and points when the coördinates are properly re-interpreted. In the new interpretation the previous coördinates of points and lines are now thought of as representing lines and points respectively. All the algebraic operations and results remain the same, but their interpretation gives the dual counterpart of the original theorem. It is to be noted that this duality does not hold in the ordinary plane of two coördinates X, Y, since the equation of a straight line in ordinary coördinates

$$aX + bY + c = 0$$

is not symmetrical in X, Y and a, b, c. Only by including the points and the line at infinity is the principle of duality perfectly established.

To pass from the homogeneous coördinates x, y, z of an ordinary point P in the plane π to ordinary rectangular coördinates, we simply set $X = x/z$, $Y = y/z$. Then X, Y represent the distances from the point P to two perpendicular axes in π, parallel to the x- and y-axes, as shown in Figure 93. We know that an equation of the form

$$aX + bY + c = 0$$

will represent a straight line in π. On substituting $X = x/z$, $Y = y/z$ and multiplying through by z we find that the equation of the same line in homogeneous coördinates is, as stated on page 195,

$$ax + by + cz = 0.$$

Thus the equation of the line $2x - 3y + z = 0$ in ordinary rectangular coördinates X, Y is $2X - 3Y + 1 = 0$. Of course, the latter equation fails for the point at infinity on this line, one set of whose homogeneous coördinates is $(3, 2, 0)$.

One thing remains to be said. We have succeeded in giving a purely analytic definition of point and line, but what of the equally important concept of projective transformation? It may be proved that a projective transformation of one plane onto another as defined on page 169 is given analytically by a set of linear equations,

$$\begin{aligned} x' &= a_1x + b_1y + c_1z, \\ y' &= a_2x + b_2y + c_2z, \\ z' &= a_3x + b_3y + c_3z, \end{aligned} \tag{4}$$

connecting the homogeneous coördinates x', y', z' of the points in the plane π' with the homogeneous coördinates x, y, z of the points in the plane π. From our present point of view we may now *define* a projective transformation as one given by any set of linear equations of the form (4). The theorems of projective geometry then become theorems on the behavior of number triples (x, y, z) under such transformations. For example, the proof that the cross-ratio of four points on a line is unchanged by such transformations becomes simply an exercise in the algebra of linear transformations. We cannot go further into the details of this analytic procedure. Instead we shall return to the more intuitive aspects of projective geometry.

§7. PROBLEMS ON CONSTRUCTIONS WITH THE STRAIGHTEDGE ALONE

In the constructions below it is understood that only the straightedge is admitted as tool.

Problems 1 to 18 are contained in a paper by J. Steiner in which he proves that the compass can be dispensed with as a tool for geometrical constructions if a fixed circle with its center is given (see Chapt. III, p. 151). The reader is advised to solve these problems in the order given.

A set of four lines a, b, c, d through a point P is called *harmonic*, if the cross-ratio $(abcd)$ equals -1. a and b are said to be *conjugate* with respect to c and d, and vice versa.

1) Prove: If, in a set of four harmonic lines a, b, c, d, the ray a bisects the angle between c and d, then b is perpendicular to a.

2) Construct the fourth harmonic line to three given lines through a point. (Hint: Use the theorem on the complete quadrilateral.)

3) Construct the fourth harmonic point to three points on a line.

4) If a given right angle and a given arbitrary angle have their vertex and one side in common, double the given arbitrary angle.

5) Given an angle and its bisector b. Construct a perpendicular to b through the vertex P of the given angle.

6) Prove: If the lines $l_1, l_2, l_3, \cdots, l_n$ through a point P intersect the straight line a in the points $A_1, A_2, \cdots, A_n$ and intersect the line b in the points $B_1, B_2, \cdots, B_n$, then all the intersections of the pairs of lines A_iB_k and A_kB_i $(i \neq k; i, k = 1, 2, \cdots, n)$ lie on a straight line.

7) Prove: If a parallel to the side BC of the triangle ABC intersects AB in B' and AC in C', then the line joining A with the intersection D of $B'C$ and $C'B$ bisects BC.

7a) Formulate and prove the converse of 7.

8) On a straight line l three point P, Q, R are given, such that Q is the midpoint of the segment PR. Construct a parallel to l through a given point S.

9) Given two parallel lines l_1 and l_2; bisect a given segment AB on l_1.

10) Draw a parallel through a given point P to two given parallel lines l_1 and l_2. (Hint: Reduce 9 to 7 using 8.)

11) Steiner gives the following solution to the problem of doubling a given line segment AB when a parallel l to AB is given: Through a point C not on l nor on the line AB draw CA intersecting l at A_1, CB intersecting l at B_1. Then (see 10) draw a parallel to l through C, which meets BA_1 at D. If DB_1 meets AB at E, then $AE = 2 \cdot AB$.

Prove the last statement.

12) Divide a segment AB into n equal parts if a parallel l to AB is given. (Hint: Construct first the n-fold of an arbitrary segment on l, using 11.)

13) Given a parallelogram $ABCD$, draw a parallel through a point P to a straight line l. (Hint: Apply 10 to the center of the parallelogram and use 8.)

14) Given a parallelogram, multiply a given segment by n. (Hint: Use 13 and 11.)

15) Given a parallelogram, divide a given segment into n parts.

16) If a fixed circle and its center are given, draw a parallel to a given straight line through a given point. (Hint: Use 13.)

17) If a fixed circle and its center are given, multiply and divide a given segment by n. (Hint: Use 13.)

18) Given a fixed circle and its center, draw a perpendicular to a given line through a given point. (Hint: Using a rectangle inscribed in the fixed circle and having two sides parallel to the given line, reduce to previous exercised.)

19) Using the results of problems 1–18, which basic construction problems can you solve if your tool is a ruler with two parallel edges?

20) Two given straight lines l_1 and l_2 intersect at a point P *outside* the given sheet of paper. Construct the line joining a given point Q with P. (Hint: Complete the given elements to the figure of Desargues's theorem for the plane in such a way that P and Q become intersections of corresponding sides of the two triangles in Desargues's Theorem.)

21) Construct the line joining two given points whose distance is greater than the length of the straightedge used. (Hint: Use 20.)

22) Two points P and Q outside the given sheet of paper are determined by two pairs of straight lines l_1, l_2 and m_1, m_2 through P and Q, respectively. Construct that part of the line PQ that lies on the given sheet of paper. (Hint: To obtain a point of PQ complete the given elements to a figure of Desargues's theorem in such a way that one triangle has two sides on l_1 and m_1 and the other one corresponding sides on l_2 and m_2.)

23) Solve 20 by means of Pascal's theorem (p. 188). (Hint: Complete the given elements to a figure of Pascal's theorem, using l_1, l_2 as a pair of opposite sides of the hexagon and Q as point of intersection of another pair of opposite sides.)

*24) Two straight lines entirely outside the given sheet of paper are each given by two pairs of straight lines intersecting at points of the lines outside the paper. Determine their point of intersection by a pair of lines through it.

§8. CONICS AND QUADRIC SURFACES

1. Elementary Metric Geometry of Conics

Until now we have been concerned only with points, lines, planes, and figures formed by a number of these. If projective geometry were nothing but the study of such "linear" figures, it would be of relatively little interest. It is a fact of fundamental importance that projective geometry is *not* confined to the study of linear figures, but includes also the whole field of conic sections and their generalizations in higher dimensions Apollonius' metric treatment of the conic sections—ellipses, hyperbolas, and parabolas—was one of the great mathematical achievements of antiquity. The importance of conic sections for pure and applied mathematics (for example, the orbits of the planets and of the electrons in the hydrogen atom are conic sections) can hardly be overestimated. It is little wonder that the classical Greek theory of conic sections is still an indispensable part of mathematical instruction. But Greek geometry was by no means final. Two thousand years later the important projective properties of the conics were discovered. In spite of the simplicity and beauty of these properties, academic inertia has so far prevented their introduction into the high school curriculum.

We shall begin by recalling the metric definitions of the conic sections. There are various such definitions whose equivalence is shown in elementary geometry. The usual ones refer to the *foci*. An *ellipse* is defined as the geometrical locus of all points P in the plane the sum of whose distances, r_1, r_2, from two fixed points F_1, F_2, the foci, has a constant value. (If the two foci coincide the figure is a circle.) The *hyperbola* is defined as the locus of all points P in the plane for which the absolute value of the difference $r_1 - r_2$ is equal to a fixed constant. The *parabola* is defined as the geometrical locus of all points P for which the distance r to a fixed point F is equal to the distance to a given line l.

In terms of analytic geometry these curves can all be expressed by equations of the second degree in the coördinates x, y. It is not hard to prove, conversely, that any curve defined analytically by an equation of the second degree:

$$ax^2 + by^2 + cxy + dx + ey + f = 0,$$

is either one of the three conics, a straight line, a pair of straight lines, a point, or imaginary. This is usually proved by introducing a new and suitable coördinate system, as is done in any course in analytic geometry.

These definitions of the conic sections are essentially metric, since they make use of the concept of distance. But there is another definition that establishes the place of the conic sections in projective geometry: *The conic sections are simply the projections of a circle on a plane.* If we project a circle C from a point O, then the projecting lines will form an infinite double cone, and the intersection of this cone with a plane π will be the projection of C. This intersection will be an ellipse or a hyperbola according as the plane cuts one or both portions of the cone. The intermediate case of the parabola occurs if π is parallel to one of the lines through O (see Fig. 94).

The projecting cone need not be a right circular cone with its vertex O perpendicularly above the center of the circle C; it may also be oblique. In all cases, as we shall here accept without proof, the intersection of the cone with a plane will be a curve whose equation is of second degree; and conversely, every curve of second degree can be obtained from a circle by such a projection. It is for this reason that the curves of second degree are called conic sections.

When the plane intersects only one portion of a right circular cone we have stated that the curve of intersection E is an ellipse. We may prove that E satisfies the usual focal definition of the ellipse, as given above, by a simple but beautiful argument given in 1822 by the Belgian mathematician G. P. Dandelin. The proof is based on the introduction of the two spheres S_1 and S_2 (Fig. 95), which are tangent to π at the points F_1 and F_2, respectively, and which touch the cone along the parallel circles K_1 and K_2 respectively. We join an arbitrary point

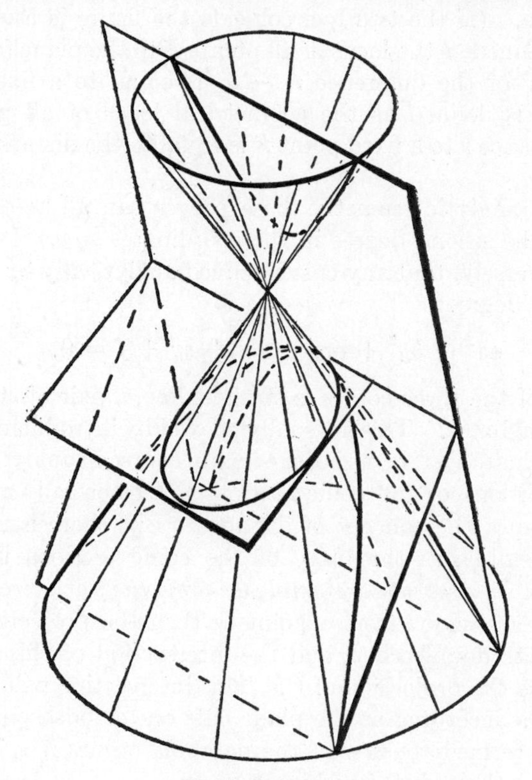

Fig. 94. Conic sections.

P of E with F_1 and F_2 and draw the line joining P to the vertex O of the cone. This line lies entirely on the surface of the cone, and intersects the circles K_1 and K_2 in the points Q_1 and Q_2 respectively. Now PF_1 and PQ_1 are two tangents from P to S_1, so that

$$PF_1 = PQ_1.$$

Similarly,

$$PF_2 = PQ_2.$$

Adding these two equations we obtain

$$PF_1 + PF_2 = PQ_1 + PQ_2.$$

But $PQ_1 + PQ_2 = Q_1Q_2$ is just the distance along the surface of the cone between the parallel circles K_1 and K_2 and is therefore independent of the particular choice of the point P on E. The resulting equation,

$$PF_1 + PF_2 = \text{constant}$$

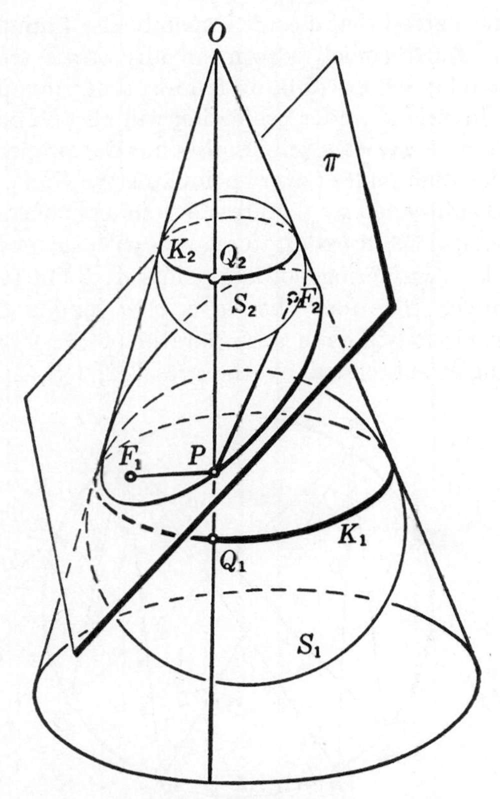

Fig. 95. Dandelin's spheres.

for all points P of E, is precisely the focal definition of an ellipse. E is therefore an ellipse and F_1, F_2 are its foci.

Exercise: When a plane cuts both portions of the cone, the curve of intersection is a hyperbola. Prove this fact, using one sphere in each portion of the cone.

2. Projective Properties of Conics

On the basis of the facts stated in the preceding section we shall adopt the tentative definition: a conic is the projection of a circle on a plane. This definition is more in keeping with the spirit of projective geometry than is the usual focal definition, since the latter is entirely based on the metric notion of distance. Even our present definition is not free from this defect, since "circle" is also a concept of metric geometry. We shall in a moment arrive at a purely projective definition of the conics.

Since we have agreed that a conic is merely the projection of a circle (i.e., that the word "conic" is to mean any curve in the projective class of the circle; see p. 186), it follows that any property of the circle that is invariant under projection will also be possessed by any conic. Now a circle has the well-known (metric) property that a given arc subtends the same angle at every point O on the circle. In Figure 96, the angle AOB subtended by the arc AB is independent of the position of O. This fact can be brought into relation with the projective concept of cross-ratio by considering not two points A, B but four points A, B, C, D on the circle. The four lines a, b, c, d joining them to a fifth point O on the circle will have a cross-ratio $(a\ b\ c\ d)$ which depends only on the angles subtended by the arcs CA, CB, DA, DB. If we

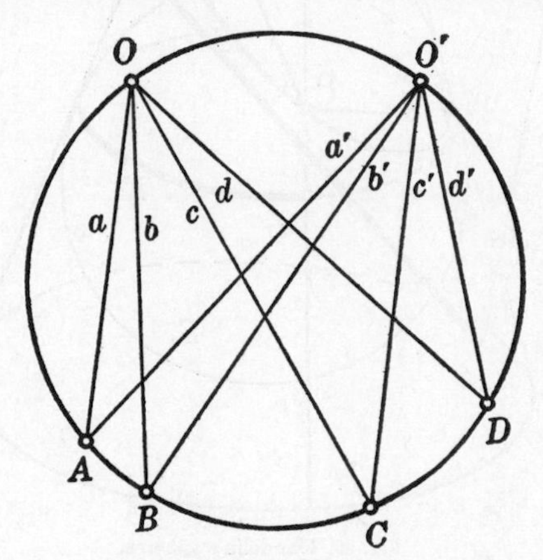

Fig. 96. Cross-ratios on a circle.

join A, B, C, D to another point O' on the circle, we obtain four rays a', b', c', d'. From the property of the circle just mentioned, the two quadruples of rays will be "congruent."† Hence they will have the same cross-ratio: $(a'\ b'\ c'\ d') = (a\ b\ c\ d)$. If we now project the

† A set of four concurrent lines a, b, c, d is said to be congruent to another set a', b', c', d' if the angles between every pair of lines of the first set are equal and have the same sense as the angles between corresponding lines of the second set.

circle into any conic K, we shall obtain on K four points, again called A, B, C, D, two other points O, O', and the two quadruples of lines a, b, c, d and a', b', c', d'. These quadruples will not be congruent, since equality of angles is in general destroyed by projection. But since cross-ratio is invariant under projection, the equality $(a\ b\ c\ d) = (a'\ b'\ c'\ d')$ will still hold. This leads to a fundamental theorem: *If any four given points* A, B, C, D *of a conic* K *are joined to a fifth point* O *of K by lines* a, b, c, d, *then the value of the cross-ratio* (a b c d) *is independent of the position of* O *on* K (Fig. 97).

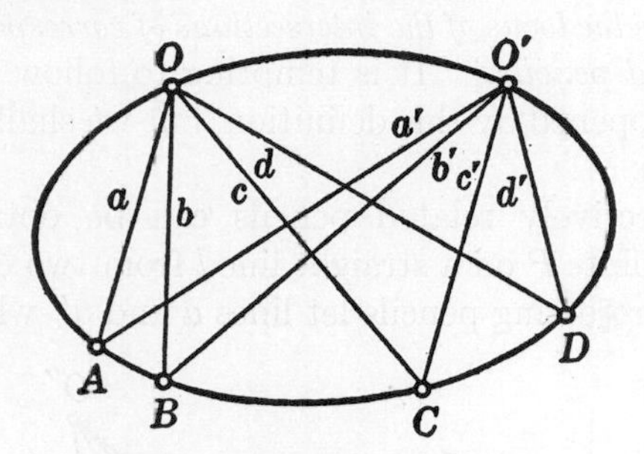

Fig. 97. Cross-ratios on an ellipse.

This is indeed a remarkable result. We already knew that any four given points on a straight line appear under the same cross-ratio from any fifth point O. This theorem on cross-ratios is the basic fact of projective geometry. Now we learn that the same is true of four points on a conic, with one important restriction: the fifth point is no longer absolutely free in the plane, but is still free to move on the given conic.

It is not difficult to prove a converse of this result in the following form: if there are two points O, O' on a curve K such that every quadruple of four points A, B, C, D on K appears under the same cross-ratio from both O and O', then K is a conic, (and therefore A, B, C, D appear under the same cross-ratio from any third point O'' of K). The proof is omitted here.

These projective properties of the conics suggest a general method for constructing such curves. By a *pencil of lines* we shall mean the set of all straight lines in a plane which pass through a given point O. Now consider the pencils through two points O and O' which are chosen to lie on a conic K. Between the lines of pencil O and those of pencil O' we may establish a biunique correspondence by coupling a line a of O with a line a' of O' whenever a and a' meet in a point A of the conic K. Then any four lines a, b, c, d of the pencil O will have the same cross-ratio as the four corresponding lines a', b', c', d' of O'. Any biunique correspondence between two pencils of lines which has this property is called a *projective correspondence.* (This definition is obviously the dual of the definition given on p. 178 of a projective correspondence between the points on two lines.) Pencils between which there is defined a projective correspondence are said to be projectively related. With this definition we can now state: The conic K is the locus of the intersections of corresponding lines of two projectively related pencils. This theorem provides the basis for a purely projective definition of the conics: *A conic is the locus of the intersections of corresponding lines in two projectively related pencils.*† It is tempting to follow the path into the theory of conics opened by this definition, but we shall confine ourselves to a few remarks.

Pairs of projectively related pencils can be obtained as follows. Project all the points P on a straight line l from two different centers O and O''; in the projecting pencils let lines a and a'' which intersect on l

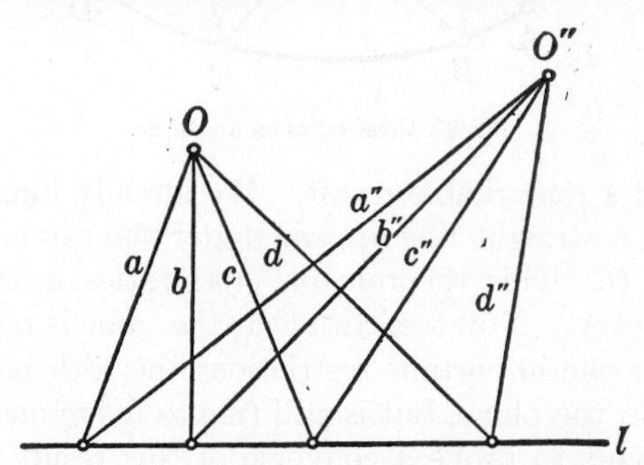

Fig. 98. Preliminary to construction of projectively related pencils.

correspond to each other. Then the two pencils will be projectively related. Now take the pencil O'' and transport it rigidly into any position O'. The resulting pencil O' will be projectively related to O. Moreover, any projective correspondence between two pencils can be so obtained. (This fact is the dual of Ex. 1 on p. 179.) If the pencils O and O' are congruent, we obtain a circle. If angles are equal but with opposite sense, the conic is an equilateral hyperbola (see Fig. 99).

Note that this definition of conic may yield a locus which is a straight line, as in Figure 98. In this case the line $O\,O''$ corresponds to itself,

† This locus may, under certain circumstances, degenerate into a straight line; see Fig. 98.

and all its points are counted as belonging to the locus. Hence the conic degenerates into a pair of lines, which agrees with the fact there are sections of a cone (those obtained by planes through the vertex) which consist of two lines.

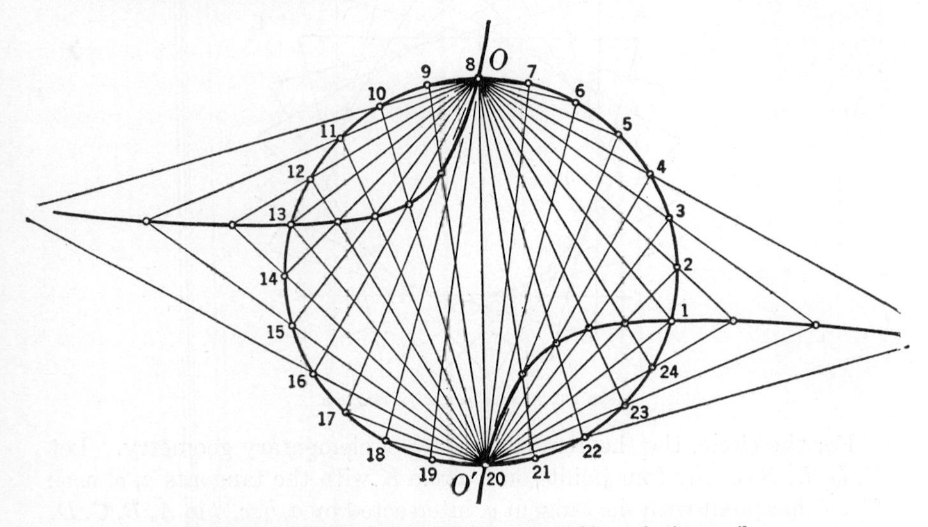

Fig. 99. Circle and equilateral hyperbola generated by projective pencils.

Exercises: 1) Draw ellipses, hyperbolas, and parabolas by means of projective pencils. (The reader is strongly urged to experiment with such constructions. They will contribute greatly to his understanding.)

2) Given five points, O, O', A, B, C, of an unknown conic K. It is required to construct the point D where a given line d through O intersects K. (Hint: Consider through O the rays a, b, c given by OA, OB, OC, and similarly through O' the rays a', b', c'. Draw through O the ray d and construct through O' the ray d' such that $(a, b, c, d) = (a', b', c', d')$. Then the intersection of d and d' is necessarily a point of K.)

3. Conics as Line Curves

The concept of tangent to a conic belongs to projective geometry, for a tangent to a conic is a straight line that touches the conic in only one point, and this property is unchanged by projection. The projective properties of tangents to conics are based on the following fundamental theorem: *The cross-ratio of the points of intersection of any four fixed tangents to a conic with a fifth tangent is the same for every position of the fifth tangent.*

The proof of this theorem is very simple. Since a conic is a projection of a circle, and since the theorem concerns only properties which are invariant under projection, a proof for the case of the circle will suffice to establish the theorem in general.

Fig. 100. A circle as a set of tangents.

For the circle, the theorem is a matter of elementary geometry. Let P, Q, R, S be any four points on a circle K with the tangents a, b, c, d; T another point with the tangent o, intersected by a, b, c, d in A, B, C, D. If M is the center of the circle, then obviously $\measuredangle\ TMA = \frac{1}{2} \measuredangle\ TMP$,

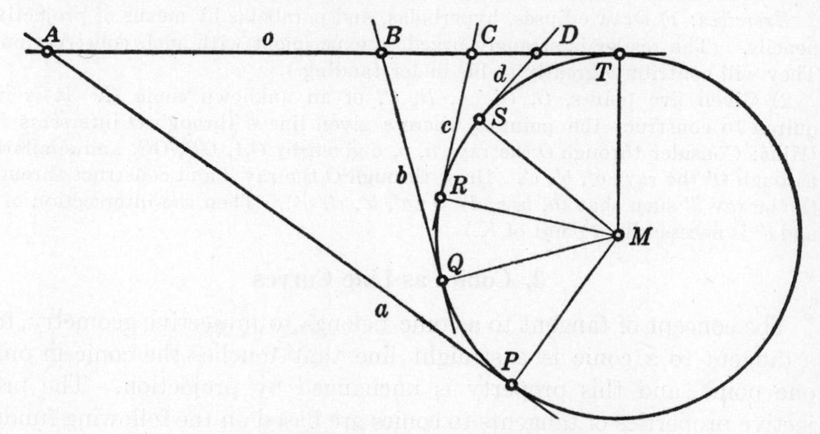

Fig. 101. The tangent property of the circle.

and $\frac{1}{2} \measuredangle\ TMP$ is equal to the angle subtended by the arc TP at a point of K. Similarly, $\measuredangle\ TMB$ is the angle subtended by the arc TQ at a point of K. Therefore $\measuredangle\ AMB = \frac{1}{2}\widehat{PQ}$, where $\frac{1}{2}\widehat{PQ}$ is the angle subtended by the arc PQ at a point of K. Hence the points A, B, C, D are projected from M by four rays whose angles are given by the fixed positions of P, Q, R, S. It follows that the cross-ratio $(A\ B\ C\ D)$ depends only on the four tangents a, b, c, d and not on the particular position of the fifth tangent o. This is exactly the theorem that we had to prove.

In the preceding section we have seen that a conic may be constructed by marking the points of intersection of corresponding lines in two projectively related pencils. The theorem just proved enables us to dualize this construction. Let us take two tangents a and a' of a conic K. A third tangent t will intersect a and a' in two points A and A' respectively. If we allow t to move along the conic, this will set up a correspondence

$$A \leftrightarrow A'$$

between the points of a and those of a'. This correspondence between the points of a and those of a' will be projective, for by our theorem any four points of a will have the same cross-ratio as the corresponding four points of a'. Hence it appears that *a conic* K, regarded as the set of its tangents, *consists of the lines which join corresponding points of the two projectively related ranges*† *of points on* a *and* a′.

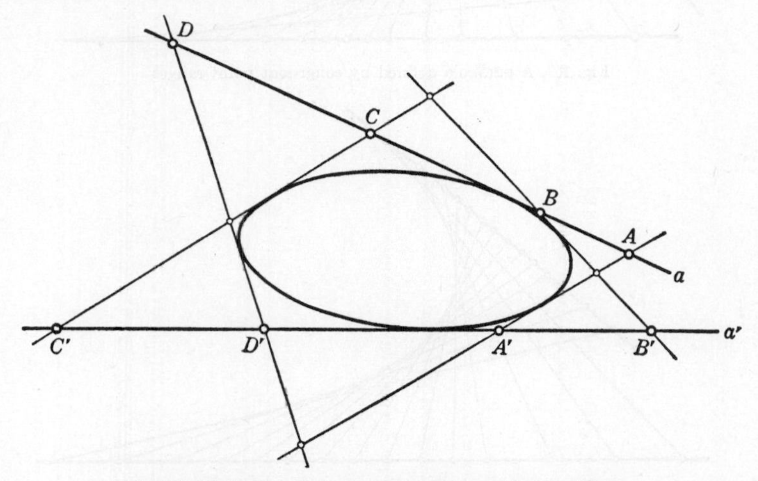

Fig. 102. Projective point ranges on two tangents of an ellipse.

† The set of points on a straight line is called a *range of points*. This is the dual of a pencil of lines.

This fact may be used to give a projective definition of a conic as a "line curve." Let us compare it with the projective definition of a conic given in the preceding section:

I	II
A conic as a set of *points* consists of the *points of intersection of corresponding lines* in two projectively related *pencils of lines.*	A conic as a set of *lines* consists of the *lines joining corresponding points* in two projectively related *ranges of points.*

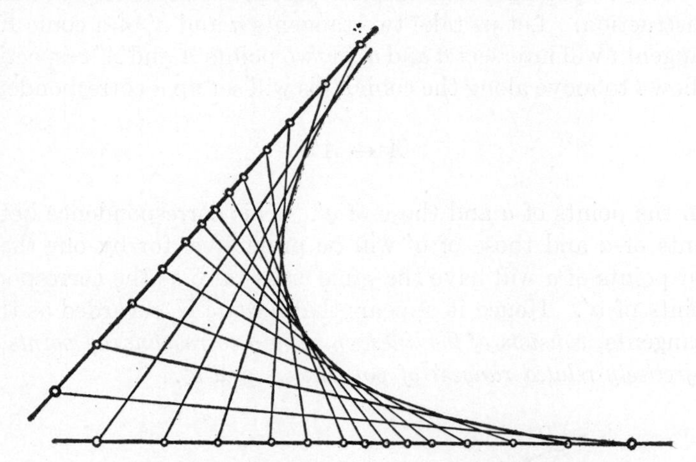

Fig. 103. A parabola defined by congruent point ranges.

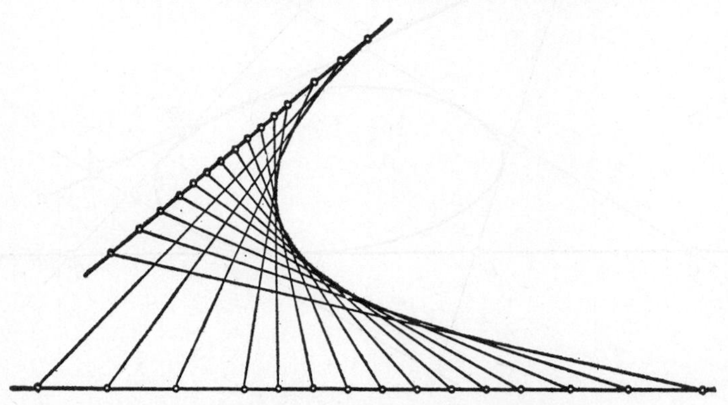

Fig. 104. A parabola defined by similar point ranges.

If we regard the tangent to a conic at a point as the dual element to the point itself, and if we consider a "line curve" (the set of all its tangents) as the dual of a "point curve" (the set of all its points), then the complete duality between these two statements is apparent. In the translation from one statement to the other, replacing each concept by its dual, the word "conic" remains the same: in one case it is a "point conic," defined by its points; in the other a "line conic," defined by its tangents. (See Fig. 100, p. 206.)

An important consequence of this fact is that the principle of duality in plane projective geometry, originally stated for points and lines only, may now be extended to cover conics. *If, in the statement of any theorem concerning points, lines, and conics, each element is replaced by its dual* (keeping in mind that the dual of a point on a conic is a tangent to the conic), *the result will also be a true theorem.* An example of the working of this principle will be found in Article 4 of this section.

The construction of conics as line curves is shown in Figures 103–104. If, on the two projectively related point ranges, the two points at infinity correspond to each other (as must be the case with congruent or similar† ranges), the conic will be a parabola; the converse is also true.

Exercise: Prove the converse theorem: On any two fixed tangents of a parabola a moving tangent cuts out two similar point ranges.

4. Pascal's and Brianchon's General Theorems for Conics

One of the best illustrations of the duality principle for conics is the relation between the general theorems of Pascal and of Brianchon. The first was discovered in 1640, the second only in 1806. Yet one is an immediate consequence of the other, since any theorem involving only conics, straight lines, and points must remain true if replaced by its dual statement.

The theorems stated in §5 under the same name are degenerate cases of the following more general theorems:

Pascal's theorem: The opposite edges of a hexagon inscribed in a conic meet in three collinear points.

Brianchon's theorem: The three diagonals joining opposite vertices of a hexagon circumscribed about a conic are concurrent.

Both theorems are clearly of a projective character. Their dual nature becomes obvious if they are formulated as follows:

Pascal's theorem: Given six points, 1, 2, 3, 4, 5, 6, on a conic. Join successive points by the lines (1, 2), (2, 3), (3, 4), (4, 5), (5, 6), (6, 1).

† It is obvious what is meant by a "congruent" or a "similar" correspondence between two ranges of points.

Mark the points of intersection of (1, 2) with (4, 5), (2, 3) with (5, 6), and (3, 4) with (6, 1). Then these three points of intersection lie on a straight line.

Brianchon's theorem: Given six tangents, 1, 2, 3, 4, 5, 6, to a conic. Successive tangents intersect in the points, (1, 2), (2, 3), (3, 4), (4, 5), (5, 6), (6, 1). Draw the lines joining (1, 2) with (4, 5), (2, 3) with (5, 6), and (3, 4) with (6, 1). Then these lines go through a point.

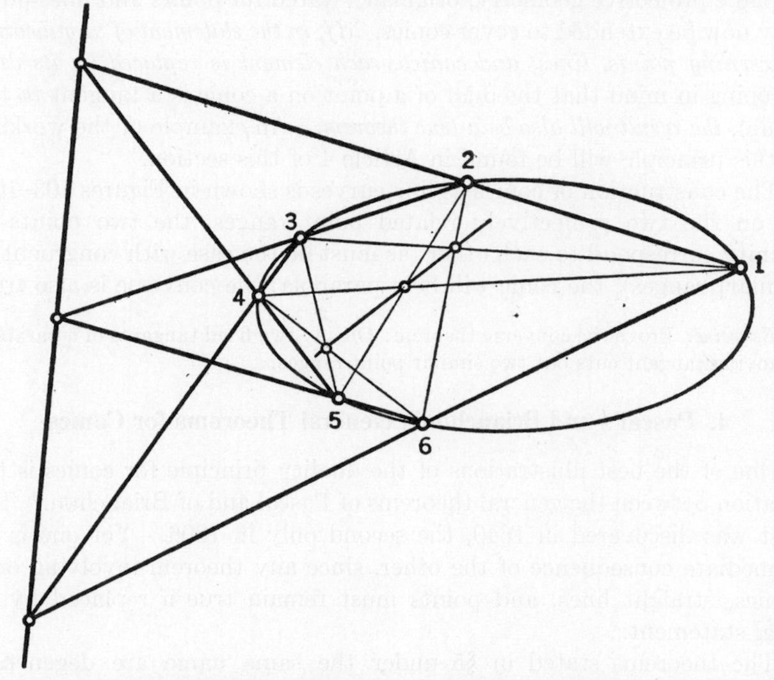

Fig 105. Pascal's general configuration. Two cases are illustrated; one for the hexagon 1, 2, 3, 4, 5, 6, and one for the hexagon 1, 3, 5, 2, 6, 4.

The proofs can be given by a specialization similar to that used in the degenerate cases. To prove Pascal's theorem, let A, B, C, D, E, F be the vertices of a hexagon inscribed in a conic K. By projection we can make AB parallel to ED and FA parallel to CD, so that we obtain the configuration of Figure 107. (For convenience in representation the hexagon is taken as self-intersecting, although this is not necessary.) Pascal's theorem now reduces to the simple statement that CB is parallel to FE; in other words, the line on which the opposite edges of the hexagon meet is the line at infinity. To prove this, let us consider

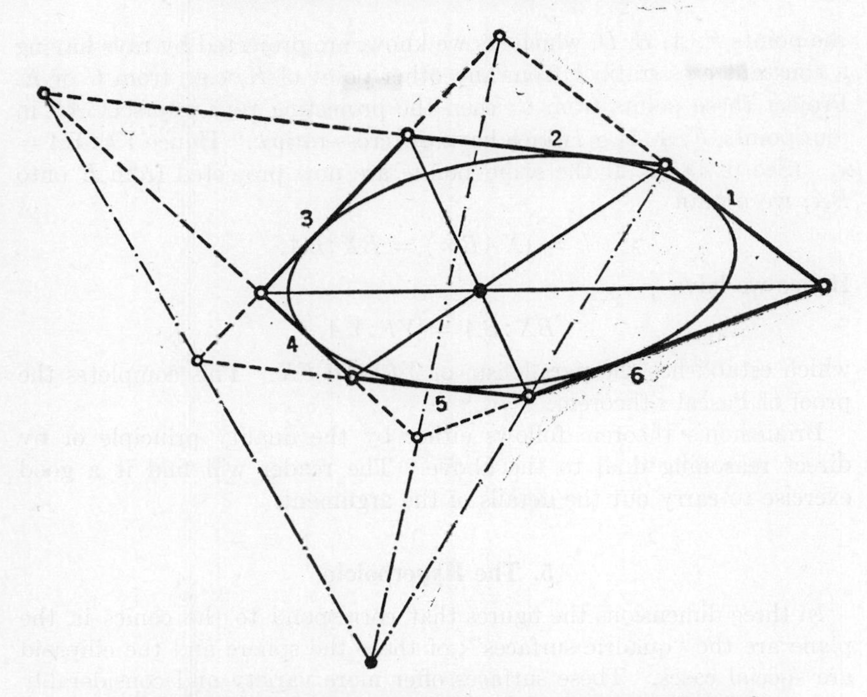

Fig. 106. Brianchon's general configuration. Again two cases are illustrated.

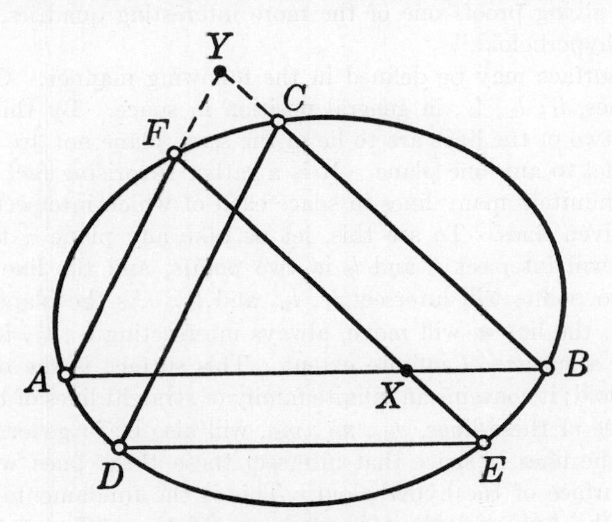

Fig. 107. Proof of Pascal's theorem.

the points F, A, B, D, which, as we know, are projected by rays having a constant cross-ratio k from any other point of K, e.g., from C or E. Project these points from C; then the projecting rays intersect AF in four points, F, A, Y, ∞, which have the cross-ratio k. Hence $YF:YA = k$. (See p. 185.) If the same points are now projected from E onto BA, we obtain

$$k = (XAB\infty) = BX:BA.$$

Hence we have

$$BX:BA = YF:YA,$$

which establishes the parallelism of YB and FX. This completes the proof of Pascal's theorem.

Brianchon's theorem follows either by the duality principle or by direct reasoning dual to the above. The reader will find it a good exercise to carry out the details of the argument.

5. The Hyperboloid

In three dimensions the figures that correspond to the conics in the plane are the "quadric surfaces"; of these the sphere and the ellipsoid are special cases. These surfaces offer more variety and considerably more difficulty than do the conics. Here we shall discuss briefly and without giving proofs one of the more interesting quadrics, the "one-sheeted hyperboloid."

This surface may be defined in the following manner. Choose any three lines, l_1, l_2, l_3, in general position in space. By this we mean that no two of the lines are to lie in the same plane nor are they all to be parallel to any one plane. It is a rather surprising fact that there will be infinitely many lines in space each of which intersects all three of the given lines. To see this, let us take any plane π through l_1. Then π will intersect l_2 and l_3 in two points, and the line m joining these two points will intersect l_1, l_2, and l_3. As the plane π rotates about l_1, the line m will move, always intersecting l_1, l_2, l_3, and will generate a surface of infinite extent. This surface is the one-sheeted hyperboloid; it contains an infinite family of straight lines of the type m. Any three of these lines, m_1, m_2, m_3, will also be in general position, and all the lines in space that intersect these three lines will also lie in the surface of the hyperboloid. This is the fundamental fact concerning the hyperboloid: it is made up of two different families of straight lines; every three lines of the same family are in general position, while each line of one family intersects all the lines of the other family.

An important projective property of the hyperboloid is that the cross-

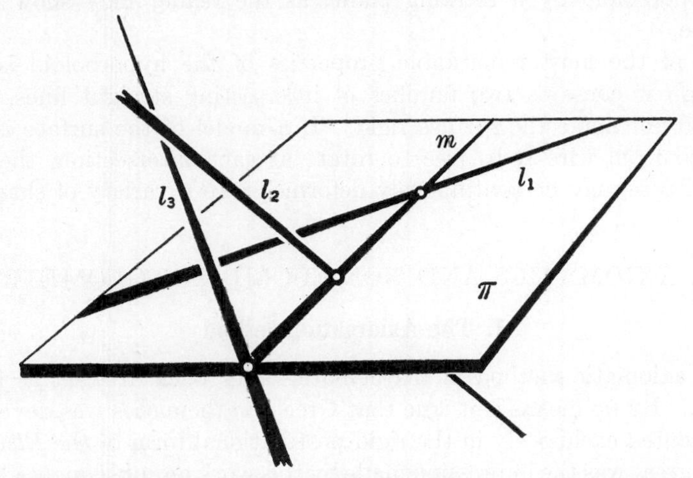

Fig. 108. Construction of lines intersecting three fixed lines in general position.

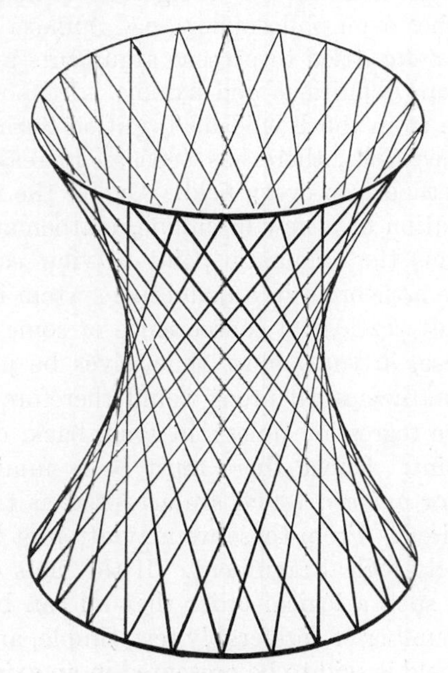

Fig. 109. The hyperboloid.

ratio of the four points where any four given lines of one family intersect a given line of the other family is independent of the position of the latter line. This follows directly from the method of construction of the hyperboloid by a rotating plane, as the reader may show as an exercise.

One of the most remarkable properties of the hyperboloid is that although it contains two families of intersecting straight lines, these lines do not make the surface rigid. If a model of the surface is constructed from wire rods, free to rotate at each intersection, then the whole figure may be continuously deformed into a variety of shapes.

§9. AXIOMATICS AND NON-EUCLIDEAN GEOMETRY

1. The Axiomatic Method

The axiomatic method in mathematics goes back at least as far as Euclid. By no means is it true that Greek mathematics was developed or presented exclusively in the rigid postulational form of the *Elements*. But so great was the impression made by this work on subsequent generations that it became a model for all rigorous demonstration in mathematics. Sometimes even philosophers, e.g. Spinoza in his *Ethica, more geometrico demonstrata,* tried to present arguments in the form of theorems deduced from definitions and axioms. In modern mathematics, after a departure from the Euclidean tradition during the seventeenth and eighteenth centuries, there has been an increasing penetration of the axiomatic method into every field. One of the most recent results has been the creation of a new discipline, mathematical logic.

In general terms the axiomatic point of view can be described as follows: To prove a theorem in a deductive system is to show that the theorem is a necessary logical consequence of some previously proved propositions; these, in turn, must themselves be proved; and so on. The process of mathematical proof would therefore be the impossible task of an infinite regression unless, in going back, one is permitted to stop at some point. Hence there must be a number of statements, called *postulates* or *axioms,* which are accepted as true, and for which proof is not required. From these we may attempt to deduce all other theorems by purely logical argument. If the facts of a scientific field are brought into such a logical order that all can be shown to follow from a selected number of (preferably few, simple, and plausible) statements, then the field is said to be presented in an axiomatic form. The choice of the propositions selected as axioms is to a large extent arbitrary. But little is gained by the axiomatic method unless the postulates are simple and not too great in number. Moreover, the postulates must be *consistent,* in the sense that no two theorems deducible from them can be mutually contradictory, and *complete,* so that every theorem of the system is deducible from them. For reasons of economy it is also desirable that the postulates be *independent,* in the sense that no one of them is a logical consequence of the others. The question of the consistency and of the completeness of a set of axioms has been the subject of much controversy. Different philosophical convictions concerning the ultimate roots of human knowledge have led to apparently irreconcilable views on the foundations of mathematics. If mathematical entities are considered as substantial objects in a realm of "pure intuition", independent of definitions and of individual acts of the human mind, then of course there can be no contradictions, since mathematical facts are objectively true statements describing existing realities. From this "Kantian" point of view there is no problem of consistency. Unfortunately, however, the actual body of mathematics cannot be fitted into such a simple philosophical framework. The modern mathematical intuitionists do not rely on pure intuition in the broad Kantian sense. They accept the denumerably infinite as the legitimate child of intuition, and they admit only constructive properties; but thus basic concepts such as the number continuum would be banished, important parts of actual mathematics excluded, and the rest almost hopelessly complicated.

Quite different is the view taken by the "formalists." They do not attribute an intuitive reality to mathematical objects, nor do they claim that axioms express obvious truths concerning the realities of pure intuition; their concern is only with the formal logical procedure of reasoning on the basis of postulates. This attitude has a definite advantage over intuitionism, since it grants to mathematics all the freedom necessary for theory and applications. But it imposes on the formalist the necessity of proving that his axioms, now appearing as arbitrary creations of the human mind, cannot possibly lead to a contradiction. Great efforts have been made during the last twenty years to find such consistency proofs, at least for the axioms of arithmetic and algebra and for the concept of the number continuum. The results are highly significant, but success is still far off. Indeed, recent results indicate that such efforts cannot be completely successful, in the sense that proofs for consistency and completeness are not possible within strictly closed systems of concepts. Remarkably enough, all these arguments on foundations proceed by methods that in themselves are thoroughly constructive and directed by intuitive patterns.

Accentuated by the paradoxes of set theory (see p. 87), the clash between the intuitionists and the formalists has been much publicized by passionate partisans of these schools. The mathematical world has resounded with a cry about the "crisis in the foundations." But the alarm was not, and must not be, taken too seriously. With all credit to the achievements produced in the struggle for clarification of the foundations, it would be completely unjustified to infer that the living body of mathematics is in the least threatened by such differences of opinion or by the paradoxes inherent in an uncontrolled drift towards boundless generality.

Quite apart from philosophical considerations and from interest in foundations, the axiomatic approach to a mathematical subject is the natural way to unravel the network of interconnections between the various facts and to exhibit the essential logical skeleton of the structure. It sometimes happens that such a concentration on the formal structure rather than on the intuitive meaning of the concepts makes it easier to find generalizations and applications that might have been overlooked in a more intuitive approach. But a significant discovery or an illuminating insight is rarely obtained by an exclusively axiomatic procedure. Constructive thinking, guided by the intuition, is the true source of mathematical dynamics. Although the axiomatic form is an ideal, it is a dangerous fallacy to believe that axiomatics constitutes *the* essence of mathematics. The constructive intuition of the mathematician brings to mathematics a non-deductive and irrational element which makes it comparable to music and art.

Since the days of Euclid, geometry has been the prototype of an axiomatized discipline. For centuries Euclid's set of axioms has been the object of intensive study. But only recently has it become apparent that his postulates must be modified and completed if all of elementary geometry is to be deducible from them. Late in the nineteenth century, for example, Pasch discovered that the ordering of points on a line, the notion of "betweenness," requires a special postulate. Pasch formulated the following statement as an axiom: A straight line that intersects one side of a triangle in any point other than a vertex must also intersect another side of the triangle. (Lack of regard for such details leads to many apparent paradoxes in which absurd consequences —e.g. the well-known "proof" that every triangle is isosceles—seem to be deduced rigorously from Euclid's axioms. This is usually done on the basis of an improperly drawn figure whose lines seem to intersect inside or outside certain triangles or circles, whereas they really do not.)

In his famous book, *Grundlagen der Geometrie* (first edition published in 1901), Hilbert gave a satisfactory set of axioms for geometry and at the same time made an exhaustive study of their mutual independence, consistency, and completeness.

Into any set of axioms there must enter certain undefined concepts, such as "point" and "line" in geometry. Their "meaning" or connection with objects of the physical world is *mathematically* unessential. They can be regarded as purely abstract entities whose mathematical properties in a deductive system are given entirely by the relations that hold among them as stated by the axioms. For example, in projective geometry we might begin with the undefined concepts of "point," "line," and "incidence," and with the two dual axioms: "Each two distinct points are incident with a unique line" and "Each two distinct lines are incident with a unique point." From the point of view of axiomatics, the dual form of such axioms is the very source of the principle of duality in projective geometry. Any theorem which contains in its statement and proof only elements connected by dual axioms must admit of dualization. For the proof of the original theorem consists in the successive application of certain axioms, and the application of the dual axioms in the same order will provide a proof for the dual theorem.

The totality of axioms of geometry provides the *implicit definition* of all "undefined" geometrical terms such as "line," "point," "incident," etc. For applications it is important that the concepts and axioms of geometry correspond well with physically verifiable statements about "real," tangible objects. The physical reality behind the concept of "point" is that of a very small object, such as a pencil dot, while a "straight line" is an abstraction from a stretched thread or a ray of light. The properties of these physical points and straight lines are found by experience to agree more or less with the formal axioms of geometry. Quite conceivably more precise experiments might necessitate modification of these axioms if they are adequately to describe physical phenomena. But if the formal axioms did not agree more or less with the properties of physical objects, then geometry would be of little interest. Thus, even for the formalist, there is an authority other than the human mind, that decides the direction of mathematical thought.

2. Hyperbolic Non-Euclidean Geometry

There is one axiom of Euclidean geometry whose "truth," that is, whose correspondence with empirical data about stretched threads or light rays, is by no means obvious. This is the famous *postulate of the unique parallel*, which states that through any point not on a given line *one and only one* line can be drawn parallel to the given line. The remarkable feature of this axiom is that it makes an assertion about the *whole* extent of a straight line, imagined as extending indefinitely in either direction; for to say that two lines are parallel is to say that they never intersect, no matter how far they may be produced. It goes without saying that there are many lines through a point which do not intersect a given line *within any fixed finite distance*, however large. Since the maximum possible length of an actual ruler, thread, or even a light ray visible to a telescope is certainly finite, and since within any finite circle there are infinitely many straight lines through a given point and not intersecting a given line inside the circle, it follows that this axiom can never be verified by experiment. All the other axioms of Euclidean geometry have a finite character in that they deal with finite portions of lines and with plane figures of finite extent. The fact that the parallel axiom is not experimentally verifiable raises the question of whether or not it is *independent* of the other axioms. If it were a necessary logical consequence of the others, then it would be possible to strike it out as an axiom and to give a proof of it in terms of the other Euclidean axioms. For centuries mathematicians tried to find such a proof, because of the widespread feeling among students of geometry that the parallel postulate is of a character essentially different from the others, lacking the sort of compelling plausibility which an axiom of geometry should possess. One of the first attempts of this nature was made by Proclus (fourth century A.D.), a commentator on Euclid, who tried to dispense with the need for a special parallel postulate by *defining* the parallel to a given line to be the locus of all points at a given fixed distance from the line. In this he failed to observe that the difficulty was only shifted to another place, for it would then be necessary to prove that the locus of such points is in fact a straight line. Since Proclus could not prove this, he would have to accept it instead of the parallel axiom as a postulate, and nothing would be gained, for the two are easily seen to be equivalent. The Jesuit Saccheri (1667–1733), and later Lambert (1728–1777), tried to prove the parallel postulate by the indirect method of assuming the contrary and drawing absurd consequences. Far from being absurd, their conclusions really amounted to theorems of the non-Euclidean geometry developed later. Had they regarded them not as absurdities, but rather as self-consistent statements, they would have been the discoverers of non-Euclidean geometry.

At that time, any geometrical system not absolutely in accordance with Euclid's would have been considered as obvious nonsense. Kant, the most influential philosopher of the period, formulated this attitude in the statement that Euclid's axioms are inherent in the human mind, and therefore have an objective validity for "real" space. This belief in the axioms of Euclidean geometry as unalterable truths, existing in the realm of pure intuition, was one of the basic tenets of Kant's philosophy. But in the long run, neither old habits of thinking nor philosophical authority could suppress the conviction that the unending record of failure in the search for a proof of the parallel postulate was due not to any lack of ingenuity, but rather to the fact that the parallel postulate is really *independent* of the others. (In much the same way, the lack of success in proving that the general equation of the fifth degree could be solved by radicals led to the suspicion, later verified, that such a solution is impossible.) The Hungarian Bolyai (1802–1860) and the Russian Lobachevsky (1793–1856), settled the question by constructing in all detail a geometry in which the parallel axiom does not hold. When the enthusiastic young genius Bolyai submitted his paper to Gauss, the "prince of mathematicians," for the recognition he so eagerly expected, he was informed that his work had been anticipated by Gauss himself, but that the latter had not cared to publish his results because he dreaded noisy publicity.

What does the independence of the parallel postulate mean? Simply that it is possible to construct a consistent system of "geometrical" statements dealing with points, lines, etc., by deduction from a set of axioms in which the parallel postulate is replaced by a contrary postulate. Such a system is called a non-Euclidean geometry. It required the intellectual courage of Gauss, Bolyai, and Lobachevsky to realize that such a geometry, based on a non-Euclidean system of axioms, can be perfectly consistent.

To show the consistency of the new geometry, it is not enough to deduce a large body of non-Euclidean theorems, as Bolyai and Lobachevsky did. Instead, we have learned to build "models" of such a geometry which satisfy all the axioms of Euclid except for the parallel postulate. The simplest such model was given by Felix Klein, whose work in the field was stimulated by the ideas of the English geometer Cayley (1821–1895). In this model, infinitely many "straight lines" can be drawn "parallel" to a given line through an external point. Such a geometry is called Bolyai-Lobachevskian or "hyperbolic" geometry. (The reason for the latter name will be found on p. 226.)

Klein's model is constructed by first considering objects of ordinary Euclidean geometry and then *renaming* certain of these objects and the relations between them in such a way that a non-Euclidean geometry arises. This must, *eo ipso*, be just as consistent as the original Euclidean geometry, because it is presented to us, seen from another point of view and described with other words, as a body of facts of ordinary Euclidean geometry. This model can be easily understood by means of some concepts of projective geometry.

If we subject the plane to a projective transformation onto another plane, or rather onto itself (by afterwards making the image plane coincide with the original plane), then, in general, a circle and its interior will be transformed into a conic section. But one can easily show (the proof is omitted here) that there exist infinitely many projective transformations of the plane onto itself such that a given circle plus its interior is transformed into itself. By such transformations points of the interior or of the boundary are in general shifted to other positions, but remain inside or on the boundary of the circle. (As a matter of fact, one can move the center of the circle into any other interior point.) Let us consider the totality of such transformations. Certainly they will not leave the shapes of figures invariant, and are therefore not rigid displacements in the usual sense. But now we take the decisive step of *calling* them "non-Euclidean displacements" in the geometry to be constructed. By means of these "displacements" we are able to define congruence—two figures being *called* congruent if there exists a non-Euclidean displacement transforming one into the other.

The Klein model of hyperbolic geometry is then the following: The "plane" consists only of the points interior to the circle; points outside are disregarded. Each point inside the circle is *called* a non-Euclidean "point"; each chord of the circle is *called* a non-Euclidean "straight line"; "displacement" and "congruence" are defined as above; joining "points" and finding the intersection of "straight lines" in the non-Euclidean sense remain the same as in Euclidean geometry. It is an easy matter to show that the new system satisfies all the postulates of Euclidean geometry, with the one exception of the parallel postulate. That the parallel postulate does not hold in the new system is shown by the fact that through any "point" not on a "straight line" infinitely many "straight lines" can be drawn having no "point" in common with the given "line." The first "straight line" is a Euclidean chord of the circle, while the second "straight line" may be any one of the chords which pass through the given "point" and do not intersect the first "line" inside the circle. This simple model is quite sufficient to settle the fundamental question which gave rise to non-Euclidean geometry; it proves that the parallel postulate cannot be deduced from the other axioms of Euclidean geometry. For if it could be so deduced, it would be a true theorem in the geometry of Klein's model, and we have seen that it is not.

Strictly speaking, this argument is based on the assumption that the geometry of Klein's model is consistent, so that a theorem together with its contrary cannot

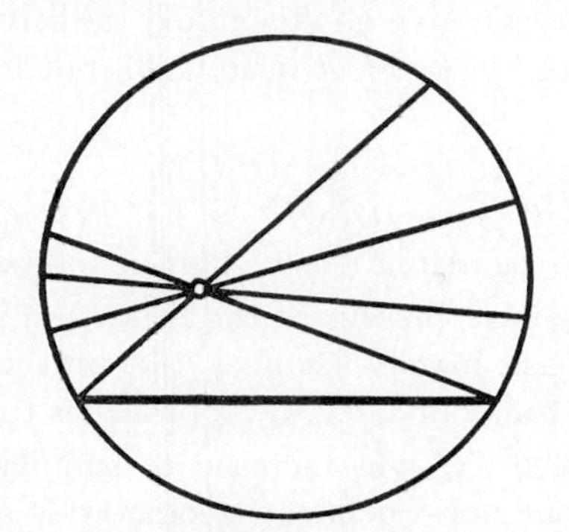

Fig. 110. Klein's non-Euclidean model.

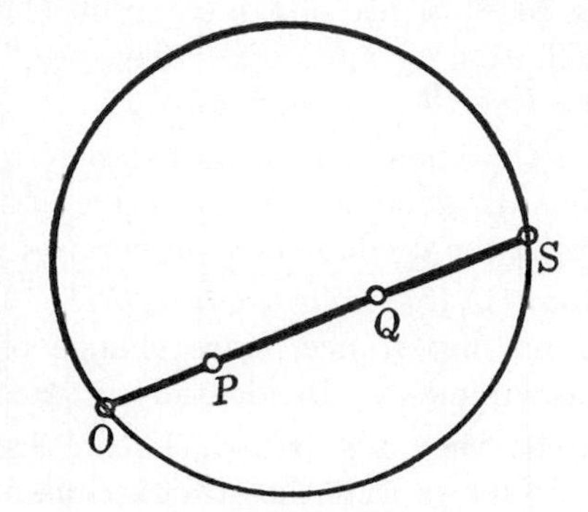

Fig. 111. Non-Euclidean distance.

be proved. But the geometry of Klein's model is certainly as consistent as ordinary Euclidean geometry, since statements concerning "points," "lines," etc. in Klein's model are merely different ways of phrasing certain theorems of Euclidean geometry. A satisfactory proof of the consistency of the axioms of Euclidean geometry has never been given, except by referring back to the concepts of analytic geometry and hence ultimately to the number continuum, whose consistency is again an open question.

* One detail which goes beyond the immediate objective should be mentioned here, namely, how to define non-Euclidean "distance" in Klein's model. This "distance" is required to be invariant under any non-Euclidean "displacement"; for displacement should leave distances invariant. We know that cross-ratios are invariant under projection. A cross-ratio involving two arbitrary points P and Q inside the circle presents itself immediately if the segment PQ is extended to meet the

circle in O and S. The cross-ratio $(OSQP)$ of these four points is a (positive) number, which one might hope to take as the definition of the "distance" PQ between P and Q. But this definition must be modified slightly to make it workable. For if the three points P, Q, R are on a line, it should be true that $\overline{PQ} + \overline{QR} = \overline{PR}$. Now in general

$$(OSQP) + (OSRQ) \neq (OSRP).$$

Instead, we have the relation

$$(OSQP)(OSRQ) = (OSRP), \tag{1}$$

as is seen from the equations

$$(OSQP)(OSRQ) = \frac{QO/QS}{PO/PS} \cdot \frac{RO/RS}{QO/QS} = \frac{RO/RS}{PO/PS} = (OSRP).$$

In consequence of the equation (1) we can give a satisfactory additive definition by measuring "distance," not by the cross-ratio itself, but by the *logarithm of the cross-ratio*:

$$\overline{PQ} = \text{non-Euclidean distance from } P \text{ to } Q = \log\,(OSQP).$$

This distance will be a positive number, since $(OSQP) > 1$ if $P \neq Q$. Using the fundamental property of the logarithm (see p. 444), it follows from (1) that $\overline{PQ} + \overline{QR} = \overline{PR}$. The base chosen for the logarithm is of no importance, since change of base merely changes the unit of measurement. Incidentally, if one of the points, e.g. Q, approaches the circle, then the non-Euclidean distance $\overline{PQ}$ will increase to infinity. This shows that the straight line of our non-Euclidean geometry is of infinite non-Euclidean length, although in the ordinary Euclidean sense it is only a finite segment of a straight line.

3. Geometry and Reality

The Klein model shows that hyperbolic geometry, viewed as a formal deductive system, is as consistent as the classical Euclidean geometry. The question then arises, which of the two is to be preferred as a description of the geometry of the physical world? As we have already seen, experiment can never decide whether there is but one or whether there are infinitely many straight lines through a point and parallel to a given line. In Euclidean geometry, however, the sum of the angles of any triangle is 180°, while it can be shown that in hyperbolic geometry the sum is less than 180°. Gauss accordingly performed an experiment to settle the question. He accurately measured the angles in a triangle formed by three fairly distant mountain peaks, and found the angle-sum to be 180°, within the limits of experimental error. Had the result been noticeably less than 180°, the consequence would have been that hyperbolic geometry is preferable to describe physical reality. But, as it turned out, nothing was settled by this experiment, since for small triangles whose sides are only a few miles in length the deviation from 180° in the hyperbolic geometry might be so small as to have been undetectable by Gauss's instruments. Thus, although the experiment was inconclusive, it showed that the Euclidean and hyperbolic geometries, which differ widely *in the large,* coincide so closely for relatively small figures that they are experimentally equivalent. Therefore, as long as purely *local* properties of space are under consideration, the choice between the two geometries is to be made solely on the basis of simplicity and convenience. Since the Euclidean system is rather simpler to deal with, we are justified in using it exclusively, as long as fairly small distances (of a few million miles!) are under consideration. But we should not necessarily expect it to be suitable for describing the universe as a whole, in its largest aspects. The situation here is precisely analogous to that which exists in physics, where the systems of Newton and Einstein give the same results for small distances and velocities, but diverge when very large magnitudes are involved.

The revolutionary importance of the discovery of non-Euclidean geometry lay in the fact that it demolished the notion of the axioms of Euclid as the immutable mathematical framework into which our experimental knowledge of physical reality must be fitted.

4. Poincaré's Model

The mathematician is free to consider a "geometry" as defined by any set of consistent axioms about "points," "straight lines," etc.; his investigations will be useful to the physicist only if these axioms correspond to the physical behavior of objects in the real world. From this point of view we wish to examine the meaning of the statement "light travels in a straight line." If this is regarded as the *physical definition* of "straight line," then the axioms of geometry must be so chosen as to correspond with the behavior of light rays. Let us imagine, with Poincaré, a world composed of the interior of a circle C, and such that the velocity of light at any point inside the circle is equal to the distance of that point from the circumference. It can be proved that rays of light will then take the form of circular arcs perpendicular at their extremities to the circumference C. In such a world, the geometrical properties of "straight lines" (defined as light rays) will differ from the Euclidean properties of straight lines. In particular, the parallel axiom will not hold, since there will be infinitely many "straight lines" through any point which do not intersect a given "straight line." As a matter of fact, the "points" and "straight lines" in this world will have exactly the geometrical properties of the "points" and "lines" of the Klein model. In other words, we shall have a different model of a hyperbolic geometry. But Euclidean geometry will also apply in this world; instead of being non-Euclidean "straight lines," the light rays would be Euclidean circles perpendicular to C. Thus we see that different systems of geometry can describe the same physical situation, provided

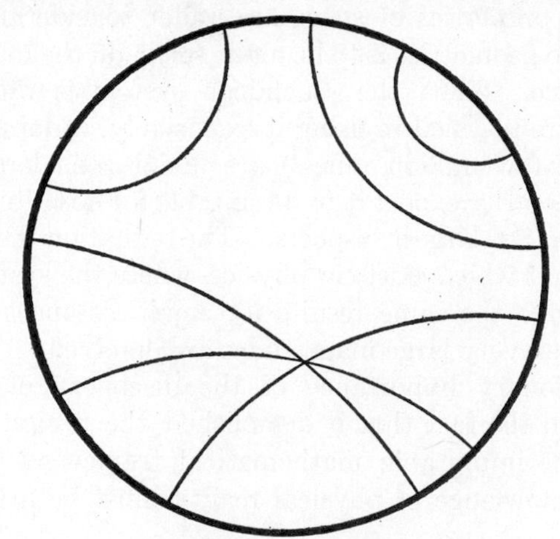

Fig. 112. Poincaré's non-Euclidean model.

that the physical objects (in this case, light rays) are correlated with different concepts of the two systems:

light ray → "straight line"—hyperbolic geometry

light ray → "circle"—Euclidean geometry.

Since the concept of a straight line in Euclidean geometry corresponds to the behavior of a light ray in a homogeneous medium, we would say that the geometry of the region inside C is hyperbolic, meaning only that the physical properties of light rays in this world correspond to the properties of the "straight lines" of hyperbolic geometry.

5. Elliptic or Riemannian Geometry

In Euclidean geometry, as well as in the hyperbolic or Bolyai-Lobachevskian geometry, the tacit assumption is made that the line is infinite (the infinite extent of the line is essentially tied up with the concept and the axioms of "betweenness"). But after hyperbolic geometry had opened the way for freedom in constructing geometries, it was only natural to ask whether different non-Euclidean geometries could be constructed in which a straight line is not infinite but finite and closed. Of course, in such geometries not only the parallel postulate, but also the axioms of "betweenness" will have to be abandoned. Modern developments have brought out the physical importance of these geometries. They were first considered in the inaugural address delivered in 1851 by Riemann upon his admission as an unpaid instructor ("Privat-Docent") at the University of Goettingen. Geometries with closed finite lines can be constructed in a completely consistent

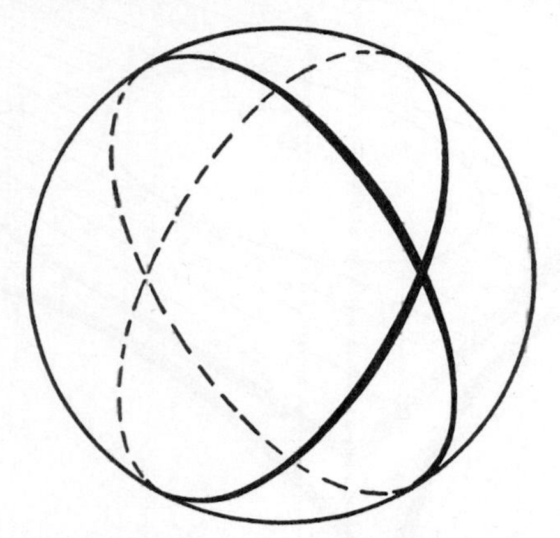

Fig. 113. "Straight lines" in a Riemannian geometry.

way. Let us imagine a two-dimensional world consisting of the surface S of a sphere, in which we define "straight line" to mean great circle of the sphere. This would be the natural way to describe the world of a navigator, since the arcs of great circles are the curves of shortest length between two points on a sphere and this is a characteristic property of straight lines in the plane. In such a world, *every* two "straight lines" intersect, so that from an external point *no* line can be drawn parallel to (i.e. not intersecting) a given "straight line." The geometry of "straight lines" in this world is called an *elliptic geometry*. In this geometry, the distance between two points is measured simply by the distance along the shorter arc of the great circle connecting the points. Angles are measured as in Euclidean geometry. We generally consider as typical of an elliptic geometry the fact that no parallel exists to a line.

Following Riemann, we can generalize this geometry as follows. Let us consider a world consisting of a curved surface in space, not necessarily a sphere, and let us define the "straight line" joining any two points to be the *curve of shortest length* or "geodesic" joining these points. The points of the surface can be divided into two classes:—1. Points in the neighborhood of which the surface is like a sphere in that it lies wholly on one side of the tangent plane at the point. 2. Points in the neighborhood of which the surface is saddle-shaped, and lies on both sides of the tangent plane at the point. Points of the first kind

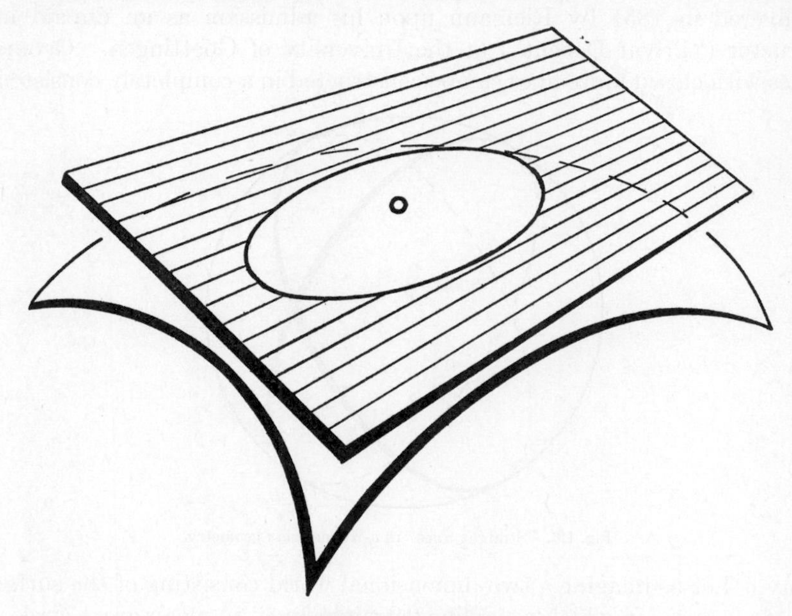

Fig. 114. Elliptic point.

are called elliptic points of the surface, since, if the tangent plane is shifted slightly parallel to itself, it intersects the surface in an elliptical curve; while points of the second kind are called hyperbolic, since, if the tangent plane is shifted slightly parallel to itself, it intersects the surface in a curve resembling a hyperbola. The geometry of the geodesic "straight lines" in the neighborhood of a point of the surface is elliptic or hyperbolic according as the point is an elliptic or hyperbolic point. In such a model of non-Euclidean geometry, angles are measured by their ordinary Euclidean value.

This idea was developed by Riemann, who considered a geometry of space analogous to this geometry of a surface, in which the "curvature" of space may change the character of the geometry from point to point. The "straight lines" in a Riemannian geometry are the geodesics. In Einstein's general theory of relativity the geometry of space is a Riemannian geometry, light travels along geodesics, and the curvature of space is determined by the nature of the matter that fills it.

From its origin in the study of axiomatics, non-Euclidean geometry has developed into an exceedingly useful instrument for application to the physical world. In the theory of relativity, in optics, and in the general theory of wave propagation, a non-Euclidean description of phenomena is sometimes far more adequate than a Euclidean one.

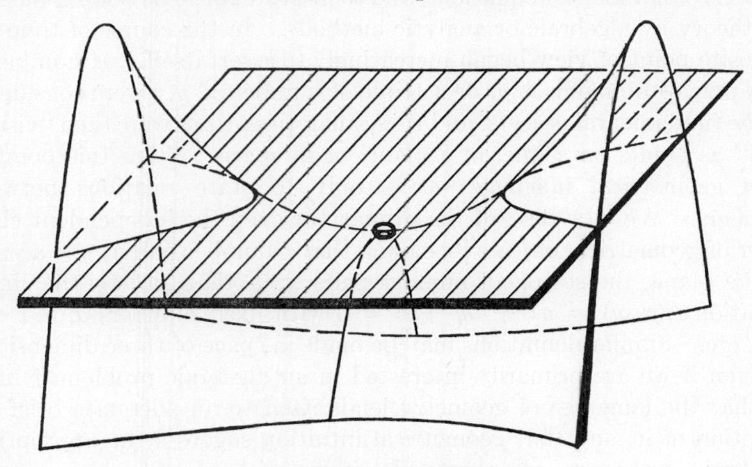

Fig. 115. Hyperbolic point.

APPENDIX

*GEOMETRY IN MORE THAN THREE DIMENSIONS

1. Introduction

The "real space" that is the medium of our physical experience has three dimensions, the plane has two dimensions, and the line one. Our spatial intuition in its ordinary sense is definitely limited to three dimensions. Still, on many occasions it is quite convenient to speak of "spaces" having four or more dimensions. What is the meaning of an n-dimensional space when n is greater than three, and what purposes can it serve? An answer can be given from the analytic as well as from the purely geometric point of view. The terminology of n-dimensional space may be regarded merely as a suggestive geometric language for mathematical ideas that are no longer within reach of ordinary geometric intuition. We shall give a brief indication of the simple considerations that motivate and justify this language.

2. Analytic Approach

We have already remarked on the inversion of meaning which came about in the course of development of analytic geometry. Points, lines, curves, etc. were originally considered to be purely "geometrical" entities, and the task of analytic geometry was merely to assign to them systems of numbers or equations, and to interpret or to develop geometrical theory by algebraic or analytic methods. In the course of time the opposite point of view began increasingly to assert itself. A number x, or a pair of numbers x, y, or a triple of numbers x, y, z were considered as the fundamental objects, and these analytic entities were then "visualized" as points on a line, in a plane, or in space. From this point of view geometrical language serves only to state relations between numbers. We may discard the primary or even the independent character of geometrical objects by saying that a number pair x, y *is* a point in the plane, the set of all number pairs x, y that satisfy the linear equation $L(x, y) = ax + by + c = 0$ with fixed numbers a, b, c *is* a line, etc. Similar definitions may be made in space of three dimensions.

Even if we are primarily interested in an algebraic problem, it may be that the language of geometry lends itself to an adequate brief description of it, and that geometrical intuition suggests the appropriate algebraic procedure. For example, if we wish to solve three simultaneous linear equations for three unknown quantities x, y, z:

$$L(x, y, z) = ax + by + cz + d = 0$$

$$L'(x, y, z) = a'x + b'y + c'z + d' = 0$$

$$L''(x, y, z) = a''x + b''y + c''z + d'' = 0,$$

we may visualize the problem as that of finding the point of intersection in three dimensional space R_3 of the three planes defined by the equations $L = 0$, $L' = 0$, $L'' = 0$. Again, if we are considering only the number pairs x, y for which $x > 0$, we may visualize them as the half-plane to the right of the x-axis. More generally, the totality of number pairs x, y for which

$$L(x, y) = ax + by + d > 0$$

may be visualized as a half-plane on one side of the line $L = 0$, and the totality of number triples x, y, z for which

$$L(x, y, z) = ax + by + cz + d > 0$$

may be visualized as the "half-space" on one side of the plane $L(x, y, z) = 0$.

The introduction of a "four-dimensional space" or even an "n-dimensional space" is now quite natural. Let us consider a quadruple of numbers x, y, z, t. Such a quadruple is said to be represented by, or simply, to *be* a point in four-dimensional space R_4. More generally, a point of n-dimensional space R_n is by definition simply an ordered set of n real numbers $x_1, x_2, \cdots, x_n$. It does not matter that we cannot visualize such a point. The geometrical language remains just as suggestive for algebraic properties involving four or n variables. The reason for this is that many of the algebraic properties of linear equations, etc. are essentially independent of the number of variables involved, or, as we may say, of the dimension of the space of the variables. For example, we call "hyperplane" the totality of all points $x_1, x_2, \cdots, x_n$ in the n-dimensional space R_n which satisfy a linear equation

$$L(x_1, x_2, \cdots, x_n) = a_1x_1 + a_2x_2 + \cdots + a_nx_n + b = 0.$$

Then the fundamental algebraic problem of solving a system of n linear equations in n unknowns,

$$\begin{aligned} L_1(x_1, x_2, \cdots, x_n) &= 0 \\ L_2(x_1, x_2, \cdots, x_n) &= 0 \\ \cdots\cdots\cdots\cdots&\cdots\cdots \\ L_n(x_1, x_2, \cdots, x_n) &= 0, \end{aligned}$$

is stated in geometrical language as that of finding the point of intersection of the n hyperplanes $L_1 = 0$, $L_2 = 0$, $\cdots$, $L_n = 0$.

The advantage of this geometrical mode of expression is only that it emphasizes certain algebraic features which are independent of n *and which are capable of visualization for* n ≤ 3. In many applications the use of such a terminology has the advantage of abbreviating, facilitating, and directing the intrinsically analytic considerations. The theory of relativity may be mentioned as an example where important progress was attained by uniting the space coördinates x, y, z and the time coördinate t of an "event" into a four-dimensional "space-time" manifold of number quadruples x, y, z, t. By the introduction of a non-Euclidean hyperbolic geometry into this analytic framework, it became possible to describe many otherwise complex situations with remarkable simplicity. Similar advantages have accrued in mechanics and statistical physics, as well as in purely mathematical fields.

Here are some examples from mathematics. The totality of all circles in the plane forms a three-dimensional manifold, because a circle with center x, y and radius t can be represented by a point with the coördinates x, y, t. Since the radius of a circle is a positive number, the totality of points representing circles fills out a half-space. In the same way, the totality of all spheres in ordinary three-dimensional space forms a four-dimensional manifold, since each sphere with center x, y, z and radius t can be represented by a point with coördinates x, y, z, t. A cube in three-dimensional space with edge of length 2, sides parallel to the coördinate planes, and center at the origin, consists of the totality of all points x_1, x_2, x_3 for which $|x_1| \leq 1$, $|x_2| \leq 1$, $|x_3| \leq 1$. In the same way a "cube" in n-dimensional space R_n with edge 2, sides parallel to the coördinate planes, and center at the origin, is defined as the totality of points $x_1, x_2, \cdots, x_n$ for which simultaneously

$$|x_1| \leq 1, \qquad |x_2| \leq 1, \cdots, |x_n| \leq 1.$$

The "surface" of this cube consists of all points for which at least one equality sign holds. The surface elements of dimension $n - 2$ consist of those points where at least *two* equality signs hold, etc.

Exercise: Describe the surface of such a cube in the three-, four-, and n-dimensional cases.

*3. Geometrical or Combinatorial Approach

While the analytical approach to n-dimensional geometry is simple and well adapted to most applications, there is another method of procedure which is purely geometrical in character. It is based on a reduction from n- to $(n - 1)$-dimensional data that enables us to define geometry in higher dimensions by a process of mathematical induction.

Let us start with the boundary of a triangle ABC in two dimensions. By cutting the closed polygon at the point C and then rotating AC and BC into the line AB we obtain the simple straight figure of Figure 116 in which the point C appears twice. This one-dimensional figure gives a complete representation of the boundary of the two dimensional triangle. By bending the segments AC and BC together in a plane, we can make the two points C coincide again. But, and this is the important point, we need not do this bending. We need only agree to "identify," i.e. not to distinguish between, the two points C in Figure 116, even though they do not actually coincide as geometrical entities in the naive sense. We may even go a step farther by taking the three segments apart at the points A and B, obtaining a set of three segments CA, AB, BC which can be put together again to form a "real" triangle by making the identified pairs of points coincide. This idea of identifying different points in a set of segments to form a polygon (in this case a triangle) is sometimes very practical. If we wish to ship a

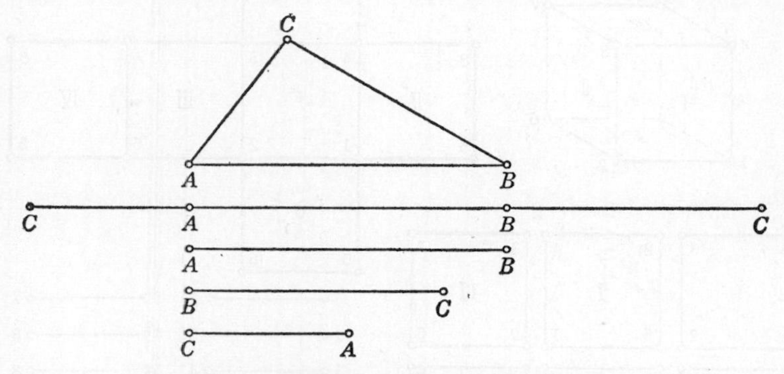

Fig. 116. Triangle defined by segments with coördinated ends.

complicated framework of steel bars, such as the framework of a bridge, we ship it in single bars and mark by the same symbol those endpoints which are to be connected when the framework is put together in space. The system of bars with marked endpoints is a complete equivalent of the spatial framework. This remark suggests the way to reduce a two-dimensional polyhedron in three-dimensional space to figures of lower dimensions. Let us take, for example, the surface of a cube (Fig. 117). It can be immediately reduced to a system of six plane squares whose boundary segments are appropriately identified, and in another step to a system of 12 straight segments with their endpoints properly identified.

In general, any polyhedron in three-dimensional space R_3 can be reduced in this way either to a system of plane polygons, or to a system of straight segments.

Exercise: Carry out this reduction for all the regular polyhedra (see p. 237).

It is now quite clear that we can invert our reasoning, *defining* a polygon in the plane by a system of straight segments, and a polyhedron in R_3 by a system of polygons in R_2 or again, with a further reduction, by a system of straight segments. Hence it is natural to define a "polyhedron" in four-dimensional space R_4 by a system of polyhedra in R_3 with proper identification of their two-dimensional faces; polyhedra in R_5 by systems of polyhedra in R_4, and so on. Ultimately we can reduce every polyhedron in R_n to a system of straight segments.

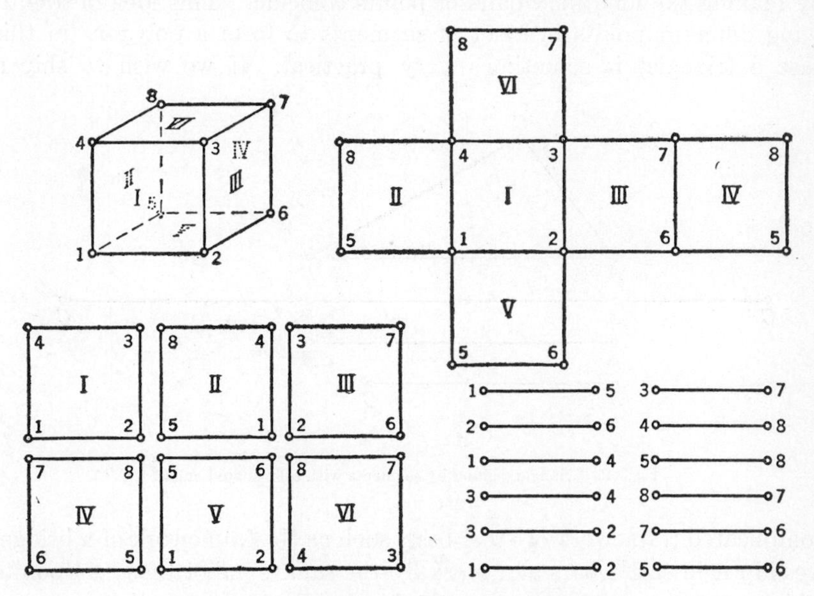

Fig. 117. Cube defined by coördination of vertices and edges.

It is not possible here to develop this subject much further. Only a few remarks without proof may be added. A cube in R_4 is bounded by 8 three-dimensional cubes, each identified with a "neighbor" along a two-dimensional face. The cube in R_4 has 16 vertices, in each of which four of the 32 straight edges meet. In R_4 there are six regular polyhedra. Besides the "cube" there is one bounded by 5 regular tetrahedra, one bounded by 16 tetrahedra, one bounded by 24 octahedra, one bounded by 120 dodecahedra, and one bounded by 600 tetrahedra. For $n > 4$ dimensions it has been proved that only 3 regular polyhedra are possible: one with $n + 1$ vertices bounded by $n + 1$ polyhedra in R_{n-1} with n sides of $(n - 2)$ dimensions; one with 2^n vertices bounded by $2n$ polyhedra in R_{n-1} with $2n - 2$ sides; and one with $2n$ vertices and 2^n polyhedra of n sides in R_{n-1} as boundaries.

* *Exercise:* Compare the definition of the cube in R_4 given in Article 2 with the definition given in this article, and show that the "analytical" definition of the surface of the cube of Article 2 is equivalent to the "combinatorial" definition of this article.

From the structural, or "combinatorial," point of view, the simplest geometrical figures of dimension 0, 1, 2, 3 are the point, the segment, the triangle, and the tetrahedron, respectively. In the interests of a uniform notation let us denote these figures by the symbols T_0, T_1,

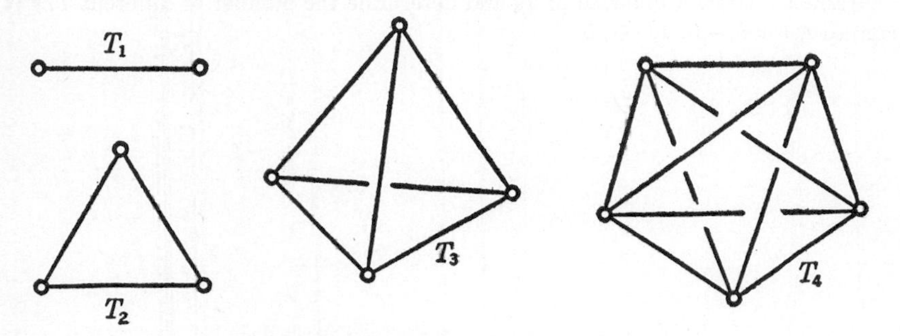

Fig. 118. The simplest elements in 1, 2, 3, 4 dimensions.

T_2, T_3, respectively. (The subscripts denote the dimension.) The structure of each of these figures is described by the statement that each T_n contains $n + 1$ vertices and that each subset of $i + 1$ vertices of a T_n $(i = 0, 1, \cdots, n)$ determines a T_i. For example, the three-dimensional tetrahedron T_3 contains 4 vertices, 6 segments, and 4 triangles.

It is clear how to proceed. We define a four-dimensional "tetrahedron" T_4 as a set of five vertices such that each subset of four vertices determines a T_3, each subset of three vertices determines a T_2, etc. The schematic diagram of T_4 is shown in Figure 118. We see that T_4 contains 5 vertices, 10 segments, 10 triangles, and 5 tetrahedra.

The generalization to n dimensions is immediate. From the theory of combinations it is known that there are exactly $C_i^r = \dfrac{r!}{i!(r - i)!}$

different subsets of i objects each that can be formed from a given set of r objects. Hence an n-dimensional "tetrahedron" contains

$$C_1^{n+1} = n + 1 \quad \text{vertices} \quad (T_0\text{'s}),$$

$$C_2^{n+1} = \frac{(n+1)!}{2!(n-1)!} \quad \text{segments} \quad (T_1\text{'s}),$$

$$C_3^{n+1} = \frac{(n+1)!}{3!(n-2)!} \quad \text{triangles} \quad (T_2\text{'s}),$$

$$C_4^{n+1} = \frac{(n+1)!}{4!(n-3)!} \quad T_3\text{'s},$$

$$\cdots\cdots\cdots\cdots\cdots\cdots$$

$$C_{n+1}^{n+1} = 1 \quad T_n\text{'s}.$$

Exercise: Draw a diagram of T_5 and determine the number of different T_i's it contains, for $i = 0, 1, \cdots, 5$.

CHAPTER V

TOPOLOGY

INTRODUCTION

In the middle of the nineteenth century there began a completely new development in geometry that was soon to become one of the great forces in modern mathematics. The new subject, called analysis situs or topology, has as its object the study of the properties of geometrical figures that persist even when the figures are subjected to deformations so drastic that all their metric and projective properties are lost.

One of the great geometers of the time was A. F. Moebius (1790–1868), a man whose lack of self-assertion destined him to the career of an insignificant astronomer in a second-rate German observatory. At the age of sixty-eight he submitted to the Paris Academy a memoir on "one-sided" surfaces that contained some of the most surprising facts of this new kind of geometry. Like other important contributions before it, his paper lay buried for years in the files of the Academy until it was eventually made public by the author. Independently of Moebius, the astronomer J. B. Listing (1808–1882) in Goettingen had made similar discoveries, and at the suggestion of Gauss had published in 1847 a little book, *Vorstudien zur Topologie*. When Bernhard Riemann (1826–1866) came to Goettingen as a student, he found the mathematical atmosphere of that university town filled with keen interest in these strange new geometrical ideas. Soon he realized that here was the key to the understanding of the deepest properties of analytic functions of a complex variable. Nothing, perhaps, has given more impetus to the later development of topology than the great structure of Riemann's theory of functions, in which topological concepts are absolutely fundamental.

At first, the novelty of the methods in the new field left mathematicians no time to present their results in the traditional postulational form of elementary geometry. Instead, the pioneers, such as Poincaré, were forced to rely largely upon geometrical intuition. Even today a student of topology will find that by too much insistence on a rigorous form of presentation he may easily lose sight of the essential geometrical content in a mass of formal detail. Still, it is a great merit of recent work to have brought topology within the framework of rigorous mathematics, where intuition remains the source but not the final validation of truth. During this process, started by L. E. J. Brouwer, the significance of topology for almost the whole of mathematics has steadily increased. American mathematicians, in particular O. Veblen, J. W. Alexander, and S. Lefschetz, have made important contributions to the subject.

While topology is definitely a creation of the last hundred years, there were a few isolated earlier discoveries that later found their place in the modern systematic development. By far the most important of these is a formula, relating the numbers of vertices, edges, and faces of a simple polyhedron, observed as early as 1640 by Descartes, and rediscovered and used by Euler in 1752. The typical character of this relation as a topological theorem became apparent much later, after Poincaré had recognized "Euler's formula" and its generalizations as one of the central theorems of topology. So, for reasons both historical and intrinsic, we shall begin our discussion of topology with Euler's formula. Since the ideal of perfect rigor is neither necessary nor desirable during one's first steps in an unfamiliar field, we shall not hesitate from time to time to appeal to the reader's geometrical intuition.

§1. EULER'S FORMULA FOR POLYHEDRA

Although the study of polyhedra held a central place in Greek geometry, it remained for Descartes and Euler to discover the following fact: In a simple polyhedron let V denote the number of vertices, E the number of edges, and F the number of faces; then always

$$V - E + F = 2. \tag{1}$$

By a *polyhedron* is meant a solid whose surface consists of a number of polygonal faces. In the case of the regular solids, all the polygons are congruent and all the angles at vertices are equal. A polyhedron is *simple* if there are no "holes" in it, so that its surface can be deformed continuously into the surface of a sphere. Figure 120 shows a simple polyhedron which is not regular, while Figure 121 shows a polyhedron which is not simple.

The reader should check the fact that Euler's formula holds for the simple polyhedra of Figures 119 and 120, but does not hold for the polyhedron of Figure 121.

To prove Euler's formula, let us imagine the given simple polyhedron to be hollow, with a surface made of thin rubber. Then if we cut out one of the faces of the hollow polyhedron, we can deform the remaining surface until it stretches out flat on a plane. Of course, the areas of the

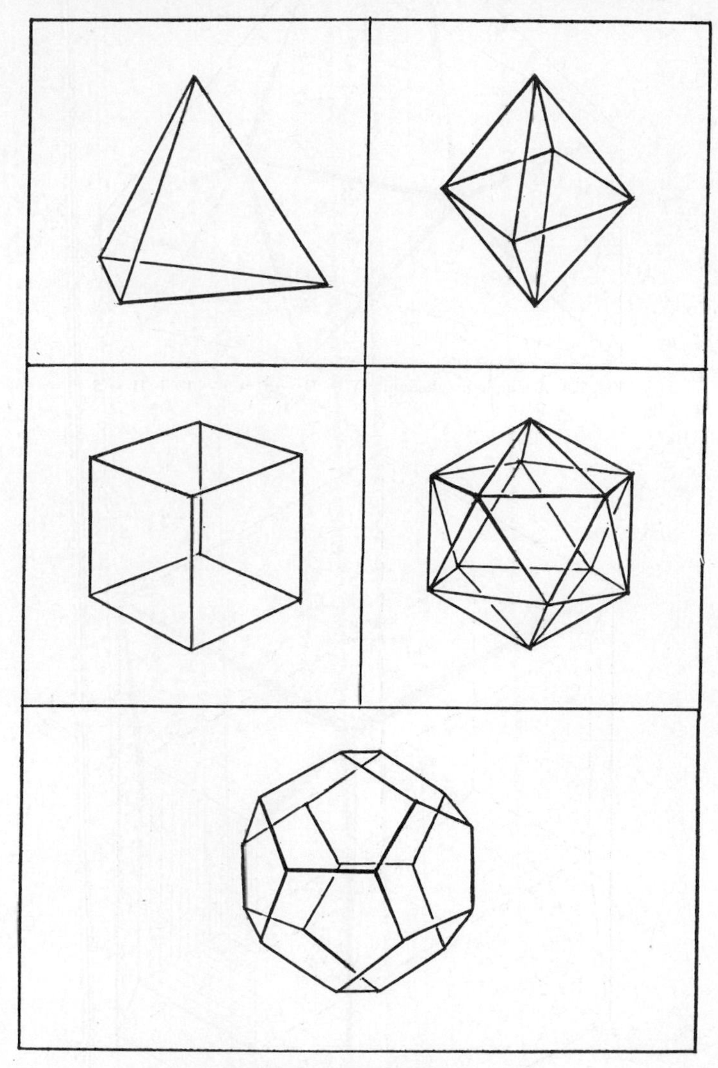

Fig. 119. The regular polyhedra.

faces and the angles between the edges of the polyhedron will have changed in this process. But the network of vertices and edges in the plane will contain the same number of vertices and edges as did the

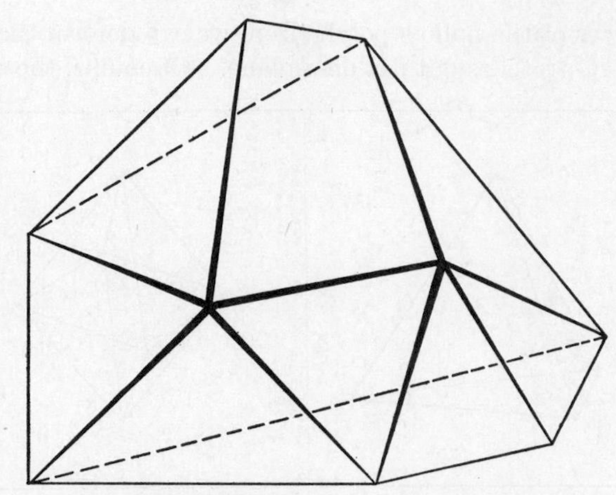

Fig. 120. A simple polyhedron. $V - E + F = 9 - 18 + 11 = 2$

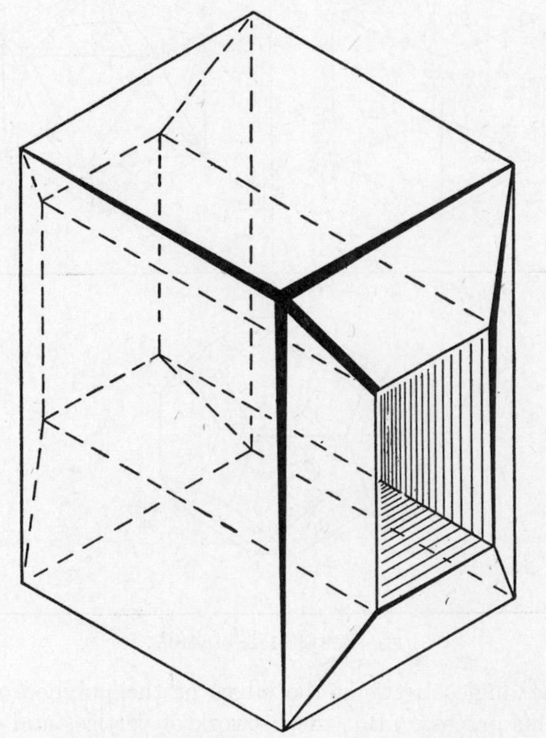

Fig. 121. A non-simple polyhedron. $V - E + F = 16 - 32 + 16 = 0.$

original polyhedron, while the number of polygons will be one less than in the original polyhedron, since one face was removed. We shall now show that for the plane network, $V - E + F = 1$, so that, if the removed face is counted, the result is $V - E + F = 2$ for the original polyhedron.

First we "triangulate" the plane network in the following way: In some polygon of the network which is not already a triangle we draw a diagonal. The effect of this is to increase both E and F by 1, thus preserving the value of $V - E + F$. We continue drawing diagonals joining pairs of points (Fig. 122) until the figure consists entirely of triangles, as it must eventually. In the triangulated network, $V - E + F$ has the value that it had before the division into tri-

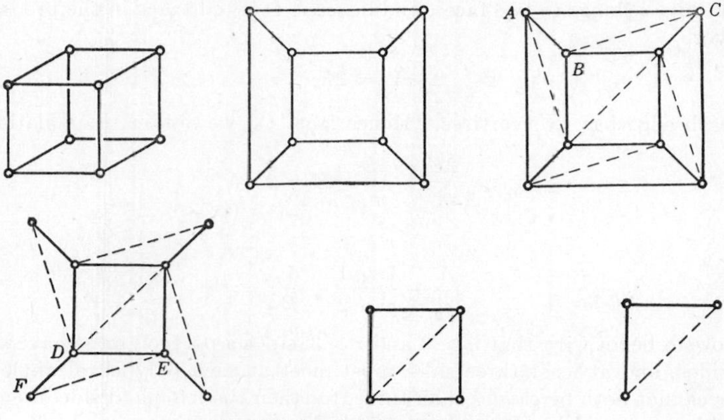

Fig. 122. Proof of Euler's theorem.

angles, since the drawing of diagonals has not changed it. Some of the triangles have edges on the boundary of the plane network. Of these some, such as ABC, have only one edge on the boundary, while other triangles may have two edges on the boundary. We take any boundary triangle and remove that part of it which does not also belong to some other triangle. Thus, from ABC we remove the edge AC and the face, leaving the vertices A, B, C and the two edges AB and BC; while from DEF we remove the face, the two edges DF and FE, and the vertex F. The removal of a triangle of type ABC decreases E and F by 1, while V is unaffected, so that $V - E + F$ remains the same. The removal of a triangle of type DEF decreases V by 1, E by 2, and F by 1, so that $V - E + F$ again remains the same. By a properly chosen sequence of these operations we can remove triangles with edges on the boundary (which changes with each removal), until finally only one triangle remains, with its three edges, three vertices, and one face. For this simple network, $V - E + F = 3 - 3 + 1 = 1$. But we have seen that by constantly erasing triangles $V - E + F$ was not altered. Therefore in the original plane network $V - E + F$ must equal 1 also, and thus equals 1 for the polyhedron with one face missing. We conclude that $V - E + F = 2$ for the complete polyhedron. This completes the proof of Euler's formula. (See (56), (57), pp. 496–7.)

On the basis of Euler's formula it is easy to show that there are no more than five regular polyhedra. For suppose that a regular polyhedron has F faces, each of which is an n-sided regular polygon, and that r edges meet at each vertex. Counting edges by faces and vertices, we see that

$$nF = 2E; \tag{2}$$

for each edge belongs to two faces, and hence is counted twice in the product nF; moreover,

$$rV = 2E, \tag{3}$$

since each edge has two vertices. Hence from (1) we obtain the equation

$$\frac{2E}{n} + \frac{2E}{r} - E = 2$$

or

$$\frac{1}{n} + \frac{1}{r} = \frac{1}{2} + \frac{1}{E}. \tag{4}$$

We know to begin with that $n \geq 3$ and $r \geq 3$, since a polygon must have at least three sides, and at least three sides must meet at each polyhedral angle. But n and r cannot both be *greater* than three, for then the left hand side of equation (4) could not exceed $\frac{1}{2}$, which is impossible for any positive value of E. Therefore, let us see what values r may have when $n = 3$, and what values n may have when $r = 3$. The totality of polyhedra given by these two cases gives the number of possible regular polyhedra.

For $n = 3$, equation (4) becomes

$$\frac{1}{r} - \frac{1}{6} = \frac{1}{E};$$

r can thus equal 3, 4, or 5. (6, or any greater number, is obviously excluded, since $1/E$ is always positive.) For these values of n and r we get $E = 6$, 12, or 30, corresponding respectively to the tetrahedron, octahedron, and icosahedron. Likewise, for $r = 3$ we obtain the equation

$$\frac{1}{n} - \frac{1}{6} = \frac{1}{E},$$

from which it follows that $n = 3$, 4, or 5, and $E = 6$, 12, or 30, respectively. These values correspond respectively to the tetrahedron, cube, and dodecahedron. Substituting these values for n, r, and E in equations (2) and (3), we obtain the numbers of vertices and faces in the corresponding polyhedra.

§2. TOPOLOGICAL PROPERTIES OF FIGURES

1. Topological Properties

We have proved that the Euler formula holds for any simple polyhedron. But the range of validity of this formula goes far beyond the polyhedra of elementary geometry, with their flat faces and straight edges; the proof just given would apply equally well to a simple polyhedron with curved faces and edges, or to any subdivision of the surface of a sphere into regions bounded by curved arcs. Moreover, if we imagine the surface of the polyhedron or of the sphere to be made out of a thin sheet of rubber, the Euler formula will still hold if the surface is deformed by bending and stretching the rubber into any other shape, so long as the rubber is not torn in the process. For the formula is concerned only with the *numbers* of the vertices, edges, and faces, and not with lengths, areas, straightness, cross-ratios, or any of the usual concepts of elementary or projective geometry.

We recall that elementary geometry deals with the magnitudes (length, angle, and area) that are unchanged by the rigid motions, while projective geometry deals with the concepts (point, line, incidence, and cross-ratio) which are unchanged by the still larger group of projective transformations. But the rigid motions and the projections are both very special cases of what are called *topological transformations*: a topological transformation of one geometrical figure A into another figure A' is given by any correspondence

$$p \leftrightarrow p'$$

between the points p of A and the points p' of A' which has the following two properties:

1. *The correspondence is biunique.* This means that to each point p of A corresponds just one point p' of A', and conversely.

2. *The correspondence is continuous in both directions.* This means that if we take any two points p, q of A and move p so that the distance between it and q approaches zero, then the distance between the corresponding points p', q' of A' will also approach zero, and conversely.

Any property of a geometrical figure A that holds as well for every figure into which A may be transformed by a topological transformation is called a *topological property* of A, and *topology* is the branch of geometry which deals only with the topological properties of figures. Imagine a figure to be copied "free-hand" by a conscientious but inexpert draftsman who makes straight lines curved and alters angles, distances and areas; then, although the metric and projective properties of the original figure would be lost, its topological properties would remain the same.

The most intuitive examples of general topological transformations are the *deformations*. Imagine a figure such as a sphere or a triangle to be made from or drawn upon a thin sheet of rubber, which is then stretched and twisted in any manner without tearing it and without bringing distinct points into actual coincidence. (Bringing distinct points into coincidence would violate condition 1. Tearing the sheet of rubber would violate condition 2, since two points of the original figure which tend toward coincidence from opposite sides of a line along which the sheet is torn would not tend towards coincidence in the torn figure.) The final position of the figure will then be a topological image of the original. A triangle can be deformed into any other triangle or

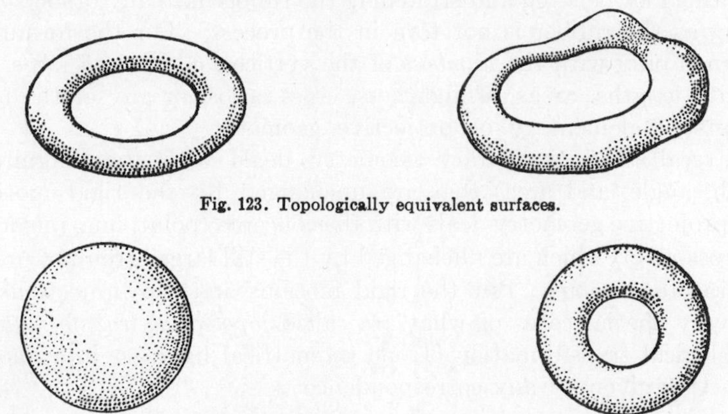

Fig. 123. Topologically equivalent surfaces.

Fig. 124. Topologically non-equivalent surfaces.

into a circle or an ellipse, and hence these figures have exactly the same topological properties. But one cannot deform a circle into a line segment, nor the surface of a sphere into the surface of an inner tube.

The general concept of topological transformation is wider than the concept of deformation. For example, if a figure is cut during a deformation and the edges of the cut sewn together after the deformation in exactly the same way as before, the process still defines a topological transformation of the original figure, although it is not a deformation. Thus the two curves of Figure 134 (p. 256) are topologically equivalent to each other or to a circle, since they may be cut, untwisted, and the cut sewn up. But it is impossible to deform one curve into the other or into a circle without first cutting the curve.

Topological properties of figures (such as are given by Euler's theorem and others to be discussed in this section) are of the greatest interest and importance in many mathematical investigations. They are in a sense the deepest and most fundamental of all geometrical properties, since they persist under the most drastic changes of shape.

2. Connectivity

As another example of two figures that are not topologically equivalent we may consider the plane domains of Figure 125. The first of

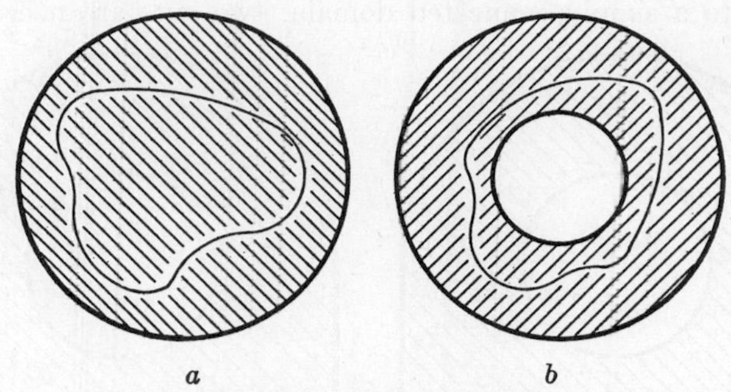

Fig. 125. Simple and double connectivity.

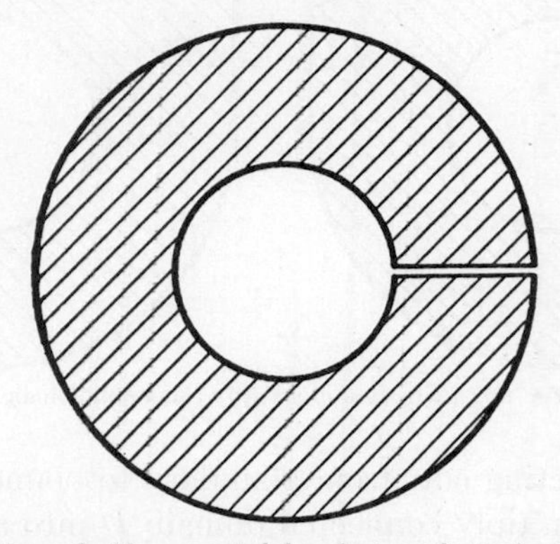

Fig. 126. Cutting a doubly connected domain to make it simply connected.

these consists of all points interior to a circle, while the second consists of all points contained between two concentric circles. Any closed curve lying in the domain a can be continuously deformed or "shrunk" down to a single point *within the domain.* A domain with this property is said to be *simply connected.* The domain b is not simply connected. For example, a circle concentric with the two boundary circles and midway between them cannot be shrunk to a single point within the domain, since during this process the curve would necessarily pass over the center of the circles, which is not a point of the domain. A domain which is not simply connected is said to be *multiply connected.* If the multiply connected domain b is cut along a radius, as in Figure 126, the resulting domain is simply connected.

More generally, we can construct domains with two, three, or more "holes," such as the domain of Figure 127. In order to convert this domain into a simply connected domain, two cuts are necessary. If

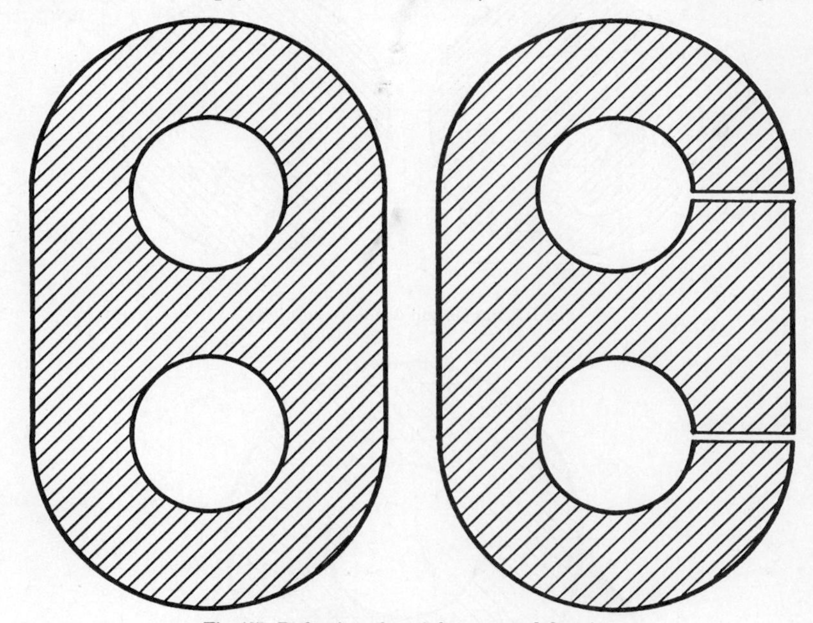

Fig. 127. Reduction of a triply connected domain.

$n - 1$ non-intersecting cuts from boundary to boundary are needed to convert a given multiply connected domain D into a simply connected domain, the domain D is said to be n-tuply connected. The degree of connectivity of a domain in the plane is an important topological invariant of the domain.

§3. OTHER EXAMPLES OF TOPOLOGICAL THEOREMS

1. The Jordan Curve Theorem

A simple closed curve (one that does not intersect itself) is drawn in the plane. What property of this figure persists even if the plane is regarded as a sheet of rubber that can be deformed in any way? The length of the curve and the area that it encloses can be changed by a deformation. But there is a topological property of the configuration which is so simple that it may seem trivial: *A simple closed curve* C *in the plane divides the plane into exactly two domains, an inside and an outside.* By this is meant that the points of the plane fall into two classes—A, the outside of the curve, and B, the inside—such that any pair of points of the same class can be joined by a curve which does not cross C, while any curve joining a pair of points belonging to different classes must cross C. This statement is obviously true for a circle or an ellipse, but the self-evidence fades a little if one contemplates a complicated curve like the twisted polygon in Figure 128.

Fig. 128. Which points of the plane are inside this polygon?

This theorem was first stated by Camille Jordan (1838–1922) in his famous *Cours d'Analyse*, from which a whole generation of mathematicians learned the modern concept of rigor in analysis. Strangely enough, the proof given by Jordan was neither short nor simple, and the surprise was even greater when it turned out that Jordan's proof was invalid and that considerable effort was necessary to fill the gaps in his reasoning. The first rigorous proofs of the theorem were quite complicated and hard to understand, even for many well-trained mathematicians. Only recently have comparatively simple proofs been found. One reason for the difficulty lies in the generality of the concept of "simple closed curve," which is not restricted to the class of polygons or "smooth" curves, but includes all curves which are topological images of a circle. On the other hand, many concepts such as "inside," "outside," etc., which are so clear to the intuition, must be made precise before a rigorous proof is possible. It is of the highest theoretical importance to analyze such concepts in their fullest generality, and much of modern topology is devoted to this task. But one should never forget that in the great majority of cases that arise from the study of concrete geometrical phenomena it is quite beside the point to work with concepts whose extreme generality creates unnecessary difficulties. As a matter of fact, the Jordan curve theorem is quite simple to prove for the reasonably well-behaved curves, such as polygons or curves with continuously turning tangents, which occur in most important problems. We shall prove the theorem for polygons in the appendix to this chapter.

2. The Four Color Problem

From the example of the Jordan curve theorem one might suppose that topology is concerned with providing rigorous proofs for the sort of obvious assertions that no sane person would doubt. On the contrary, there are many topological questions, some of them quite simple in form, to which the intuition gives no satisfactory answer. An example of this kind is the renowned "four color problem."

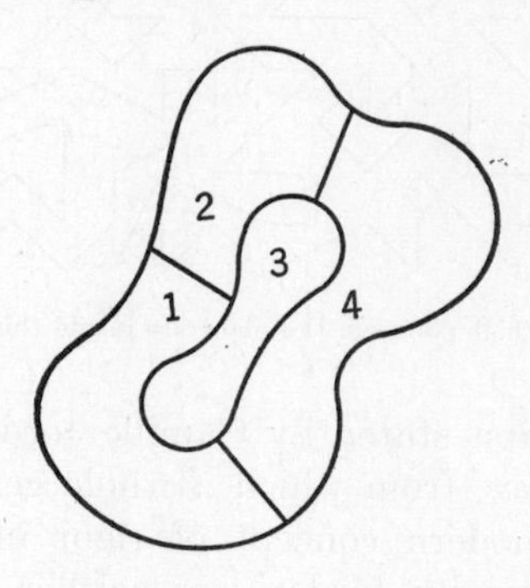

Fig. 129. Coloring a map.

In coloring a geographical map it is customary to give different colors to any two countries that have a portion of their boundary in common. It has been found empirically that any map, no matter how many countries it contains nor how they are situated, can be so colored by using only *four* different colors. It is easy to see that no smaller number of colors will suffice for all cases. Figure 129 shows an island in the sea that certainly cannot be properly colored with less than four colors, since it contains four countries, each of which touches the other three.

The fact that no map has yet been found whose coloring requires more than four colors suggests the following mathematical theorem: *For any subdivision of the plane into non-overlapping regions, it is always possible to mark the regions with one of the numbers* 1, 2, 3, 4 *in such a way that no two adjacent regions receive the same number.* By "adjacent" regions we mean regions with a whole segment of boundary in common; two regions which meet at a single point only or at a finite number of points (such as the states of Colorado and Arizona) will not be called adjacent, since no confusion would arise if they were colored with the same color.

The problem of proving this theorem seems to have been first proposed by Moebius in 1840, later by DeMorgan in 1850, and again by Cayley in 1878. A "proof" was published by Kempe in 1879, but in 1890 Heawood found an error in Kempe's reasoning. By a revision of Kempe's proof, Heawood was able to show that *five* colors are always sufficient. (A proof of the five color theorem is given in the appendix to this chapter.) Despite the efforts of many famous mathematicians, the matter essentially rests with this more modest result: It has been *proved* that five colors suffice for all maps and it is *conjectured* that four will likewise suffice. But, as in the case of the famous Fermat theorem (see p. 42), neither a proof of this conjecture nor an example contradicting it has been produced, and it remains one of the great unsolved problems in mathematics. The four color theorem has indeed been proved for all maps containing less than thirty-eight regions. In view of this fact it appears that even if the general theorem is false it cannot be disproved by any very simple example.

In the four color problem the maps may be drawn either in the plane or on the surface of a sphere. The two cases are equivalent: any map on the sphere may be represented on the plane by boring a small hole through the interior of one of the regions A and deforming the resulting surface until it is flat, as in the proof of Euler's theorem. The resulting map in the plane will be that of an "island" consisting of the remaining regions, surrounded by a "sea" consisting of the region A. Conversely, by a reversal of this process, any map in the plane may be represented on the sphere. We may therefore confine ourselves to maps on the sphere. Furthermore, since deformations of the regions and their boundary lines do not affect the problem, we may suppose that the boundary of each region is a simple closed polygon composed of circular arcs. Even thus "regularized," the problem remains unsolved; the difficulties here, unlike those involved in the Jordan curve theorem, do not reside in the generality of the concepts of region and curve.

A remarkable fact connected with the four color problem is that for surfaces more complicated than the plane or the sphere the corresponding theorems have actually been proved, so that, paradoxically enough, the analysis of more complicated geometrical surfaces appears in this respect to be easier than that of the simplest cases. For example, on the surface of a torus (see Figure 123), whose shape is that of a doughnut or an inflated inner tube, it has been shown that any map may be colored by using seven colors, while maps may be constructed containing seven regions, each of which touches the other six.

*3. The Concept of Dimension

The concept of dimension presents no great difficulty so long as one deals only with simple geometric figures such as points, lines, triangles, and polyhedra. A single point or any *finite* set of points has dimension zero, a line segment is one-dimensional, and the surface of a triangle or of a sphere two-dimensional. The set of points in a solid cube is three-dimensional. But when one attempts to extend this concept to more general point sets, the need for a precise definition arises. What dimension should be assigned to the point set R consisting of all points on the x-axis whose coördinates are *rational* numbers? The set of rational points is dense on the line and might therefore be considered to be one-dimensional, like the line itself. On the other hand, there are irrational gaps between any pair of rational points, as between any two points of a finite point set, so that the dimension of the set R might also be considered to be zero.

An even more knotty problem arises if one tries to assign a dimension to the following curious point set, first considered by Cantor. From the unit segment remove the middle third, consisting of all points x such that $1/3 < x < 2/3$. Call the remaining set of points C_1. Now from C_1 remove the middle third of each of its two segments, leaving a set which we call C_2. Repeat this process by removing the middle third of each of the four intervals of C_2, leaving a set C_3,

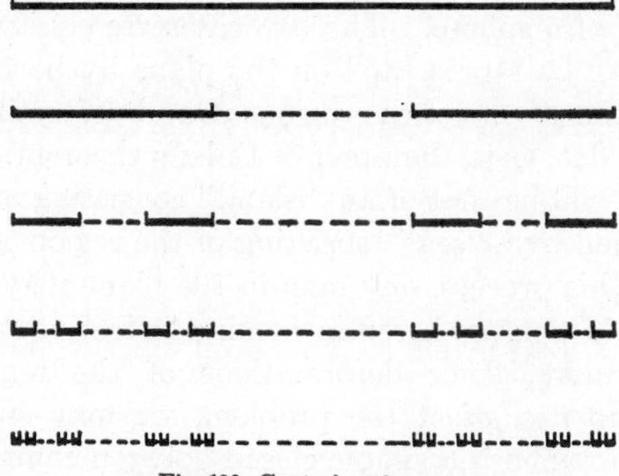

Fig. 130. Cantor's point set.

and proceed in this manner to form sets $C_4, C_5, C_6, \cdots$. Denote by C the set of points on the unit segment that are left after all these intervals have been removed, i.e. C is the set of points common to the infinite sequence of sets $C_1, C_2, \cdots$. Since one interval, of length 1/3, was removed at the first step; two intervals, each of length $1/3^2$, at the second step; etc.; the total length of the segments removed is

$$1 \cdot \frac{1}{3} + 2 \cdot \frac{1}{3^2} + 2^2 \cdot \frac{1}{3^3} + \cdots = \frac{1}{3}\left(1 + \left(\frac{2}{3}\right) + \left(\frac{2}{3}\right)^2 + \cdots\right).$$

The infinite series in parentheses is a geometrical series whose sum is $1/(1 - 2/3) = 3$; hence the total length of the segments removed is 1. Still there remain points in the set C. Such, for example, are the points 1/3, 2/3, 1/9, 2/9, 7/9, 8/9, $\cdots$, by which the successive segments are trisected. As a matter of fact, it is easy to show that C will consist precisely of all those points x whose expansions in the form of infinite triadic fractions can be written in the form

$$x = \frac{a_1}{3} + \frac{a_2}{3^2} + \frac{a_3}{3^3} + \cdots + \frac{a_n}{3^n} + \cdots,$$

where each a_i is either 0 or 2, while the triadic expansion of every point removed will have at least one of the numbers a_i equal to 1.

What shall be the dimension of the set C? The diagonal process used to prove the non-denumerability of the set of all real numbers can be so modified as to yield the same result for the set C. It would seem, therefore, that the set C should be one-dimensional. Yet C contains no complete interval, no matter how small, so that C might also be thought of as zero-dimensional, like a finite set of points. In the same spirit, we might ask whether the set of points of the *plane*, obtained by erecting at each rational point or at each point of the Cantor set C a segment of unit length, should be considered to be one-dimensional or two-dimensional.

It was Poincaré who (in 1912) first called attention to the need for a deeper analysis and a precise definition of the concept of dimensionality. Poincaré observed that the line is one-dimensional because we may separate any two points on it by cutting it at a single point (which is of dimension 0), while the plane is two-dimensional because in order to separate a pair of points in the plane we must cut out a whole closed curve (of dimension 1). This suggests the inductive nature of dimensionality: a space is n-dimensional if any two points may be separated by removing an $(n - 1)$-dimensional subset, and if a lower-dimensional subset will not always suffice. An inductive definition of dimensionality is also contained implicitly in Euclid's *Elements*, where a one-dimensional figure is something whose boundary consists of points, a two-dimensional figure one whose boundary consists of curves, and a three-dimensional figure one whose boundary consists of surfaces.

In recent years an extensive theory of dimension has been developed. One definition of dimension begins by making precise the concept "point set of dimension 0." Any *finite* set of points has the property that each point of the set can be enclosed in a region of space which can be made as small as we please, and which contains no points of the set on its boundary. This property is now taken as the definition of 0-dimensionality. For convenience, we say that an empty set, containing no points at all, has dimension -1. Then a point set S is of dimension 0 if it is not of dimension -1 (i.e. if S contains at least one point), and if each point of S can be enclosed within an arbitrarily small region whose boundary intersects S in a set of dimension -1 (i.e. whose boundary contains no points of S). For example, the set of rational points on the line is of dimension 0, since each rational point can be made the center of an arbitrarily small interval with irrational endpoints. The Cantor set C is also seen to be of dimension 0, since, like the set of rational points, it is formed by removing a dense set of points from the line.

So far we have defined only the concepts of dimension -1 and dimension 0. The definition of dimension 1 suggests itself at once: a set S of points is of dimension 1 if it is not of dimension -1 or 0, and if each point of S can be enclosed within an arbitrarily small region whose boundary intersects S in a set of dimension 0. A line segment has this property, since the boundary of any interval is a pair of points, which is a set of dimension 0 according to the preceding definition. Moreover, by proceeding in the same manner, we can successively define the concepts of dimension 2, 3, 4, 5, $\cdots$, each resting on the previous definitions. Thus a set S will be of dimension n if it is not of any lower dimension, and if each point of S can be enclosed within an arbitrarily small region whose boundary intersects S in a set of dimension $n - 1$. For example, the plane is of dimension 2, since each point of the plane can be enclosed within an arbitrarily small circle, whose circumference is of dimension 1.† No point set in ordinary space can have dimension higher than 3, since each point of space can be made the center of an arbitrarily small sphere whose surface is of dimension 2. But in modern mathematics the word "space" is used to denote any system of objects for which a notion of "distance" or "neighborhood" is defined (see p. 316), and these abstract "spaces" may have dimensions higher than 3. A simple example is *Cartesian n-space*, whose "points" are ordered arrays of n real numbers:

$$P = (x_1, x_2, x_3, \cdots, x_n),$$
$$Q = (y_1, y_2, y_3, \cdots, y_n);$$

with the "distance" between the points P and Q defined as

$$d(P, Q) = \sqrt{(x_1 - y_1)^2 + (x_2 - y_2)^2 + \cdots + (x_n - y_n)^2}.$$

This space may be shown to have dimension n. A space which does not have dimension n for any integer n is said to be of dimension infinity. Many examples of such spaces are known.

One of the most interesting facts of dimension theory is the following characteristic property of two-, three- or, in general, n-dimensional figures. Consider first the two-dimensional case. If any simple two-dimensional figure is subdivided into sufficiently small regions (each of which is regarded as including its

† This does not purport to be a rigorous proof that the plane is of dimension 2 according to our definition, since it assumes that the circumference of a circle is known to be of dimension 1, and that the plane is known not to be of dimension 0 or 1. But a proof can be given for these facts and for their analogs in higher dimensions. This proof shows that the definition of the dimension of a general point set does not contradict ordinary usage for simple sets.

boundary), then there will necessarily be points where *three or more* of these regions meet, *no matter what the shapes of the regions.* In addition, *there will exist subdivisions* of the figure in which each point belongs to *at most* three regions of the subdivision. Thus, if the two-dimensional figure is a square, as in Figure 131, then a point will belong to the three regions, 1, 2, and 3, while for this particular subdivision no point belongs to more than three regions. Similarly, in the three-dimensional case it may be proved that, if a volume is covered by sufficiently small volumes, there always exist points common to at least four of the latter, while for a properly chosen subdivision no more than four will have a point in common.

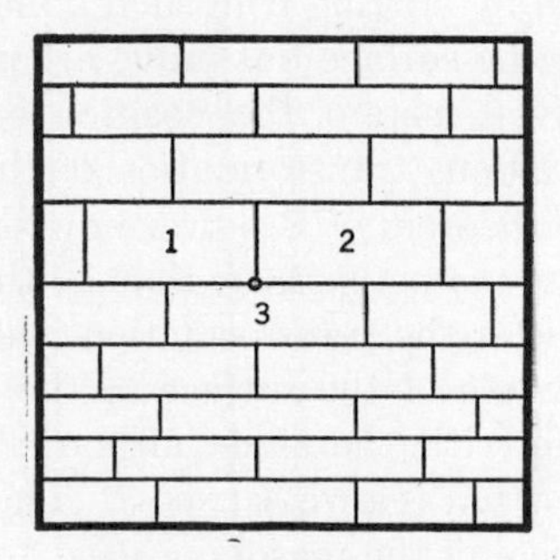

Fig. 131. The tiling theorem.

These observations suggest the following theorem, due to Lebesgue and Brouwer: If an n-dimensional figure is covered in any way by sufficiently small subregions, then there will exist points which belong to at least $n + 1$ of these subregions; moreover, it is always possible to find a covering by arbitrarily small regions for which no point will belong to more than $n + 1$ regions. Because of the method of covering considered here, this is known as the "tiling" theorem. It characterizes the dimension of any geometrical figure: those figures for which the theorem holds are n-dimensional, while all others are of some other dimension. For this reason it may be taken as the *definition* of dimensionality, as is done by some authors.

The dimension of any set is a topological feature of the set; no two figures of different dimensions can be topologically equivalent. This is the famous topological theorem of "invariance of dimensionality," which gains in significance by comparison with the fact stated on page 85, that the set of points in a square has the same cardinal number as the set of points on a line segment. The correspondence there defined is not topological because the continuity conditions are violated.

*4. A Fixed Point Theorem

In the applications of topology to other branches of mathematics, "fixed point" theorems play an important rôle. A typical example is the following theorem of Brouwer. It is much less obvious to the intuition than most topological facts.

We consider a circular disk in the plane. By this we mean the region consisting of the interior of some circle, together with its circumference. Let us suppose that the points of this disk are subjected to any continuous transformation (which need not even be biunique) in which each point remains within the circle, although differently situated. For example, a thin rubber disk might be shrunk, turned, folded, stretched, or deformed in any way, so long as the final position of each point of the disk lies within its original circumference. Again, if the liquid in a glass is set into motion by stirring it in such a manner that particles on the surface remain on the surface but move around on it to other positions, then at any given instant the position of the particles on the surface defines a continuous transformation of the original distribution of the particles. The theorem of Brouwer now states: *Each such transformation leaves at least one point fixed*; that is, there exists at least one point whose position after the transformation is the same as its original position. (In the example of the surface of the liquid, the fixed point will in general change with the time, although for a simple circular rotation it is the center that is always fixed.) The proof of the existence of a fixed point is typical of the reasoning used to establish many topological theorems.

Consider the disk before and after the transformation, and assume, contrary to the statement of the theorem, that *no* point remains fixed, so that under the transformation each point moves to another point

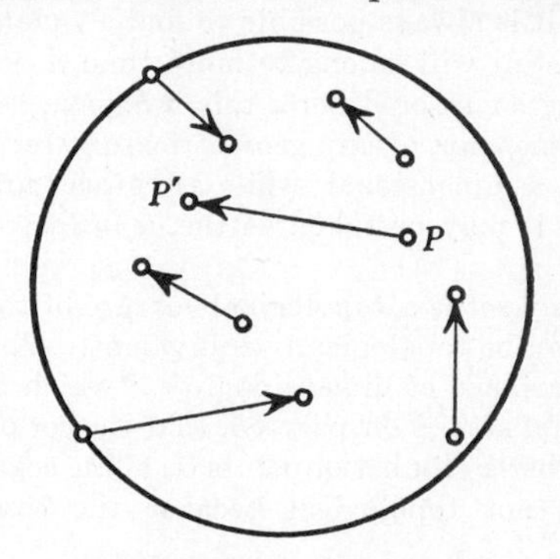

Fig. 132. Transformation vectors.

inside or on the circle. To each point P of the original disk attach a little arrow or "vector" pointing in the direction PP', where P' is the image of P under the transformation. At every point of the disk there is such an arrow, for every point was assumed to move somewhere else.

Now consider the points on the boundary of the circle, with their associated vectors. All of these vectors point into the circle, since, by assumption, no points are transformed into points outside the circle. Let us begin at some point P_1 on the boundary and travel in the counter-clockwise direction around the circle. As we do so, the direction of the vector will change, for the points on the boundary have variously pointed vectors associated with them. The directions of these vectors may be shown by drawing parallel arrows that issue from a single point in the plane. We notice that in traversing the circle once from P_1

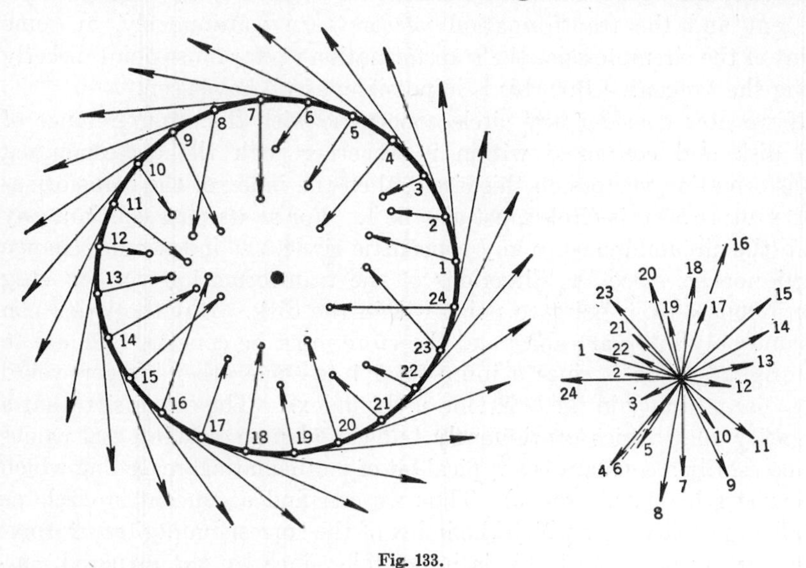

Fig. 133.

around to P_1, the vector turns around and comes back to its original position. Let us call the number of complete revolutions made by this vector the "index" of the vectors on the circle; more precisely, we define the index as the *algebraic sum* of the various changes in angle of the vectors, so that each clockwise portion of a revolution is taken with a negative sign, while each counter-clockwise portion is regarded as positive. The index is the net result, which may *a priori* be any one of the numbers $0, \pm 1, \pm 2, \pm 3, \cdots$, corresponding to a total change in angle of $0, \pm 360, \pm 720, \cdots$ degrees. We now assert that *the index equals* 1; that is, the total change in the direction of the arrow amounts to exactly one positive revolution. To show this, we recall that the transformation vector at any point P on the circle is always directed inside the circle and never along the tangent. Now, if this transformation vector turns through a total angle different from the total angle through which the *tangent* vector turns (which is 360°, because the tangent vector obviously makes one complete positive revolution), then the difference between the total angles through which the tangent vector and the transformation vector turn will be some non-zero multiple of 360°, since each makes an integral number of revolutions. Hence the transformation vector must turn completely around the tangent at least once during the complete circuit from P_1 back to P_1, and since the tangent and the transformation vectors turn continuously, at some point of the circumference the transformation vector must point directly along the tangent. But this is impossible, as we have seen.

If we now consider any circle concentric with the circumference of the disk and contained within it, together with the corresponding transformation vectors on this circle, then the index of the transformation vectors on this circle must also be 1. For as we pass continuously from the circumference to any concentric circle, the index must change continuously, since the directions of the transformation vectors vary continuously from point to point within the disk. But the index can assume only integral values and therefore must be constantly equal to its original value 1, since a jump from 1 to some other integer would be a discontinuity in the behavior of the index. (The conclusion that a quantity that varies continuously but can assume only integral values is necessarily a constant is a typical bit of mathematical reasoning which intervenes in many proofs.) Thus we can find a concentric circle as small as we please for which the index of the corresponding transformation vectors is 1. But this is impossible, since by the assumed continuity of the transformation the vectors on a sufficiently small circle will all point in approximately the same direction as the vector at the center of the circle. Thus the total net change of their angles can be made as small as we please, less than 10°, say, by taking a small enough circle. Hence the index, which must be an integer, will be zero. This contradiction shows our initial hypothesis that there is no fixed point under the transformation to be untenable, and completes the proof.

The theorem just proved holds not only for a disk but also for a triangular or square region or any other surface that is the image of a disk under a topological transformation. For if A is any figure correlated with a disk by a biunique and continuous transformation, then a continuous transformation of A into itself which had no fixed point would define a continuous transformation of the disk into itself without a fixed point, which we have proved to be impossible. The theorem also holds in three dimensions for solid spheres or cubes, but the proof is not so simple.

Although the Brouwer fixed point theorem for the disk is not very obvious to the intuition, it is easy to show that it is an immediate consequence of the following fact, the truth of which is intuitively evident: It is impossible to transform continuously a circular disk into its circumference alone so that each point of the circumference remains fixed. We shall show that the existence of a fixed-point-free transformation of a disk into itself would contradict this fact. Suppose $P \to P'$ were such a transformation; for each point P of the disk we could draw an arrow starting at P' and continuing through P until it reached the circumference at some point P^*. Then the transformation $P \to P^*$ would be a continuous transformation of the whole disk into its circumference alone and would leave each point of the circumference fixed, contrary to the assumption that such a transformation is impossible. Similar reasoning may be used to establish the Brouwer theorem in three dimensions for the solid sphere or cube.

It is easy to see that some geometrical figures do admit continuous fixed-point-free transformations into themselves. For example, the ring-shaped region between two concentric circles admits as a continuous fixed-point-free transformation a rotation through any angle not a multiple of 360° about its center. The surface of a sphere admits the continuous fixed-point-free transformation that takes each point into its diametrically opposite point. But it may be proved, by reasoning analogous to that which we have used for the disk, that any continuous transformation which carries no point into its diametrically opposite point (e.g., any small deformation) has a fixed point.

Fixed point theorems such as these provide a powerful method for the proof of many mathematical "existence theorems" which at first sight may not seem to be of a geometrical character. A famous example is a fixed point theorem conjectured by Poincaré in 1912, shortly before his death. This theorem has as an immediate consequence the existence of an infinite number of periodic orbits in the restricted problem of three bodies. Poincaré was unable to confirm his conjecture, and it was a major achievement of American mathematics when in the following year G. D. Birkhoff succeeded in giving a proof. Since then topological methods have been applied with great success to the study of the qualitative behaviour of dynamical systems.

5. Knots

As a final example it may be pointed out that the study of knots presents difficult mathematical problems of a topological character. A knot is formed by first looping and interlacing a piece of string and then joining the ends together. The resulting closed curve represents a geometrical figure that remains essentially the same even if it is deformed by pulling or twisting without breaking the string. But how is it possible to give an intrinsic characterization that will distinguish a knotted closed curve in space from an unknotted curve such as the circle? The answer is by no means simple, and still less so is the complete mathematical analysis of the various kinds of knots and the differences between them. Even for the simplest case this has proved to be a sizable task. Consider the two trefoil knots shown in Figure 134. These two knots are completely symmetrical "mirror images" of one another, and are topologically equivalent, but they are not congruent. The problem arises whether it is possible to deform one of these knots into the other in a continuous way. The answer is in the negative, but the proof of this fact requires considerably more knowledge of the technique of topology and group theory than can be presented here.

Fig. 134. Topologically equivalent knots that are not deformable into one another.

§4. THE TOPOLOGICAL CLASSIFICATION OF SURFACES

1. The Genus of a Surface

Many simple but important topological facts arise in the study of two-dimensional surfaces. For example, let us compare the surface of a sphere with that of a torus. It is clear from Figure 135 that the two surfaces differ in a fundamental way: on the sphere, as in the plane, every simple closed curve such as C separates the surface into two parts. But on the torus there exist closed curves such as C' that do not

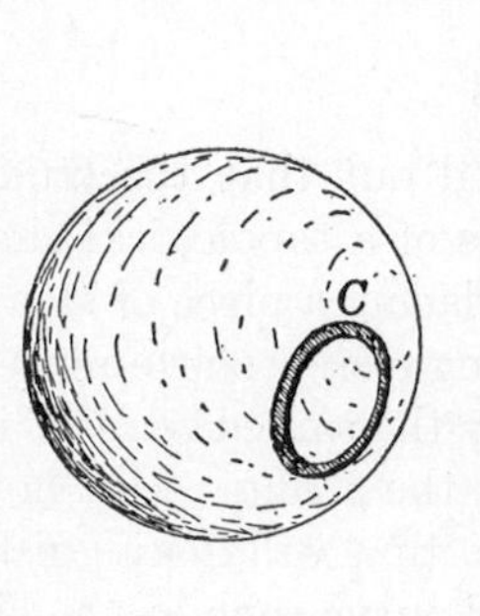

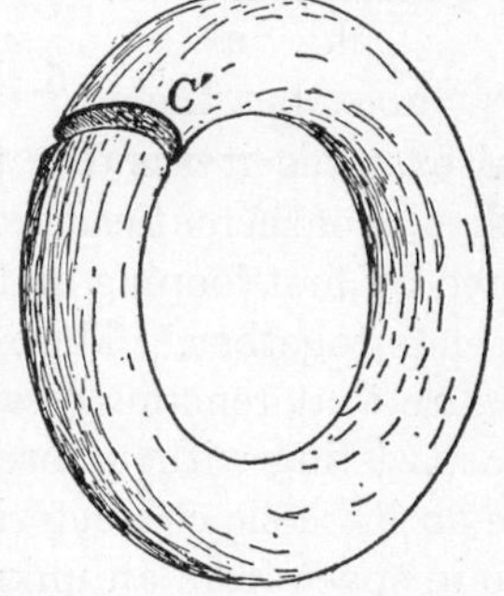

Fig. 135. Cuts on sphere and torus.

separate the surface into two parts. To say that C separates the sphere into two parts means that if the sphere is cut along C it will fall into two distinct and unconnected pieces, or, what amounts to the same thing, that we can find two points on the sphere such that any curve on the sphere which joins them must intersect C. On the other hand, if the torus is cut along the closed curve C', the resulting surface still hangs together: any point of the surface can be joined to any other point by a curve that does not intersect C'. This difference between the sphere and the torus marks the two types of surfaces as topologically distinct, and shows that it is impossible to deform one into the other in a continuous way.

Next let us consider the surface with two holes shown in Figure 136. On this surface we can draw *two* non-intersecting closed curves A and B which do not separate the surface. The torus is always separated into two parts by any two such curves. On the other hand, *three* closed non-intersecting curves always separate the surface with two holes.

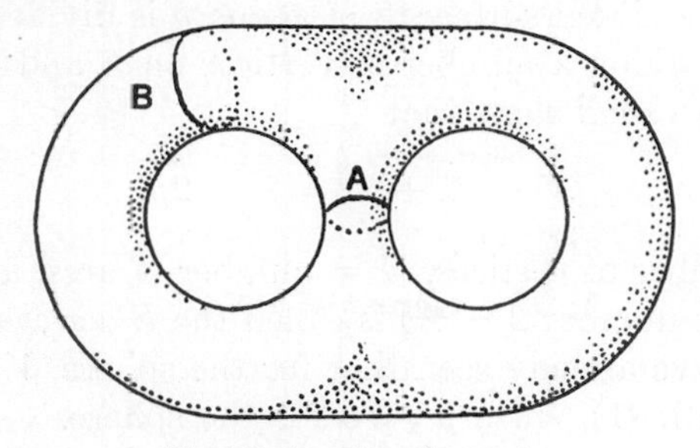

Fig. 136. A surface of genus 2.

These facts suggest that we define the *genus* of a surface as the largest number of non-intersecting simple closed curves that can be drawn on the surface without separating it. The genus of the sphere is 0, that of the torus is 1, while that of the surface in Figure 136 is 2. A similar surface with p holes has the genus p. The genus is a topological property of a surface and remains the same if the surface is deformed. Conversely, it may be shown (we omit the proof) that if two closed surfaces have the same genus, then one may be deformed into the other, so that the genus $p = 0, 1, 2, \cdots$ of a closed surface characterizes it completely from the topological point of view. (We are assuming that the surfaces considered are ordinary "two-sided" closed surfaces. In Article 3 of this section we shall consider "one-sided" surfaces.) For example, the two-holed doughnut and the sphere with two "handles" of Figure 137 are both closed surfaces of genus 2, and it is clear that either of these surfaces may be continuously deformed into the other. Since the doughnut with p holes, or its equivalent, the sphere with p handles, is

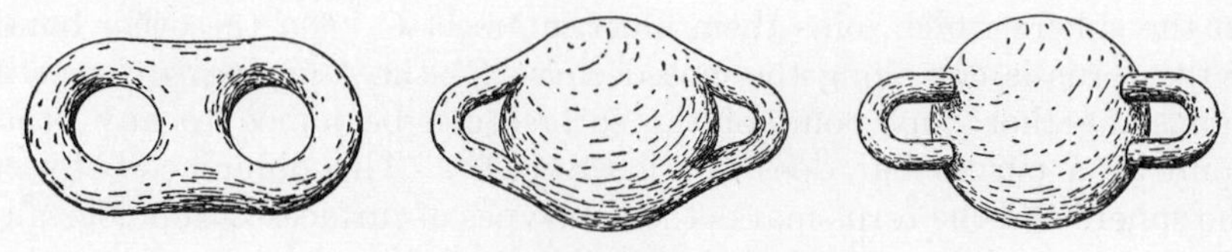

Fig. 137 Surfaces of genus 2.

of genus p, we may take either of these surfaces as the topological representative of all closed surfaces of genus p.

*2. The Euler Characteristic of a Surface

Suppose that a closed surface S of genus p is divided into a number of regions by marking a number of vertices on S and joining them by curved arcs. We shall show that

$$V - E + F = 2 - 2p, \tag{1}$$

where V = number of vertices, E = number of arcs, and F = number of regions. The number $2 - 2p$ is called the *Euler characteristic* of the surface. We have already seen that for the sphere, $V - E + F = 2$, which agrees with (1), since $p = 0$ for the sphere.

To prove the general formula (1), we may assume that S is a sphere with p handles. For, as we have stated, any surface of genus p may be continuously deformed into such a surface, and during this deformation the numbers $V - E + F$ and $2 - 2p$ will not change. We shall choose the deformation so as to ensure that the closed curves $A_1, A_2, B_1, B_2, \cdots$ where the handles join the sphere consist of arcs of the given subdivision. (We refer to Fig. 138, which illustrates the proof for the case $p = 2$.)

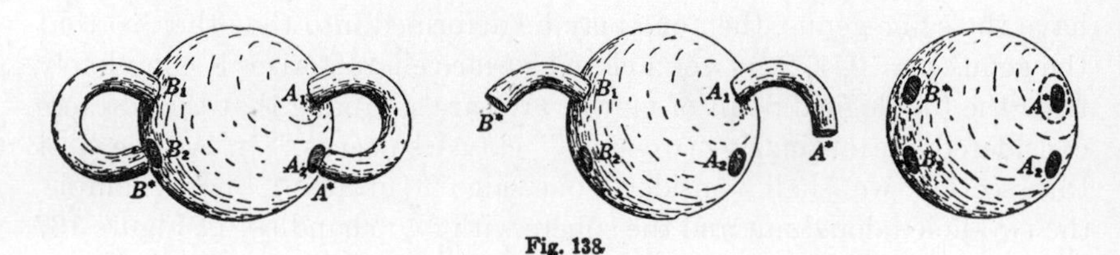

Fig. 138

Now let us cut the surface S along the curves A_2, B_2, $\cdots$ and straighten the handles out. Each handle will have a free edge bounded by a new curve A^*, B^*, $\cdots$ with the same number of vertices and arcs as A_2, B_2, $\cdots$ respectively. Hence $V - E + F$ will not change, since the additional vertices exactly counterbalance the additional arcs, while no new regions are created. Next, we deform the surface by flattening out the projecting handles, until the resulting surface is simply a sphere from which $2p$ regions have been removed. Since $V - E + F$ is known to equal 2 for any subdivision of the whole sphere, we have

$$V - E + F = 2 - 2p$$

for the sphere with $2p$ regions removed, and hence for the original sphere with p handles, as was to be proved.

Figure 121 illustrates the application of formula (1) to a surface S consisting of flat polygons. This surface may be continuously deformed into a torus, so that the genus p is 1 and $2 - 2p = 2 - 2 = 0$. As predicted by formula (1),

$$V - E + F = 16 - 32 + 16 = 0.$$

Exercise: Subdivide the doughnut with two holes of Figure 137 into regions, and show that $V - E + F = -2$.

3. One-Sided Surfaces

An ordinary surface has two sides. This applies both to closed surfaces like the sphere or the torus and to surfaces with boundary curves, such as the disk or a torus from which a piece has been removed. The two sides of such a surface could be painted with different colors to distinguish them. If the surface is closed, the two colors never meet. If the surface has boundary curves, the two colors meet only along these curves. A bug crawling along such a surface and prevented from crossing boundary curves, if any exist, would always remain on the same side.

Moebius made the surprising discovery that there are surfaces with only *one* side. The simplest such surface is the so-called Moebius strip, formed by taking a long rectangular strip of paper and pasting its two ends together after giving one a half-twist, as in Figure 139. A bug crawling along this surface, keeping always to the middle of the strip, will return to its original position upside down. The Moebius strip has only one edge, for its boundary consists of a single closed curve. The ordinary two-sided surface formed by pasting together the two ends of a rectangle without twisting has two distinct boundary curves. If the latter strip is cut along the center line it falls apart into two different strips of the same kind. But if the Moebius strip is cut along this line (shown in Figure 139) we find that it remains in one piece. It is rare for anyone not familiar with the Moebius strip to predict this

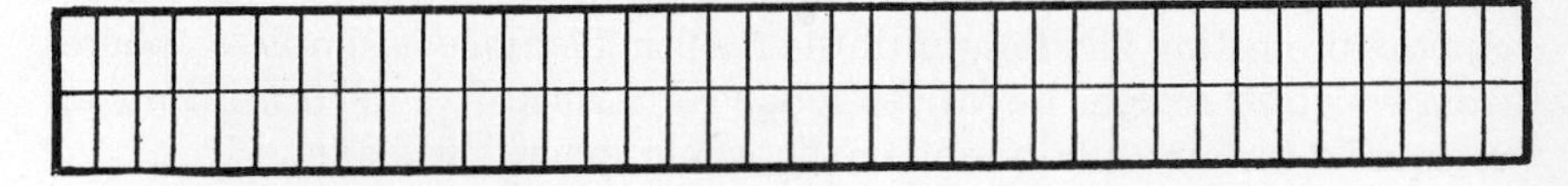

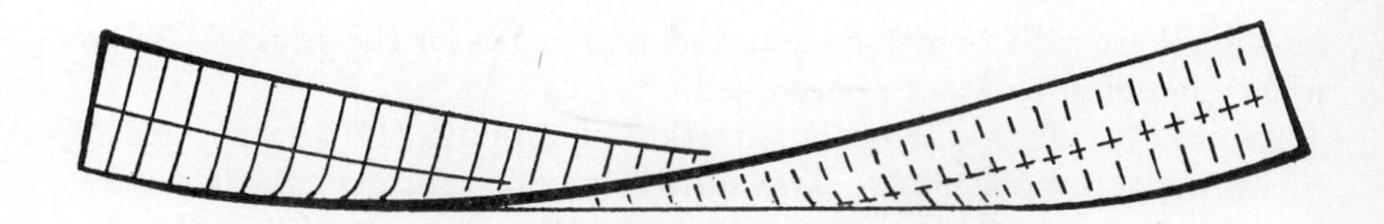

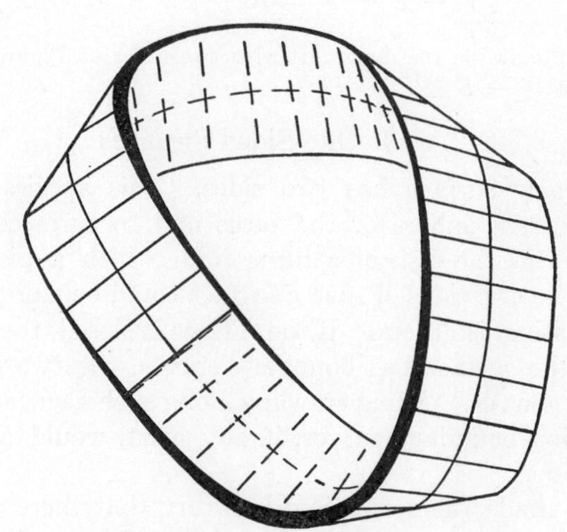

Fig. 139. Forming a Moebius strip.

behavior, so contrary to one's intuition of what "should" occur. If the surface that results from cutting the Moebius strip along the middle is again cut along its middle, two separate but intertwined strips are formed.

It is fascinating to play with such strips by cutting them along lines parallel to a boundary curve and 1/2, 1/3, etc. of the distance across.

The boundary of a Moebius strip is an unknotted closed curve which can be deformed into a flat one e.g. a circle. During the deformation, the strip may be allowed to intersect itself so that a onesided selfintersecting surface results as in Figure 140 known as a cross-cap. The locus of selfintersection is regarded as two different lines, each belonging to

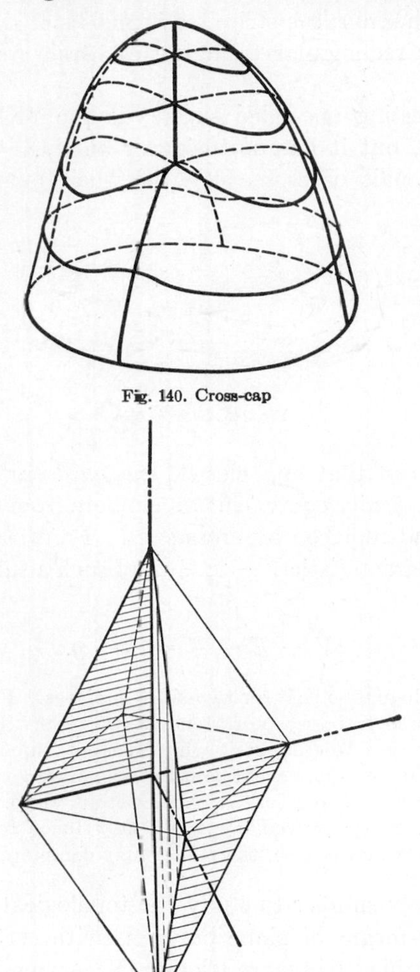

Fig. 140. Cross-cap

Fig. 141. Moebius strip with plane boundary.

one of the two portions of the surface which intersect there. The onesidedness of the Moebius strip is preserved because this property is topological; a one-sided surface cannot be continuously deformed into a two-sided surface. Strikingly enough it is even possible to conduct the deformation in such a way that the boundary of the Moebius strip becomes flat, e.g. triangular, while the strip remains free from selfintersections. Figure 141 indicates such a model, due to Dr. B. Tuckermann; the boundary is a triangle defining one half of one diagonal square of a regular octahedron; the strip itself consists of six faces of the octahedron and four rectangular triangles, each one fourth of a diagonal plane.

Another interesting one-sided surface is the "Klein bottle." This surface is closed, but it has no inside or outside. It is topologically equivalent to a pair of cross-caps with their boundaries coinciding.

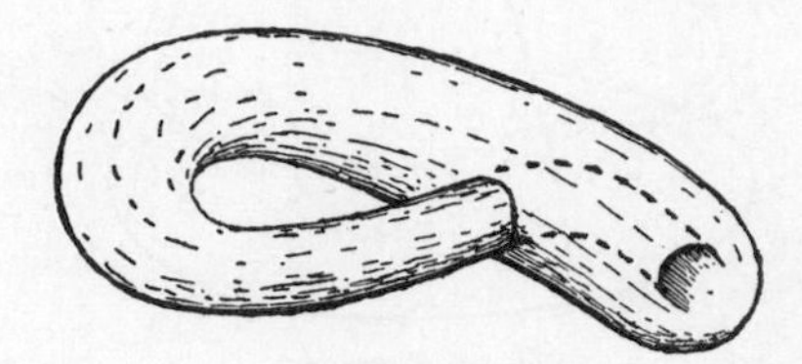

Fig. 142. Klein bottle.

It may be shown that any closed, *one-sided* surface of genus $p = 1, 2, \cdots$ is topologically equivalent to a sphere from which p disks have been removed and replaced by cross-caps. From this it easily follows that the Euler characteristic $V - E + F$ of such a surface is related to p by the equation

$$V - E + F = 2 - p.$$

The proof is analogous to that for two-sided surfaces. First we show that the Euler characteristic of a cross-cap or Moebius strip is 0. To do this we observe that, by cutting across a Moebius strip which has been subdivided into a number of regions, we obtain a rectangle that contains two more vertices, one more edge, and the same number of regions as the Moebius strip. For the rectangle, $V - E + F = 1$, as we proved on page 239. Hence for the Moebius strip $V - E + F = 0$. As an exercise, the reader may complete the proof.

It is considerably simpler to study the topological nature of surfaces such as these by means of plane polygons with certain pairs of edges conceptually identified (compare Chapt. IV, Appendix, Article 3). In the diagrams of Figure 143, parallel arrows are to be brought into coincidence—actual or conceptual—in position and direction.

This method of identification may also be used to define three-dimensional closed manifolds, analogous to the two-dimensional closed surfaces. For example, if we identify corresponding points of opposite

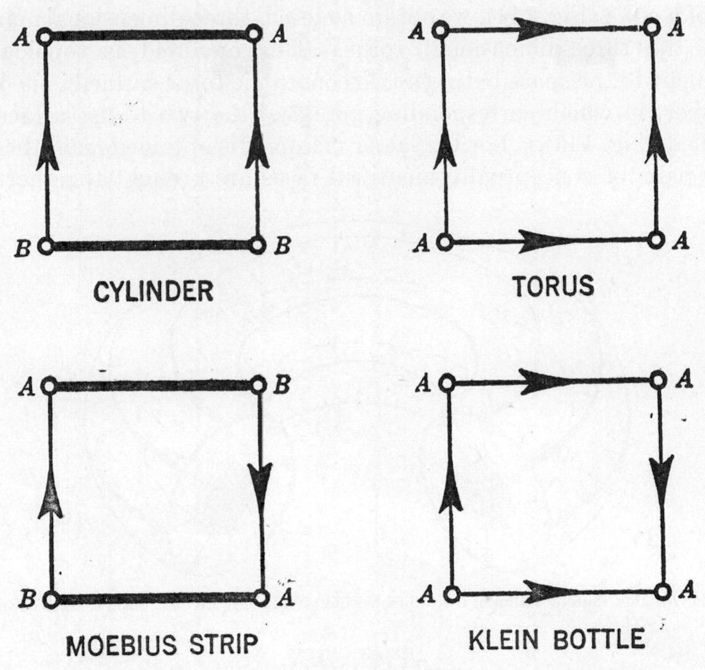

Fig. 143. Closed surfaces defined by coördination of edges in plane figure.

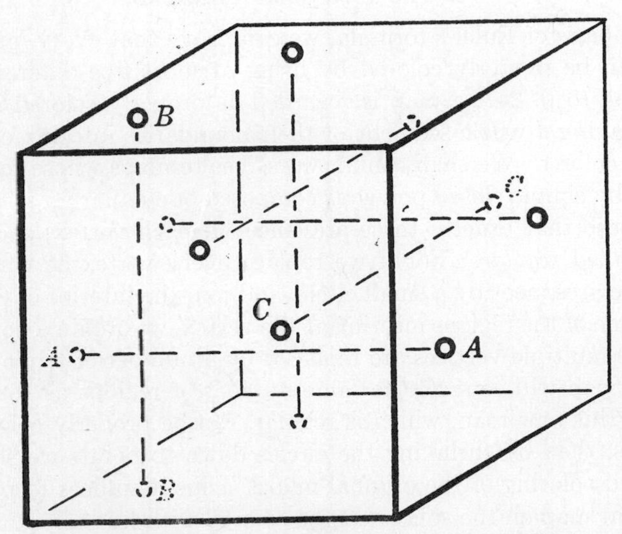

Fig. 144. Three-dimensional torus defined by boundary identification.

faces of a cube (Fig. 144), we obtain a closed, three-dimensional manifold called the three-dimensional torus. This manifold is topologically equivalent to the space between two concentric torus surfaces, one inside the other, in which corresponding points of the two torus surfaces are identified (Fig. 145). For the latter manifold is obtained from the cube if two pairs of conceptually identified faces are brought together.

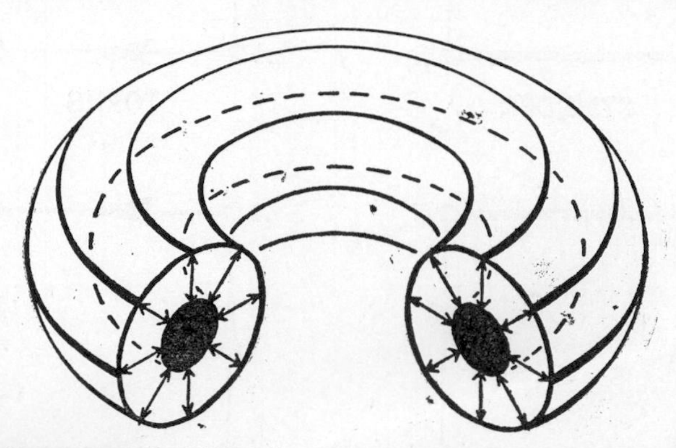

Fig. 145. Another representation of three-dimensional torus. (Figure cut to show identification.)

APPENDIX

*1. The Five Color Theorem

On the basis of Euler's formula, we can prove that every map on the sphere can be properly colored by using at most five different colors. (According to p. 247, a map is regarded as properly colored if no two regions having a whole segment of their boundaries in common receive the same color.) We shall confine ourselves to maps whose regions are bounded by simple closed polygons composed of circular arcs. We may also suppose that exactly three arcs meet at each vertex; such a map will be called *regular*. For if we replace every vertex at which more than three arcs meet by a small circle, and join the interior of each such circle to one of the regions meeting at the vertex, we obtain a new map in which the multiple vertices are replaced by a number of triple vertices. The new map will contain the same number of regions as the original map. If this new map, which is regular, can be properly colored with five colors, then by shrinking the circles down to points we shall have the desired coloring of the original map. Thus it suffices to prove that any regular map on the sphere can be colored with five colors.

First we show that every regular map must contain at least one polygon with fewer than six sides. Denote by F_n the number of regions of n sides in a regular map; then, if F denotes the total number of regions,

$$F = F_2 + F_3 + F_4 + \cdots . \tag{1}$$

Each arc has two ends, and three arcs end at each vertex. Hence, if E denotes the number of arcs in the map, and V the number of vertices,

$$2E = 3V. \tag{2}$$

Furthermore, a region bounded by n arcs has n vertices, and each vertex belongs to three regions, so that

$$2E = 3V = 2F_2 + 3F_3 + 4F_4 + \cdots . \tag{3}$$

By Euler's formula, we have

$$V - E + F = 2, \quad \text{or} \quad 6V - 6E + 6F = 12.$$

From (2), we see that $6V = 4E$, so that $6F - 2E = 12$.

Hence, from (1) and (3),

$$6(F_2 + F_3 + F_4 + \cdots) - (2F_2 + 3F_3 + 4F_4 + \cdots) = 12,$$

or

$$(6-2)F_2 + (6-3)F_3 + (6-4)F_4 + (6-5)F_5 + (6-6)F_6 + (6-7)F_7 + \cdots = 12.$$

Hence at least one of the terms on the left must be positive, so that at least one of the numbers F_2, F_3, F_4, F_5 is positive, as we wished to show.

Now to prove the five color theorem. Let M be any regular map on the sphere with n regions in all. We know that at least one of these regions has fewer than six sides.

Case 1. M contains a region A with 2, 3, or 4 sides. In this case, remove the boundary between A and one of the regions adjoining it. (If A has 4 sides, one region may come around and touch two non-adjacent sides of A. In this case, by the Jordan curve theorem, the regions touching the other two sides of A will be distinct, and we remove the boundary between A and one of the latter regions.)

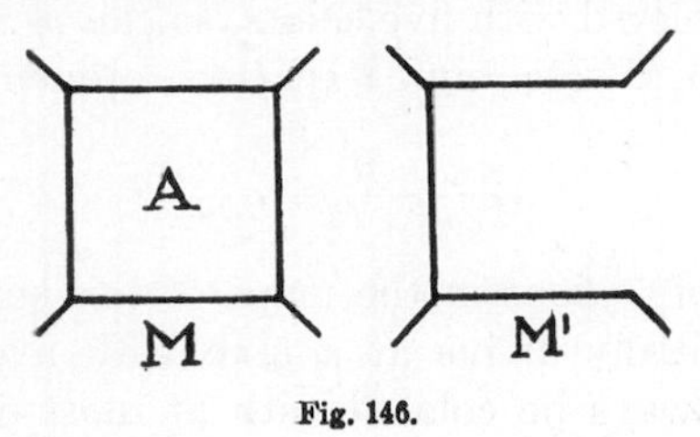

Fig. 146.

The resulting map M' will be a regular map with $n - 1$ regions. *If M' can be properly colored with 5 colors, so can M.* For since at most four regions of M adjoin A, we can always find a fifth color for A.

Case 2. M contains a region A with five sides. Consider the five regions adjoining A, and call them B, C, D, E, and F. We can always find a pair among these which do not touch each other; for if, say, B and D touch, they will prevent C from touching either E or F, since any path leading from C to E or F will have to go through at least one of the regions A, B, and D (Fig. 147). (It is clear that this fact, too, depends essentially on the Jordan curve theorem, which holds for the plane or sphere. It is not true on the torus, for example.) We may therefore assume, say, that C and F do not touch. We remove the sides of A

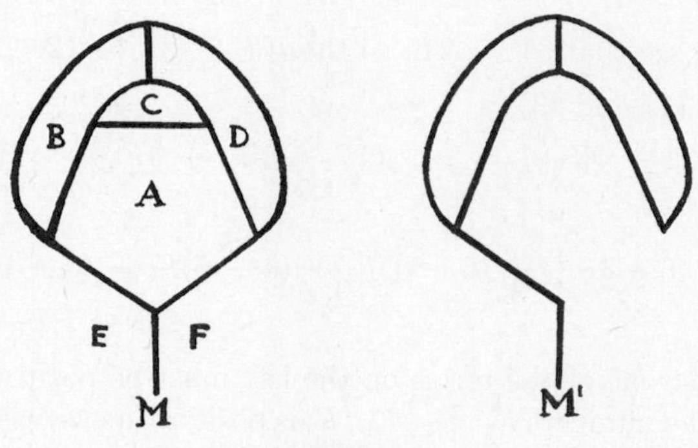

Fig. 147.

adjoining C and F, forming a new map M' with $n - 2$ regions, which is also regular. *If the new map can be properly colored with five colors, then so can the original map M.* For when the boundaries are restored, A will be in contact with no more than four different colors, since C and F have the same color, and we can therefore find a fifth color for A.

Thus in either case if M is a regular map with n regions, we can construct a new regular map M' having $n - 1$ or $n - 2$ regions, and such that if M' can be colored with five colors, so can M. This process may again be applied to M' etc., and leads to a sequence of maps derived from M:

$$M, M', M'', \cdots.$$

Since the number of regions in the maps of this sequence steadily decreases, we must finally arrive at a map with five or fewer regions. Such a map can always be colored with at most five colors. Hence, returning step by step to M, we see that M itself can be colored with five colors. This completes the proof. Note that this proof is constructive, in that it gives a perfectly practicable, although wearisome, method of actually coloring any map with n regions in a finite number of steps.

2. The Jordan Curve Theorem for Polygons

The Jordan curve theorem states that any simple closed curve C divides the points of the plane not on C into two distinct domains (with no points in common) of which C is the common boundary. We shall give a proof of this theorem for the case where C is a closed *polygon* P.

We shall show that the points of the plane not on P fall into two classes, A and B, such that any two points of the same class can be joined by a polygonal path which does not cross P, while any path joining a point of A to a point of B must cross P. The class A will form the "outside" of the polygon, while the class B will form the "inside."

We begin the proof by choosing a fixed direction in the plane, not parallel to any of the sides of P. Since P has but a finite number of sides, this is always possible. We now define the classes A and B as follows:

The point p belongs to A if the ray through p in the fixed direction intersects P in an *even* number, 0, 2, 4, 6, $\cdots$, of points. The point p belongs to B if the ray through p in the fixed direction intersects P in an *odd* number, 1, 3, 5, $\cdots$, of points.

With regard to rays that intersect P at vertices, we shall not count an intersection at a vertex where both edges of P meeting at the vertex are on the same side of the ray, but we shall count an intersection at a vertex where the two edges are on opposite sides of the ray. We shall say that two points p and q have the same "parity" if they belong to the same class, A or B.

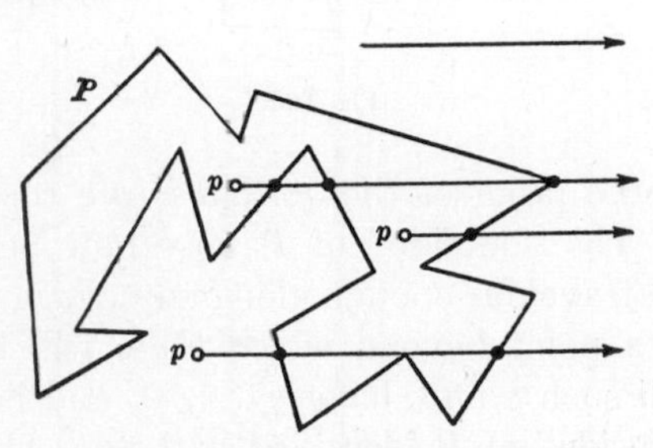

Fig. 148. Counting intersections.

First we observe that all the points on any line segment not intersecting P have the same parity. For the parity of a point p moving along such a segment can only change when the ray in the fixed direction through p passes through a vertex of P, and in neither of the two possible cases will the parity actually change, because of the agreement made in the preceding paragraph. From this it follows that *if any point* p_1 *of* A *is joined to a point* p_2 *of* B *by a polygonal path, then this path must intersect* P, for otherwise the parity of all the points of the path, and in particular of p_1 and p_2, would be the same. Moreover, we can show that *any two points of the same class,* A *or* B, *can be joined by a polygonal path which does not intersect* P. Call the two points p and q. If the straight segment pq joining p to q does not intersect P it is the desired path. Otherwise, let p' be the first point of intersection of this segment with P, and let q' be the last such point (Fig. 149). Construct the path starting from p along the segment pp', then turning off just before p' and following along P until P returns to pq at q'. If we can prove that this path will intersect pq between q' and q, rather than between p' and q', then the path may be continued to q along $q'q$ without intersecting P. It is clear that any two points r and s near enough to each other, but on opposite sides of some segment of P, must have different parity, for the ray through r will intersect P in one more point than will the ray through s. Thus we see that the parity changes as we cross the point q' along the segment pq. It follows that the dotted path crosses pq between q' and q, since p and q (and hence every point on the dotted path) have the same parity.

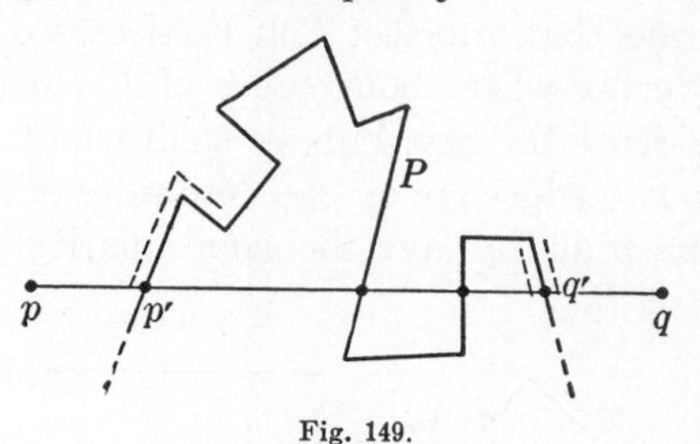

Fig. 149.

This completes the proof of the Jordan curve theorem for the case of a polygon P. The "outside" of P may now be identified as the class A, since if we travel far enough along any ray in the fixed direction we shall come to a point beyond which there will be no intersection with P, so that all such points have parity 0, and hence belong to A. This leaves the "inside" of P identified with the class B. No matter how twisted the simple closed polygon P, we can always determine whether a given point p of the plane is inside or outside P by drawing a ray and counting the number of intersections of the ray with P. If this number is odd, then the point p is imprisoned within P, and cannot escape without crossing P at some point. If the number is even, then the point p is outside P. (Try this for Figure 128.)

*One may also prove the Jordan curve theorem for polygons in the following way: Define the *order* of a point p_0 with respect to any closed curve C which does not pass through p_0 as the net number of complete revolutions made by an arrow joining p_0 to a moving point p on the curve as p traverses the curve once. Let

A = all points p_0 not on P and with *even* order with respect to P,

B = all points p_0 not on P and with *odd* order with respect to P.

Then A and B, thus defined, form the outside and inside of P respectively. The carrying out of the details of this proof is left as an exercise.

**3. The Fundamental Theorem of Algebra

The "fundamental theorem of algebra" states that if

$$f(z) = z^n + a_{n-1}z^{n-1} + a_{n-2}z^{n-2} + \cdots + a_1 z + a_0, \tag{1}$$

where $n \geq 1$, and $a_{n-1}, a_{n-2}, \cdots, a_0$ are any complex numbers, then there exists a complex number α such that $f(\alpha) = 0$. In other words, *in the field of complex numbers every polynomial equation has a root.* (On p. 102 we drew the conclusion that $f(z)$ can be factored into n linear factors:

$$f(z) = (z - \alpha_1)(z - \alpha_2) \cdots (z - \alpha_n),$$

where $\alpha_1, \alpha_2, \cdots, \alpha_n$ are the zeros of $f(z)$.) It is remarkable that this theorem can be proved by considerations of a topological character, related to those used in proving the Brouwer fixed point theorem.

The reader will recall that a complex number is a symbol $x + yi$, where x and y are real numbers and i has the property that $i^2 = -1$. The complex number $x + yi$ may be represented by the point in the plane whose coördinates with respect to a pair of perpendicular axes are x, y. If we introduce polar coördinates in this plane, taking the origin and the positive direction of the x-axis as pole and prime direction respectively, we may write

$$z = x + yi = r(\cos\theta + i\sin\theta),$$

where $r = \sqrt{x^2 + y^2}$. It follows from De Moivre's formula that

$$z^n = r^n(\cos n\theta + i\sin n\theta).$$

(See p. 96.) Thus, if we allow the complex number z to describe a circle of radius r about the origin, z^n will describe n complete times a circle of radius r^n as z describes its circle once. We also recall that r, the modulus of z, written $|z|$, gives the distance of z from O, and that if $z' = x' + iy'$, then $|z - z'|$ is the distance between z and z'. With these preliminaries we may proceed to the proof of the theorem.

Let us suppose that the polynomial (1) has no root, so that for every complex number z

$$f(z) \neq 0.$$

On this assumption, if we now allow z to describe any closed curve in the x, y-plane, $f(z)$ will describe a closed curve Γ which never passes

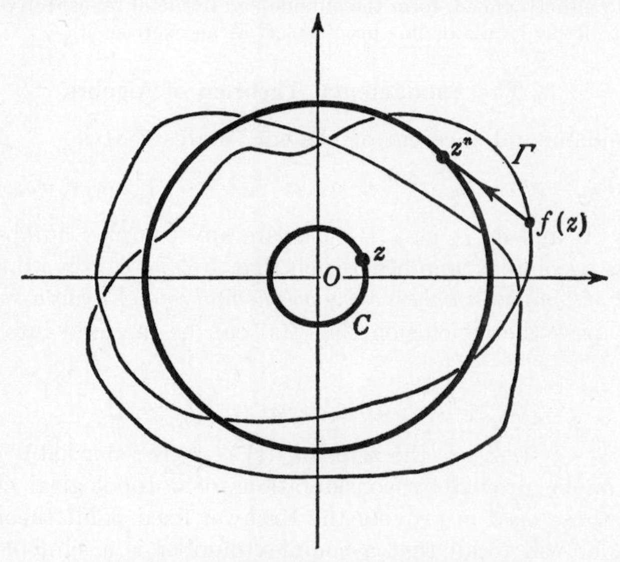

Fig. 150. Proof of fundamental theorem of algebra.

through the origin (Fig. 150). We may, therefore, define the *order* of the origin O with respect to the function $f(z)$ for any closed curve C as *the net number of complete revolutions made by an arrow joining* O *to a point on the curve* Γ *traced out by the point representing* f(z) as z traces out the curve C. As the curve C we shall take a circle with O as center and with radius t, and we define the function $\phi(t)$ to be the order of O with respect to the function $f(z)$ for the circle about O with radius t. Clearly $\phi(0) = 0$, since a circle with radius 0 is a single point, and the curve Γ reduces to the point $f(0) \neq O$. We shall show in the next paragraph that $\phi(t) = n$ for large values of t. But the order $\phi(t)$ depends continuously on t, since $f(z)$ is a continuous function of z. Hence we shall have a contradiction, for the function $\phi(t)$ can assume only integral values and therefore cannot pass continuously from the value 0 to the value n.

It remains only to show that $\phi(t) = n$ for large values of t. We observe that on a circle of radius $z = t$ so large that

$$t > 1 \quad \text{and} \quad t > |a_0| + |a_1| + \cdots + |a_{n-1}|,$$

we have the inequality

$$\begin{aligned} |f(z) - z^n| &= |a_{n-1}z^{n-1} + \cdots + a_0| \\ &\leq |a_{n-1}|\cdot|z|^{n-1} + |a_{n-2}|\cdot|z|^{n-2} + \cdots + |a_0| \\ &= t^{n-1}\left[|a_{n-1}| + \cdots + \frac{|a_0|}{t^{n-1}}\right] \\ &\leq t^{n-1}[|a_{n-1}| + |a_{n-2}| + \cdots + |a_0|] < t^n = |z^n|. \end{aligned}$$

Since the expression on the left is the distance between the two points z^n and $f(z)$, while the last expression on the right is the distance of the point z^n from the origin, we see that the straight line segment joining the two points $f(z)$ and z^n cannot pass through the origin so long as z is on the circle of radius t about the origin. This being so, we may continuously deform the curve traced out by $f(z)$ into the curve traced out by z^n without ever passing through the origin, simply by pushing each point $f(z)$ along the segment joining it to z^n. Since the order of the origin will vary continuously and can assume only integral values during this deformation, it must be the same for both curves. Since the order for z^n is n, the order for $f(z)$ must also be n. This completes the proof.

CHAPTER VI

FUNCTIONS AND LIMITS

INTRODUCTION

The main body of modern mathematics centers around the concepts of function and limit. In this chapter we shall analyze these notions systematically.

An expression such as

$$x^2 + 2x - 3$$

has no definite numerical value until the value of x is assigned. We say that the value of this expression is a *function* of the value of x, and write

$$x^2 + 2x - 3 = f(x).$$

For example, when $x = 2$ then $2^2 + 2 \cdot 2 - 3 = 5$, so that $f(2) = 5$. In the same way we may find by direct substitution the value of $f(x)$ for any integral, fractional, irrational, or even complex number x.

The number of primes less than n is a function $\pi(n)$ of the integer n. When a value of n is given, the value $\pi(n)$ is determined, even though no algebraic expression for computing it is known. The area of a triangle is a function of the lengths of its three sides; it varies as the lengths of the sides vary and is determined when these lengths are given definite values. If a plane is subjected to a projective or a topological transformation, then the coördinates of a point after the transformation depend on, i.e. are functions of, the original coördinates of the point. The concept of function enters whenever quantities are connected by a definite physical relationship. The volume of a gas enclosed in a cylinder is a function of the temperature and of the pressure on the piston. The atmospheric pressure as observed in a balloon is a function of the altitude above sea level. The whole domain of periodic phenomena—the motion of the tides, the vibrations of a plucked string, the emission of light waves from an incandescent filament—is governed by the simple trigonometric functions $\sin x$ and $\cos x$.

To Leibniz (1646–1716), who first used the word "function," and to the mathematicians of the eighteenth century, the idea of a functional relationship was more or less identified with the existence of a simple mathematical formula expressing the exact nature of the relationship. This concept proved too narrow for the requirements of mathematical physics, and the idea of a function, together with the related notion of limit, was subjected to a long process of generalization and clarification, of which we shall give an account in this chapter.

§1. VARIABLE AND FUNCTION

1. Definitions and Examples

Often mathematical objects occur which we are free to choose arbitrarily from a whole set S of objects. Then we call such an object a *variable* within the *range* or *domain* S. It is customary to use letters from the latter portion of the alphabet for variables. Thus if S denotes the set of all integers, the variable X with the domain S denotes an arbitrary integer. We say, "the variable X ranges over the set S," meaning that we are free to identify the symbol X with any member of the set S. The use of variables is convenient when we wish to make statements involving objects chosen at will from a whole set. For example, if S again denotes the set of integers and X and Y are both variables with the domain S, the statement

$$X + Y = Y + X$$

is a convenient symbolic expression of the fact that the sum of any two integers is independent of the order in which they are taken. A particular case is expressed by the equation

$$2 + 3 = 3 + 2,$$

involving constants, but to express the general law, valid for all pairs of numbers, symbols having the meaning of variables are needed.

It is by no means necessary that the domain S of a variable X be a set of numbers. For example, S might be the set of all circles in the plane; then X would denote any individual circle. Or S might be the set of all closed polygons in the plane, and X any individual polygon. Nor is it necessary that the domain of a variable contain an infinite number of elements. For example, X might denote any member of the population S of a given city at a given time. Or X might denote any one of the possible remainders when an integer is divided by 5; in this case the domain S would consist of the five numbers 0, 1, 2, 3, 4.

The most important case of a numerical variable—in this case we customarily use a small letter x—is that in which the domain of variability S is an interval $a \leq x \leq b$ of the real number axis. We then call x a *continuous variable* in the interval. The domain of variability of a continuous variable may be extended to infinity. Thus S may be the set of all positive real numbers, $x > 0$, or even the set of all real numbers without exception. In a similar way we may consider a variable X whose values are the points in a plane or in some given domain of the plane, such as the interior of a rectangle or of a circle. Since each point of the plane is defined by its two coördinates, x, y, with respect to a fixed pair of axes, we often say in this case that we have a *pair of continuous variables*, x and y.

It may be that with each value of a variable X there is associated a definite value of another variable U. Then U is called a *function* of X. The way in which U is related to X is expressed by a symbol such as

$$U = F(X) \qquad \text{(read, "}F\text{ of }X\text{")}.$$

If X ranges over the set S, then the variable U will range over another set, say T. For example, if S is the set of all triangles X in the plane, a function $F(X)$ may be defined by assigning to each triangle X the length, $U = F(X)$, of its perimeter; T will be the set of all positive numbers. Here we note that two different triangles, X_1 and X_2, may have the same perimeter, so that the equation $F(X_1) = F(X_2)$ is possible even though $X_1 \neq X_2$. A projective transformation of one plane, S, onto another, T, assigns to each point X of S a single point U of T according to a definite rule which we may express by the functional symbol $U = F(X)$. In this case $F(X_1) \neq F(X_2)$ whenever $X_1 \neq X_2$, and we say that the mapping of S onto T is *biunique* (see p. 78).

Functions of a continuous variable are often defined by algebraic expressions. Examples are the functions

$$u = x^2, \qquad u = \frac{1}{x}, \qquad u = \frac{1}{1 + x^2}.$$

In the first and last of these expressions, x may range over the whole set of real numbers; while in the second, x may range over the set of real numbers with the exception of 0—the value 0 being excluded since $1/0$ is not a number.

The number $B(n)$ of prime factors of n is a function of n, where n ranges over the domain of all natural numbers. More generally, any sequence of numbers, $a_1, a_2, a_3, \cdots$, may be regarded as the set of values of a function, $u = F(n)$, where the domain of the independent variable n is the set of natural numbers. It is only for brevity that we write a_n for the nth term of the sequence, instead of the more explicit functional notation $F(n)$ The expressions discussed in Chapter I,

$$S_1(n) = 1 + 2 + \cdots + n = \frac{n(n+1)}{2},$$

$$S_2(n) = 1^2 + 2^2 + \cdots + n^2 = \frac{n(n+1)(2n+1)}{6},$$

$$S_3(n) = 1^3 + 2^3 + \cdots + n^3 = \frac{n^2(n+1)^2}{4},$$

are functions of the integral variable n.

If $U = F(X)$ we usually reserve for X the name *independent variable*, while U is called the *dependent variable*, since its value depends on the value chosen for X.

It may happen that the same value of U is assigned to all values of X, so that the set T consists of one element only. We then have the special case where the value U of the function does not actually vary; that is, U is *constant*. We shall include this case under the general concept of function, even though this might seem strange to a beginner, for whom the emphasis naturally seems to lie in the idea that U varies when X does. But it will do no harm—and will in fact be useful—to regard a constant as the special case of a variable whose "domain of variation" consists of a single element only.

The concept of function is of the greatest importance, not only in pure mathematics but also in practical applications. Physical laws are nothing but statements concerning the way in which certain quantities depend on others when some of these are permitted to vary. Thus the pitch of the note emitted by a plucked string depends on the length, weight, and tension of the string, the pressure of the atmosphere depends on the altitude, and the energy of a bullet depends on its mass and velocity. The task of the physicist is to determine the exact or approximate nature of this functional dependence.

The function concept permits an exact mathematical characterization of motion. If a moving particle is concentrated at a point in space with rectangular coördinates x, y, z, and if t measures the time, then the motion of the particle is completely described by giving its coördinates x, y, z as functions of t:

$$x = f(t), \qquad y = g(t), \qquad z = h(t).$$

Thus, if a particle falls freely along the vertical z-axis under the influence of gravity alone,

$$x = 0, \qquad y = 0, \qquad z = -\tfrac{1}{2}gt^2,$$

where g is the acceleration due to gravity. If a particle rotates uniformly on a circle of unit radius in the x, y-plane, its motion is characterized by the functions

$$x = \cos \omega t, \qquad y = \sin \omega t,$$

where ω is a constant, the so-called angular velocity of the motion.

A mathematical function is simply a law governing the interdependence of variable quantities. It does not imply the existence of any relationship of "cause and effect" between them. Although in ordinary language the word "function" is often used with the latter connotation, we shall avoid all such philosophical interpretations. For example, Boyle's law for a gas contained in an enclosure at constant temperature states that the product of the pressure p and the volume v is a constant c (whose value in turn depends on the temperature):

$$pv = c.$$

This relation may be solved for either p or v as a function of the other variable,

$$p = \frac{c}{v} \quad \text{or} \quad v = \frac{c}{p},$$

without implying that a change in volume is the "cause" of a change in pressure any more than that the change in pressure is the "cause" of the change in volume It is only the form of the *connection* between the two variables which is relevant to the mathematician.

Mathematicians and physicists differ sometimes as to the aspect of the function concept on which they put the emphasis. The former usually stresses the *law of correspondence*, the mathematical operation that is applied to the independent variable x to obtain the value of the dependent variable u. In this sense $f(\ )$ is a symbol for a *mathematical operation*; the value $u = f(x)$ is the result of applying the operation $f(\ )$ to the number x. On the other hand, the physicist is often more interested in the *quantity* u as such than in any mathematical procedure by which the values of u can be computed from those of x. Thus the resistance u of the air to a moving object depends on the velocity v and can be found by experiment, whether or not an explicit mathematical formula for computing $u = f(v)$ is known. It is the actual resistance which primarily interests the physicist and not any particular mathematical formula $f(v)$, except insofar as the study of such a formula may aid in analyzing the behavior of the quantity u. This is the attitude ordinarily taken if one *applies* mathematics to physics or engineering. In more advanced calculations with functions confusion can sometimes be avoided only by knowing exactly whether one means the operation $f(\ )$ which assigns to x a quantity $u = f(x)$, or the quantity u itself, which may also be considered to depend, in a quite different manner, on some other variable, z. For example, the area of a circle is given by the function $u = f(x) = \pi x^2$, where x is the radius, and also by the function $u = g(z) = z^2/4\pi$, where z is the circumference.

Perhaps the simplest types of mathematical functions of one variable are the *polynomials*, of the form

$$u = f(x) = a_0 + a_1 x + a_2 x^2 + \cdots + a_n x^n,$$

with constant "coefficients," $a_0, a_1, \cdots, a_n$. Next come the *rational functions*, such as

$$u = \frac{1}{x}, \qquad u = \frac{1}{1 + x^2}, \qquad u = \frac{2x + 1}{x^4 + 3x^2 + 5},$$

which are quotients of polynomials, and the *trigonometric functions*, $\cos x$, $\sin x$, and $\tan x = \sin x/\cos x$, which are best defined by reference to the unit circle in the ξ, η-plane, $\xi^2 + \eta^2 = 1$. If the point $P(\xi, \eta)$ moves on the circumference of this circle, and if x is the directed angle through which the positive ξ-axis must be rotated in order to coincide with OP, then $\cos x$ and $\sin x$ are the coördinates of P: $\cos x = \xi$, $\sin x = \eta$.

2. Radian Measure of Angles

For all practical purposes angles are measured in units obtained by subdividing a right angle into a number of equal parts. If this number is 90, then the unit is the familiar "degree." A subdivision into 100 parts would be better adapted to our decimal system, but would represent the same principle of measuring. For theoretical purposes, however, it is advantageous to use an essentially different method of characterizing the size of an angle, the so-called radian measure. Many important formulas involving the trigonometric functions of angles have a simpler form in this system than if the angles are measured in degrees.

To find the radian measure of an angle we describe a circle of radius 1 about the vertex of the angle. The angle will cut out an arc s on the circumference of this circle, and we define the length of this arc as the *radian measure* of the angle. Since the total circumference of a circle with radius 1 has the length 2π, the full angle of 360° has the radian measure 2π. It follows that if x denotes the radian measure of an angle and y its degree measure, then x and y are connected by the relation $y/360 = x/2\pi$ or

$$\pi y = 180x.$$

Thus an angle of 90° ($y = 90$) has the radian measure $x = 90\pi/180 = \pi/2$, etc. On the other hand, an angle of 1 radian (the angle with radian measure $x = 1$) is the angle that cuts out an arc equal to the radius of the circle; in degrees this will be an angle of $y = 180/\pi = 57.2957 \cdots$ degrees. We must always multiply the radian measure x of an angle by the factor $180/\pi$ to obtain its degree measure y.

The radian measure x of an angle is also equal to twice the area A of the sector of the unit circle cut out by the angle; for this area bears to the whole area of the circle the ratio which the arc along the circumference bears to the whole circumference: $x/2\pi = A/\pi$, $x = 2A$.

Henceforth the angle x will mean the angle whose radian measure is x. An angle of x degrees will be written $x°$, to avoid ambiguity.

It will become apparent that radian measure is very convenient for analytic operations. For practical use, however, it would be rather inconvenient. Since π is irrational, we shall never return to the same point of the circle if we mark off repeatedly the unit angle, i.e. the angle of radian measure 1. The ordinary measure is so devised that after marking off 1 degree 360 times, or 90 degrees 4 times, we return to the same position.

3. The Graph of a Function. Inverse Functions

The character of a function is often most clearly shown by a simple geometrical graph. If x, u are coördinates in a plane with respect to a pair of perpendicular axes, then linear functions such as

$$u = ax + b$$

are represented by straight lines; quadratic functions such as

$$u = ax^2 + bx + c$$

by parabolas; the function

$$u = \frac{1}{x}$$

by a hyperbola, etc. By definition, the *graph* of any function $u = f(x)$ consists of all the points in the plane whose coördinates x, u are in the relationship $u = f(x)$. The functions sin x, cos x, tan x, are represented by the curves in Figures 151 and 152. These graphs show clearly how the values of the functions increase or decrease as x varies.

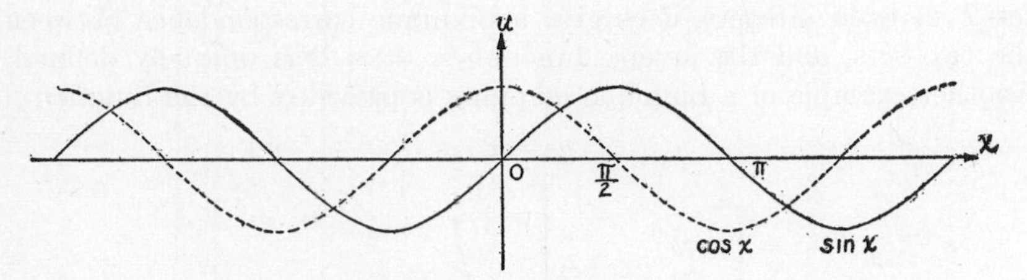

Fig. 151. Graphs of sin x and cos x.

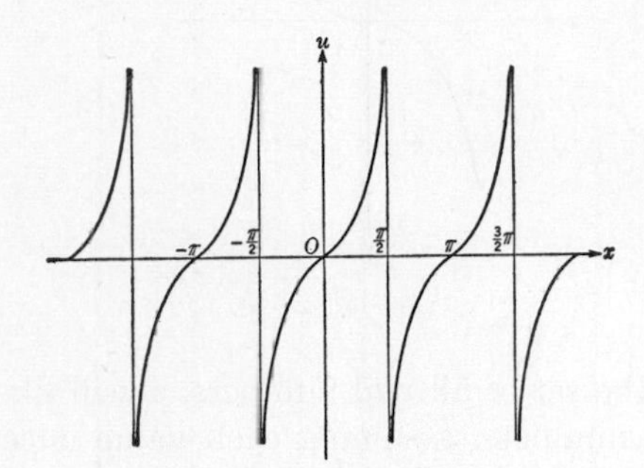

Fig. 152. $u = \tan x$.

An important method for introducing new functions is the following. Beginning with a known function, $F(X)$, we may try to solve the equation $U = F(X)$ for X, so that X will appear as a function of U:

$$X = G(U).$$

The function $G(U)$ is then called an *inverse function* of $F(X)$. This process leads to a unique result only if the function $U = F(X)$ defines a biunique mapping of the domain of X onto that of U, i.e. if the inequality $X_1 \neq X_2$ always implies the inequality $F(X_1) \neq F(X_2)$, for only then will there be a uniquely defined X correlated with each U. Our previous example in which X denoted any triangle in the plane and $U = F(X)$ was its perimeter is a case in point. Obviously this mapping of the set S of triangles onto the set T of positive real numbers is not biunique, since there are infinitely many different triangles with the same perimeter. Hence in this case the relation $U = F(X)$ does not serve to define a unique inverse function. On the other hand, the function $m = 2n$, where n ranges over the set S of integers and m over the set T of even integers, does give a biunique correspondence between the two sets, and the inverse function $n = m/2$ is uniquely defined. Another example of a biunique mapping is provided by the function

$$u = x^3.$$

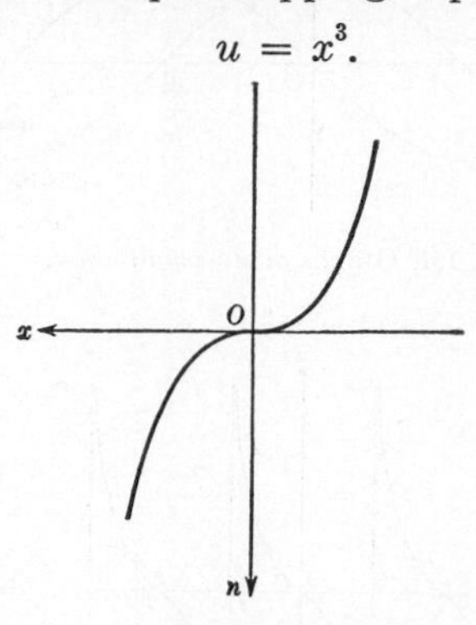

Fig. 153. $u = x^3$.

As x ranges over the set of all real numbers, u will likewise range over the set of all real numbers, assuming each value once and only once. The uniquely defined inverse function is

$$x = \sqrt[3]{u}.$$

In the case of the function

$$u = x^2,$$

an inverse function is not uniquely determined. For since $u = x^2 = (-x)^2$, each positive value of u will have *two* antecedents. But if, as is customary, we define the symbol $\sqrt{u}$ to mean the *positive* number whose square is u, then the inverse function

$$x = \sqrt{u}$$

exists, so long as we restrict x and u to positive values.

The existence of a unique inverse of a function of one variable, $u = f(x)$, can be seen by a glance at the graph of the function. The inverse function will be uniquely defined only if to each value of u there corresponds but one value of x. In terms of the graph, this means that no parallel to the x-axis intersects the graph in more than one point. This will certainly be the case if the function $u = f(x)$ is *monotone*, i.e. steadily increasing or steadily decreasing as x increases. For example, if $u = f(x)$ is steadily increasing, then for $x_1 < x_2$ we always have $u_1 = f(x_1) < u_2 = f(x_2)$. Hence for a given value of u there can be at most one x such that $u = f(x)$, and the inverse function will be uniquely defined. The graph of the inverse function $x = g(u)$ is obtained merely by rotating the original graph through an angle of 180° about the dotted line (Fig. 154), so that the positions of the x-axis and the u-axis are interchanged. The new position of the graph will depict x as a function of u. In its original position the graph shows u as the height above the horizontal x-axis, while after the rotation the same graph shows x as the height above the horizontal u-axis.

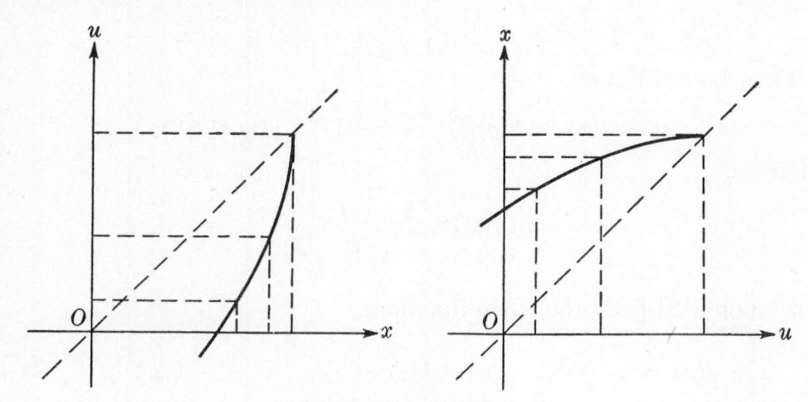

Fig. 154. Inverse functions.

The considerations of the preceding paragraph may be illustrated for the case of the function

$$u = \tan x.$$

This function is monotone for $-\pi/2 < x < \pi/2$ (Fig. 152). The values of u, which increase steadily with x, range from $-\infty$ to $+\infty$; hence the inverse function,

$$x = g(u),$$

is defined for all values of u. This function is denoted by $\tan^{-1} u$ or $\text{arc} \tan u$. Thus $\text{arc} \tan(1) = \pi/4$, since $\tan \pi/4 = 1$. Its graph is shown in Figure 155.

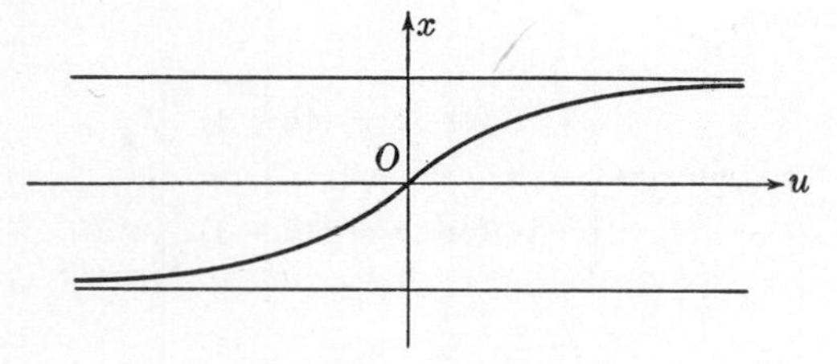

Fig. 155. $x = \text{arc} \tan u$.

4. Compound Functions

A second important method for creating new functions from two or more given ones is the *compounding* of functions. For example, the function

$$u = f(x) = \sqrt{1 + x^2}$$

is "compounded" from the two simpler functions

$$z = g(x) = 1 + x^2, \qquad u = h(z) = \sqrt{z},$$

and can be written as

$$u = f(x) = h(g[x]) \qquad \text{(read, "}h\text{ of }g\text{ of }x\text{")}.$$

Likewise,

$$u = f(x) = \frac{1}{\sqrt{1 - x^2}}$$

is compounded from the three functions

$$z = g(x) = 1 - x^2, \qquad w = h(z) = \sqrt{z}, \qquad u = k(w) = \frac{1}{w},$$

so that

$$u = f(x) = k(h[g(x)]).$$

The function

$$u = f(x) = \sin\frac{1}{x}$$

is compounded from the two functions

$$z = g(x) = \frac{1}{x}, \qquad u = h(z) = \sin z.$$

The function $f(x)$ is not defined for $x = 0$, since for $x = 0$ the expression $1/x$ has no meaning. The graph of this remarkable function is obtained from that of the sine. We know that $\sin z = 0$ for $z = k\pi$, where k is any positive or negative integer. Furthermore,

$$\sin z = \begin{cases} 1 & \text{for} \quad z = (4k + 1)\dfrac{\pi}{2}, \\ -1 & \text{for} \quad z = (4k - 1)\dfrac{\pi}{2}, \end{cases}$$

if k is any integer. Hence

$$\sin\frac{1}{x} = \begin{cases} 0 & \text{for} \quad x = \dfrac{1}{k\pi}, \\ 1 & \text{for} \quad x = \dfrac{2}{(4k+1)\pi}, \\ -1 & \text{for} \quad x = \dfrac{2}{(4k-1)\pi}. \end{cases}$$

If we set successively

$$k = 1, 2, 3, 4, \cdots,$$

then, since the denominators of these fractions increase without limit, the values of x for which the function $\sin(1/x)$ has the values $1, -1, 0$, will cluster nearer and nearer to the point $x = 0$. Between any such point and the origin there will still be an infinite number of oscillations of the function. The graph of the function is shown in Figure 156.

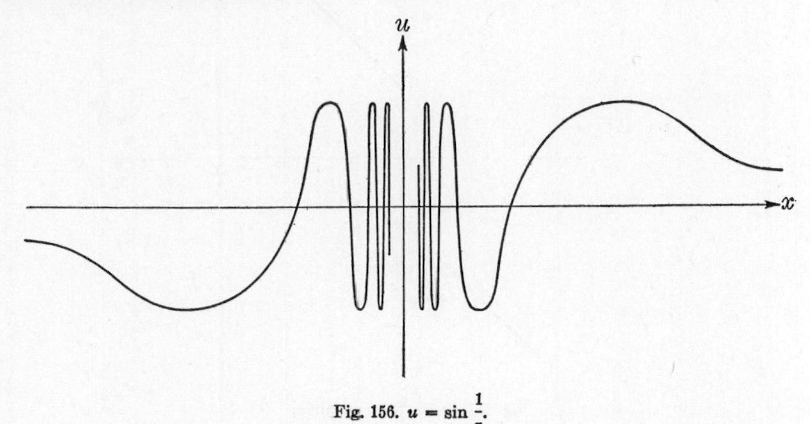

Fig. 156. $u = \sin\frac{1}{x}$.

5. Continuity

The graphs of the functions so far considered give an intuitive idea of the property of continuity. We shall give a precise analysis of this concept in §4, after the limit concept has been put on a rigorous basis. But roughly speaking, we say that a function is continuous if its graph is an uninterrupted curve (see p. 310). A given function $u = f(x)$ may be tested for continuity by letting the independent variable x move continuously from the right side and from the left side towards any specified value x_1. Unless the function $u = f(x)$ is constant in the neighborhood of x_1, its value will also change. If the value $f(x)$ approaches as a limit the value $f(x_1)$ of the function at the specified point $x = x_1$, *no matter whether we approach* x_1 *from one side or the other*, then the function is said to be *continuous at* x_1. If this holds for every point x_1 of a certain interval, then the function is said to be *continuous in the interval.*

Although every function represented by an unbroken graph is continuous, it is quite easy to define functions that are not everywhere continuous. For example, the function of Figure 157, defined for all values of x by setting

$$f(x) = 1 + x \qquad \text{for} \quad x > 0$$

$$f(x) = -1 + x \qquad \text{for} \quad x \leq 0$$

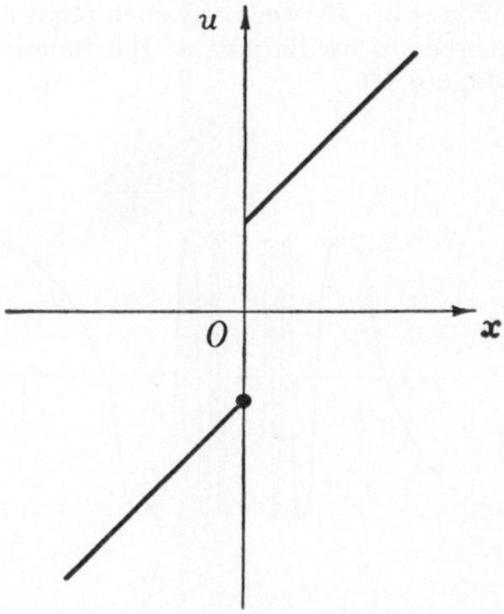

Fig. 157. Jump discontinuity.

is discontinuous at the point $x_1 = 0$, where it has the value -1. If we try to draw a graph of this function, we shall have to lift our pencil from the paper at this point. If we approach the value $x_1 = 0$ from the right side, then $f(x)$ approaches $+1$. But this value differs from the actual value, -1, at this point. The fact that -1 is approached by $f(x)$ as x tends to zero from the *left* side does not suffice to establish continuity.

The function $f(x)$ defined for all x by setting

$$f(x) = 0 \quad \text{for} \quad x \neq 0, \qquad f(0) = 1,$$

presents a discontinuity of a different sort at the point $x_1 = 0$. Here both right- and left-hand limits exist and are equal as x approaches 0, but this common limiting value differs from $f(0)$.

Another type of discontinuity is shown by the function of Figure 158,

$$u = f(x) = \frac{1}{x^2},$$

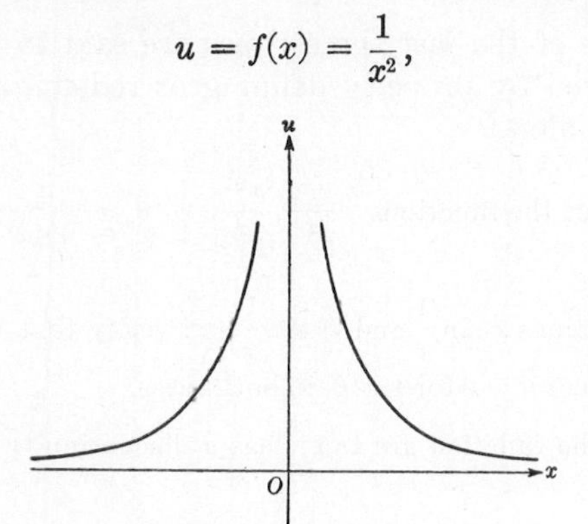

Fig. 158. Infinite discontinuity.

at the point $x = 0$. If x is allowed to approach zero from either side, u tends to infinity; the graph of the function is broken at this point, and small changes of x in the neighborhood of $x = 0$ may produce very large changes in u. Strictly speaking, the value of the function is not defined for $x = 0$, since we do not admit infinity as a number and therefore we cannot say that $f(x)$ *is* infinite when $x = 0$. Hence we say only that $f(x)$ "tends to infinity" as x approaches zero.

A still different type of discontinuity appears in the function $u = \sin(1/x)$ at the point $x = 0$, as is apparent from the graph of that function (Fig. 156).

The preceding examples exhibit several ways in which a function can fail to be continuous at a point $x = x_1$:

1) It may be possible to make the function continuous at $x = x_1$ by properly defining or redefining its value when $x = x_1$. For example, the function $u = x/x$ is constantly equal to 1 when $x \neq 0$; it is not defined for $x = 0$, since $0/0$ is a meaningless symbol. But if we agree in this case that the value $u = 1$ shall also correspond to the value $x = 0$, then the function so extended becomes continuous for every value of x without exception. The same effect is produced if we redefine $f(0) = 0$ for the function defined at the bottom of the preceding page. A discontinuity of this kind is said to be *removable.*

2) Different limits may be approached by the function as x approaches x_1 from the right and from the left, as in Figure 157.

3) Even one-sided limits may not exist, as in Figure 156.

4) The function may tend to infinity as x approaches x_1, as in Figure 158.

Discontinuities of the last three types are said to be *essential*; they cannot be removed by properly defining or redefining the function at the point $x = x_1$ alone.

Exercises: 1) Plot the functions $\frac{x-1}{x^2}, \frac{x^2-1}{x^2+1}, \frac{x}{(x^2-1)(x^2+1)}$ and find their discontinuities.

2) Plot the functions $x \sin \frac{1}{x}$ and $x^2 \sin \frac{1}{x}$ and verify that they are continuous at $x = 0$, if one defines $u = 0$ for $x = 0$, in both cases.

*3) Show that the function arc tan $\frac{1}{x}$ has a discontinuity of the second type (jump) at $x = 0$.

*6. Functions of Several Variables

We return to our systematic discussion of the function concept. If the independent variable P is a point in the plane with coördinates x, y, and if to each such point P corresponds a single number u—for example, u might be the distance of the point P from the origin—then we usually write

$$u = f(x, y).$$

This notation is also used if, as often happens, two quantities x and y appear from the outset as independent variables. For example, the pressure u of a gas is a function of the volume x and the temperature y, and the area u of a triangle is a function $u = f(x, y, z)$ of the lengths x, y, and z of its three sides.

In the same way that a graph gives a geometrical representation of a function of one variable, a geometrical representation of a function $u = f(x, y)$ of two variables is afforded by a surface in the three-dimensional space with x, y, u as coördinates. To each point x, y in the x, y-plane we assign the point in space whose coördinates are x, y, and $u = f(x, y)$. Thus the function $u = \sqrt{1 - x^2 - y^2}$ is represented by a spherical surface with the equation $u^2 + x^2 + y^2 = 1$, the linear function $u = ax + by + c$ by a plane, the function $u = xy$ by a hyperbolic paraboloid, etc.

A different representation of the function $u = f(x, y)$ may be given in the x, y-plane alone by means of *contour lines*. Instead of considering the three-dimensional "landscape" $u = f(x, y)$, we draw, as on a contour map, the level curves of the function, indicating the projections on the x, y-plane of all points with equal vertical elevation u. These level curves are simply the curves $f(x, y) = c$, where c remains constant for each curve. Thus the function $u = x + y$ is characterized by Figure

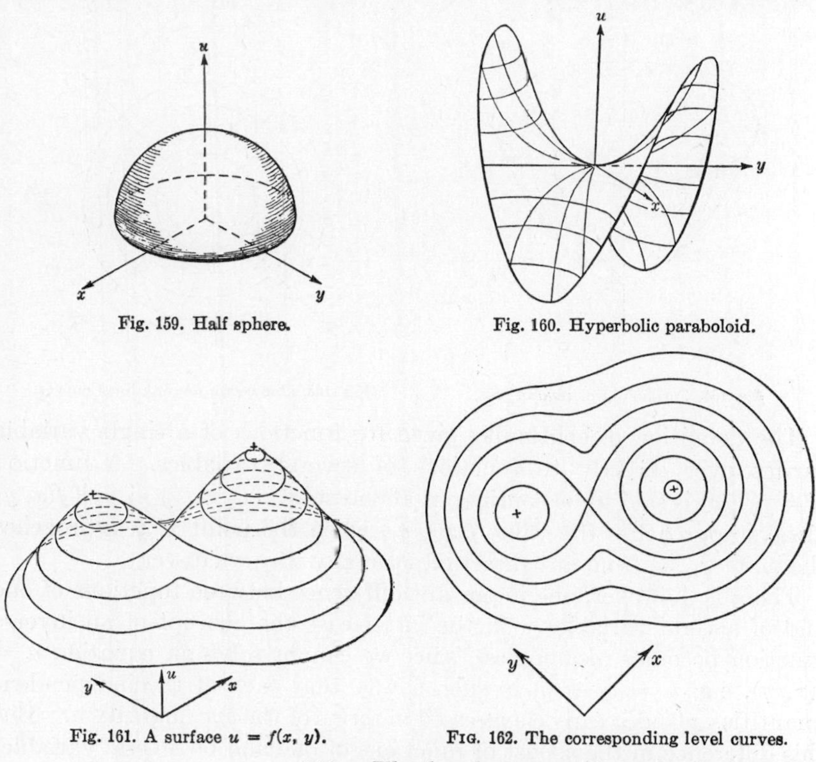

Fig. 159. Half sphere.

Fig. 160. Hyperbolic paraboloid.

Fig. 161. A surface $u = f(x, y)$.

Fig. 162. The corresponding level curves.

Fig. 163. Level curves of $u = x + y$.

163. The level curves of a spherical surface are a set of concentric circles. The function $u = x^2 + y^2$ representing a paraboloid of revolution is likewise characterized by circles (Fig. 165). By numbers attached to the different curves one may indicate the height $u = c$.

Functions of several variables occur in physics when the motion of a continuous substance is to be described. For example, suppose a string is stretched between two points on the x-axis and then deformed so that the particle with the position x is moved a certain distance perpendicularly to the axis. If the string is then released, it will vibrate in such a way that the particle with the original coördinate x will have at the time t a distance $u = f(x, t)$ from the x-axis. The motion is completely described as soon as the function $u = f(x, t)$ is known.

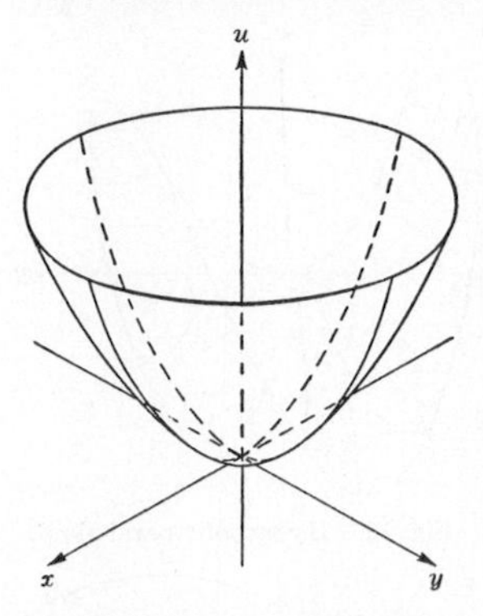

Fig. 164. Paraboloid of revolution.

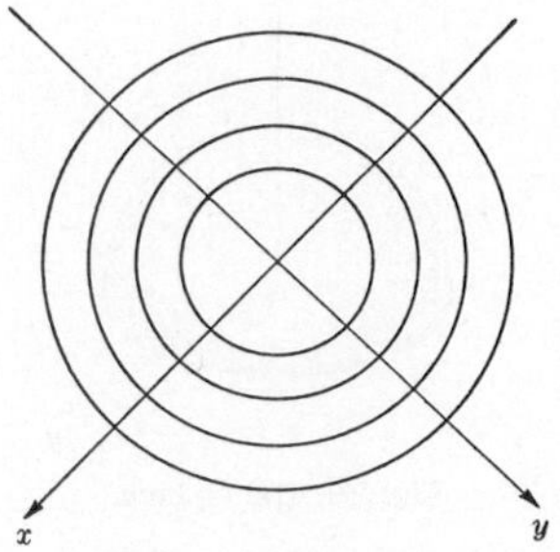

Fig. 165. The corresponding level curves.

The definition of continuity given for functions of a single variable carries over directly to functions of several variables. A function $u = f(x, y)$ is said to be continuous at the point $x = x_1$, $y = y_1$ if $f(x, y)$ always approaches the value $f(x_1, y_1)$ when the point x, y approaches the point x_1, y_1 from any direction or in any way whatever.

There is, however, one important difference between functions of one and of several variables. In the latter case the concept of an inverse function becomes meaningless, since we cannot solve an equation $u = f(x, y)$, e.g. $u = x + y$, in such a way that each of the independent quantities x and y can be expressed in terms of the *one* quantity u. But this difference in the aspect of functions of one and of several variables disappears if we emphasize the idea of a function as defining a mapping or transformation.

*7. Functions and Transformations

A correspondence between the points of one line l, characterized by a coördinate x along the line, and the points of another line l', characterized by a coördinate x', is simply a function $x' = f(x)$. In case the correspondence is biunique we also have an inverse function $x = g(x')$. The simplest example is a transformation by projection, which—we state here without proof—is characterized in general by a function of the form $x' = f(x) = (ax + b)/(cx + d)$, where a, b, c, d, are constants. In this case, the inverse function is $x = g(x') = (-dx' + b)/(cx' - a)$.

Mappings in two dimensions from a plane π with coördinates x, y

onto a plane π' with coördinates x', y' cannot be represented by a single function $x' = f(x)$, but require two functions of two variables:

$$x' = f(x, y),$$
$$y' = g(x, y).$$

For example, a projective transformation is given by a function system,

$$x' = \frac{ax + by + c}{gx + hy + k},$$
$$y' = \frac{dx + ey + f}{gx + hy + k},$$

where $a, b, \cdots, k$ are constants, and where x, y and x', y' are coördinates in the two planes respectively. From this point of view the idea of an inverse transformation makes sense. We simply have to *solve this system of equations* for x and y in terms of x' and y'. Geometrically, this amounts to finding the inverse mapping of π' onto π. This will be uniquely defined, provided the correspondence between the points of the two planes is biunique.

The transformations of the plane studied in topology are given, not by simple algebraic equations, but by any system of functions,

$$x' = f(x, y),$$
$$y' = g(x, y),$$

that define a biunique and bicontinuous transformation.

Exercises: *1) Show that the transformation of inversion (Chapter III, p. 141) in the unit circle is given analytically by the equations $x' = x/(x^2 + y^2)$, $y' = y/(x^2 + y^2)$. Find the inverse transformation. Prove analytically that inversion transforms the totality of lines and circles into lines and circles.

2) Prove that by a transformation $x' = (ax + b)/(cx + d)$ four points of the x-axis are transformed into four points of the x'-axis with the same cross-ratio. (See p. 175.)

§2. LIMITS

1. The Limit of a Sequence a_n

As we have seen in §1, the description of the continuity of a function is based on the limit concept. Up to now we have used this concept in a more or less intuitive form. In this and the following sections we shall consider it in a more systematic way. Since sequences are rather simpler than functions of a continuous variable, we shall begin with a study of sequences.

In Chapter II we encountered sequences a_n of numbers and studied their limits as n increases indefinitely or "tends to infinity." For example, the sequence whose nth term is $a_n = 1/n$,

$$1, \frac{1}{2}, \frac{1}{3}, \cdots, \frac{1}{n}, \cdots, \tag{1}$$

has the limit 0 for increasing n:

$$\frac{1}{n} \to 0 \quad \text{as} \quad n \to \infty. \tag{2}$$

Let us try to state exactly what is meant by this. As we go out farther and farther in the sequence, the terms become smaller and smaller. After the 100th term all the terms are smaller than 1/100, after the 1000th term all the terms are smaller than 1/1000, and so on. None of the terms is actually equal to 0. But if we go out *far enough* in the sequence (1), we can be sure that each of its terms will differ from 0 by *as little as we please.*

The only trouble with this explanation is that the meaning of the italicized phrases is not entirely clear. How far is "far enough," and how little is "as little as we please"? If we can attach a precise meaning to these phrases then we can give a precise meaning to the limiting relation (2).

A geometric interpretation will help to make the situation clearer. If we represent the terms of the sequence (1) by their corresponding points on the number axis we observe that the terms of the sequence appear to cluster around the point 0. Let us choose any interval I on the number axis with center at the point 0 and total width 2ϵ, so that the interval extends a distance ϵ on each side of the point 0. If we choose $\epsilon = 10$, then, of course, *all* the terms $a_n = 1/n$ of the sequence will lie inside the interval I. If we choose $\epsilon = 1/10$, then the first few terms of the sequence will lie outside I, but all the terms from a_{11} cn,

$$\frac{1}{11}, \frac{1}{12}, \frac{1}{13}, \frac{1}{14}, \cdots,$$

will lie within I. Even if we choose $\epsilon = 1/1000$, only the first thousand terms of the sequence will fail to lie within I, while from the term a_{1001} on, all the infinitely many terms

$$a_{1001}, a_{1002}, a_{1003}, \cdots$$

will lie within I. Clearly, this reasoning holds for any positive number ϵ: as soon as a positive ϵ is chosen, no matter how small it may be, we can then find an integer N so large that

$$\frac{1}{N} < \epsilon.$$

From this it follows that all the terms a_n of the sequence for which $n \geq N$ will lie within I, and only the finite number of terms $a_1, a_2, \cdots, a_{N-1}$ can lie outside. The important point is this: *First* the width of the interval I is assigned at pleasure by choosing ϵ. *Then* a suitable integer N can be found. This process of first choosing a number ϵ and then finding a suitable integer N can be carried out for any positive number ϵ, no matter how small, and gives a precise meaning to the statement that all the terms of the sequence (1) will differ from 0 by as little as we please, provided we go out far enough in the sequence.

To summarize: Let ϵ be any positive number. Then we can find an integer N such that all the terms a_n of the sequence (1) for which $n \geq N$ will lie within the interval I of total width 2ϵ and with center at the point 0. This is the precise meaning of the limiting relation (2).

On the basis of this example we are now ready to give an exact definition of the general statement: "The sequence of real numbers $a_1, a_2, a_3, \cdots$ has the limit a." We include a in the interior of an interval I of the number axis: if the interval is small, some of the numbers a_n may lie outside the interval, but as soon as n becomes sufficiently large, say greater than or equal to some integer N, then all the numbers a_n for which $n \geq N$ must lie within the interval I. Of course, the integer N may have to be taken very large if a very small interval I is chosen, but no matter how small the interval I, such an integer N must exist if the sequence is to have a as its limit.

The fact that a sequence a_n has the limit a is expressed symbolically by writing

$$\lim a_n = a \qquad \text{as } n \to \infty,$$

or simply

$$a_n \to a \qquad \text{as } n \to \infty$$

(read: *a_n tends to a*, or *converges to a*). The definition of the convergence of a sequence a_n to a may be formulated more concisely as follows: *The sequence* $a_1, a_2, a_3, \cdots$ *has the limit* a *as* n *tends to infinity if, corre-*

sponding to any positive number ϵ, *no matter how small, there may be found an integer* N (*depending on* ϵ), *such that*

$$|a - a_n| < \epsilon \tag{3}$$

for all

$$n \geq N.$$

This is the abstract formulation of the notion of the limit of a sequence. Small wonder that when confronted with it for the first time one may not fathom it in a few minutes. There is an unfortunate, almost snobbish attitude on the part of some writers of textbooks, who present the reader with this definition without a thorough preparation, as though an explanation were beneath the dignity of a mathematician.

The definition suggests a contest between two persons, A and B. A sets the requirement that the fixed quantity a should be approached by a_n with a degree of accuracy better than a chosen margin $\epsilon = \epsilon_1$; B meets the requirement by demonstrating that there is a certain integer $N = N_1$ such that all the a_n after the element a_{N_1} satisfy the ϵ_1-requirement. Then A may become more exacting and set a new, smaller, margin, $\epsilon = \epsilon_2$. B again meets his demand by finding a (perhaps much larger) integer $N = N_2$. *If* B *can satisfy* A *no matter how small* A *sets his margin, then we have the situation expressed by* $a_n \rightarrow a$.

There is a definite psychological difficulty in grasping this precise definition of limit. Our intuition suggests a "dynamic" idea of a limit as the result of a process of "motion": we move on through the row of integers 1, 2, 3, $\cdots$, n, $\cdots$ and then observe the behavior of the sequence a_n. We feel that the approach $a_n \rightarrow a$ should be observable. But this "natural" attitude is not capable of clear mathematical formulation. To arrive at a precise definition we must *reverse* the order of steps; instead of first looking at the independent variable n and then at the dependent variable a_n, we must base our definition on what we have to do if we wish actually to check the statement $a_n \rightarrow a$. In such a procedure we must first choose an arbitrarily small margin around a and then determine whether we can meet this condition by taking the independent variable n sufficiently large. Then, by giving symbolic names, ϵ and N, to the phrases "arbitrarily small margin" and "sufficiently large n," we are led to the precise definition of limit.

As another example, let us consider the sequence

$$\frac{1}{2}, \frac{2}{3}, \frac{3}{4}, \frac{4}{5}, \cdots, \frac{n}{n+1}, \cdots,$$

where $a_n = \frac{n}{n+1}$. I state that $\lim a_n = 1$. If you choose an interval whose center is the point 1 and for which $\epsilon = 1/10$, then I can satisfy your requirement (3) by choosing $N = 10$; for

$$0 < 1 - \frac{n}{n+1} = \frac{n+1-n}{n+1} = \frac{1}{n+1} < \frac{1}{10}$$

as soon as $n \geq 10$. If you strengthen your demand by choosing $\epsilon = 1/1000$, then again I can meet it by choosing $N = 1000$; and similarly for any positive number ϵ, no matter how small, which you may choose; in fact, I need only choose any integer N greater than $1/\epsilon$. This process of assigning an arbitrarily small margin ϵ about the number a and then proving that the terms of the sequence a_n are all within a distance ϵ of a if we go far enough out in the sequence, is the detailed description of the fact that $\lim a_n = a$.

If the members of the sequence $a_1, a_2, a_3, \cdots$ are expressed as infinite decimals, then the statement $\lim a_n = a$ simply means that for any positive integer m the first m digits of a_n coincide with the first m digits of the infinite decimal expansion of the fixed number a, provided that n is chosen sufficiently large, say greater than or equal to some value N (depending on m). This merely corresponds to choices of ϵ of the form 10^{-m}.

There is another, quite suggestive, way of expressing the limit concept. If $\lim a_n = a$, and if we enclose a in the interior of an interval I, then no matter how small I may be, all the numbers a_n for which n is greater than or equal to some integer N will lie within I, so that at most *a finite number*, N—1, *of terms* at the beginning of the sequence,

$$a_1, a_2, \cdots, a_{N-1},$$

can lie outside I. If I is very small, N may be very large, say a hundred or even a thousand billion; still only a finite number of terms of the sequence will lie outside I, while the infinitely many remaining terms will lie within I.

We may say of the members of any infinite sequence that "almost all" have a certain property if only a finite number, no matter how great, do not have the property. For example "almost all" positive integers are greater than 1,000,000,000,000. Using this terminology, the statement $\lim a_n = a$ is equivalent to the statement: *If* I *is any interval with* a *as its center, then almost all of the numbers* a_n *lie within* I.

It should be noted in passing that it is not necessarily assumed that all the terms a_n of a sequence have different values. It is permissible for some, infinitely many, or even *all* the numbers a_n to be equal to the limit value a. For example, the sequence for which $a_1 = 0, a_2 = 0, \cdots, a_n = 0, \cdots$ is a legitimate sequence, and its limit, of course, is 0.

A sequence a_n with a limit a is called *convergent*. A sequence a_n without a limit is called *divergent*.

Exercises: Prove:

1. The sequence for which $a_n = \dfrac{n}{n^2 + 1}$ has the limit 0. (Hint: $a_n = \dfrac{1}{n + \frac{1}{n}}$ is less than $\dfrac{1}{n}$ and greater than 0.)

2. The sequence $a_n = \dfrac{n^2 + 1}{n^3 + 1}$ has the limit 0. (Hint: $a_n = \dfrac{1 + \frac{1}{n^2}}{n + \frac{1}{n^2}}$ lies between 0 and $\dfrac{2}{n}$.)

3. The sequence 1, 2, 3, 4, $\cdots$ and the oscillating sequences

$$1, 2, 1, 2, 1, 2, \cdots,$$

$$-1, 1, -1, 1, -1, \cdots \quad (\text{i.e. } a_n = (-1)^n),$$

$$\text{and } 1, \tfrac{1}{2}, 1, \tfrac{1}{3}, 1, \tfrac{1}{4}, 1, \tfrac{1}{5}, \cdots$$

do *not* have limits.

If in a sequence a_n the members become so large that eventually a_n is larger than any preassigned number K, then we say that a_n *tends to infinity* and write $\lim a_n = \infty$, or $a_n \to \infty$. For example, $n^2 \to \infty$ and $2^n \to \infty$. This terminology is useful, though perhaps not quite consistent, because ∞ is not considered to be a number a. *A sequence tending to infinity is still called divergent.*

Exercise: Prove that the sequence $a_n = \dfrac{n^2 + 1}{n}$ tends to infinity; similarly for $a_n = \dfrac{n^2 + 1}{n + 1}$, $a_n = \dfrac{n^3 - 1}{n + 1}$, and $a_n = \dfrac{n^n}{n^2 + 1}$.

Beginners sometimes fall into the error of thinking that a passage to the limit as $n \to \infty$ may be performed simply by substituting $n = \infty$ in the expression for a_n. For example, $1/n \to 0$ because "$1/\infty = 0$." But the symbol ∞ is not a number, and its use in the expression $1/\infty$ is illegitimate. Trying to imagine the limit of a sequence as the "ultimate" or "last" term a_n when $n = \infty$ misses the point and obscures the issue.

2. Monotone Sequences

In the general definition of page 291, no specific type of approach of a convergent sequence $a_1, a_2, a_3, \cdots$ to its limit a is required. The simplest type is exhibited by a so-called monotone sequence, such as the sequence

$$\frac{1}{2}, \frac{2}{3}, \frac{3}{4}, \cdots, \frac{n}{n+1}, \cdots.$$

Each term of this sequence is greater than the preceding term. For $a_{n+1} = \frac{n+1}{n+2} = 1 - \frac{1}{n+2} > 1 - \frac{1}{n+1} = \frac{n}{n+1} = a_n$. A sequence of this sort, where $a_{n+1} > a_n$, is said to be *monotone increasing*. Similarly, a sequence for which $a_n > a_{n+1}$, such as the sequence 1, 1/2, 1/3, $\cdots$, is called *monotone decreasing*. Such sequences can approach their limits from one side only. In contrast to these, there are sequences that oscillate, such as the sequence $-1, +1/2, -1/3, +1/4, \cdots$. This sequence approaches its limit 0 from both sides (see Fig. 11, p. 69).

The behavior of a monotone sequence is especially easy to determine. Such a sequence may have no limit, but run away completely, like the sequence

$$1, 2, 3, 4, \cdots,$$

where $a_n = n$, or the sequence

$$2, 3, 5, 7, 11, 13, \cdots,$$

where a_n is the nth prime number, p_n. In this case the sequence tends to infinity. But if the terms of a monotone increasing sequence remain bounded—that is, if every term is less than an upper bound B, known in advance—then it is intuitively clear that the sequence must tend to a certain limit a which will be less than or at most equal to B. We

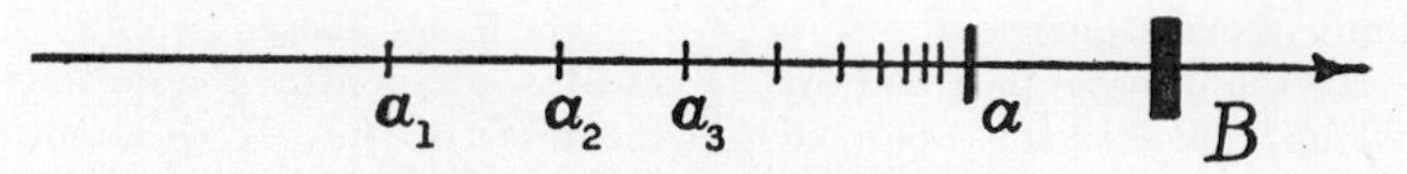

Fig. 166. Monotone bounded sequence.

formulate this as the *Principle of Monotone Sequences: Any monotone increasing sequence that has an upper bound must converge to a limit.* (A similar statement holds for any monotone *decreasing* sequence with a *lower* bound.) It is remarkable that the value of the limit a need not be given or known in advance; the theorem states that under the prescribed conditions the limit *exists*. Of course, this theorem depends on the introduction of irrational numbers and would otherwise not always be true; for, as we have seen in Chapter II, any irrational number (such as $\sqrt{2}$) is the limit of the monotone increasing and bounded sequence of rational decimal fractions obtained by breaking off a certain infinite decimal at the nth digit.

* Although the principle of monotone sequences appeals to the intuition as an obvious truth, it will be instructive to give a rigorous proof in the modern fashion. To do this we must show that the principle is a logical consequence of the definitions of real number and limit.

Suppose that the numbers $a_1, a_2, a_3, \cdots$ form a monotone increasing but bounded sequence. We can express the terms of this sequence as infinite decimals,

$$a_1 = A_1.p_1p_2p_3\cdots,$$
$$a_2 = A_2.q_1q_2q_3\cdots,$$
$$a_3 = A_3.r_1r_2r_3\cdots,$$
$$\cdots\cdots\cdots\cdots\cdots\cdots$$

where the A_i are integers and the p_i, q_i, etc. are digits from 0 to 9. Now run down the column of integers $A_1, A_2, A_3, \cdots$. Since the sequence $a_1, a_2, a_3, \cdots$ is *bounded*, these integers cannot increase indefinitely, and since the sequence is *monotone increasing*, the sequence of integers $A_1, A_2, A_3, \cdots$ *will remain constant after attaining its maximum value.* Call this maximum value A, and suppose that it is attained at the N_0th row. Now run down the second column $p_1, q_1, r_1, \cdots$, confining attention to the terms of the N_0th and subsequent rows. If x_1 is the largest digit to appear in this column after the N_0th row, then x_1 will appear *constantly* after its first appearance, which we may suppose to occur in the N_1th row, where $N_1 \geq N_0$. For if the digit in this column decreased at any time thereafter, the sequence $a_1, a_2, a_3, \cdots$ would not be monotone increasing. Next we consider the digits $p_2, q_2, r_2, \cdots$ of the third column. A similar argument shows that after a certain integer $N_2 \geq N_1$ the digits of the third column are constantly equal to some digit x_2. If we repeat this process for the 4th, 5th, $\cdots$ columns we obtain digits $x_3, x_4, x_5, \cdots$ and corresponding integers $N_3, N_4, N_5, \cdots$. It is easy to see that the number

$$a = A.x_1x_2x_3x_4\cdots$$

is the limit of the sequence $a_1, a_2, a_3, \cdots$. For if ϵ is chosen $\geq 10^{-m}$, then for all $n \geq N_m$ the integral part and first m places of digits after the decimal point in a_n will coincide with those of a, so that the difference $|a - a_n|$ cannot exceed 10^{-m}. Since this can be done for any positive ϵ, however small, by choosing m sufficiently large, the theorem is proved.

It is also possible to prove this theorem on the basis of any one of the other definitions of real numbers given in Chapter II; for example, the definition by nested intervals or by Dedekind cuts. Such proofs are to be found in most texts on advanced calculus.

The principle of monotone sequences could have been used in Chapter II to define the sum and product of two positive infinite decimals,

$$a = A.a_1a_2a_3 \cdots,$$
$$b = B.b_1b_2b_3 \cdots.$$

Two such expressions cannot be added or multiplied in the ordinary way, starting from the right-hand end, for there is no such end. (As an example, the reader may try to add the two infinite decimals 0.333333 ··· and 0.989898 ··· .) But if x_n denotes the *finite* decimal fraction obtained by breaking off the expressions for a and b at the nth place and adding in the ordinary way, then the sequence $x_1, x_2, x_3, \cdots$ will be monotone increasing and bounded (by the integer $A + B + 2$, for example). Hence this sequence has a limit, and we may define $a + b = \lim x_n$. A similar process serves to define the product ab. These definitions can then be extended by the ordinary rules of arithmetic to cover all cases, where a and b are positive or negative.

Exercise: Show in this way that the sum of the two infinite decimals considered above is the real number $1.323232 \cdots = 131/99$.

The importance of the limit concept in mathematics lies in the fact that *many numbers are defined only as limits*—often as limits of monotone bounded sequences. This is why the field of rational numbers, in which such limits may not exist, is too narrow for the needs of mathematics.

3. Euler's Number e

The number e has had an established place in mathematics alongside the Archimedean number π ever since the publication in 1748 of Euler's *Introductio in Analysin Infinitorum.* It provides an excellent illustration of how the principle of monotone sequences can serve to define a new real number. Using the abbreviation

$$n! = 1 \cdot 2 \cdot 3 \cdot 4 \cdots n$$

for the product of the first n integers, we consider the sequence $a_1, a_2, a_3, \cdots$, where

$$a_n = 1 + \frac{1}{1!} + \frac{1}{2!} + \cdots + \frac{1}{n!}. \tag{4}$$

The terms a_n form a monotone increasing sequence, since a_{n+1} originates from a_n by the addition of the positive increment $\frac{1}{(n+1)!}$. Moreover, the values of a_n are bounded above:

$$a_n < B = 3. \tag{5}$$

For we have $$\frac{1}{s!} = \frac{1}{2}\frac{1}{3}\cdots\frac{1}{s} < \frac{1}{2}\frac{1}{2}\cdots\frac{1}{2} = \frac{1}{2^{s-1}},$$

and hence

$$a_n < 1 + 1 + \frac{1}{2} + \frac{1}{2^2} + \frac{1}{2^3} + \cdots + \frac{1}{2^{n-1}} = 1 + \frac{1 - (\frac{1}{2})^n}{1 - \frac{1}{2}}$$
$$= 1 + 2(1 - (\tfrac{1}{2})^n) < 3,$$

using the formula given on page 13 for the sum of the first n terms of a geometric series. Hence, by the principle of monotone sequences, a_n must approach a limit as n tends to infinity, and *this limit we call* e. To express the fact that $e = \lim a_n$, we may write e as the "infinite series"

$$e = 1 + \frac{1}{1!} + \frac{1}{2!} + \frac{1}{3!} + \cdots + \frac{1}{n!} + \cdots. \tag{6}$$

This "equality," with a row of dots at the end, is simply another way of expressing the content of the two statements

$$a_n = 1 + \frac{1}{1!} + \frac{1}{2!} + \cdots + \frac{1}{n!}$$

and

$$a_n \to e \text{ as } n \to \infty.$$

The series (6) permits the calculation of e to any desired degree of accuracy. For example, the sum (to nine digits) of the terms in (6) up to and including $1/12!$ is $\Sigma = 2.71828183 \cdots$. (The reader should check this result.) The "error," i.e. the difference between this value and the true value of e can easily be appraised. We have for the difference $(e - \Sigma)$ the expression

$$\frac{1}{13!} + \frac{1}{14!} + \cdots < \frac{1}{13!}\left(1 + \frac{1}{13} + \frac{1}{13^2} + \cdots\right)$$
$$= \frac{1}{13!} \cdot \frac{1}{1 - \frac{1}{13}} = \frac{1}{12 \cdot 12!}.$$

This is so small that it cannot affect the ninth digit of Σ. Hence, allowing for a possible error in the last figure of the value given above, we have $e = 2.7182818$, to eight digits.

* *The number* e *is irrational.* To prove this we shall proceed indirectly by assuming that $e = p/q$, where p and q are integers, and then deducing an absurdity from this assumption. Since we know that $2 < e < 3$, e cannot be an integer,

and therefore q must be at least equal to 2. Now we multiply both sides of (6) by $q! = 2 \cdot 3 \cdots q$, obtaining

$$\begin{aligned} e \cdot q! &= p \cdot 2 \cdot 3 \cdots (q-1) \\ &= [q! + q! + 3 \cdot 4 \cdots q + 4 \cdot 5 \cdots q + \cdots + (q-1)q + q + 1] \\ &\quad + \frac{1}{(q+1)} + \frac{1}{(q+1)(q+2)} + \cdots . \end{aligned} \tag{7}$$

On the left side we obviously have an integer. On the right side, the term in brackets is likewise an integer. The remainder of the right side, however, is a positive number that is less than $\frac{1}{2}$ and hence no integer. For $q \geq 2$, and hence the terms of the series $1/(q+1) + \cdots$ are respectively not greater than the corresponding terms of the geometrical series $1/3 + 1/3^2 + 1/3^3 + \cdots$, whose sum is $1/3[1/(1 - 1/3)] = \frac{1}{2}$. Hence (7) presents a contradiction: the integer on the left side cannot be equal to the number on the right side; for this latter number, being the sum of an integer and a positive number less than $\frac{1}{2}$, is not an integer.

4. The Number π

As is known from school mathematics, the length of the circumference of a circle of unit radius can be defined as the limit of a sequence of lengths of regular polygons with an increasing number of sides. The length of the circumference so defined is denoted by 2π. More precisely, if p_n denotes the length of the inscribed, and q_n the length of the circumscribed regular n-sided polygon, then $p_n < 2\pi < q_n$. Moreover, as n increases, each of the sequences p_n, q_n approaches 2π monotonically, and with each step we obtain a smaller margin for the error in the approximation of 2π given by p_n or q_n.

On page 124 we found the expression

$$p_{2^m} = 2^m\sqrt{2 - \sqrt{2 + \sqrt{2 + \cdots}}}$$

containing $m - 1$ nested square root signs. This formula can be used to compute the approximate value of 2π.

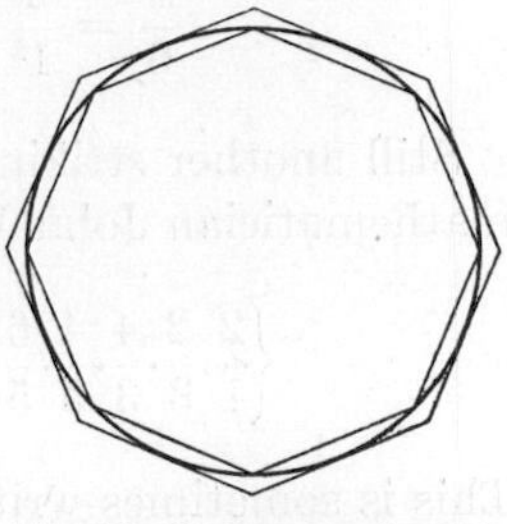

Fig. 167. Circle approximated by polygons.

Exercises: 1. Find the approximate value of π given by p_4, p_8, and p_{16}.

* 2. Find a formula for q_{2^m}.

* 3. Use this formula to find q_4, q_8, and q_{16}. From a knowledge of p_{16} and q_{16}, state bounds between which π must lie.

What *is* the number π? The inequality $p_n < 2\pi < q_n$ gives the complete answer by setting up a sequence of nested intervals which close down on the point 2π. Still, this answer leaves something to be desired, for it gives no information about the nature of π as a real number: is it rational or irrational, algebraic or transcendental? As we have mentioned on page 140, π is in fact a transcendental number, and hence irrational. In contrast to the proof for e, the proof of the irrationality of π, first given by J. H. Lambert (1728–1777), is rather difficult and will not be undertaken here. However, other information about π is within our reach. Recalling the statement that the integers are the basic material of mathematics, we may ask whether the number π has any simple relationship to the integers. The decimal expansion of π, although it has been calculated to several hundred places, reveals no trace of regularity. This is not surprising, since π and 10 have nothing to do with one another. But in the eighteenth century Euler and others found beautiful expressions linking π to the integers by means of infinite series and products. Perhaps the simplest such formula is the following:

$$\frac{\pi}{4} = 1 - \frac{1}{3} + \frac{1}{5} - \frac{1}{7} + \cdots,$$

expressing $\pi/4$ as the limit for increasing n of the partial sums

$$s_n = 1 - \frac{1}{3} + \frac{1}{5} - \cdots + (-1)^n \frac{1}{2n+1}.$$

We shall derive this formula in Chapter VIII. Another infinite series for π is

$$\frac{\pi^2}{6} = \frac{1}{1^2} + \frac{1}{2^2} + \frac{1}{3^2} + \frac{1}{4^2} + \frac{1}{5^2} + \frac{1}{6^2} + \cdots.$$

Still another striking expression for π was discovered by the English mathematician John Wallis (1616–1703). His formula states that

$$\left\{\frac{2}{1}\cdot\frac{2}{3}\cdot\frac{4}{3}\cdot\frac{4}{5}\cdot\frac{6}{5}\cdot\frac{6}{7}\cdots\frac{2n}{2n-1}\cdot\frac{2n}{2n+1}\right\} \to \frac{\pi}{2} \text{ as } n \to \infty.$$

This is sometimes written in the abbreviated form

$$\frac{\pi}{2} = \frac{2}{1}\cdot\frac{2}{3}\cdot\frac{4}{3}\cdot\frac{4}{5}\cdot\frac{6}{5}\cdot\frac{6}{7}\cdot\frac{8}{7}\cdot\frac{8}{9}\cdots,$$

the expression on the right being called an *infinite product.*

A proof of the last two formulas will be found in any comprehensive book on the calculus (see p. 482 and pp. 509–510).

*5. Continued Fractions

Interesting limiting processes occur in connection with continued fractions. A finite continued fraction, such as

$$\frac{57}{17} = 3 + \cfrac{1}{2 + \cfrac{1}{1 + \cfrac{1}{5}}},$$

represents a rational number. On page 49 we showed that every rational number can be written in this form by means of the Euclidean algorithm. For irrational numbers, however, the algorithm does not stop after a finite number of steps. Instead, it leads to a sequence of fractions of increasing length, each representing a rational number. In particular, all real algebraic numbers (see p. 103) of degree 2 may be expressed in this way. Consider, for example, the number $x = \sqrt{2} - 1$, which is a root of the quadratic equation

$$x^2 + 2x = 1, \qquad \text{or} \qquad x = \frac{1}{2 + x}.$$

If on the right side x is again replaced by $1/(2 + x)$ this yields the expression

$$x = \cfrac{1}{2 + \cfrac{1}{2 + x}},$$

and then

$$x = \cfrac{1}{2 + \cfrac{1}{2 + \cfrac{1}{2 + x}}},$$

and so on, so that after n steps we obtain the equation

$$x = \left.\cfrac{1}{2 + \cfrac{1}{2 + \cfrac{1}{2 + \ddots \cfrac{1}{2 + x}}}}\right\} n \text{ steps.}$$

As n tends to infinity, we obtain the "infinite continued fraction"

$$\sqrt{2} = 1 + \cfrac{1}{2 + \cfrac{1}{2 + \cfrac{1}{2 + \cfrac{1}{2 + \ddots}}}}.$$

This remarkable formula connects $\sqrt{2}$ with the integers in a much more striking way than does the decimal expansion of $\sqrt{2}$, which displays no regularity in the succession of its digits.

For the positive root of any quadratic equation of the form

$$x^2 = ax + 1, \qquad \text{or} \qquad x = a + \frac{1}{x},$$

we obtain the expansion

$$x = a + \cfrac{1}{a + \cfrac{1}{a + \cfrac{1}{a + \ddots}}}.$$

For example, setting $a = 1$, we find

$$x = \tfrac{1}{2}(1 + \sqrt{5}) = 1 + \cfrac{1}{1 + \cfrac{1}{1 + \cfrac{1}{1 + \ddots}}}$$

(cf. p. 123). These examples are special cases of a general theorem which states that *the real roots of quadratic equations with integral coefficients have periodic continued fraction developments*, just as rational numbers have periodic decimal expansions.

Euler was able to find almost equally simple infinite continued fractions for e and π. The following are exhibited without proof:

$$e = 2 + \cfrac{1}{1 + \cfrac{1}{2 + \cfrac{1}{1 + \cfrac{1}{1 + \cfrac{1}{4 + \cfrac{1}{1 + \cfrac{1}{1 + \cfrac{1}{6 + \ddots}}}}}}}};$$

$$e = 2 + \cfrac{1}{1 + \cfrac{1}{2 + \cfrac{2}{3 + \cfrac{3}{4 + \cfrac{4}{5 + \ddots}}}}};$$

$$\frac{\pi}{4} = \cfrac{1}{1 + \cfrac{1^2}{2 + \cfrac{3^2}{2 + \cfrac{5^2}{2 + \cfrac{7^2}{2 + \cfrac{9^2}{2 + \ddots}}}}}}.$$

§3. LIMITS BY CONTINUOUS APPROACH

1. Introduction. General Definition

In §2, Article 1 we succeeded in giving a precise formulation of the statement, "The sequence a_n (i.e. the function $a_n = F(n)$ of the integral variable n) has the limit a as n tends to infinity." We shall now give a corresponding definition of the statement, "The function $u = f(x)$ of the continuous variable x has the limit a as x tends to the value x_1." In an intuitive form this concept of limit by continuous approach of the independent variable x was used in §1, Article 5 to test the continuity of the function $f(x)$.

Again let us begin with a particular example. The function $f(x) = \frac{(x + x^3)}{x}$ is defined for all values of x other than $x = 0$, where the denominator vanishes. If we draw a graph of the function $u = f(x)$ for values of x in the neighborhood of 0, it is evident that as x "approaches" 0 from either side the corresponding value of $u = f(x)$ "approaches" the limit 1. In order to give a precise description of this fact, let us find an explicit formula for the difference between the value $f(x)$ and the fixed number 1:

$$f(x) - 1 = \frac{x + x^3}{x} - 1 = \frac{x + x^3 - x}{x} = \frac{x^3}{x}.$$

If we agree to consider only values of x near 0, but not the value $x = 0$ itself (for which $f(x)$ is not even defined), we may divide both numerator and denominator of the expression on the right side of this equation by x, obtaining the simpler formula

$$f(x) - 1 = x^2.$$

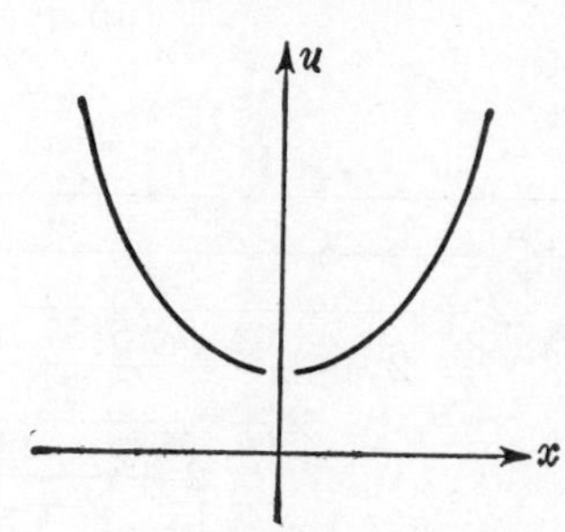

Fig. 168. $u = (x + x^3)/x$.

Clearly, we can make this difference *as small as we please* by confining x to a *sufficiently small* neighborhood of the value 0. Thus for $x = \pm\frac{1}{10}$, $f(x) - 1 = \frac{1}{100}$; for $x = \pm\frac{1}{100}$, $f(x) - 1 = \frac{1}{10{,}000}$; and so on. More generally, if ϵ is any positive number, no matter how small, then the difference between $f(x)$ and 1 will be smaller than ϵ, provided only that the distance of x from 0 is less than the number $\delta = \sqrt{\epsilon}$. For if

$$|x| < \sqrt{\epsilon}$$

then

$$|f(x) - 1| = |x^2| < \epsilon.$$

The analogy with our definition of limit for a sequence is complete. On page 291 we made the definition, "The sequence a_n has the limit a as n tends to infinity if, corresponding to every positive number ϵ, no matter how small, there may be found an integer N (depending on ϵ) such that

$$|a_n - a| < \epsilon$$

for all n satisfying the inequality

$$n \geq N."$$

In the case of a function $f(x)$ of a continuous variable x as x tends to a finite value x_1, we merely replace the "sufficiently large" n given by

N by the "sufficiently near" x_1 given by a number δ, and arrive at the following definition of limit by continuous approach, first given by Cauchy around 1820: *The function* f(x) *has the limit* a *as* x *tends to the value* x_1 *if, corresponding to every positive number* ϵ, *no matter how small, there may be found a positive number* δ *(depending on* ϵ*) such that*

$$|f(x) - a| < \epsilon$$

for all $x \neq x_1$ *satisfying the inequality*

$$|x - x_1| < \delta.$$

When this is the case we write

$$f(x) \to a \quad \text{as} \quad x \to x_1.$$

In the case of the function $f(x) = (x + x^3)/x$ we showed above that $f(x)$ has the limit 1 as x tends to the value $x_1 = 0$. In this case it was sufficient always to choose $\delta = \sqrt{\epsilon}$.

2. Remarks on the Limit Concept

The (ϵ, δ)-definition of limit is the result of more than a hundred years of trial and error, and embodies in a few words the result of persistent effort to put this concept on a sound mathematical basis. Only by limiting processes can the fundamental notions of the calculus—derivative and integral—be defined. But a clear understanding and a precise definition of limits had long been blocked by an apparently insurmountable difficulty.

In their study of motion and change the mathematicians of the seventeenth and eighteenth centuries accepted as a matter of course the concept of a quantity x steadily changing and moving in a continuous flow toward a limiting value x_1. Associated with this primary flow of time or of a quantity x behaving like time they considered a secondary value $u = f(x)$ that followed the motion of x. The problem was to attach a precise mathematical meaning to the idea that $f(x)$ "tends to" or "approaches" a fixed value a as x moves toward x_1.

But from the time of Zeno and his paradoxes the intuitive physical or metaphysical concept of continuous motion has eluded all attempts at an exact mathematical formulation. There is no difficulty in proceeding step by step through a discrete sequence of values $a_1, a_2, a_3, \ldots$. But in dealing with a continuous variable x that ranges over a whole interval of the number axis it is impossible to say how x shall "approach" the fixed value x_1 in such a way as to assume consecutively and in their order of magnitude all the values in the interval. For the points on a line form a dense set, and there is no "next" point after a given point has been reached. Certainly, the intuitive idea of a continuum has a psychological reality in the human mind. But it cannot be called upon to resolve a mathematical impossibility; there must remain a discrepancy between the intuitive idea and the mathematical language designed to describe the scientifically relevant features of our intuition in exact logical terms. Zeno's paradoxes are a pointed indication of this discrepancy.

Cauchy's achievement was to realize that, as far as the mathematical concepts are concerned, any reference to a prior intuitive idea of continuous motion may and even must be omitted. As happens so often, the path to scientific progress was opened by resigning an attempt in a metaphysical direction and instead operating solely with notions that in principle correspond to "observable" phenomena. If we analyze what we really mean by the words "continuous approach," how we must proceed to verify it in a specific case, then we are forced to accept a definition such as Cauchy's. This definition is *static*; it does not presuppose the intuitive idea of motion. On the contrary, only such a static definition makes possible a precise mathematical analysis of continuous motion in time, and disposes of Zeno's paradoxes as far as mathematical science is concerned.

In the (ϵ, δ)-definition the independent variable does not move; it does not "tend to" or "approach" a limit x_1 in any physical sense. These phrases and the symbol $\rightarrow$ still remain, and no mathematician need or should lose the suggestive intuitive feeling that they express. But when it comes to checking the existence of a limit in actual scientific procedure it is the (ϵ, δ)-definition that must be applied. Whether this definition corresponds satisfactorily with the intuitive "dynamic" notion of approach is a question of the same sort as whether the axioms of geometry provide a satisfactory description of the intuitive concept of space. Both formulations leave out something that is real to the intuition, but they provide an adequate mathematical framework for expressing our knowledge of these concepts.

As in the case of sequential limit, the key to Cauchy's definition lies in the reversal of the "natural" order in which the variables are considered. First we fix our attention on a margin ϵ for the dependent variable, and then we seek to determine a suitable margin δ for the independent variable. The statement "$f(x) \rightarrow a$ as $x \rightarrow x_1$" is only a brief way of saying that this can be done for every positive number ϵ. In particular, no *part* of this statement, e.g. "$x \rightarrow x_1$" has a meaning by itself.

One more point should be stressed. In letting x "tend to" x_1 we may permit x to be greater than or less than x_1, but we expressly exclude equality by requiring that $x \neq x_1$: x tends to x_1, but never actually *assumes* the value x_1. Thus we can apply our definition to functions that are not defined for $x = x_1$, but have definite limits as x tends to x_1; e.g. the function $f(x) = \dfrac{x + x^3}{x}$ considered on page 303. Excluding $x = x_1$ corresponds to the fact that, for limits of sequences a_n as $n \to \infty$, e.g. $a_n = 1/n$, we never substitute $n = \infty$ in the formula.

However, as x tends to x_1, $f(x)$ may approach the limit a in such a way that there are values $x \neq x_1$ for which $f(x) = a$. For example, in considering the function $f(x) = x/x$ as x tends to 0 we never allow x to equal 0, but $f(x) = 1$ for all $x \neq 0$ and the limit a exists and is equal to 1 according to our definition.

3. The Limit of $\frac{\sin x}{x}$

If x denotes the radian measure of an angle, then the expression $\dfrac{\sin x}{x}$ is defined for all x except $x = 0$, where it becomes the meaningless symbol 0/0. The reader with access to a table of trigonometric functions will be able to compute the value of $\dfrac{\sin x}{x}$ for small values of x. These tables are commonly given in terms of the degree measure of angles; we recall from §1, Article 2 that the degree measure x is related to the radian measure y by the relation $x = \dfrac{\pi}{180} y = 0.01745\, y$, to 5 places. From a four-place table we find that for an angle of

10°,	$x = 0.1745$,	$\sin x = 0.1736$,	$\dfrac{\sin x}{x} = 0.9948$
5°,	0.0873,	0.0872,	0.9988
2°,	0.0349,	0.0349,	1.0000
1°,	0.0175,	0.0175,	1.0000.

Although these figures are stated to be accurate only to four places, it would appear that

$$\sin x/x \to 1 \qquad \text{as } x \to 0. \tag{1}$$

We shall now give a rigorous proof of this limiting relation.

From the unit circle definition of the trigonometric functions, if x is the radian measure of angle BOC, for $0 < x < \frac{\pi}{2}$ we have

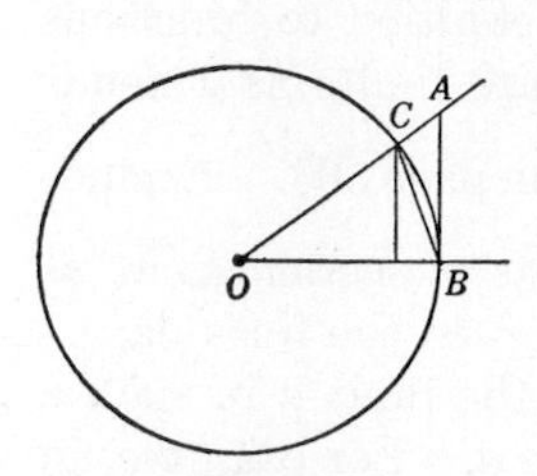

Fig. 169.

area of triangle $OBC = \frac{1}{2} \cdot 1 \cdot \sin x$

area of circular sector $OBC = \frac{1}{2} \cdot x$ (see p. 278)

area of triangle $OBA = \frac{1}{2} \cdot 1 \cdot \tan x$.

Hence

$$\sin x < x < \tan x.$$

Dividing by $\sin x$ we obtain

$$1 < \frac{x}{\sin x} < \frac{1}{\cos x},$$

or

$$\cos x < \frac{\sin x}{x} < 1. \tag{2}$$

Now $1 - \cos x = (1 - \cos x)\dfrac{1 + \cos x}{1 + \cos x} = \dfrac{1 - \cos^2 x}{1 + \cos x} = \dfrac{\sin^2 x}{1 + \cos x} < \sin^2 x$. Since $\sin x < x$, this shows that

$$1 - \cos x < x^2, \tag{3}$$

or

$$1 - x^2 < \cos x.$$

Together with (2), this yields the final inequality

$$1 - x^2 < \frac{\sin x}{x} < 1. \tag{4}$$

Although we have been assuming that $0 < x < \frac{\pi}{2}$, this inequality is also true for $\frac{-\pi}{2} < x < 0$, since $\dfrac{\sin(-x)}{(-x)} = \dfrac{-\sin x}{-x} = \dfrac{\sin x}{x}$, and $(-x)^2 = x^2$.

From (4) the limit relation (1) is an immediate consequence. For the difference between $\dfrac{\sin x}{x}$ and 1 is less than x^2, and this can be made less than any number ϵ by choosing $|x| < \delta = \sqrt{\epsilon}$.

Exercises: 1) From the inequality (3) deduce the limiting relation $\dfrac{1 - \cos x}{x} \to 0$ as $x \to 0$.

Find the limits as $x \to 0$ of the following functions:

2) $\frac{\sin^2 x}{x}$. 3) $\frac{\sin x}{x(x-1)}$. 4) $\frac{\tan x}{x}$. 5) $\frac{\sin ax}{x}$.

6) $\frac{\sin ax}{\sin bx}$. 7) $\frac{x \sin x}{1 - \cos x}$. 8) $\frac{\sin x}{x}$, if x is measured in degrees.

9) $\frac{1}{x} - \frac{1}{\tan x}$. 10) $\frac{1}{\sin x} - \frac{1}{\tan x}$.

4. Limits as $x \to \infty$

If the variable x is sufficiently large, then the function $f(x) = 1/x$ becomes arbitrarily small, or "tends to 0." In fact, the behavior of this function as x increases is essentially the same as that of the sequence $1/n$ as n increases. We give the general definition: *The function* f(x) *has the limit* a *as* x *tends to infinity, written*

$$f(x) \to a \qquad \text{as } x \to \infty,$$

if, corresponding to each positive number ϵ, *no matter how small, there can be found a positive number* K *(depending on* ϵ*) such that*

$$|f(x) - a| < \epsilon$$

provided only that $|x| > K$. (Compare with the corresponding definition on p. 305.)

In the case of the function $f(x) = 1/x$, for which $a = 0$, it suffices to choose $K = 1/\epsilon$, as the reader may at once verify.

Exercises: 1. Show that the foregoing definition of the statement

$$f(x) \to a \qquad \text{as} \qquad x \to \infty$$

is equivalent to the statement

$$f(x) \to a \qquad \text{as} \qquad 1/x \to 0.$$

Prove that the following limit relations hold:

2. $\frac{x+1}{x-1} \to 1$ as $x \to \infty$. 3. $\frac{x^2+x+1}{x^2-x-1} \to 1$ as $x \to \infty$.

4. $\frac{\sin x}{x} \to 0$ as $x \to \infty$. 5. $\frac{x+1}{x^2+1} \to 0$ as $x \to \infty$.

6. $\frac{\sin x}{x + \cos x} \to 0$ as $x \to \infty$. 7. $\frac{\sin x}{\cos x}$ has no limit as $x \to \infty$

8. Define: "$f(x) \to \infty$ as $x \to \infty$." Give an example.

There is one difference between the case of a function $f(x)$ and a sequence a_n. In the case of a sequence, n can tend to infinity only by increasing, but for a function we may allow x to become infinite either positively or negatively. If it is desired to restrict attention to the behavior of $f(x)$ when x assumes large *positive* values only, we may replace the condition $|x| > K$ by the condition $x > K$; for large negative values of x we use the condition $x < -K$. To symbolize these two methods of "one-sided" approach to infinity we write

$$x \to +\infty, \qquad x \to -\infty$$

respectively.

§4. PRECISE DEFINITION OF CONTINUITY

In §1, Article 5 we stated what amounts to the following criterion for the continuity of a function: "A function $f(x)$ is continuous at the point $x = x_1$ if, when x approaches x_1, the quantity $f(x)$ approaches the value $f(x_1)$ as a limit." If we analyze this definition we see that it consists of two different requirements:

a) the limit a of $f(x)$ must exist as x tends to x_1,
b) this limit a must be equal to the value $f(x_1)$.

If in the limit definition of page 305 we set $a = f(x_1)$, then the condition for continuity takes the following form: *The function* f(x) *is continuous for the value* x = x₁ *if, corresponding to every positive number* ϵ, *no matter how small, there may be found a positive number* δ *(depending on* ϵ*) such that*

$$|f(x) - f(x_1)| < \epsilon$$

for all x *satisfying the inequality*

$$|x - x_1| < \delta.$$

(The restriction $x \neq x_1$ imposed in the limit definition is unnecessary here, since the inequality $|f(x_1) - f(x_1)| < \epsilon$ is automatically satisfied.)

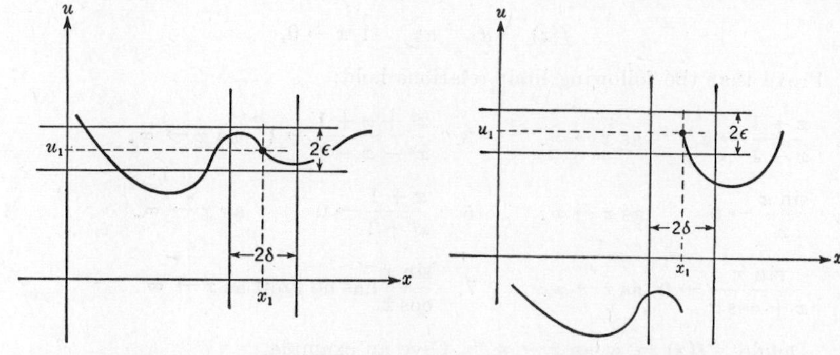

Fig. 170. A function continuous at $x = x_1$.   Fig. 171. A function discontinuous at $x = x_1$.

As an example, let us check the continuity of the function $f(x) = x^3$ at the point $x_1 = 0$, say. We have

$$f(x_1) = 0^3 = 0.$$

Now let us assign any positive value to ϵ, for example $\epsilon = \frac{1}{1000}$. Then we must show that by confining x to values sufficiently near $x_1 = 0$, the corresponding values of $f(x)$ will not differ from 0 by more than $\frac{1}{1000}$, i.e. will lie between $\frac{-1}{1000}$ and $\frac{1}{1000}$. We see immediately that this margin is not exceeded if we restrict x to values differing from $x_1 = 0$ by less than $\delta = \sqrt[3]{\frac{1}{1000}} = \frac{1}{10}$; for if $|x| < \frac{1}{10}$, then $|f(x)| = x^3 < \frac{1}{1000}$. In the same way we can replace $\epsilon = \frac{1}{1000}$ by $\epsilon = 10^{-4}, 10^{-5}$, or whatever margin we desire; $\delta = \sqrt[3]{\epsilon}$ will always satisfy the requirement, since if $|x| < \sqrt[3]{\epsilon}$, then $|f(x)| = x^3 < \epsilon$.

On the basis of the (ϵ, δ)-definition of continuity one can show in a similar way that all polynomials, rational functions, and trigonometric functions are continuous, except for isolated values of x where the functions may become infinite.

In terms of the graph of a function $u = f(x)$, the definition of continuity takes the following geometrical form. Choose any positive number ϵ and draw parallels to the x-axis at a height $f(x_1) - \epsilon$ and $f(x_1) + \epsilon$ above it. Then it must be possible to find a positive number δ such that the whole portion of the graph which lies within the vertical band of width 2δ about x_1 is also contained within the horizontal band of width 2ϵ about $f(x_1)$. Figure 170 shows a function which is continuous at x_1, while Figure 171 shows a function which is not. In the latter case, no matter how narrow we make the vertical band about x_1, it will always include a portion of the graph that lies outside the horizontal band corresponding to the choice of ϵ.

If I assert that a given function $u = f(x)$ is continuous for the value $x = x_1$, it means that I am prepared to fulfill the following contract with you. You may choose any positive number ϵ, as small as you please, but fixed. Then I must produce a positive number δ such that $|x - x_1| < \delta$ implies $|f(x) - f(x_1)| < \epsilon$. I do *not* contract to produce at the outset a number δ that will suffice for whatever ϵ you may subsequently choose; *my* choice of δ will depend on *your* choice of ϵ. If you can produce but one value ϵ for which I cannot provide a suitable δ, then my assertion is contradicted. Hence to prove that I can fulfill my contract in any concrete case of a function $u = f(x)$, I usually construct an explicit positive function

$$\delta = \varphi(\epsilon),$$

defined for every positive number ϵ, for which I can prove that $|x - x_1| < \delta$ implies always $|f(x) - f(x_1)| < \epsilon$. In the case of the function $u = f(x) = x^3$ at the value $x_1 = 0$, the function $\delta = \varphi(\epsilon)$ was $\delta = \sqrt[3]{\epsilon}$.

Exercises: 1) Prove that $\sin x$, $\cos x$ are continuous functions.

2) Prove the continuity of $1/(1 + x^4)$ and of $\sqrt{1 + x^2}$.

It should now be clear that the (ϵ, δ)-definition of continuity agrees with what might be called the observable facts concerning a function. As such it is in line with the general principle of modern science that sets up as the criterion for the usefulness of a concept or for the "scientific existence" of a phenomenon the possibility of its observation (at least in principle) or of its reduction to observable facts.

§5. TWO FUNDAMENTAL THEOREMS ON CONTINUOUS FUNCTIONS

1. Bolzano's Theorem

Bernard Bolzano (1781–1848), a Catholic priest trained in scholastic philosophy, was one of the first to introduce the modern concept of rigor into mathematical analysis. His important booklet, *Paradoxien des Unendlichen*, appeared in 1850. Here for the first time it was recognized that many apparently obvious statements concerning continuous functions can and must be proved if they are to be used in full generality. The following theorem on continuous functions of one variable is an example.

A continuous function of a variable x *which is positive for some value of* x *and negative for some other value of* x *in a closed interval* $a \leq x \leq b$ *of continuity must have the value zero for some intermediate value of* x. Thus, if $f(x)$ is continuous as x varies from a to b, while $f(a) < 0$ and $f(b) > 0$, then there will exist a value α of x such that $a < \alpha < b$ and $f(\alpha) = 0$.

Bolzano's theorem corresponds perfectly with our intuitive idea of a continuous curve, which, in order to get from a point below the x-axis to a point above, must somewhere cross the axis. That this need *not* be true of discontinuous functions is shown by Figure 157 on page 284.

*2. Proof of Bolzano's Theorem

A rigorous proof of this theorem will be given. (Like Gauss and other great mathematicians, one may accept and use the fact without proof.) Our objective is to reduce the theorem to fundamental properties of the real number system, in particular to the Dedekind-Cantor postulate concerning nested intervals

(p. 68). To this end we consider the interval I, $a \leq x \leq b$, in which the function $f(x)$ is defined, and bisect it by marking the mid-point, $x_1 = \frac{a + b}{2}$. If at

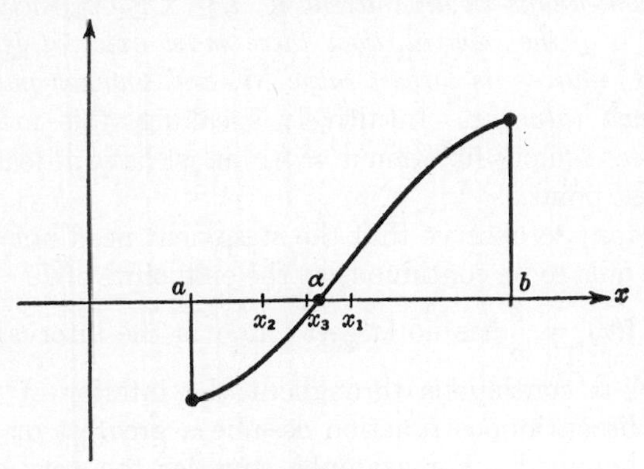

Fig. 172. Bolzano's theorem.

this mid-point we find that $f(x_1) = 0$, then there remains nothing further to prove. If, however, $f(x_1) \neq 0$, then $f(x_1)$ must be either greater than or less than zero. In either case one of the halves of I will again have the property that the sign of $f(x)$ is different at its two extremes. Let us call this interval I_1. We continue the process by bisecting I_1; then either $f(x) = 0$ at the midpoint of I_1, or we can choose an interval I_2, half of I_1, with the property that the sign of $f(x)$ is different at its two extremes. Repeating this procedure, either we shall find after a finite number of bisections a point for which $f(x) = 0$, or we shall obtain a sequence of nested intervals $I_1, I_2, I_3, \cdots$. In the latter case, the Dedekind-Cantor postulate assures the existence of a point α in I common to all these intervals. We assert that $f(\alpha) = 0$, so that α is the point whose existence proves the theorem.

So far the assumption of continuity has not been used. It now serves to clinch the argument by a bit of indirect reasoning. We shall prove that $f(\alpha) = 0$ by assuming the contrary and deducing a contradiction. Suppose that $f(\alpha) \neq 0$, e.g. that $f(\alpha) = 2\epsilon > 0$. Since $f(x)$ is continuous, we can find a (perhaps very small) interval J of length 2δ with α as midpoint, such that the value of $f(x)$ everywhere in J differs from $f(\alpha)$ by less than ϵ. Hence, since $f(\alpha) = 2\epsilon$, we can be sure that $f(x) > \epsilon$ everywhere in J, so that $f(x) > 0$ in J. But the interval J is fixed, and if n is sufficiently large the little interval I_n must necessarily fall within J, since the sequence I_n shrinks to zero. This yields the contradiction; for it follows from the way I_n was chosen that the function $f(x)$ has opposite signs at the two endpoints of every I_n, so that $f(x)$ must have negative values somewhere in J. Thus the absurdity of $f(\alpha) > 0$ and (in the same way) of $f(\alpha) < 0$ proves that $f(\alpha) = 0$.

3. Weierstrass' Theorem on Extreme Values

Another important and intuitively plausible fact about continuous functions was formulated by Karl Weierstrass (1815–1897), who, perhaps more than anyone else, was responsible for the modern trend towards rigor in mathematical analysis. This theorem states: *If a function* f(x) *is continuous in an interval* I, a≤ x ≤ b, *including the endpoints* a *and* b *of the interval, then there must exist at least one point in* I *where* f(x) *attains its largest value* M, *and another point where* f(x) *attains its least value* m. Intuitively speaking, this means that the graph of the continuous function $u = f(x)$ must have at least one highest and one lowest point.

It is important to observe that the statement need not be true if the function $f(x)$ fails to be continuous at the endpoints of I. For example, the function $f(x) = \frac{1}{x}$ has no largest value in the interval $0 < x \leq 1$, although $f(x)$ is continuous throughout the interior of this interval. Nor need a discontinuous function assume a greatest or a least value even if it is bounded. For example, consider the very discontinuous function $f(x)$ defined by setting

$$f(x) = x \text{ for irrational } x,$$

$$f(x) = \tfrac{1}{2} \text{ for rational } x,$$

in the interval $0 \leq x \leq 1$. This function always takes on values between 0 and 1, in fact values as near to 1 and 0 as we may wish, if x is chosen as an irrational number sufficiently near to 0 or 1. But $f(x)$ can never be *equal* to 0 or 1, since for rational x we have $f(x) = \frac{1}{2}$, and for irrational x we have $f(x) = x$. Hence 0 and 1 are never attained.

* Weierstrass' theorem can be proved in much the same way as Bolzano's theorem. We divide I into two closed half-intervals I' and I'' and fix our attention on I' as the interval in which the greatest value of $f(x)$ must be sought, *unless there is a point* α *in* I'' *such that* $f(\alpha)$ *exceeds all the values of* $f(x)$ *in* I'; in the latter case we select I''. The interval so selected we call I_1. Now we proceed with I_1 in the same say as we did with I, obtaining an interval I_2, and so on. This process will define a sequence $I_1, I_2, \cdots, I_n, \cdots$ of nested intervals all containing a point z. We shall prove that the value $f(z) = M$ is the largest attained by $f(x)$ in I, i.e. that there cannot be a point s in I for which $f(s) > M$. Suppose there were a point s with $f(s) = M + 2\epsilon$, where ϵ is a (perhaps very small) positive number. Around z as center we can, because of the continuity of $f(x)$, mark off a small interval K, leaving s outside, and such that in K the values of $f(x)$ differ from $f(z) = M$ by less than ϵ, so that we certainly have $f(x) < M + \epsilon$ in K. But for sufficiently large n the interval I_n lies inside K, and I_n was so defined that no value of $f(x)$ for x outside I_n can exceed all the values of $f(x)$ for x in I_n. Since s is outside I_n and $f(s) > M + \epsilon$, while in K, and hence in I_n, we have $f(x) < M + \epsilon$, we have arrived at a contradiction.

The existence of a least value m may be proved in the same way, or it follows directly from what has already been proved, since the least value of $f(x)$ is the greatest value of $g(x) = -f(x)$.

Weierstrass' theorem can be proved in a similar way for continuous functions of two or more variables $x, y, \cdots$. Instead of an interval with its endpoints we have to consider a *closed* domain, e.g. a rectangle in the x, y-plane which includes its boundary.

Exercise: Where, in the proofs of Bolzano's and Weierstrass' theorems, did we use the fact that $f(x)$ was assumed to be defined and continuous in the whole *closed* interval $a \leq x \leq b$ and not merely in the interval $a < x \leq b$ or $a < x < b$?

The proofs of Bolzano's and Weierstrass' theorems have a decidedly non-constructive character. They do not provide a method for actually finding the location of a zero or of the greatest or smallest value of a function with a prescribed degree of precision in a finite number of steps. Only the mere existence, or rather the absurdity of the non-existence, of the desired values is proved. This is another important instance where the "intuitionists" (see p. 86) have raised objections; some have even insisted that such theorems be eliminated from mathematics. The student of mathematics should take this no more seriously than did most of the critics.

*4. A Theorem on Sequences. Compact Sets

Let $x_1, x_2, x_3, \cdots$ be any infinite sequence of numbers, distinct or not, all contained in the *closed interval* I, $a \leq x \leq b$. The sequence may or may not tend to a limit. But in any case, *it is always possible to extract from such a sequence, by omitting certain of its terms, an infinite subsequence,* $y_1, y_2, y_3, \cdots$, *which tends to some limit* y *contained in the interval* I.

To prove this theorem we divide the interval I into two closed subintervals I' and I'' by marking the midpoint $\frac{a+b}{2}$ of I:

$$I': \ a \leq x \leq \frac{a+b}{2},$$

$$I'': \frac{a+b}{2} \leq x \leq b.$$

In at least one of these, which we may call I_1, there must lie infinitely many terms x_n of the original sequence. Choose any one of these terms, say x_{n_1}, and call it y_1. Now proceed in the same way with the interval I_1. Since there are infinitely many terms x_n in I_1, there must be infinitely many terms in at least one of the halves of I_1, which we may call I_2. Hence we can certainly find a term x_n in I_2 for which $n > n_1$.

Choose some one of these, and call it y_2. Proceeding in this way, we can find a sequence $I_1, I_2, I_3, \cdots$ of nested intervals and a subsequence $y_1, y_2, y_3, \cdots$ of the original sequence, such that y_n lies in I_n for every n. This sequence of intervals closes down on a point y of I, and it is clear that the sequence $y_1, y_2, y_3, \cdots$ has the limit a, as was to be proved.

* These considerations are capable of the type of generalization that is typical of modern mathematics. Let us consider a variable X ranging over a general set S in which some notion of "distance" is defined. S may be a set of points in the plane or in space. But this is not necessary; for example, S might be the set of all triangles in the plane. If X and Y are two triangles, with vertices A, B, C and A', B', C' respectively, then we can define the "distance" between the two triangles as the number

$$d(X, Y) = AA' + BB' + CC',$$

where AA', etc. denotes the ordinary distance between the points A and A'. Whenever there exists such a notion of "distance" in a set S we may define the concept of a sequence of elements $X_1, X_2, X_3, \cdots$ tending to a limit element X of S. By this we mean that $d(X, X_n) \to 0$ as $n \to \infty$. We shall now say that *the set* S *is compact if from any sequence* $X_1, X_2, X_3, \cdots$ *of elements of* S *we can always extract a subsequence which tends to some limit element* X *of* S. We have shown in the preceding paragraph that a closed interval $a \leq x \leq b$ is compact in this sense. Hence the concept of a compact set may be regarded as a generalization of a *closed interval* of the number axis. Note that the number axis *as a whole* is not compact, since the sequence of integers 1, 2, 3, 4, 5, $\cdots$ neither tends to a limit nor contains any subsequence that does. Nor is an open interval such as $0 < x < 1$, not including its endpoints, compact, since the sequence $\frac{1}{2}$, $\frac{1}{3}$, $\frac{1}{4}$, $\cdots$ or any subsequence of it tends to the limit 0, which is not a point of the open interval. In the same way it may be shown that the region of the plane consisting of the points interior to a square or rectangle is not compact, but becomes compact if the boundary points are added. Furthermore, the set of all triangles whose vertices lie within or on the circumference of a given circle is compact.

We may also extend the notion of continuity to the case where the variable X ranges over any set S in which the notion of limit is defined. The function $u = F(X)$, where u is a real number, is said to be continuous at the element X if, for any sequence of elements $X_1, X_2, X_3, \cdots$ which tends to X as limit, the corresponding sequence of numbers $F(X_1)$, $F(X_2)$, $\cdots$ tends to the limit $F(X)$. (An equivalent (ϵ, δ)-definition could also be given.) It is quite easy to show that Weierstrass' theorem also holds in the general case of a continuous function defined over the elements of any compact set:

If $u = F(X)$ *is any continuous function defined on a compact set* S, *then there always exists an element of* S *for which* F(X) *attains its largest value, and also one for which it attains its smallest value.*

The proof is simple once one has grasped the general concepts involved, but we shall not go further into this subject. It will appear in Chapter VII that the general theorem of Weierstrass is of great importance in the theory of maxima and minima.

§6. SOME APPLICATIONS OF BOLZANO'S THEOREM

1. Geometrical Applications

Bolzano's simple yet general theorem may be used to prove many facts which are not at all obvious at first sight. We begin by proving: *If* A *and* B *are any two areas in the plane, then there exists a straight line in the plane which bisects* A *and* B *simultaneously.* By an "area" we mean any portion of the plane included within a simple closed curve.

Let us begin by choosing some fixed point P in the plane, and drawing from P a directed ray PR from which to measure angles. If we take any ray PS which makes an angle x with PR, there will exist a directed straight line in the plane *bisecting the area* A, and with the same direction as the ray PS. For if we take a directed line l_1 with the direction of PS and lying wholly on one side of A and move this line parallel to itself until it is in position l_2 (see Fig. 173), wholly on the other side of A, then the function whose value is defined to be the area of A to the right of the line (the east direction if the arrow on the line points north) minus the area of A to the left of the line will be positive for position l_1 and negative for position l_2. Since this function is continuous, by Bolzano's theorem it must be zero for some intermediate position l_x, which therefore bisects A. For each value of x from $x = 0°$ to $x = 360°$, the line l_x which bisects A is uniquely defined.

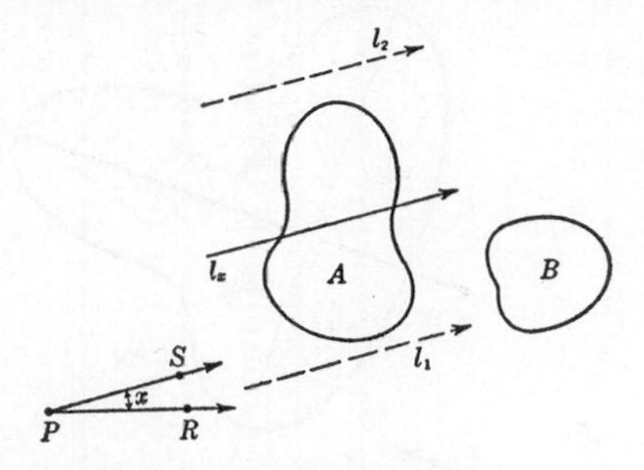

Fig. 173. Simultaneous bisection of two areas.

Now let the function $y = f(x)$ be defined as the area of B to the right of l_x minus the area of B to the left of l_x. Suppose that the line l_0 which bisects A and has the direction of PR has more of B to the right than to the left; then for $x = 0°$, y is positive. Let x increase to 180°, then the line l_{180} with direction RP which bisects A is the same as l_0 but oppositely directed, with right and left interchanged; hence the value of y for $x = 180°$ is the same numerically as for $x = 0°$, but with opposite sign, and therefore negative. Since y is a continuous function of x as l_x turns around, there exists some value α of x between 0° and 180° for which y is zero. It follows that the line l_α bisects A and B simultaneously. This completes the proof.

Note that although we have proved the *existence* of a line with the desired property, we have given no definite procedure for *constructing* it; this exhibits again the distinguishing feature of mathematical existence proofs as compared with constructions.

A similar problem is the following: Given a single area in the plane, it is desired to cut it into *four* equal pieces by two *perpendicular* lines. In order to prove that this is always possible, we return to our previous problem at the stage where we had defined l_x for any angle x, but we forget about the area B. Instead, we take the line l_{x+90} which is perpendicular to l_x and which also bisects A. If we number the four pieces of A as shown in Figure 174, then we have

$$A_1 + A_2 = A_3 + A_4$$

and

$$A_2 + A_3 = A_1 + A_4,$$

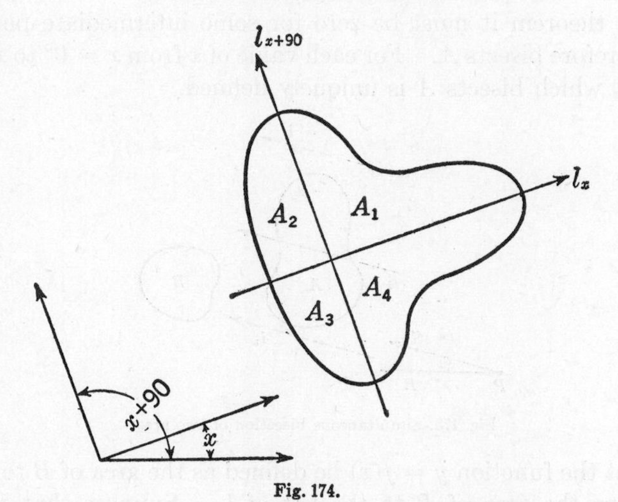

Fig. 174.

from which it follows, on subtracting the second equation from the first, that

$$A_1 - A_3 = A_3 - A_1,$$

i.e.

$$A_1 = A_3,$$

and hence

$$A_2 = A_4.$$

Thus if we can show the existence of an angle α such that for l_α

$$A_1(\alpha) = A_2(\alpha),$$

then our theorem will be proved, since for such an angle all four areas will be equal. To do this, we define a function $y = f(x)$ by drawing l_x and setting

$$f(x) = A_1(x) - A_2(x).$$

For $x = 0°$, $f(0) = A_1(0) - A_2(0)$ may be positive. In that case, for $x = 90°$, $A_1(90) - A_2(90) = A_2(0) - A_3(0) = A_2(0) - A_1(0)$ will be negative. Therefore, since $f(x)$ varies continuously as x increases from 0° to 90°, there will be some value α between 0° and 90° for which $f(\alpha) = A_1(\alpha) - A_2(\alpha) = 0$. The lines l_α and $l_{\alpha+90}$ then divide the area into four equal pieces.

It is interesting to observe that these problems may be generalized to three and higher dimensions. In three dimensions the first problem becomes: Given three volumes in space, to find a plane which bisects all three simultaneously. The proof that this is always possible again depends on Bolzano's theorem. In more than three dimensions the theorem is still true but the proof requires more advanced methods.

*2. Application to a Problem of Mechanics

We shall conclude this section by discussing an apparently difficult problem in mechanics that is easily answered by an argument based on continuity concepts. (This problem was suggested by H. Whitney.) Suppose a train travels from station A to station B along a straight section of track. The journey need not be of uniform speed or acceleration. The train may act in any manner, speeding up, slowing down, coming to a halt, or even backing up for a while, before reaching B. But the exact motion of the train is supposed to be known in advance; that is, the function $s = f(t)$ is given, where s is the distance of the train from station A, and t is the time, measured from the instant of departure. On the floor of one of the cars a rod is pivoted so that it may move without friction either forward or backward until it touches the floor. If it does touch the floor, we assume that it remains on the floor henceforth; this will be the case if the rod does not bounce. Is it possible to place the rod in such a position that, if it is released at the instant when the train starts and allowed to move solely under the influence of gravity and the motion of the train, it will not fall to the floor during the entire journey from A to B?

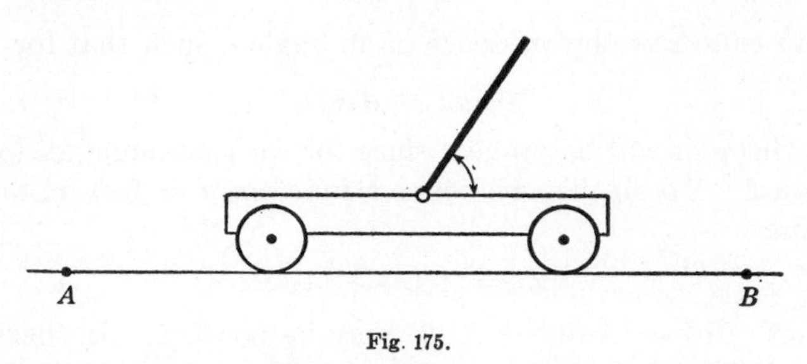

Fig. 175.

It might seem quite unlikely that for any given schedule of motion the interplay of gravity and reaction forces will always permit such a maintenance of balance under the single condition that the initial position of the rod is suitably chosen. Yet we state that such a position always exists.

Paradoxical as this assertion might seem at first sight, it can be proved easily once one concentrates on its essentially topological character. No detailed knowledge of the laws of dynamics is needed; only the following simple assumption of a physical nature need be granted: *The motion of the rod depends continuously on its initial position.* Let us characterize the initial position of the rod by the initial angle x which it makes with the floor, and by y the angle which the rod makes with the floor at the end of the journey, when the train reaches the point B. If the rod has fallen to the floor we have either $y = 0$ or $y = \pi$. For a given initial position x the end position y is, according to our assumption, uniquely determined as a function $y = g(x)$ which is continuous and has the values $y = 0$ for $x = 0$ and $y = \pi$ for $x = \pi$ (the latter assertion simply expressing that the rod will remain flat on the floor if it starts in this position). Now we recall that $g(x)$, as a continuous function in the interval $0 \leq x \leq \pi$, assumes all the values between $g(0) = 0$ and $g(\pi) = \pi$; consequently, for any such values y, e.g. for the value $y = \frac{\pi}{2}$, there exists a specific value of x such that $g(x) = y$; in particular, there exists an initial position for which the end position of the rod at B is perpendicular to the floor. (Note: In this argument it should not be forgotten that the motion of the train is fixed once for all.)

Of course, the reasoning is entirely theoretical. If the journey is of long duration or if the train schedule, expressed by $s = f(t)$, is very erratic, then the range of initial positions x for which the end position

$g(x)$ differs from 0 or π will be exceedingly small, as is known to anyone who has tried to balance a needle upright on a plate for an appreciable time. Still, our reasoning should be of value even to a practical mind inasmuch as it shows how qualitative results in dynamics may be obtained by simple arguments without technical manipulation.

Exercises: 1. Using the theorem of page 315, show that the reasoning above may be generalized to the case where the journey is of infinite duration.

2. Generalize to the case where the motion of the train is along any curve in the plane and the rod may fall in any direction. (Hint: It is not possible to map a circular disk continuously onto its circumference alone by a mapping which eaves every point of the circumference fixed (see p. 255)).

3. Show that the time required for the rod to fall to the floor, if the car is stationary and the rod is released at an angle ϵ from the vertical position, tends to infinity as ϵ tends to zero.

SUPPLEMENT TO CHAPTER VI

MORE EXAMPLES ON LIMITS AND CONTINUITY

§1. EXAMPLES OF LIMITS

1. General Remarks

In many cases the convergence of a sequence a_n can be proved by an argument of the following sort. We find two other sequences, b_n and c_n, whose terms have a simpler structure than those of the original sequence, and such that

$$b_n \leq a_n \leq c_n \tag{1}$$

for every n. *Then if we can show that the sequences* b_n *and* c_n *both converge to the same limit* α, *it follows that* a_n *also converges to the limit* α. We shall leave the formal proof of the statement to the reader.

It is clear that applications of this procedure will involve the use of inequalities. It is therefore appropriate to recall a few elementary rules which govern arithmetical operations with inequalities.

1. If $a > b$, then $a + c > b + c$ (any number may be added to both sides of an inequality).
2. If $a > b$ and the number c is *positive*, then $ac > bc$ (an inequality may be multiplied by any positive number).
3. If $a < b$, then $-b < -a$ (the sense of an inequality is reversed if both sides are multiplied by -1). Thus $2 < 3$ but $-3 < -2$.
4. If a and b have the same sign, and if $a < b$, then $1/a > 1/b$.
5. $|a + b| \leq |a| + |b|$.

2. The Limit of q^n

If q is a number greater than 1, the sequence q^n will increase beyond any bound, as does the sequence $2, 2^2, 2^3, \cdots$ for $q = 2$. The sequence "tends to infinity" (see p. 294). The proof in the general case is based on the important inequality (proved on p. 15)

$$(1 + h)^n \geq 1 + nh > nh, \tag{2}$$

where h is any positive number. We set $q = 1 + h$, where $h > 0$; then

$$q^n = (1 + h)^n > nh.$$

If k is any positive number, no matter how large, then for all $n > k/h$ it follows that

$$q^n > nh > k;$$

hence $q^n \to \infty$.

If $q = 1$, then the members of the sequence q^n are all equal to 1, and 1 is therefore the limit of the sequence. If q is negative, then q^n will alternate between positive and negative values, and will have no limit if $q \leq -1$.

Exercise: Give a rigorous proof of the last statement.

On page 64 we showed that if $-1 < q < 1$, then $q^n \to 0$. We may give another and very simple proof of this fact. First we consider the case where $0 < q < 1$. Then the numbers $q, q^2, q^3 \cdots$ form a monotone decreasing sequence bounded below by 0. Hence, according to page 295, the sequence must approach a limit: $q^n \to a$. Multiplying both sides of this relation by q we obtain $q^{n+1} \to aq$.

Now q^{n+1} must have the same limit as q^n, since the name, n or $n + 1$, of the increasing exponent, does not matter. Hence $aq = a$, or $a(q - 1) = 0$. Since $1 - q \neq 0$, this implies that $a = 0$.

If $q = 0$, the statement $q^n \to 0$ is trivial. If $-1 < q < 0$, then $0 < |q| < 1$; hence $|q^n| = |q|^n \to 0$ by the preceding argument From this it follows that always $q^n \to 0$ for $|q| < 1$. This complete the proof.

Exercises: Prove that for $n \to \infty$:

1) $(x^2/1 + x^2)^n \to 0$;
2) $(x/1 + x^2)^n \to 0$;
3) $(x^3/4 + x^2)^n$ tends to infinity for $x > 2$, to 0 for $|x| < 2$.

3. The Limit of $\sqrt[n]{p}$

The sequence $a_n = \sqrt[n]{p}$, i.e. the sequence $p, \sqrt{p}, \sqrt[3]{p}, \sqrt[4]{p}, \cdots$, has the limit 1 for any fixed positive number p:

$$\sqrt[n]{p} \to 1 \text{ as } n \to \infty. \tag{3}$$

(By the symbol $\sqrt[n]{p}$ we mean, as always, the positive nth root. For negative numbers p there are no real nth roots when n is even.)

To prove the relation (3), we first suppose that $p > 1$; then $\sqrt[n]{p}$ will also be greater than 1. Thus we may set

$$\sqrt[n]{p} = 1 + h_n,$$

where h_n is a positive quantity depending on n. The inequality (2) then shows that

$$p = (1 + h_n)^n > nh_n.$$

On dividing by n we see that

$$0 < h_n < p/n.$$

Since the sequences $b_n = 0$ and $c_n = p/n$ both have the limit 0, it follows by the argument of Article 1 that h_n also has the limit 0 as n increases, and our assertion is proved for $p > 1$. Here we have a typical instance where a limiting relation, in this case $h_n \to 0$, is recognized by enclosing h_n between two bounds whose limits are more easily obtained.

Incidentally, we have derived an estimate for the difference h_n between $\sqrt[n]{p}$ and 1; this difference must always be less than p/n.

If $0 < p < 1$, then $\sqrt[n]{p} < 1$, and we may set

$$\sqrt[n]{p} = \frac{1}{1 + h_n},$$

where h_n is again a positive number depending on n. It follows that

$$p = \frac{1}{(1 + h_n)^n} < \frac{1}{nh_n},$$

so that

$$0 < h_n < \frac{1}{np}.$$

From this we conclude that h_n tends to 0 as n increases. Hence, since $\sqrt[n]{p} = 1/(1 + h_n)$, it follows that $\sqrt[n]{p} \to 1$.

The equalizing effect of nth root extraction, which tends to push every positive number towards 1 as n increases, is even strong enough to do this in some cases if the radicand does not remain constant. We shall prove that the sequence $1, \sqrt{2}, \sqrt[3]{3}, \sqrt[4]{4}, \sqrt[5]{5}, \cdots$ tends to 1, i.e. that

$$\sqrt[n]{n} \to 1$$

as n increases. By a little device this can again be shown to follow from the inequality (2). Instead of the nth root of n, we take the nth root of $\sqrt{n}$. If we set $\sqrt[n]{\sqrt{n}} = 1 + k_n$, where k_n is a positive number depending on n, then the inequality yields $\sqrt{n} = (1 + k_n)^n > nk_n$, so that

$$k_n < \frac{\sqrt{n}}{n} = \frac{1}{\sqrt{n}}.$$

Hence

$$1 < \sqrt[n]{n} = (1 + k_n)^2 = 1 + 2k_n + k_n^2 < 1 + \frac{2}{\sqrt{n}} + \frac{1}{n}.$$

The right side of this inequality tends to 1 as n increases, so that $\sqrt[n]{n}$ must also tend to 1.

4. Discontinuous Functions as Limits of Continuous Functions

We may consider limits of sequences a_n when a_n is not a fixed number but depends on a variable x: $a_n = f_n(x)$. If this sequence converges as $n \to \infty$, then the limit is again a function of x,

$$f(x) = \lim f_n(x).$$

Such representations of functions $f(x)$ as limits of others are often useful in reducing "higher" functions $f(x)$ to elementary functions $f_n(x)$.

This is true in particular of the representation of discontinuous functions by explicit formulas. For example, let us consider the sequence $f_n(x) = \dfrac{1}{1 + x^{2n}}$. For $|x| = 1$ we have $x^{2n} = 1$ and hence $f_n(x) = 1/2$ for every n, so that $f_n(x) \to 1/2$. For $|x| < 1$ we have $x^{2n} \to 0$, and hence $f_n(x) \to 1$, while for $|x| > 1$ we have $x^{2n} \to \infty$, and hence $f_n(x) \to 0$. Summarizing:

$$f(x) = \lim \frac{1}{1 + x^{2n}} = \begin{cases} 1 & \text{for } |x| < 1, \\ 1/2 & \text{for } |x| = 1, \\ 0 & \text{for } |x| > 1. \end{cases}$$

Here the discontinuous function $f(x)$ is represented as the limit of a sequence of continuous rational functions.

Another interesting example of a similar character is given by the sequence

$$f_n(x) = x^2 + \frac{x^2}{1 + x^2} + \frac{x^2}{(1 + x^2)^2} + \cdots + \frac{x^2}{(1 + x^2)^n}.$$

For $x = 0$ all the values $f_n(x)$ are zero, and therefore $f(0) = \lim f_n(0) = 0$. For $x \neq 0$ the expression $1/(1 + x^2) = q$ is positive and less than 1; our results on geometrical series guarantee the convergence of $f_n(x)$ for $n \to \infty$. The limit, i.e. the sum of the infinite geometrical series, is $\dfrac{x^2}{1 - q} = \dfrac{x^2}{1 - \dfrac{1}{1 + x^2}}$, which is equal to $1 + x^2$. Thus we see that $f_n(x)$ tends to the function $f(x) = 1 + x^2$ for $x \neq 0$, and to $f(x) = 0$ for $x = 0$. This function has a removable discontinuity at $x = 0$.

*5. Limits by Iteration

Often the terms of a sequence are such that a_{n+1} is obtained from a_n by the same procedure as a_n from a_{n-1}; the same process, indefinitely repeated, produces the whole sequence from a given initial term. In such cases we speak of a process of "iteration."

For example, the sequence

$$1, \sqrt{1+1}, \sqrt{1+\sqrt{2}}, \sqrt{1+\sqrt{1+\sqrt{2}}}, \cdots$$

has such a law of formation; each term after the first is formed by taking the square root of 1 plus its predecessor. Thus the formula

$$a_1 = 1, \; a_{n+1} = \sqrt{1 + a_n}$$

defines the whole sequence. Let us find its limit. Obviously a_n is greater than 1 for $n > 1$. Furthermore a_n is a monotone increasing sequence, for

$$a_{n+1}^2 - a_n^2 = (1 + a_n) - (1 + a_{n-1}) = a_n - a_{n-1}.$$

Hence whenever $a_n > a_{n-1}$ it will follow that $a_{n+1} > a_n$. But we know that $a_2 - a_1 = \sqrt{2} - 1 > 0$, from which we conclude by mathematical induction that $a_{n+1} > a_n$ for all n, i.e. that the sequence is monotone increasing. Moreover it is bounded; for by the previous results we have

$$a_{n+1} = \frac{1 + a_n}{a_{n+1}} < \frac{1 + a_{n+1}}{a_{n+1}} = 1 + \frac{1}{a_{n+1}} < 2.$$

By the principle of monotone sequences we conclude that for $n \to \infty$ $a_n \to a$, where a is some number between 1 and 2. We easily see that a is the positive root of the quadratic equation $x^2 = 1 + x$. For as $n \to \infty$ the equation $a_{n+1}^2 = 1 + a_n$ becomes $a^2 = 1 + a$. Solving this equation, we find that the positive root is $a = \dfrac{1+\sqrt{5}}{2}$. Thus we may solve this quadratic equation by an iteration process which gives the value of the root with any degree of approximation if we continue long enough.

We can solve many other algebraic equations by iteration in a similar way. For example, we may write the cubic equation $x^3 - 3x + 1 = 0$ in the form

$$x = \frac{1}{3 - x^2}.$$

We now choose any value for a_1, say $a_1 = 0$, and define

$$a_{n+1} = \frac{1}{3 - a_n^2},$$

obtaining the sequence $a_2 = 1/3 = .3333 \cdots$, $a_3 = 9/26 = .3461 \cdots$, $a_4 = 676/1947 = .3472 \cdots$, etc. It may be shown that the sequence a_n obtained in this way converges to a limit $a = .3473 \cdots$ which is a solution of the given cubic equation. Iteration processes such as this are highly important both in pure mathematics, where they yield "existence proofs," and in applied mathematics, where they provide approximation methods for the solution of many types of problems.

Exercises on limits. For $n \to \infty$:

1) Prove that $\sqrt{n+1} - \sqrt{n} \to 0$.

(Hint: write the difference in the form

$$\frac{\sqrt{n+1} - \sqrt{n}}{\sqrt{n+1} + \sqrt{n}} \cdot (\sqrt{n+1} + \sqrt{n}).)$$

2) Find the limit of $\sqrt{n^2 + a} - \sqrt{n^2 + b}$.
3) Find the limit of $\sqrt{n^2 + an + b} - n$.
4) Find the limit of $\dfrac{1}{\sqrt{n+1} + \sqrt{n}}$.
5) Prove that the limit of $\sqrt[n]{n+1}$ is 1.
6) What is the limit of $\sqrt[n]{a^n + b^n}$ if $a > b > 0$?
7) What is the limit of $\sqrt[n]{a^n + b^n + c^n}$ if $a > b > c > 0$?
8) What is the limit of $\sqrt[n]{a^n b^n + a^n c^n + b^n c^n}$ if $a > b > c > 0$?
9) We shall see later (p. 449) that $e = \lim (1 + 1/n)^n$. What then is $\lim (1 + 1/n^2)^n$?

§2. EXAMPLE ON CONTINUITY

To give a precise proof of the continuity of a function requires the explicit verification of the definition of page 310. Sometimes this is a lengthy procedure, and therefore it is fortunate that, as we shall see in Chapter VIII, continuity is a consequence of differentiability. Since the latter will be established systematically for all elementary functions, we may follow the usual course of omitting tedious individual proofs of continuity. But as a further illustration of the general definition we shall analyze one further example, the function $f(x) = \dfrac{1}{1 + x^2}$. We may restrict x to a fixed interval $|x| \leq M$, where M is an arbitrarily selected number. Writing

$$f(x_1) - f(x) = \frac{1}{1 + x_1^2} - \frac{1}{1 + x^2}$$

$$= \frac{x^2 - x_1^2}{(1 + x^2)(1 + x_1^2)} = (x - x_1) \frac{(x + x_1)}{(1 + x^2)(1 + x_1^2)},$$

we find for $|x| \leq M$ and $|x_1| \leq M$

$$|f(x_1) - f(x)| \leq |x - x_1||x + x_1| \leq |x - x_1| \cdot 2M.$$

Hence it is clear that the difference on the left side will be smaller than any positive number ϵ if only $|x_1 - x| < \delta = \dfrac{\epsilon}{2M}$.

It should be noted that we are being quite generous in our appraisals. For large values of x and x_1 the reader will easily see that a much larger δ would suffice.

CHAPTER VII

MAXIMA AND MINIMA

INTRODUCTION

A straight segment is the shortest connection between its endpoints. An arc of a great circle is the shortest curve joining two points on a sphere. Among all closed plane curves of the same length the circle encloses the largest area, and among all closed surfaces of the same area the sphere encloses the largest volume.

Maximum and minimum properties of this type were known to the Greeks, even though the results were often stated without a real attempt at a proof. One of the most significant Greek discoveries is ascribed to Heron, the Alexandrian scientist of the first century A.D. It had long been known that a light ray from a point P meeting a plane mirror L in a point R is reflected in the direction of a point Q such that PR and QR form equal angles with the mirror. Heron found that if R' is any other point on the mirror, then the total distance $PR' + R'Q$ is larger than the distance $PR + RQ$. This theorem, which we shall prove presently, characterizes the actual path of light PRQ between P and Q as the shortest possible path from P to Q by way of the mirror—a discovery that can be considered the germ of the theory of geometrical optics.

It is only natural that mathematicians should be interested in questions of this sort. In daily life problems of maxima and minima, of the "best" and the "worst," arise constantly. Many problems of practical importance present themselves in this form. For example, how should a boat be shaped so as to have the least possible resistance in water? What cylindrical container made from a given amount of material has a maximum volume?

Starting in the seventeenth century, the general theory of extreme values—maxima and minima—has become one of the systematic integrating principles of science. Fermat's first steps in his differential calculus were prompted by the desire to study questions of maxima and minima by general methods. In the century that followed, the scope of these methods was greatly widened by the invention of the "calculus of variations." It became increasingly apparent that the physical laws of nature are most adequately expressed in terms of a minimum principle that provides a natural access to a more or less complete solution of particular problems. One of the most remarkable achievements of contemporary mathematics is the theory of stationary values—an extension of the notion of extreme values which combines analysis and topology. Our approach to the whole subject will be quite elementary.

§1. PROBLEMS IN ELEMENTARY GEOMETRY

1. Maximum Area of a Triangle with Two Sides Given

Given two segments a and b; required to find the triangle of maximum area having a and b as sides. The solution is simply the right triangle whose two legs are a and b. For consider any triangle with a and b as sides, as in Figure 176. If h is the altitude on the base a, then the area of the triangle is $A = \frac{1}{2}ah$. Now $\frac{1}{2}ah$ is clearly a maximum when h is largest, and this occurs when h coincides with b; that is, for a right triangle. Hence the maximum area is $\frac{1}{2}ab$.

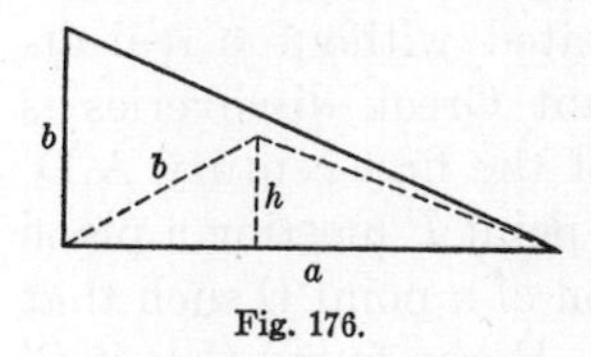

Fig. 176.

2. Heron's Theorem. Extremum Property of Light Rays

Given a line L and two points P and Q on the same side of L. For what point R on L is $PR + RQ$ the shortest path from P to L to Q? This is Heron's problem of the light ray. (If L were the bank of a stream, and someone had to go from P to Q as fast as possible, fetching a pail of water from L on the way, then he would have to solve just this problem.) To find the solution, we reflect P in L as in a mirror, obtaining the point P' such that L is the perpendicular bisector of PP'. The line $P'Q$ intersects L in the required point R. It is simple to prove that $PR + RQ$ is smaller than $PR' + R'Q$ for any other point R' on L. For $PR = P'R$ and $PR' = P'R'$; hence, $PR + RQ = P'R + RQ = P'Q$ and $PR' + R'Q = P'R' + R'Q$. But $P'R' + R'Q$ is greater than $P'Q$ (since the sum of any two sides of a triangle is greater than the third side), hence $PR' + R'Q$ is greater than $PR + RQ$, which was to be proved. In what follows we assume that neither P nor Q lies on L.

From Figure 177 we see that $\measuredangle 3 = \measuredangle 2$, and $\measuredangle 2 = \measuredangle 1$, so that $\measuredangle 1 = \measuredangle 3$. In other words, R *is the point such that* PR *and* QR *make equal angles with* L. From this it follows that a light ray reflected in

L (which is known from experiment to make equal angles of incidence and reflection) actually takes the shortest path from P to L to Q, as stated above in the introduction.

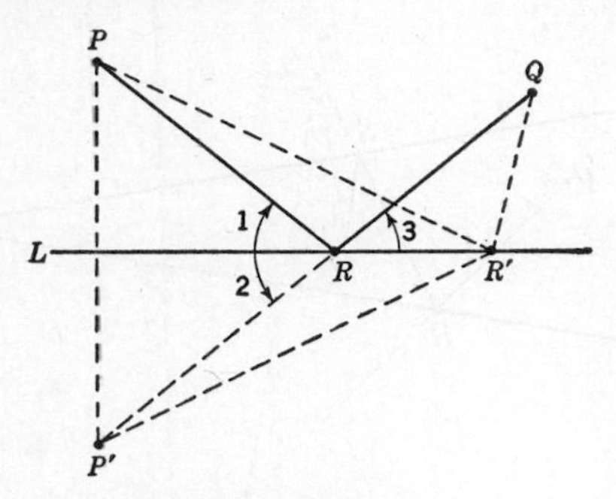

Fig. 177. Heron's theorem.

The problem can be generalized to include several lines $L, M, \cdots$. For example, consider the case where we have two lines L, M and two points P, Q situated as in Figure 178, with the problem of finding the path of minimum length from P to L, then to M, then to Q. Let Q' be the reflection of Q in M and Q'' the reflection of Q' in L. Draw

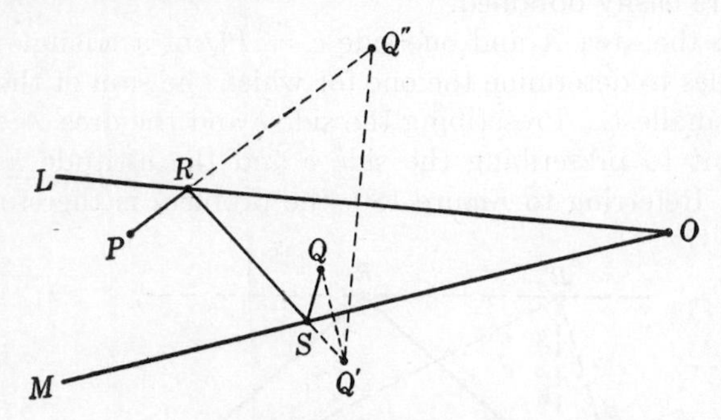

Fig. 178. Reflection in two mirrors.

PQ'' intersecting L in R and RQ' intersecting M in S; then R and S are the required points such that $PR + RS + SQ$ is the path of minimum length from P to L to M to Q. The proof of this fact is very similar to that of the previous problem, and is left as an exercise for the reader. If L and M were mirrors, a light ray from P, reflected from L to M, and there reflected to Q, would meet L at R and M at S; hence the light ray would again take the path of shortest length.

One might ask for the shortest path first from P to M, then to L, and from there to Q. This would give a path $PRSQ$ (see Fig. 179) determined in a manner similar to the previous path $PRSQ$. The length of the first path may be greater than, equal to, or less than that of the second.

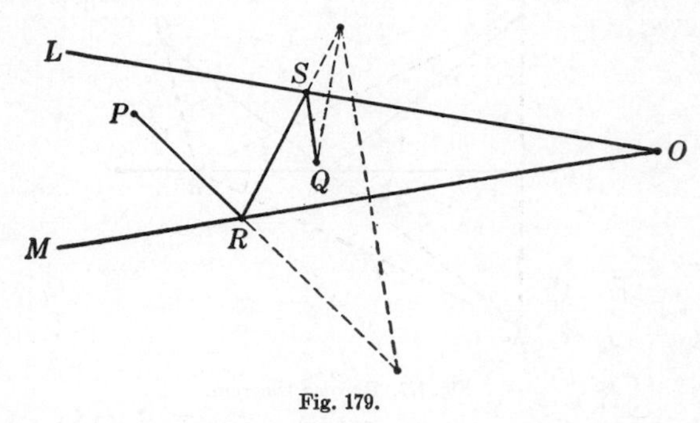

Fig. 179.

* *Exercise:* Show that the first path is smaller than the second if O and R lie on the same side of the line PQ. When will the two paths be of equal length?

3. Applications to Problems on Triangles

With the help of Heron's theorem solutions to the following two problems are easily obtained.

a) Given the area A and one side $c = PQ$ of a triangle; among all such triangles to determine the one for which the sum of the other sides a and b is smallest. Prescribing the side c and the area A of a triangle is equivalent to prescribing the side c and the altitude h on c, since $A = \frac{1}{2}hc$. Referring to Figure 180, the problem is therefore to find a

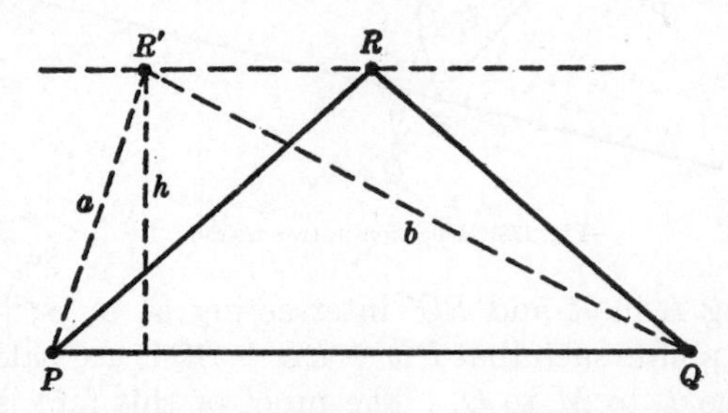

Fig. 180. Triangle of minimum perimeter with given base and area.

point R such that the distance from R to the line PQ is equal to the given h, and such that the sum $a + b$ is a minimum. From the first condition it follows that R must lie on the line parallel to PQ at a distance h. The answer is given by Heron's theorem for the special case where P and Q are equally distant from L: the required triangle PRQ is isosceles.

b) In a triangle let one side c and the sum $a + b$ of the two other sides be given; to find among all such triangles the one with the largest area. This is just the converse of problem a). The solution is again the isosceles triangle for which $a = b$. As we have just shown, this triangle has the minimum value of $a + b$ for its area; that is, any other triangle with the base c and the same area has a greater value of $a + b$. Moreover, it is clear from a) that any triangle with base c and an area greater than that of the isosceles triangle also has a greater value of $a + b$. Hence any other triangle with the same values of $a + b$ and of c must have a smaller area, so that the isosceles triangle provides the maximum area for given c and $a + b$.

4. Tangent Properties of Ellipse and Hyperbola. Corresponding Extremum Properties

The problem of Heron is connected with some important geometrical theorems. We have proved that if R is the point on L such that $PR + RQ$ is a minimum, then PR and RQ make equal angles with L. This minimum total distance we shall call $2a$. Let p and q denote the distances from any point in the plane to P and Q respectively, and consider the locus of *all* points in the plane for which $p + q = 2a$. This locus is an ellipse, with P and Q as foci, that passes through the point R on the line L. Moreover, L *must be tangent to the ellipse at* R. If L intersected the ellipse at a point other than R, there would be a

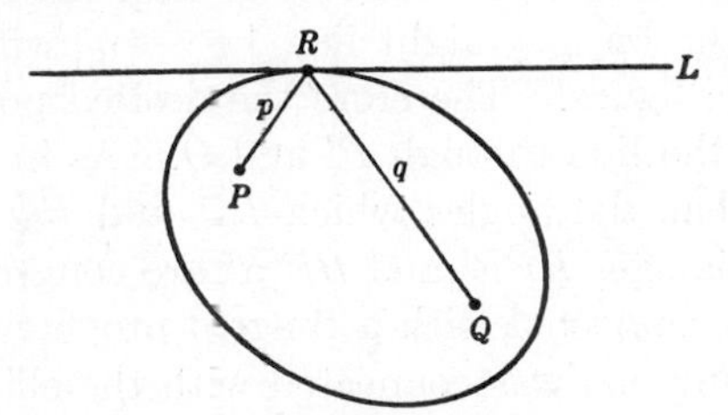

Fig. 181. Tangent property of ellipse.

segment of L lying inside the ellipse; for each point of this segment $p + q$ would be less than $2a$, since it is easily seen that $p + q$ is less than $2a$ inside the ellipse and greater than $2a$ outside. Since we know that $p + q \geq 2a$ on L, this is impossible. Hence L must be tangent to the ellipse at R. But we know that PR and RQ make equal angles with L; hence we have incidentally proved the important theorem: A tangent to an ellipse makes equal angles with the lines joining the foci to the point of tangency.

Closely related to the foregoing discussion is the following problem: Given a straight line L and two points P and Q on *opposite* sides of L (see Fig. 182), to find a point R on L such that the quantity $|p - q|$, that is, the absolute value of the *difference* of the distances from P and Q to R, is a *maximum*. (We shall assume that L is not the perpendicular bisector of PQ; for then $p - q$ would be zero for every point R on L and the problem would be meaningless.) To solve this problem, we first reflect P in L, obtaining the point P' on the same side of L as Q. For any point R' on L, we have $p = R'P = R'P'$, $q = R'Q$. Since R', Q, and P' can be regarded as the vertices of a triangle, the quantity $|p - q| = |R'P' - R'Q|$ is never greater than $P'Q$, for the difference

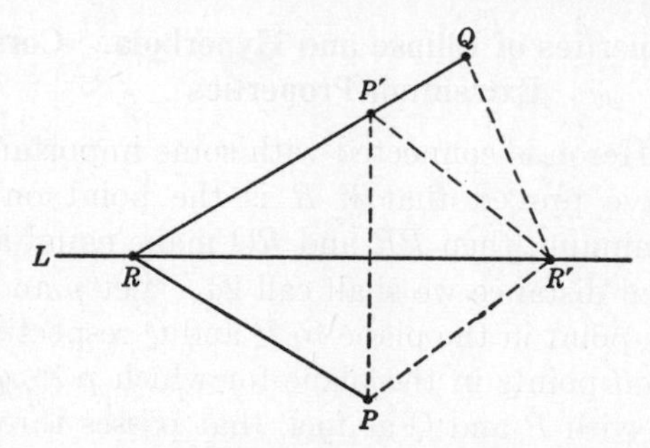

Fig. 182. $|PR - QR|$ = maximum.

between two sides of a triangle is never greater than the third side. If R', P', and Q all lie on a straight line, $|p - q|$ will be *equal* to $P'Q$, as is seen from the figure. Therefore the desired point R is the intersection of L with the line through P' and Q. As in the previous case, it is easily seen that the angles which RP and RQ make with L are equal, since the triangles RPR' and $RP'R'$ are congruent.

This problem is connected with a tangent property of the hyperbola, just as the preceding one was connected with the ellipse. If the maximum difference $|PR - QR|$ has the value $2a$, we can consider the locus of all points in the plane for which $p - q$ has the absolute value $2a$. This is a hyperbola with P and Q as its foci and passing through the point R. As is easily shown, the absolute value of $p - q$ is less than $2a$ in the region between two branches of the hyperbola, and greater than $2a$ on that side of each branch where the corresponding focus lies. It follows, by essentially the same argument as for the ellipse, that L must be tangent to the hyperbola at R. Which of the two branches is touched by L depends on whether P or Q is nearer to L; if P is nearer, the branch surrounding P will touch L, and likewise for Q (see Fig. 183).

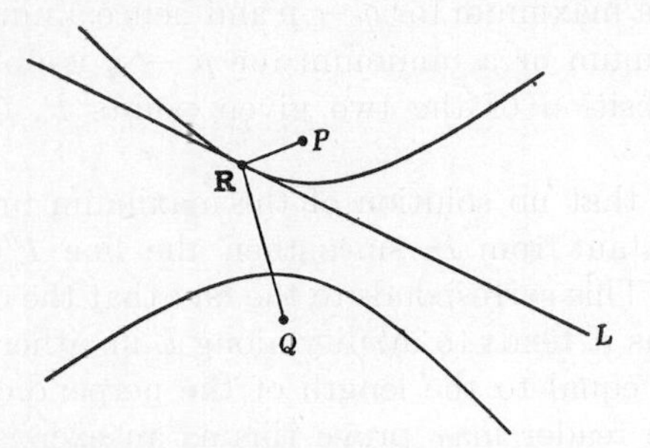

Fig. 183. Tangent property of hyperbola.

If P and Q are equidistant from L, then L will touch neither branch of the hyperbola, but will, instead, be one of the curve's asymptotes. This result becomes plausible when one observes that for this case the preceding construction will yield no (finite) point R, since the line $P'Q$ will be parallel to L.

In the same way as before this argument proves the well known theorem: A tangent to a hyperbola at any point bisects the angle subtended at that point by the foci of the hyperbola.

It may seem strange that if P and Q are on the same side of L we have a minimum problem to solve, while if they are on opposite sides of L we considered a maximum problem. That this is natural can be seen at once. In the first problem, each of the distances p, q, and therefore their sum, becomes larger without bound as we proceed along L indefinitely in either direction. Hence it would be impossible to find a maximum value for $p + q$, and a *minimum* problem is the only possibility. It is quite different in the second case, where P and Q lie on different sides of L. Here, to avoid confusion, we have to distinguish between the difference $p - q$, its negative $q - p$, and the absolute value $|p - q|$; it is the latter which was made a *maximum.* The situation is best understood if we let the point R move along the line L through different positions, R_1, R_2, R_3, $\cdots$. There is one point for which the difference $p - q$ is zero: the intersection of the perpendicular bisector of PQ with L. This point therefore gives a minimum for the absolute value $|p - q|$. But on one side of this point p is greater than q, and on the other, less; hence the quantity $p - q$ is positive on one side of the point and negative on the other. Consequently $p - q$ itself is neither a maximum nor a minimum at the point for which

$|p - q| = 0$. However, the point making $|p - q|$ a maximum is actually an extremum of $p - q$. If $p > q$, we have a maximum for $p - q$; if $q > p$, a maximum for $q - p$ and hence a minimum for $p - q$. Whether a maximum or a minimum for $p - q$ is obtainable is determined by the position of the two given points P, Q relative to the line L.

We have seen that no solution of the maximum problem exists if P and Q are equidistant from L, since then the line $P'Q$ in Figure 182 is parallel to L. This corresponds to the fact that the quantity $|p - q|$ tends to a limit as R tends to infinity along L in either direction. This limiting value is equal to the length of the perpendicular projection s of PQ on L (the reader may prove this as an exercise). If P and Q have the same distance from L, then $|p - q|$ will always be less than this limit and no maximum exists, since to each point R we can find another farther out for which $|p - q|$ is larger, but still not quite equal to s.

*5. Extreme Distances to a Given Curve

First we shall determine the *shortest* and the *longest distance* from a point P to a given curve C. For simplicity we shall suppose that C is a simple closed curve with a tangent everywhere, as in Figure 184. (The concept of tangent to a curve is here accepted on an intuitive basis that will be analyzed in the next chapter.) The answer is very simple:

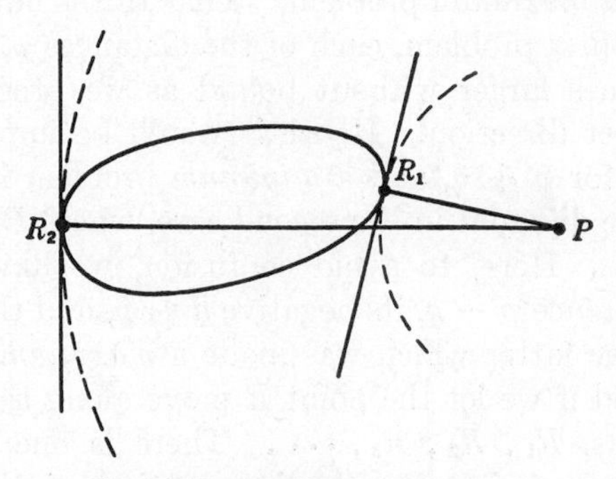

Fig. 184. Extreme distances to a curve.

A point R on C for which the distance PR has its smallest *or* its largest value must be such that the line PR is perpendicular to the tangent to C at R; in other words, PR is perpendicular to C. The proof is as follows: the circle with center at P and passing through R must be tangent to the curve. For if R is the point of minimum distance, C must lie entirely outside the circle, and therefore cannot cross it at R, while if R is the point of maximum distance, C must lie entirely inside the circle, and again cannot cross it at R. (This follows from the obvious fact that the distance from any point to P is less than RP if the point lies inside the circle, and greater than RP if the point lies outside the circle.) Hence the circle and the curve C will touch and have a common tangent at R. Now the line PR, being a radius of the circle, is perpendicular to the tangent to the circle at R, and therefore perpendicular to C at R.

Incidentally, the diameter of such a closed curve C, that is, its longest chord, must be perpendicular to C at both endpoints. The proof is left as an exercise for the reader. A similar statement should be formulated and proved in three dimensions.

Exercise: Prove that the shortest and longest segments connecting two non-intersecting closed curves are perpendicular to the curves at their endpoints.

The problems in Article 4 concerning the sum or difference of distances can now be generalized. Consider, instead of a straight line L, a simple closed curve C with a tangent at every point, and two points, P and Q, not on C. We wish to characterize the points on C for which the sum, $p + q$, and the difference, $p - q$, take on their extreme values, where p and q denote the distances from any point on C to P and Q respectively. No use can be made of the simple construction of reflection with which we solved the problems for the case where C is a straight line. But we may use the properties of the ellipse and hyperbola to solve the present problems. Since C is a closed curve and no longer a line extending to infinity, both the minimum and maximum problems make sense here, for it may be taken as granted that the quantities $p + q$ and $p - q$ have greatest *and* least values on any finite segment of a curve, in particular on a closed curve (see §7).

For the case of the sum, $p + q$, suppose R is the point on C for which $p + q$ is a maximum, and let $2a$ denote the value of $p + q$ at R. Consider the ellipse with foci at P and Q that is the locus of all points for which $p + q = 2a$. This ellipse must be tangent to C at R (the proof is left as an exercise for the reader). But we have seen that the lines PR and QR make equal angles with the ellipse at R; since the ellipse is tangent to C at R, the lines PR and QR must also make equal angles with C at R. If $p + q$ is a minimum for R, we see in the same way that

PR and QR make equal angles with C at R. Thus we have the theorem: Given a closed curve C and two points P and Q on the same side of C; then at a point R of C where the sum $p + q$ takes on its greatest or least value on C, the lines PR and QR make equal angles with the curve C (i.e. with its tangent) at R.

If P is inside C and Q outside, this theorem also holds for the greatest value of $p + q$, but fails for the least value, since the ellipse degenerates into a straight line.

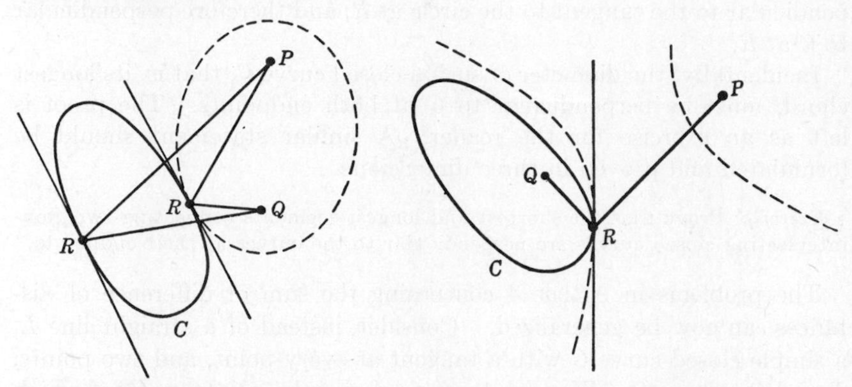

Fig. 185. Greatest and least values of $PR + QR$.

Fig. 186. Least value of $PR - QR$.

By an entirely analogous procedure, making use of properties of the hyperbola instead of the ellipse, the reader may prove the following theorem: Given a closed curve C and two points P and Q, on different sides of C; then at a point R of C where $p - q$ takes on its greatest or least value on C, the lines PR and QR make equal angles with C. Again we emphasize that the problem for a closed curve C differs from that for an infinite line inasmuch as in the latter problem the maximum of the absolute value $| p - q |$ was sought while now a maximum (as well as a minimum) of $p - q$ exits.

*§2. A GENERAL PRINCIPLE UNDERLYING EXTREME VALUE PROBLEMS

1. The Principle

The preceding problems are examples of a general question which is best formulated in analytic language. If, in the problem of finding the extreme values of $p + q$, we denote by x, y the coördinates of the point

R, by x_1, y_1 the coördinates of the fixed point P, and by x_2, y_2 those of Q, then

$$p = \sqrt{(x - x_1)^2 + (y - y_1)^2}, \qquad q = \sqrt{(x - x_2)^2 + (y - y_2)^2},$$

and the problem is to find the extreme values of the function

$$f(x, y) = p + q.$$

This function is continuous everywhere in the plane, but the point with the coördinates x, y is restricted to the given curve C. This curve will be defined by an equation $g(x, y) = 0$; e.g. $x^2 + y^2 - 1 = 0$ if it is the unit circle. Our problem then is to find the extreme values of $f(x, y)$ when x and y are restricted by the condition that $g(x, y) = 0$, and we shall consider this general type of problem.

To characterize the solutions, we consider the family of curves with the equations $f(x, y) = c$; that is, the curves given by equations of this form, where the constant c may have any value, the same for all points of any one curve of the family. Let us assume that one and only one curve of the family $f(x, y) = c$ passes through each point of the plane, at least if we restrict ourselves to the vicinity of the curve C. Then as c changes, the curve $f(x, y) = c$ will sweep out a part of the plane, and no point in this part will be touched twice in the sweeping process. (The curves $x^2 - y^2 = c$, $x + y = c$, and $x = c$ are such families.) In particular, one curve of the family will pass through the point R_1, where

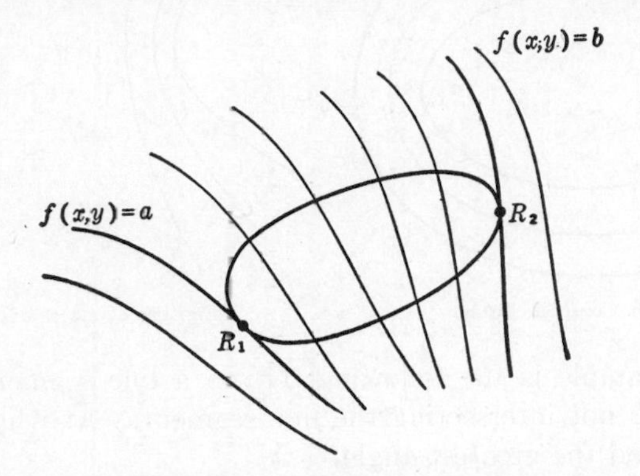

Fig. 187. Extrema of a function on a curve.

$f(x, y)$ takes on its greatest value on C, and another one through the point R_2 where $f(x, y)$ takes on its least value. Let us call the greatest value a and the least value b. On one side of the curve $f(x, y) = a$ the value of $f(x, y)$ will be less than a. and on the other side greater than a.

Since $f(x, y) \leq a$ on C, C must lie entirely on one side of the curve $f(x, y) = a$; hence it must be tangent to that curve at R_1. Similarly, C must be tangent to the curve $f(x, y) = b$ at R_2. We thus have the general theorem: *If at a point* R *on a curve* C *a function* f(x, y) *has an extreme value* a, *the curve* f(x, y) = a *is tangent to* C *at* R.

2. Examples

The results of the preceding section are easily seen to be special cases of this general theorem. If $p + q$ is to have an extreme value, the function $f(x, y)$ is $p + q$, and the curves $f(x, y) = c$ are the confocal ellipses with foci P and Q. As predicted by the general theorem, the ellipses passing through the points on C where $f(x, y)$ takes on its extreme values were seen to be tangent to C at these points. In the case where the extrema of $p - q$ are sought, the function $f(x, y)$ is $p - q$, the curves $f(x, y) = c$ are the confocal hyperbolas with P and Q as their foci, and the hyperbolas passing through the points of extreme value of $f(x, y)$ were seen to be tangent to C.

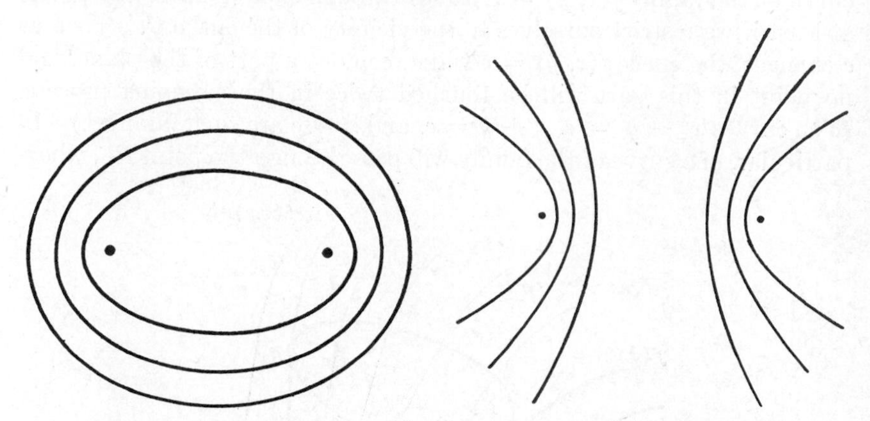

Fig. 188. Confocal ellipses. Fig. 189. Confocal hyperbolas.

Another example is the following: Given a line segment PQ and a straight line L not intersecting the line segment. At what point of L will PQ subtend the greatest angle?

The function to be maximized here is the angle θ subtended by PQ from points on L. The angle subtended by PQ from any point R in the plane is a function $\theta = f(x, y)$ of the coördinates of R. From elementary geometry we know that the family of curves $\theta = f(x, y) = c$ is the family of circles through P and Q, since a chord of a circle subtends the same angle at all points of the circumference on the same side of the chord. As is seen from Figure 190, two of these circles will, in gen-

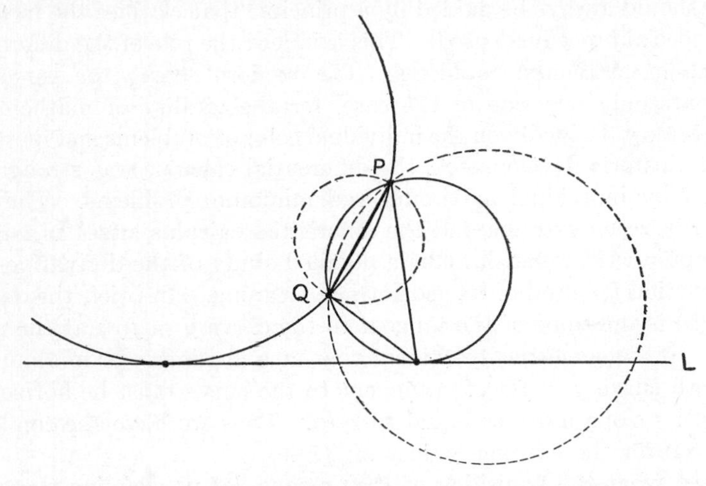

Fig. 190. Point on L from which segment PQ appears largest.

eral, be tangent to L, with centers on opposite sides of PQ. One of the points of tangency gives the absolute maximum for θ, while the other point yields a "relative" maximum (that is, the value of θ will be less *in a certain neighborhood* of this point than at the point itself. The greater of the two maxima, the absolute maximum, is given by that point of tangency which lies in the acute angle formed by the extension of PQ and L, and the smaller one by the point which lies in the obtuse angle formed by these two lines. (The point where the extension of the segment PQ intersects L gives the minimum value of θ, zero.)

As a generalization of this problem we may replace L by a curve C and seek the point R on C at which a given line segment PQ (not intersecting C) subtends the greatest or least angle. Here again, the circle through P, Q, and R must be tangent to C at R.

§3. STATIONARY POINTS AND THE DIFFERENTIAL CALCULUS

1. Extrema and Stationary Points

In the preceding arguments the technique of the differential calculus was not used. As a matter of fact, our elementary methods are far more simple and direct than those of the calculus. As a rule in scientific thinking it is better to consider the individual features of a problem rather than to rely exclusively on general methods, although individual efforts should always be guided by a principle that clarifies the meaning of the special procedures used. This is indeed the rôle of the differential calculus in extremum problems. The modern search for generality represents only one side of the case, for the vitality of mathematics depends most decidedly on the individual color of problems and methods.

In its historic development, the differential calculus was strongly influenced by individual maximum and minimum problems. The connection between extrema and the differential calculus arises as follows. In Chapter VIII we shall make a detailed study of the derivative $f'(x)$ of a function $f(x)$ and of its geometrical meaning. In brief, the derivative $f'(x)$ is the slope of the tangent to the curve $y = f(x)$ at the point (x, y). It is geometrically evident that at a maximum or minimum of a smooth curve $y = f(x)$ the tangent to the curve must be horizontal, that is, its slope must be equal to zero. Thus we have the condition $f'(x) = 0$ for the extreme values of $f(x)$.

To see what the vanishing of $f'(x)$ means, let us examine the curve of Figure 191. There are five points, A, B, C, D, E, at which the tangent

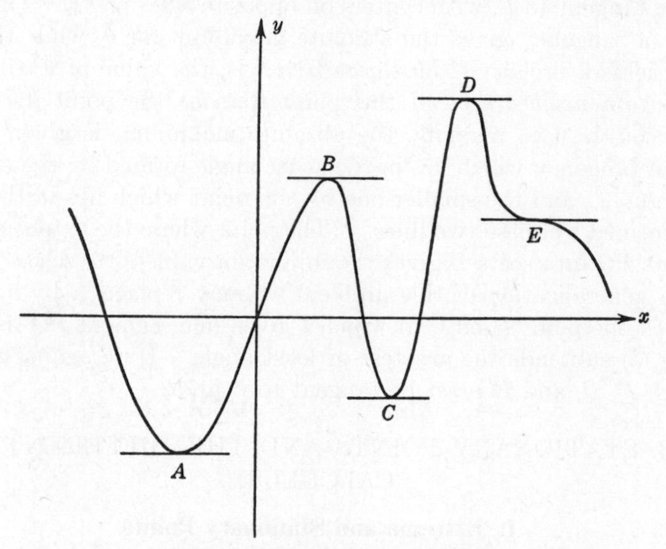

Fig. 191. Stationary points of a function.

to this curve is horizontal; let the values of $f(x)$ at these points be a, b, c, d, e respectively. The maximum of $f(x)$ in the interval pictured is at

D, the minimum at A. The point B also represents a maximum, in the sense that for all other points in the *immediate neighborhood* of B, $f(x)$ is less than b, although $f(x)$ is greater than b for points close to D. For this reason we call B a *relative maximum* of $f(x)$, while D is the *absolute maximum.* Similarly, C represents a relative minimum and A the absolute minimum. Finally, at E, $f(x)$ has neither a maximum nor a minimum, even though $f'(x) = 0$. From this it follows that the vanishing of $f'(x)$ is a *necessary*, but not a *sufficient* condition for the occurrence of an extremum of a smooth function $f(x)$; in other words, at any extremum, relative or absolute, $f'(x) = 0$, but not every point at which $f'(x) = 0$ need be an extremum. A point where the derivative vanishes, whether it is an extremum or not, is called a *stationary* point. By a more refined analysis, it is possible to arrive at more or less complicated conditions on the higher derivatives of $f(x)$ which completely characterize the maxima, minima, and other stationary points.

2. Maxima and Minima of Functions of Several Variables. Saddle Points

There are problems of maxima and minima that cannot be expressed in terms of a function $f(x)$ of one variable. The simplest such case is that of finding the extreme values of a function $z = f(x, y)$ of two variables.

We can represent $f(x, y)$ by the height z of a surface above the x, y-plane, which we may interpret, say, as a mountain landscape. A maximum of $f(x, y)$ corresponds to a mountain top; a minimum, to the bottom of a depression or of a lake. In both cases, if the surface is smooth the tangent plane to the surface will be horizontal. But there are other points besides summits and the bottoms of valleys for which the tangent plane is horizontal; these are the points given by mountain passes. Let us examine these points in more detail. Consider as in

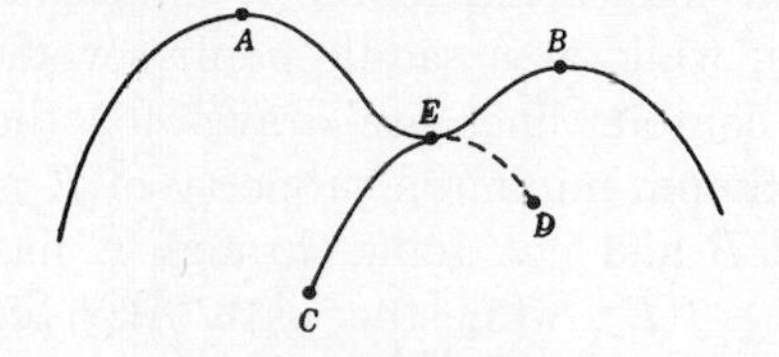

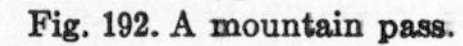

Fig. 192. A mountain pass.

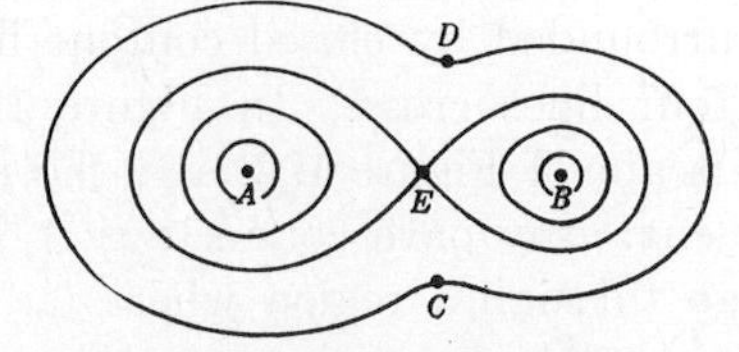

Fig. 193 The corresponding contour map.

Figure 192 two mountains A and B on a range and two points C and D on different sides of the mountain range, and suppose that we wish to go from C to D. Let us first consider only the paths leading from C to D obtained by cutting the surface with some plane through C and D. Each such path will have a highest point. By changing the position of the plane, we change the path, and there will be one path CD for which the altitude of that *highest point* is *least*. The point E of maximum altitude on this path is a mountain pass, called in mathematical language a *saddle point*. It is clear that E is neither a maximum nor a minimum, since we can find points as near E as we please which are higher and lower than E. Instead of confining ourselves to paths that lie in a plane, we might just as well consider paths without this restriction. The characterization of the saddle point E remains the same.

Similarly, if we want to proceed from the peak A to the peak B, any particular path will have a lowest point; if again we consider only plane sections, there will be one path AB for which this lowest point is highest, and the minimum for this path is again at the point E found above. This saddle point E thus has the property of being a highest minimum or a lowest maximum; that is, a *maxi-minimum* or a *mini-maximum*. The tangent plane at E is horizontal; for, since E is the minimum point of AB, the tangent line to AB at E must be horizontal, and similarly, since E is the maximum point of CD, the tangent line to CD at E must be horizontal. The tangent plane, which is the plane determined by these lines, is therefore also horizontal. Thus we find three different types of points with horizontal tangent planes: maxima, minima, and saddle points; corresponding to these we have different types of stationary values of $f(x, y)$.

Another way of representing a function $f(x, y)$ is by drawing contour lines, such as those used in maps for representing altitudes (see p. 286). A contour line is a curve in the x, y-plane along which the function $f(x, y)$ has a constant value; thus the contour lines are identical with the curves of the family $f(x, y) = c$. Through an ordinary point of the plane there passes exactly one contour line; a maximum or minimum is surrounded by closed contour lines; while at a saddle point several contour lines cross. In Figure 193 contour lines are drawn for the landscape of Figure 192, and the maximum-minimum property of E is evident: Any path connecting A and B and not going through E has to go through a region where $f(x, y) < f(E)$, while the path AEB of Figure 192 has a minimum at E. In the same way we see that the value of $f(x, y)$ at E is the smallest maximum for paths connecting C and D.

3. Minimax Points and Topology

There is an intimate connection between the general theory of stationary points and the concepts of topology. Here we can give only a brief indication of these ideas in connection with a simple example.

Let us consider the mountain landscape on a ring-shaped island B with the two boundaries C and C'. If again we represent the altitude above sea-level by $u = f(x, y)$, with $f(x, y) = 0$ on C and C' and $f(x, y) > 0$ in the interior of B, then there must exist at least one mountain pass on the island, shown in Figure 194 by the point where the contour lines cross. Intuitively, this can be seen if one tries to go from

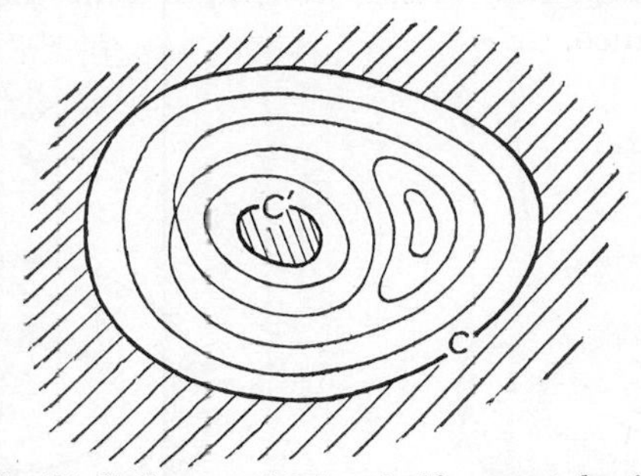

Fig. 194. Stationary points in a doubly connected region.

C to C' in such a way that one's path does not rise higher than necessary. Each path from C to C' must possess a highest point, and if we select that path whose highest point is as low as possible, then the highest point of this path is a saddle point of $u = f(x, y)$. (There is a trivial exception when a horizontal plane is tangent to the mountain crest all around the ring.) For a domain bounded by p curves there must exist, in general, at least $p - 1$ stationary points of minimax type. Similar relations have been discovered by Marston Morse to hold in higher dimensions, where there is a greater variety of topological possibilities and of types of stationary points. These relations form the basis of the modern theory of stationary points.

4. The Distance from a Point to a Surface

For the distance between a point P and a closed curve there are (at least) two stationary values, a minimum and a maximum. Nothing new occurs if we try to extend this result to three dimensions, so long as we consider a surface C topologically equivalent to a sphere, e.g. an ellipsoid. But new phenomena appear if the surface is of higher genus, e.g. a torus. There is still a shortest and a longest distance from P to a torus C, both segments being perpendicular to C. In addition we find extrema of different types representing maxima of minima or minima of maxima. To find them, we draw on the torus a closed "meridian" circle L, as in Figure 195, and we seek on L the point Q nearest to P. Then we try to move L so that the distance PQ becomes: a) a minimum. This Q is simply the point on C nearest to P. b) a maximum. This yields another stationary point. We could just as well seek on L the point farthest from P, and then find L such that this maximum distance is: c) a maximum, which will be attained at the point on C farthest from P. d) a minimum. Thus we obtain four different stationary values of the distance.

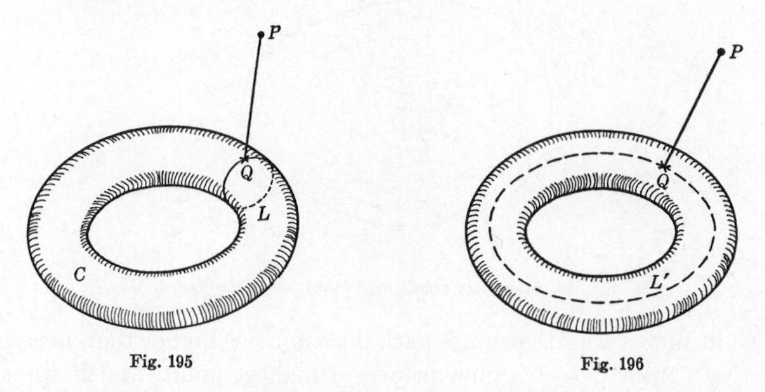

Fig. 195 Fig. 196

* *Exercise:* Repeat the reasoning with the other type L' of closed curve on C that cannot be contracted to a point, as in Figure 196.

§4. SCHWARZ'S TRIANGLE PROBLEM

1. Schwarz's Proof

Hermann Amandus Schwarz (1843–1921) was a distinguished mathematician of the University of Berlin and one of the great contributors to modern function theory and analysis. He did not disdain to write on elementary subjects, and one of his papers treats the following problem: Given an acute-angled triangle, to inscribe in it another triangle with the least possible perimeter. (By an inscribed triangle we mean one with a vertex on each side of the original triangle.) We shall see that there is exactly one such triangle, and that its vertices are the foot-points of the altitudes of the given triangle. We shall call this triangle the *altitude triangle.*

Schwarz proved the minimum property of the altitude triangle by the method of reflection, with the help of the following theorem of elementary geometry (see Fig. 197): At each vertex, P, Q, R, the two sides of the altitude triangle make equal angles with the side of the original triangle; this angle is equal to the angle at the opposite vertex of the original triangle. For example, the angles ARQ and BRP are both equal to angle C, etc.

To prove this preliminary theorem, we note that $OPBR$ is a quadrilateral that can be inscribed in a circle, since $\measuredangle OPB$ and $\measuredangle ORB$ are right angles. Consequently, $\measuredangle PBO = \measuredangle PRO$, since they subtend the same arc PO in the circumscribed circle. Now $\measuredangle PBO$ is complementary to $\measuredangle C$, since CBQ is a right triangle, and $\measuredangle PRO$ is complementary to $\measuredangle PRB$. Therefore the latter is equal to $\measuredangle C$. In the same way, using the quadrilateral $QORA$, we see that $\measuredangle QRA = \measuredangle C$, etc.

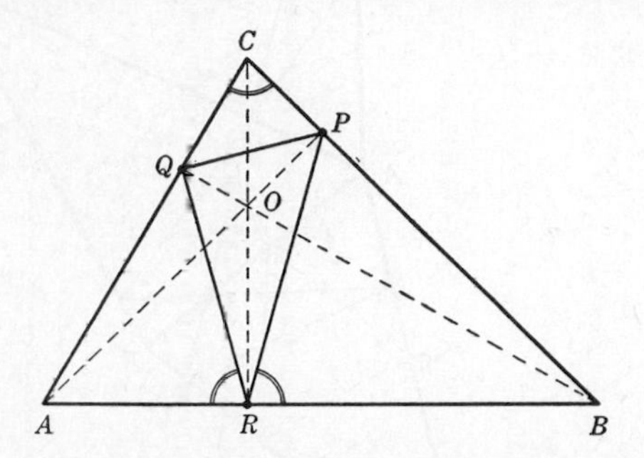

Fig. 197. Altitude triangle of ABC, showing equal angles.

This result enables us to state the following reflection property of the altitude triangle: Since, for example $\measuredangle AQR = \measuredangle CQP$, the reflection of RQ in the side AC is the continuation of PQ, and vice versa; similarly for the other sides.

We shall now prove the minimum property of the altitude triangle. In the triangle ABC consider, together with the altitude triangle, any other inscribed triangle, UVW. Reflect the whole figure first in the side AC of ABC, then reflect the resulting triangle in its side AB then in BC then again in AC, and finally in AB. In this way we obtain altogether six congruent triangles, each with the altitude triangle and the other one inscribed. The side BC of the last triangle is parallel to the original side BC. For in the first reflection, BC is rotated clockwise through an angle $2C$, then through $2B$ clockwise; in the third reflection it is not affected, in the fourth it rotates through $2C$ counterclockwise,

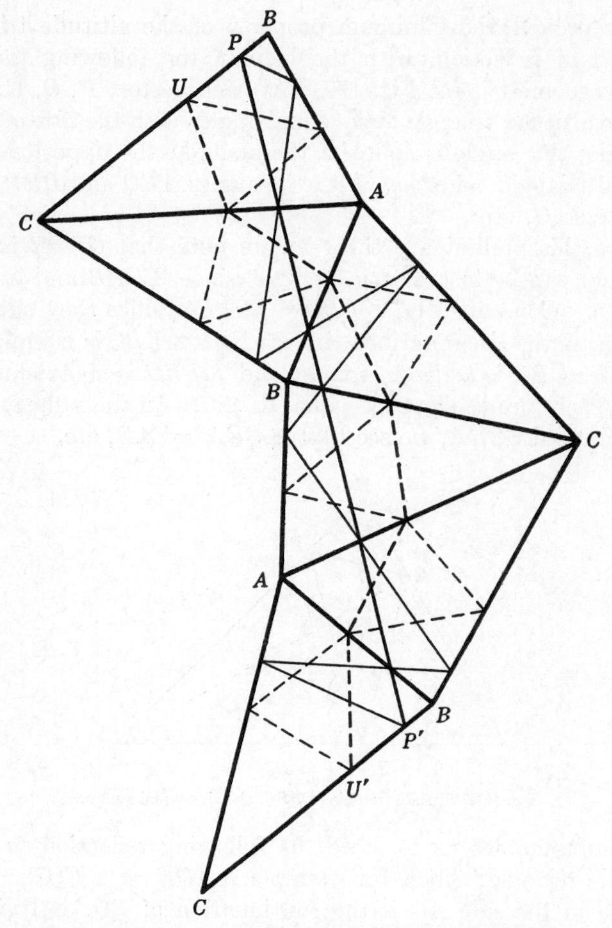

Fig. 198. Schwarz's proof that altitude triangle has least perimeter.

and in the fifth through $2B$ counterclockwise. Thus the total angle through which it has turned is zero.

Due to the reflection property of the altitude triangle, the straight line segment PP' is equal to twice the perimeter of the altitude triangle; for PP' is composed of six pieces that are, in turn, equal to the first, second, and third side of the triangle, each side occurring twice. Similarly, the broken line from U to U' is twice the perimeter of the other inscribed triangle. This line is not shorter than the straight line segment UU'. Since UU' is parallel to PP', the broken line from U to U' is not shorter than PP', and therefore the perimeter of the altitude triangle is the shortest possible for any inscribed triangle, as was to be proved. Thus we have at the same time shown that there is a minimum and that it is given by the altitude triangle. That there is no other triangle with perimeter equal to that of the altitude triangle will be seen presently.

2. Another Proof

Perhaps the simplest solution of Schwarz's problem is the following, based on the theorem proved earlier in this chapter that the sum of the distances from two points P and Q to a line L is least at that point R of L where PR and QR make the same angle with L, provided that P and Q lie on the same side of L and neither lies on L. Assume that the triangle PQR inscribed in the triangle ABC solves the minimum problem. Then R must be the point on the side AB where $p + q$ is a minimum, and therefore the angles ARQ and BRP must be equal; similarly, $\measuredangle AQR = \measuredangle CQP$, $\measuredangle BPR = \measuredangle CPQ$. Thus the minimum triangle, if it exists, must have the equal-angle property used in Schwarz's proof. It remains to be shown that the only triangle with this property is the altitude triangle. Moreover, since in the theorem on which this proof is based it is assumed that P and Q do not lie on AB, the proof does not hold in case one of the points P, Q, R is a vertex of the original triangle (in which case the minimum triangle would degenerate into twice the corresponding altitude); in order to complete the proof we must show that the perimeter of the altitude triangle is shorter than twice any altitude.

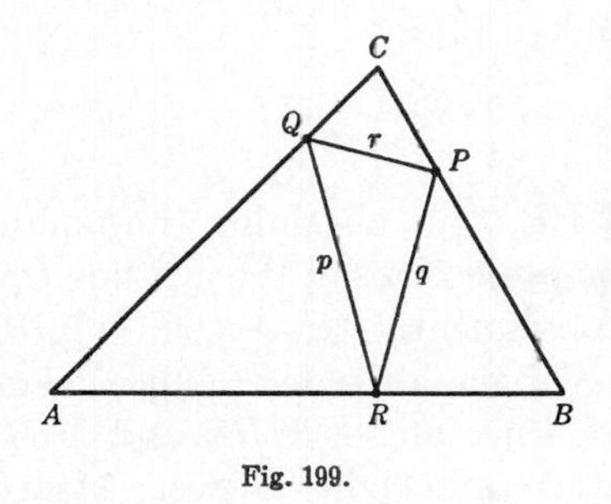

Fig. 199.

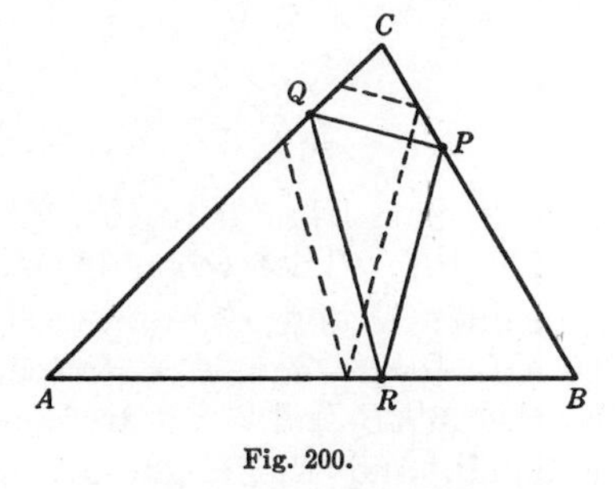

Fig. 200.

To dispose of the first point, we observe that if an inscribed triangle has the equal-angle property mentioned above, the angles at P, Q, and R must be equal to $\measuredangle A$, $\measuredangle B$, and $\measuredangle C$ respectively. For assume, say, that $\measuredangle ARQ = \measuredangle C + \delta$. Then, since the sum of the angles of a triangle is 180°, the angle at Q must be $B - \delta$, and at P, $A - \delta$, in order that the triangles ARQ and BRP may have the sum of their angles equal to 180°. But then the sum of the angles of the triangle CPQ is $A - \delta + B - \delta + C = 180° - 2\delta$; on the other hand, this sum must be 180°. Therefore δ is equal to zero. We have already seen that the altitude triangle has this equal-angle property. Any other triangle with this property would have its sides parallel to the corresponding sides of the altitude triangle; in other words, it would have to be similar to it and oriented in the same way. The reader may show that no other such triangle can be inscribed in the given triangle (see Fig. 200).

Finally, we shall show that the perimeter of the altitude triangle is less than twice any altitude, provided the angles of the original triangle are all acute. We produce the sides QP and QR and draw the

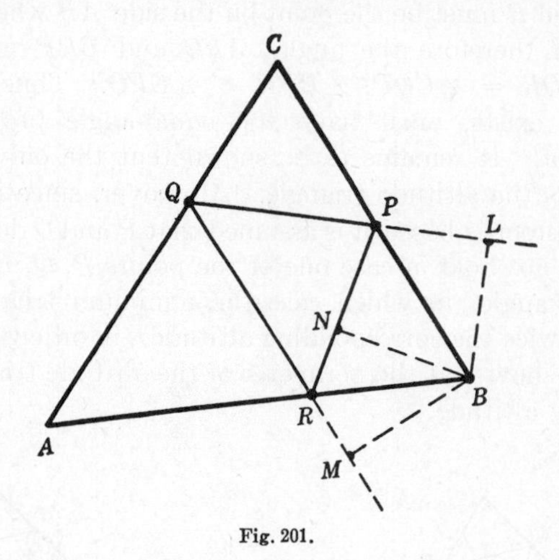

Fig. 201.

perpendiculars from B to QP, QR, and PR, thus obtaining the points L, M, and N. Then QL and QM are the projections of the altitude QB on the lines QP and QR respectively. Consequently, $QL + QM < 2QB$. Now $QL + QM$ equals p, the perimeter of the altitude triangle. For triangles MRB and NRB are congruent, since angles MRB and NRB are equal, and the angles at M and N are right angles. Hence $RM = RN$; therefore $QM = QR + RN$. In the same way, we see that $PN = PL$, so that $QL = QP + PN$. We therefore have $QL + QM = QP + QR + PN + NR = QP + QR + PR = p$. But we have shown that $2QB > QL + QM$. Therefore p is less than twice the altitude QB; by exactly the same argument, p is less than twice any altitude, as was to be proved. The minimum property of the altitude triangle is thus completely proved.

Incidentally, the preceding construction permits the direct calculation of p. We know that the angles PQC and RQA are equal to B, and therefore $PQB = RQB = 90° - B$, so that $\cos (PQB) = \sin B$. Therefore, by elementary trigonometry, $QM = QL = QB \sin B$, and $p = 2QB \sin B$. In the same way, it can be shown that $p = 2PA \sin A = 2RC \sin C$. From trigonometry, we know that $RC = a \sin B = b \sin A$, etc., which gives $p = 2a \sin B \sin C = 2b \sin C \sin A = 2c \sin A \sin B$. Finally, since $a = 2r \sin A$, $b = 2r \sin B$, $c = 2r \sin C$, where r is the radius of the circumscribed circle, we obtain the symmetrical expression, $p = 4r \sin A \sin B \sin C$.

3. Obtuse Triangles

In both of the foregoing proofs it has been assumed that the angles A, B, and C are all acute. If, say, C is obtuse, as in Figure 202, the

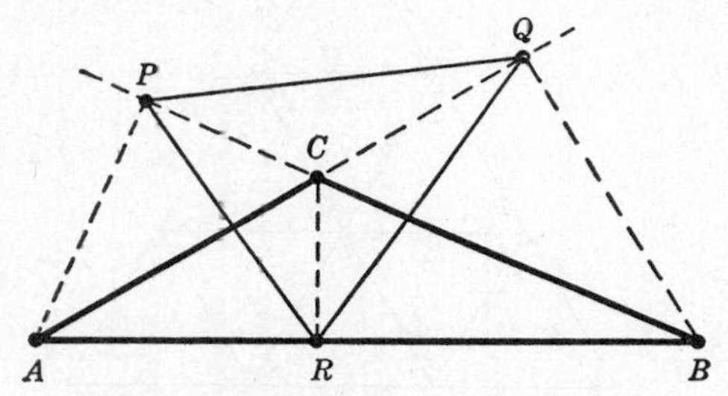

Fig. 202. Altitude triangle for obtuse triangle.

points P and Q will lie outside the triangle. Therefore the altitude triangle can no longer, strictly speaking, be said to be *in*scribed in the triangle, unless by an inscribed triangle we merely mean one whose vertices are on the sides or on the extensions of the sides of the original triangle. At any rate, the altitude triangle does not now give the minimum perimeter, for $PR > CR$ and $QR > CR$; hence $p = PR + QR + PQ > 2CR$. Since the reasoning in the first part of the last proof showed that the minimum perimeter, if not given by the altitude triangle, must be twice an altitude, we conclude that for obtuse triangles the "inscribed triangle" of smallest perimeter is the shortest altitude counted twice, although this is not properly a triangle. Still, one can find a proper triangle whose perimeter differs from twice the altitude by as little as we please. For the boundary case, the right triangle, the two solutions—twice the shortest altitude, and the altitude triangle—coincide.

The interesting question whether the altitude triangle has any sort of extremum property for obtuse triangles cannot be discussed here. Only this much may be stated: the altitude triangle gives, not a minimum for the sum of the sides, $p + q + r$, but a stationary value of minimax type for the expression $p + q - r$, where r denotes the side of the inscribed triangle opposite the obtuse angle.

4. Triangles Formed by Light Rays

If the triangle ABC represents a chamber with reflecting walls, then the altitude triangle is the only triangular light path possible in the chamber. Other closed light paths in form of polygons are not excluded, as Figure 203 shows, but the altitude triangle is the only such polygon with three sides.

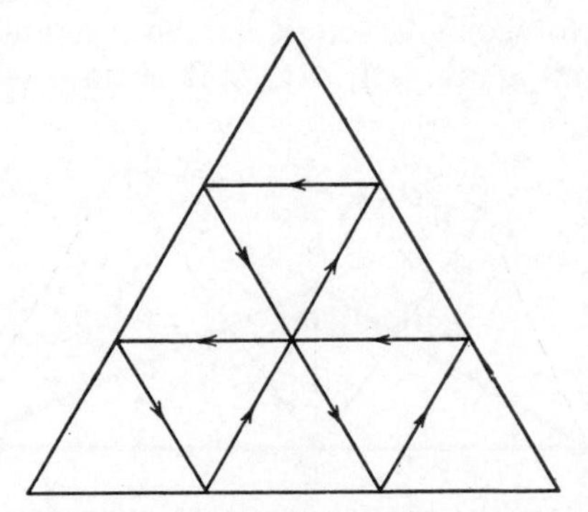

Fig. 203. Closed light path in a triangular mirror.

We may generalize this problem by asking for the possible "light triangles" in an arbitrary domain bounded by one or even several smooth curves; i.e. we ask for triangles with their vertices somewhere on the boundary curves and such that each two adjacent edges form the same angle with the curve. As we have seen in §1, the equality of angles is a condition for maximum as well as minimum total length of the two edges, so that we may, according to circumstances, find different types of light triangles. For example, if we consider the interior of a single smooth curve C, then the inscribed triangle of maximum length must be a light triangle. Or we may consider (as suggested to the authors by Marston Morse) the exterior of three smooth closed curves. A light triangle ABC may be characterized by the fact that its length has a stationary value; this value may be a minimum with respect to all three points A, B, C, it may be a minimum with respect to any of the combinations such as A and B and a maximum with respect to the third point C, it may be a minimum with respect to one point and a maximum with respect to the two others, or finally it may be a maximum with respect to all three points. Altogether the existence of at least $2^3 = 8$ light triangles is assured, since for each of the three points independently either a maximum or a minimum is possible.

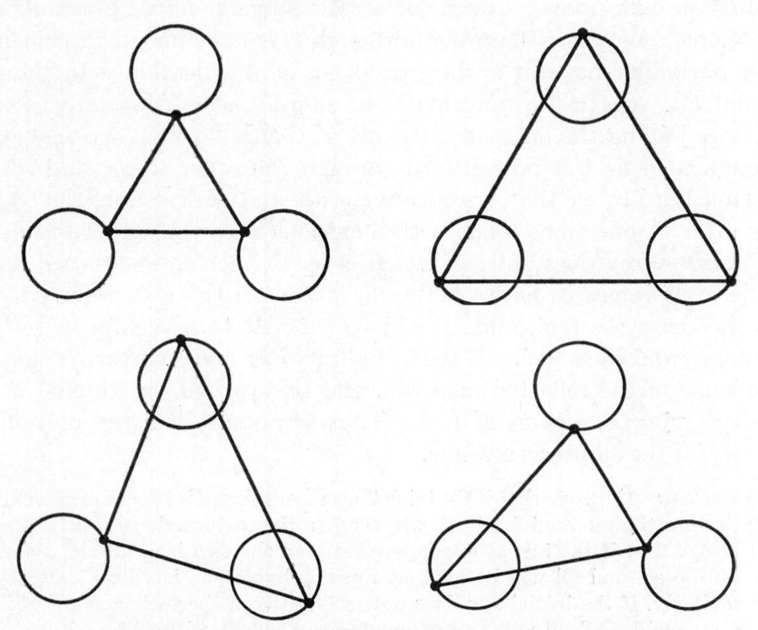

Figs. 204–7. The four types of light triangles between three circles.

*5. Remarks Concerning Problems of Reflection and Ergodic Motion

It is a problem of major interest in dynamics and optics to describe the path or "trajectory" of a particle in space or of a light ray for an unlimited length of time. If by some physical device the particle or ray is restricted to a bounded portion of space, it is of particular interest to know whether the trajectory will, in the limit, fill the space everywhere with an approximately equal distribution. Such a trajectory is called *ergodic*. The assumption of its existence is basic for statistical methods in modern dynamics and atomic theory. But very few relevant instances are known where a rigorous mathematical proof of the "ergodic hypothesis" can be given.

The simplest examples refer to the case of motion within a plane curve C, where the wall C is supposed to act as a perfect mirror, reflecting the otherwise free particle at the same angle at which it hits the boundary. For example, a rectangular box (an idealized billiard table with perfect reflection and a mass point as billiard ball) leads in general to an ergodic path; the ideal billiard ball going on for ever will reach the vicinity of every point, except for certain singular initial positions and directions. We omit the proof, although it is not difficult in principle.

Of particular interest is the case of an elliptical table with the foci F_1 and F_2. Since the tangent to an ellipse makes equal angles with the lines joining the point of tangency to the two foci, every trajectory through a focus will be reflected through the other focus, and so on. It is not hard to see that, irrespective of the initial direction, the trajectory after n reflections tends with increasing n to the major axis F_1F_2. If the initial ray does not go through a focus, then there are two possibilities. If it passes between the foci, then all the reflected rays will pass between the foci, and all will be tangent to a certain hyperbola having F_1 and F_2 as foci. If the initial ray does not separate F_1 and F_2, then none of the reflected rays will, and they will all be tangent to an ellipse having F_1 and F_2 as foci. Thus in no case will the motion be ergodic for the ellipse as a whole.

**Exercises:* 1) Prove that if the initial ray passes through a focus of the ellipse, the nth reflection of the initial ray will tend to the major axis as n increases.

2) Prove that if the initial ray passes between the two foci, all the reflected rays will do so, and all will be tangent to some hyperbola having F_1 and F_2 as foci; similarly, if the initial ray does not pass between the foci, none of the reflected rays will, and all will be tangent to some ellipse with F_1 and F_2 as foci. (Hint: Show that the ray before and after reflection at R makes equal angles with the lines RF_1 and RF_2 respectively, and then prove that tangents to confocal conics can be characterized in this way.)

§5. STEINER'S PROBLEM

1. Problem and Solution

A very simple but instructive problem was treated by Jacob Steiner, the famous representative of geometry at the University of Berlin in the early nineteenth century. Three villages A, B, C are to be joined by a system of roads of minimum total length. Mathematically, three points A, B, C are given in a plane, and a fourth point P in the plane is sought so that the sum $a + b + c$ shall be a minimum, where a, b, c denote the three distances from P to A, B, C respectively. The answer to the problem is: If in the triangle ABC all angles are less than 120°, then P is the point from which each of the three sides, AB, BC, CA,

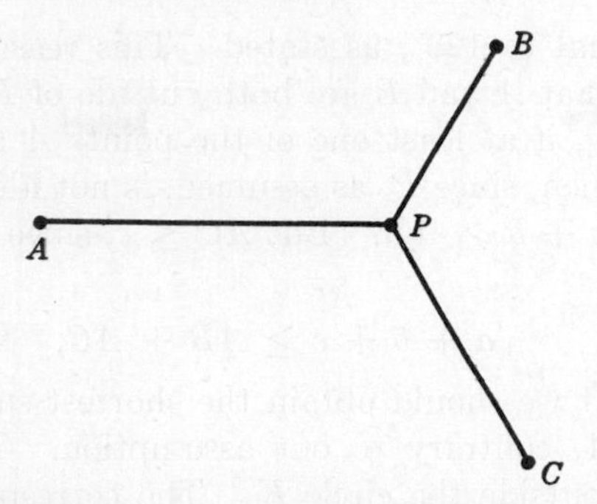

Fig. 208. Least sum of distances to three points.

subtends an angle of 120°. If, however, an angle of ABC, e.g. the angle at C, is equal to or larger than 120°, then the point P coincides with the vertex C.

It is an easy matter to obtain this solution if we use our previous results concerning extrema. Suppose P is the required minimum point. There are these alternatives: either P coincides with one of the vertices A, B, C, or P differs from these vertices. In the first case it is clear that P must be the vertex of the largest angle C of ABC, because the sum $CA + CB$ is less than any other sum of two sides of the triangle ABC. Thus, to complete the proof of our statement, we must analyze the second case. Let K be the circle with radius c around C. Then P must be the point on K such that $PA + PB$ is a minimum. If A and B

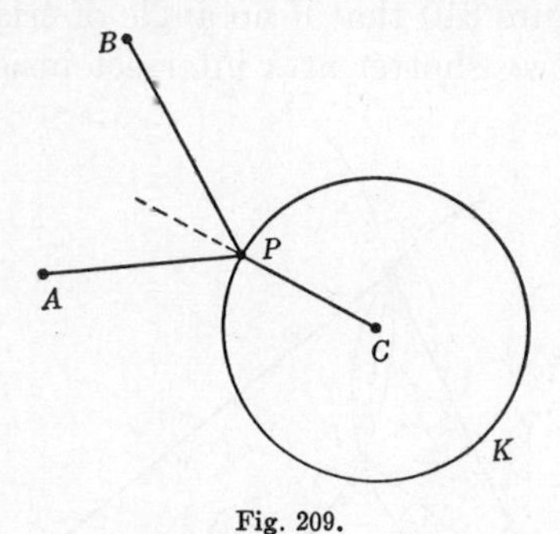

Fig. 209.

are outside of K, as in Figure 209, then, according to the result in §1, PA and PB must form equal angles with the circle K and hence with the radius PC, which is perpendicular to K. Since the same reasoning applies also to the position of P and the circle with the radius a around A, it follows that all three angles formed by PA, PB, PC are equal, and consequently equal to 120°, as stated. This reasoning was based on the assumption that A and B are both outside of K, which remains to be proved. Now, if at least one of the points A and B, say A, were on or inside K, then, since P, as assumed, is not identical with A or B, we should have $a + b \geq AB$. But $AC \leq c$, since A is not outside K. Hence

$$a + b + c \geq AB + AC,$$

which means that we should obtain the shortest sum of distances if P coincided with A, contrary to our assumption. This proves that A and B are both outside the circle K. The corresponding fact is similarly proved for the other combinations: B, C with respect to a circle of radius a about A, and A, C with respect to a circle of radius b about B.

2. Analysis of the Alternatives

To test which of the two alternatives for the point P actually occurs we must examine the construction of P. To find P, we merely draw the circles K_1, K_2 on which two of the sides, say AC and BC, subtend arcs of 120°. Then AC will subtend 120° from any point on the shorter arc into which AC divides K_1, but will subtend 60° from any point on the longer arc. The intersection of the two shorter arcs, provided such an intersection exists, gives the required point P, for not only will AC and BC subtend 120° at P, but AB will also, the sum of the three angles being 360°.

It is clear from Figure 210 that if no angle of triangle ABC is greater than 120°, then the two shorter arcs intersect inside the triangle. On

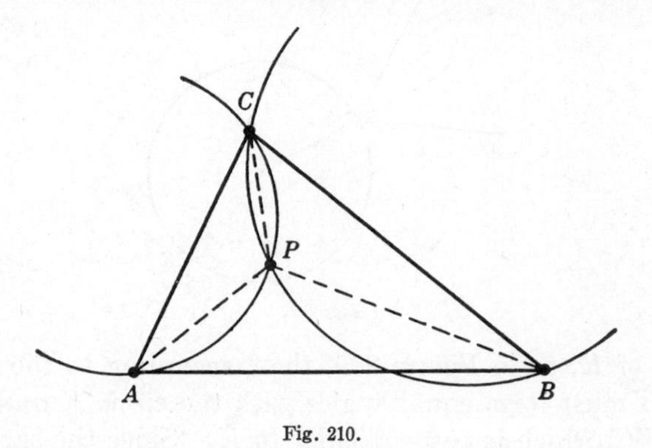

Fig. 210.

the other hand, if one angle, C, of triangle ABC is greater than 120°, then the two shorter arcs of K_1 and K_2 fail to intersect, as is shown in

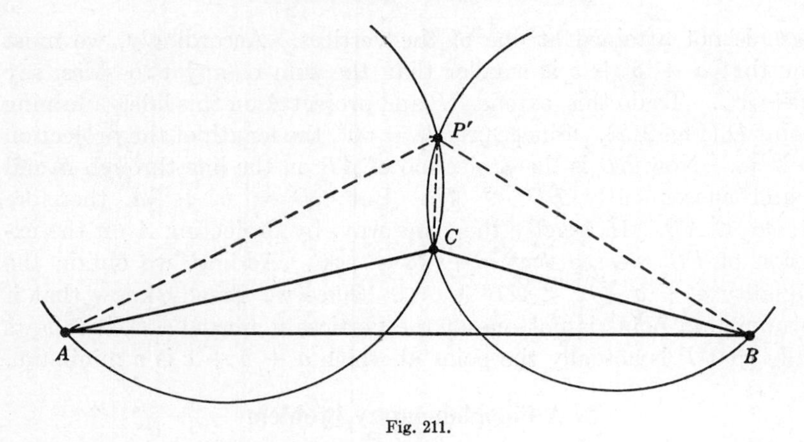

Fig. 211.

Figure 211. In this case there is no point P from which all three sides subtend 120°. However, K_1 and K_2 determine at their intersection a point P' from which AC and BC subtend angles of 60° each, while the side AB opposite the obtuse angle subtends 120°.

For a triangle ABC having an angle greater than 120° there is, then, no point at which each side subtends 120°. Hence the minimum point P must coincide with a vertex, since that was shown to be the only other alternative, and this must be the vertex at the obtuse angle. If, on the other hand, all the angles of a triangle are less than 120°, we have seen that a point P can be constructed from which each side subtends 120°. But to complete the proof of our theorem we have yet to show that $a + b + c$ will actually be less here than if P coincided with any vertex, for we have only shown that P gives a minimum *if* the smallest total

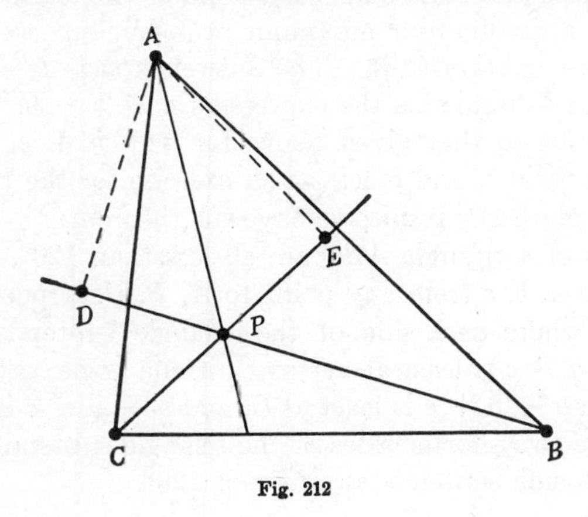

Fig. 212

length is not attained at one of the vertices. Accordingly, we must show that $a + b + c$ is smaller than the sum of any two sides, say $AB + AC$. To do this, extend BP and project A on this line, obtaining a point D (Fig. 212). Since $\measuredangle APD = 60°$, the length of the projection PD is $\frac{1}{2}a$. Now BD is the projection of AB on the line through B and P, and consequently $BD < AB$. But $BD = b + \frac{1}{2}a$, therefore $b + \frac{1}{2}a < AB$. In exactly the same way, by projecting A on the extension of PC, we see that $c + \frac{1}{2}a < AC$. Adding, we obtain the inequality $a + b + c < AB + AC$. Since we already know that if the minimum point is not one of the vertices it must be P, it follows finally that P is actually the point at which $a + b + c$ is a minimum.

3. A Complementary Problem

The formal methods of mathematics sometimes reach out beyond one's original intention. For example, if the angle at C is greater than 120° the procedure of geometrical construction produces, instead of the solution P (which in this case is the point C itself) another point P', from which the larger side AB of the triangle ABC appears under an

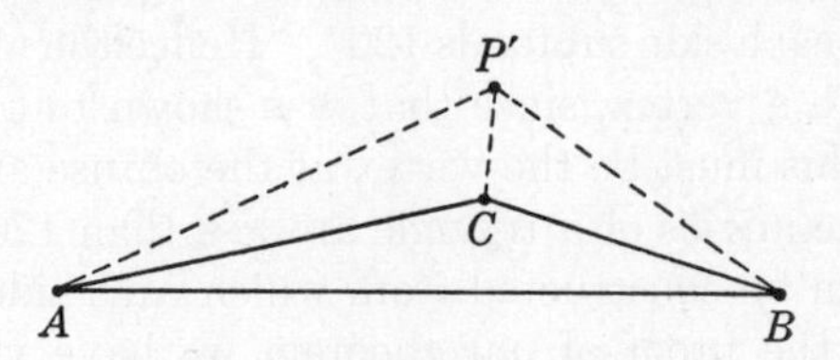

Fig. 213. $a + b - c =$ minimum.

angle of 120°, and the two smaller sides under the angle of 60°. Certainly P' does not solve our minimum problem, but we may suspect that it has some relation to it. The answer is that P' solves the following problem: to minimize the expression $a + b - c$. The proof is entirely analogous to that given above for $a + b + c$, based on the results of §1, Article 5, and is left as an exercise for the reader. Combined with the preceding result, we have the theorem:

If the angles of a triangle ABC are all less than 120°, then the sum of the distances a, b, c from any point to A, B, C, respectively, is least at that point where each side of the triangle subtends an angle of 120°, and $a + b - c$ is least at vertex C; if one angle, say C, is greater than 120°, then $a + b + c$ is least at C, and $a + b - c$ is least at that point where the two shorter sides of the triangle subtend angles of 60° and the longest side subtends an angle of 120°.

Thus, of the two minimum problems, one is always solved by the circle construction and the other by a vertex. For $\measuredangle C = 120°$ the two solutions of each problem, and indeed the solutions of the two problems, coincide, since the point obtained by the construction is then precisely the vertex C.

4. Remarks and Exercises

If from a point P inside an equilateral triangle UVW we drop three perpendicular lines PA, PB, PC as shown in Figure 214, then A, B, C, and P form the figure studied above. This remark can serve in solving Steiner's problem by starting with the points A, B, C and then finding U, V, W.

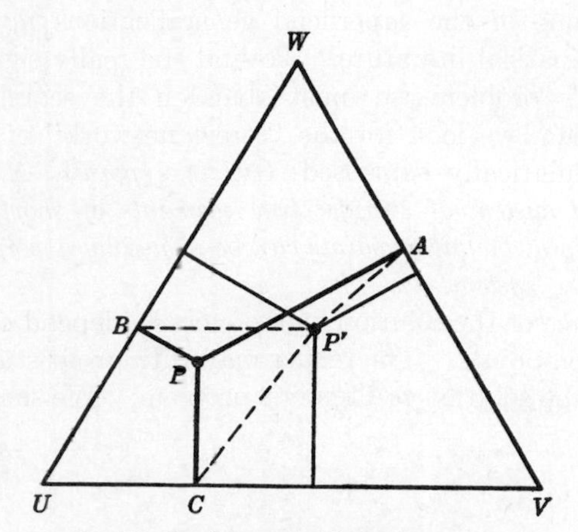

Fig. 214. Another proof of Steiner's solution.

Exercises: 1) Carry out this scheme, using the fact that from any point in an equilateral triangle the sum of the three perpendicular segments is constant and equal to the altitude.

2) Using the corresponding fact when P is outside UVW, discuss the complementary problem.

In three dimensions one might study a similar problem: Given four points A, B, C, D; to find a fifth point P such that $a + b + c + d$ is a minimum.

* *Exercise:* Investigate this problem and its complementary problem by the methods of §1, or by using a regular tetrahedron.

5. Generalization to the Street Network Problem

In Steiner's problem three fixed points A, B, C are given. It is natural to generalize this problem to the case of n given points, $A_1, A_2, \cdots, A_n$; we ask for the point P in the plane for which the sum of the distances $a_1 + a_2 + \cdots + a_n$ is a minimum, where a_i is the distance PA_i. (For four points, arranged as in Fig. 215, the point P

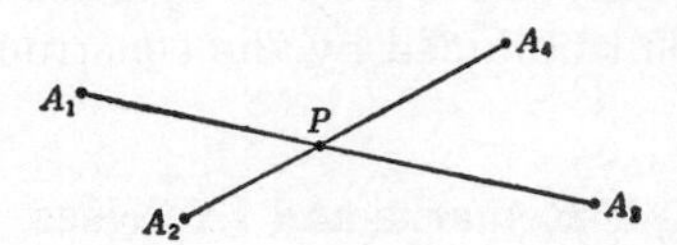

Fig. 215. Least sum of distances to four points.

is the point of intersection of the diagonals of the quadrilateral $A_1A_2A_3A_4$; the reader may prove this as an exercise.) This problem, which was also treated by Steiner, does not lead to interesting results. It is one of the superficial generalizations not infrequently found in mathematical literature. To find the really significant extension of Steiner's problem we must abandon the search for a s ngle point P. Instead, we look for the "street network" of shortest total length. Mathematically expressed: *Given* n *points,* $A_1, \cdots, A_n$ *to find a connected system of straight line segments of shortest total length such that any two of the given points can be joined by a polygon consisting of segments of the system.*

The appearance of the solution will, of course, depend on the arrangement of the given points. The reader may with profit study the subject on the basis of the solution to Steiner's problem. We shall content our-

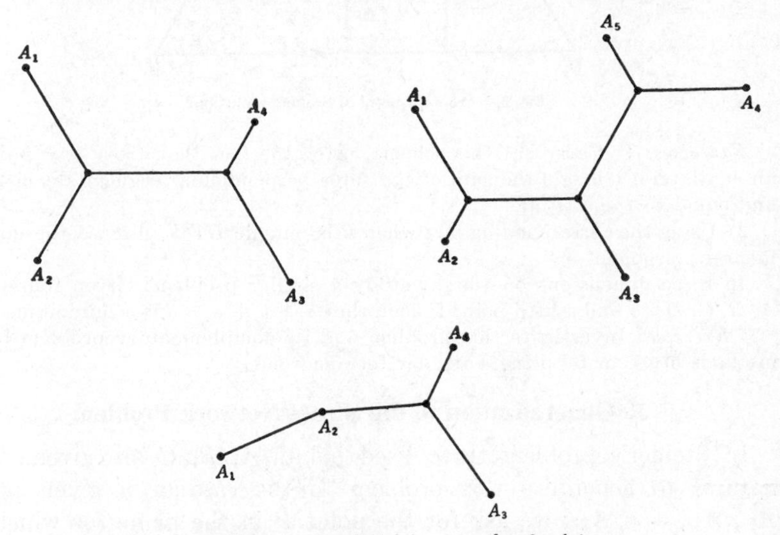

Figs. 216-8. Shortest networks joining more than 3 points.

selves here with pointing out the answer in the typical cases shown in Figures 216–8. In the first case the solution consists of five segments with two multiple intersections where three segments meet at angles of 120°. In the second case the solution contains three multiple intersections. If the points are differently arranged, figures such as these may not be possible. One or more of the multiple intersections may degenerate and be replaced by one or more of the given points, as in the third case.

In the case of n given points, there will be at most $n - 2$ multiple intersections, at each of which three segments meet at angles of 120°.

The solution of the problem is not always uniquely determined. For four points A, B, C, D forming a square we have the two equivalent solutions shown in Figures 219–20. If the points $A_1, A_2, \cdots, A_n$ are

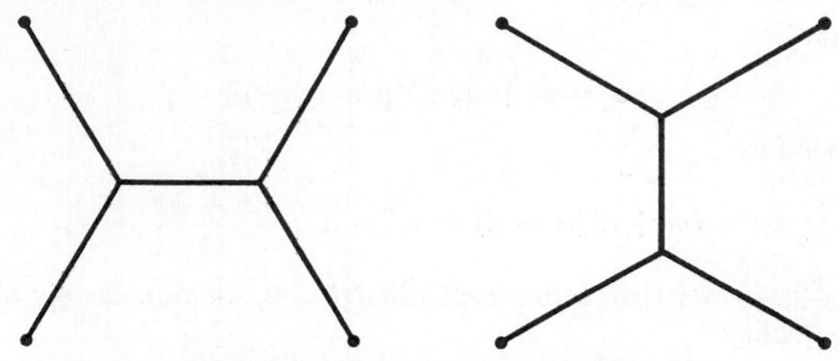

Figs. 219–20. Two shortest networks joining 4 points.

the vertices of a simple polygon with sufficiently flat angles, then the polygon itself will give the minimum.

§6. EXTREMA AND INEQUALITIES

One of the characteristic features of higher mathematics is the important rôle played by inequalities. The solution of a maximum problem always leads, in principle, to an inequality which expresses the fact that the variable quantity under consideration is less than or at most equal to the maximum value provided by the solution. In many cases such inequalities have an independent interest. As an example we shall consider the important inequality between the arithmetical and geometrical means.

1. The Arithmetical and Geometrical Mean of Two Positive Quantities

We begin with a simple maximum problem which occurs very often in pure mathematics and its applications. In geometrical language it amounts to the following: Among all rectangles with a prescribed perimeter, to find the one with largest area. The solution, as one might expect, is the square. To prove this we reason as follows. Let $2a$ bt the prescribed perimeter of the rectangle. Then the fixed sum of the lengths x and y of two adjacent edges is $x + y$, while the variable are xy is to be made as large as possible. The "arithmetical mean" of and y is simply

$$m = \frac{x+y}{2}.$$

We shall also introduce the quantity

$$d = \frac{x-y}{2},$$

so that

$$x = m + d, \qquad y = m - d,$$

and therefore

$$xy = (m+d)(m-d) = m^2 - d^2 = \frac{(x+y)^2}{4} - d^2.$$

Since d^2 is greater than zero except when $d = 0$, we immediately obtain the inequality

$$\sqrt{xy} \leq \frac{x+y}{2}, \tag{1}$$

where the equality sign holds only when $d = 0$ and $x = y = m$.

Since $x + y$ is fixed, it follows that $\sqrt{xy}$, and therefore the area xy, is a maximum when $x = y$. The expression

$$g = \sqrt{xy},$$

where the positive square root is meant, is called the "geometrical mean" of the positive quantities x and y; the inequality (1) expresses the fundamental relation between the arithmetical and geometrical means.

The inequality (1) also follows directly from the fact that the expression

$$(\sqrt{x} - \sqrt{y})^2 = x + y - 2\sqrt{xy}$$

is necessarily non-negative, being a square, and is zero only for $x = y$.

A geometrical derivation of the inequality may be given by considering the fixed straight line $x + y = 2m$ in the plane, together with the family of curves $xy = c$, where c is constant for each of these curves (hyperbolas) and varies from curve to curve. As is evident from Figure

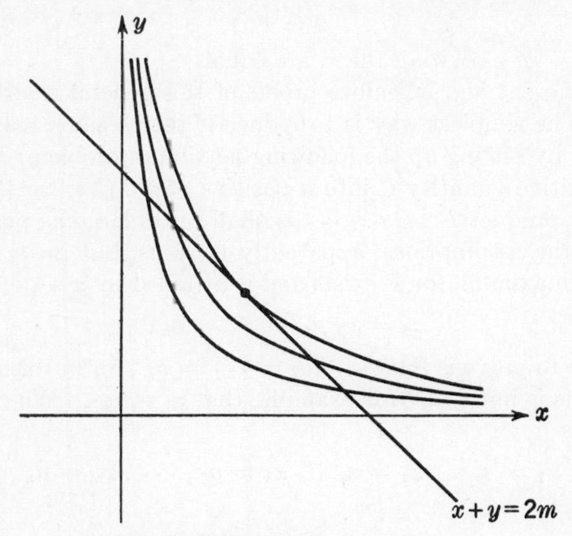

Fig. 221. Maximum xy for given $x + y$.

221, the curve with the greatest value of c having a point in common with the given straight line will be the hyperbola tangent to the line at the point $x = y = m$; for this hyperbola, therefore, $c = m^2$. Hence

$$xy \leq \left(\frac{x + y}{2}\right)^2.$$

It should be remarked that any inequality, $f(x, y) \leq g(x, y)$, can be read both ways and therefore gives rise to a maximum as well as to a minimum property. For example, (1) also expresses the fact that among all rectangles of given area the square has the least perimeter.

2. Generalization to n Variables

The inequality (1) between the arithmetical and geometrical means of two positive quantities can be extended to any number n of positive quantities, denoted by $x_1, x_2, \cdots, x_n$. We call

$$m = \frac{x_1 + x_2 + \cdots + x_n}{n}$$

their arithmetical mean, and

$$g = \sqrt[n]{x_1 x_2 \cdots x_n}$$

their geometrical mean, where the positive nth root is meant. The general theorem states that

$$g \leq m, \tag{2}$$

and that $g = m$ only if all the x_i are equal.

Many different and ingenious proofs of this general result have been devised. The simplest way is to reduce it to the same reasoning used in Article 1 by setting up the following maximum problem: To partition a given positive quantity C into n positive parts, $C = x_1 + \cdots + x_n$, so that the product $P = x_1x_2 \cdots x_2$ shall be as large as possible. We start with the assumption—apparently obvious, but analyzed later in §7—that a maximum for P exists and is attained by a set of values

$$x_1 = a_1, \cdots, x_n = a_n.$$

All we have to prove is that $a_1 = a_2 = \cdots = a_n$, for in this case $g = m$. Suppose this is not true—for example, that $a_1 \neq a_2$. We consider the n quantities

$$x_1 = s, \qquad x_2 = s, \qquad x_3 = a_3, \cdots, x_n = a_n,$$

where

$$s = \frac{a_1 + a_2}{2}.$$

In other words, we replace the quantities a_i by another set in which only the first two are changed and made equal, while the total sum C is retained. We can write

$$a_1 = s + d, \qquad a_2 = s - d,$$

where

$$d = \frac{a_1 - a_2}{2}.$$

The new product is

$$P' = s^2 \cdot a_3 \cdots a_n,$$

while the old product is

$$P = (s + d)\cdot(s - d)\cdot a_3 \cdots a_n = (s^2 - d^2)\cdot a_3 \cdots a_n,$$

so that obviously, unless $d = 0$,

$$P < P',$$

contrary to the assumption that P was the maximum. Hence $d = 0$ and $a_1 = a_2$. In the same way we can prove that $a_1 = a_i$, where a_i

is any one of the a's; it follows that all the a's are equal. Since $g = m$ when all the x_i are equal, and since we have shown that only this gives the maximum value of g, it follows that $g < m$ otherwise, as stated in the theorem.

3. The Method of Least Squares

The arithmetical mean of n numbers, $x_1, \cdots, x_n$, which need not be assumed all positive in this article, has an important minimum property. Let u be an unknown quantity that we want to determine as accurately as possible with some measuring instrument. To this end we make a number n of readings which may yield slightly different results, $x_1, \cdots, x_n$, due to various sources of experimental error. Then the question arises, what value of u is to be accepted as most trustworthy? It is customary to select for this "true" or "optimal" value the arithmetical mean $m = \frac{x_1 + \cdots + x_n}{n}$. To give a real justification for this assumption one must enter into a detailed discussion of the theory of probability. But we can at least point out a minimum property of m which makes it a reasonable choice. Let u be any possible value for the quantity measured. Then the differences $u - x_1, \cdots, u - x_n$ are the deviations of this value from the different readings. These deviations can be partly positive, partly negative, and the tendency will naturally be to assume as the optimal value for u one for which the total deviation is in some sense as small as possible. Following Gauss, it is customary to take, not the deviations themselves, but their squares, $(u - x_i)^2$, as appropriate measures of inaccuracy, and to choose as the optimal value among all the possible values for u one such that the sum of the squares of the deviations

$$(u - x_1)^2 + (u - x_2)^2 + \cdots + (u - x_n)^2$$

is as small as possible. *This optimal value for* u *is exactly the arithmetic mean* m, and it is this fact that constitutes the point of departure in Gauss's important "method of least squares." We can prove the italicized statement in an elegant way. By writing

$$(u - x_i) = (m - x_i) + (u - m),$$

we obtain

$$(u - x_i)^2 = (m - x_i)^2 + (u - m)^2 + 2(m - x_i)(u - m).$$

Now add all these equations for $i = 1, 2, \cdots, n$. The last terms yield $2(u - m)(nm - x_1 - \cdots - x_n)$, which is zero because of the definition of m; consequently we retain

$$(u - x_1)^2 + \cdots + (u - x_n)^2$$
$$= (m - x_1)^2 + \cdots + (m - x_n)^2 + n(m - u)^2.$$

This shows that

$$(u - x_1)^2 + \cdots + (u - x_n)^2 \geq (m - x_1)^2 + \cdots + (m - x_n)^2,$$

and that the equality sign holds only for $u = m$, which is exactly what we were to prove.

The general method of least squares takes this result as a guiding principle in more complicated cases when the problem is to decide on a plausible result from slightly incompatible measurements. For example, suppose we have measured the coördinates of n points x_i, y_i of a theoretically straight line, and suppose that these measured points do not lie exactly on a straight line. How shall we draw the line that best fits the n observed points? Our previous result suggests the following procedure, which, it is true, might be replaced by equally reasonable variants. Let $y = ax + b$ represent the equation of the line, so that the problem is to find the coefficients a and b. The distance in the y direction from the line to the point x_i, y_i is given by $y_i - (ax_i + b) = y_i - ax_i - b$, with a positive or negative sign according as the point is above or below the line. Hence the square of this distance is $(y_i - ax_i - b)^2$, and the method is simply to determine a and b in such a way that the expression

$$(y_1 - ax_1 - b)^2 + \cdots + (y_n - ax_n - b)^2$$

attains its least possible value. Here we have a minimum problem involving two unknown quantities, a and b. The detailed discussion of the solution, though quite simple, is omitted here.

§7. THE EXISTENCE OF AN EXTREMUM. DIRICHLET'S PRINCIPLE

1. General Remarks

In some of the previous extremum problems the solution is directly demonstrated to give a better result than any of its competitors. A striking instance is Schwarz's solution of the triangle problem, where we could see at once that no inscribed triangle has a perimeter smaller than that of the altitude triangle. Other examples are the minimum or maximum problems whose solutions depend on an explicit inequality, such as that between the arithmetical and geometrical means. But in some of our problems we followed a different path. We began with the assumption that a solution had been found; then we analyzed this assumption and drew conclusions which eventually permitted a description and construction of the solution. This was the case, for example, with the solution of Steiner's problem and with the second treatment of Schwarz's problem. The two methods are logically different. The first one is, in a way, more perfect, since it gives a more or less constructive demonstration of the solution. The second method, as we saw in the case of the triangle problem, is likely to be simpler. But it is not so direct, and it is, above all, conditional in its structure, for it starts with the assumption that a solution to the problem *exists*. It gives the solution only provided that this is granted or proved. Without this assumption it merely shows that *if* a solution exists, then it must have a certain character.†

Because of the apparent obviousness of the premise that a solution exists, mathematicians until late in the nineteenth century paid no attention to the logical point involved, and assumed the existence of a solution to extremum problems as a matter of course. Some of the greatest mathematicians of the nineteenth century—Gauss, Dirichlet, and Riemann—used this assumption indiscriminately as the basis for deep and otherwise hardly accessible theorems in mathematical physics and the theory of functions. The climax came when, in 1849, Riemann published his doctoral thesis on the foundations of the theory of functions of a complex variable. This concisely written paper, one of the great pioneering achievements of modern mathematics, was so completely unorthodox in its approach to the subject that many people would have liked to ignore it. Weierstrass was then the foremost mathematician at the University of Berlin and the acknowledged leader in the building of a rigorous function theory. Impressed but somewhat doubtful, he soon discovered a logical gap in the paper which the author had not bothered to fill. Weierstrass' shattering criticism, though it did not disturb Riemann, resulted at first in an almost general neglect of his theory. Riemann's meteoric career came to a sudden end after a few years with his death from consumption. But his ideas always found

† The logical necessity of ascertaining the existence of an extremum is illustrated by the following fallacy: 1 is the largest integer. For let us denote the largest integer by x. If $x > 1$, then $x^2 > x$, hence x could not be the largest integer. Therefore x must be equal to 1.

some enthusiastic disciples, and fifty years after the publication of his thesis Hilbert finally succeeded in opening the way for a complete answer to the questions that he had left unsettled. This whole development in mathematics and mathematical physics became one of the great triumphs in the history of modern mathematical analysis.

In Riemann's paper the point open to critical attack is the question of the existence of a minimum. Riemann based much of his theory on what he called Dirichlet's principle (Dirichlet had been Riemann's teacher at Goettingen, and had lectured but never written about this principle.) Let us suppose, for example, that part of a plane or of any surface is covered with tinfoil and that a stationary electric current is set up in the layer of tinfoil by connecting it at two points with the poles of an electric battery. There is no doubt that the physical experiment leads to a definite result. But how about the corresponding mathematical problem, which is of the utmost importance in function theory and other fields? According to the theory of electricity, the physical phenomenon is described by a "boundary value problem of a partial differential equation". It is this mathematical problem that concerns us; its solvability is made plausible by its assumed equivalence to a physical phenomenon but is by no means mathematically proved by this argument. Riemann disposed of the mathematical question in two steps. First he showed that the problem is equivalent to a minimum problem: a certain quantity expressing the energy of the electric flow is minimized by the actual flow in comparison to the other flows possible under the prescribed conditions. Then he stated as "Dirichlet's principle" that such a minimum problem has a solution. Riemann took not the slightest step towards a mathematical proof of the second assertion, and this was the point attacked by Weierstrass Not only was the existence of the minimum not at all evident, but, as it turned out, it was an extremely delicate question for which the mathematics of that time was not yet prepared and which was finally settled only after many decades of intensive research.

2. Examples

We shall illustrate the sort of difficulty involved by two examples. 1) We mark two points A and B at a distance d on a straight line L, and ask for the polygon of shortest length that starts at A in a direction perpendicular to L and ends at B. Since the straight segment AB is the shortest connection between A and B for all paths, we can be certain that any path admissible in the competition has a length greater than d, for the only path giving the value d is the straight segment AB, which violates the restriction imposed on the direction at A, and hence is not admissible under the terms of the problem. On the other hand, con-

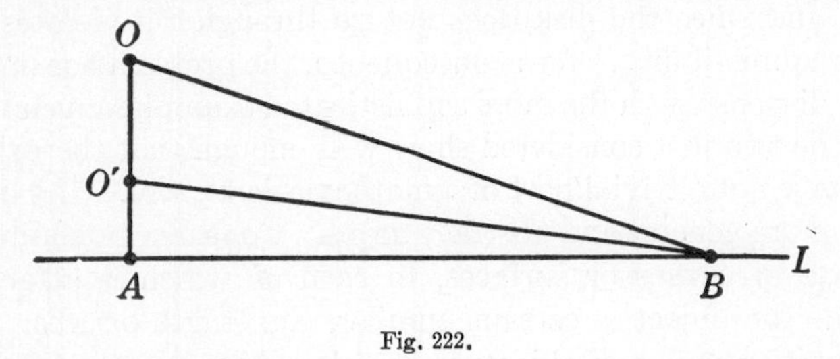

Fig. 222.

sider the admissible path AOB in Figure 222. If we replace O by a point O' near enough to A, we can obtain an admissible path with a length differing as little from d as we like; hence if a *shortest* admissible path exists, it cannot have a length exceeding d and must therefore have the exact length d. But the only path of that length is not admissible, as we saw. Hence there can exist no shortest admissible path, and the proposed minimum problem has no solution.

2) As in Figure 223, let C be a circle and S a point at a distance 1 above its center. Consider the class of all surfaces bounded by C that go through the point S and lie above C in such a way that no two different points have the same vertical projection on the plane of C. Which of these surfaces has the least area? This problem, natural as it appears, has no solution: there is no admissible surface with a minimum area. If the condition that the surface go through S had not been prescribed, the solution would obviously be the plane circular disk bounded by C. Let us denote its area by A. Any other surface bounded by C must have an area larger than A. But we can find an admissible surface whose area exceeds A by as little as we please. For this purpose we take a conical surface of height 1 and so slender that its area is less than whatever margin may have been assigned. We place this cone on top of the disk with its vertex at S, and consider the total surface formed by the surface of the cone and the part of the disk outside the base of the cone. It is immediately clear that this surface, which deviates from the plane only near the center, has an area exceeding A by less than the given margin.

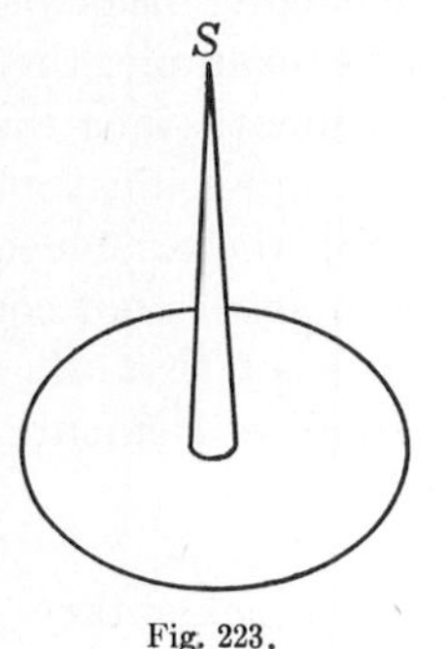

Fig. 223.

Since this margin can be chosen as small as we like, it follows again that the minimum, if it exists, cannot be other than the area A of the disk. But among all the surfaces bounded by C only the disk itself has this area, and since the disk does not go through S it violates the conditions for admissibility. As a consequence, the problem has no solution.

We can dispense with the more sophisticated examples given by Weierstrass. The two just considered show well enough that the existence of a minimum is not a trivial part of a mathematical proof. Let us put the matter in more general and abstract terms. Consider a definite class of objects. e.g. of curves or surfaces, to each of which is attached as a function of the object a certain number, e.g. length or area. If there is only a finite number of objects in the class, there must obviously be a largest and a smallest among the corresponding numbers. But if there are infinitely many objects in the class, there need be neither a largest nor a smallest number, even if all these numbers are contained between fixed bounds. In general, these numbers will form an infinite set of points on the number axis. Let us suppose, for simplicity, that all the numbers are positive. Then the set has a "greatest lower bound", that is, a point α below which no number of the set lies, and which is either itself an element of the set or is approached with any degree of accuracy by members of the set. If α belongs to the set, it is the smallest element; otherwise the set simply does not contain a smallest element. For example, the set of numbers 1, 1/2, 1/3, ... contains no smallest element, since the lower bound, 0, does not belong to the set. These examples illustrate in an abstract way the logical difficulties connected with the existence problem. The mathematical solution of a minimum problem is not complete until one has provided, explicitly or implicitlz, a proof that the set of values associated with the problem contains a smallest element.

3. Elementary Extremum Problems

In elementary problems it requires only an attentive analysis of the basic concepts involved to settle the question of the existence of a solution. In Chapter VI, §5 the genaral notion of a compact set was discussed; it was stated that a continuous function defined for the elements of a compact set always assumes a largest and a smallest value somewhere in the set. In each of the elementary problems previously discussed, the competing values can be regarded as the values of a function of one or severval ariables in a domain that is either compact or can easily be made so without essential change in the problem. In such a case the existence of a maximum and a minimum is assured. In Steiner's problem, for example, the quantity under consideration is the sum of three distances, and this depends continuously on the position of the movable point. Since the domain of this point is the whole plane, nothing is lost if we enclose the figure in a large circle and restrict the point to its interior and boundary. For as soon as the movable point is sufficiently far away from the three given points, the sum of its distances to these points will certainly exceed $AB + AC$, which is one of the admissible values of the function. Hence if there is a minimum for a point restricted to a large circle, this will also be the minimum for the unrestricted problem. But it is easy to show that the domain consisting of a circle plus its interior is compact, hence a minimum for Steiner's problem exists.

The importance of the assumption that the domain of the independent variable is compact can be shown by the following example. Given two closed curves C_1 and C_2, there always exist two points. P_1, P_2, on C_1, C_2 respectively, which have the least possible distance from each other, and points Q_1, Q_2 which have the largest possible distance. For the distance between a point A_1 on C_1 and a point A_2 on C_2 is a continuous function on the compact set consisting of the pairs A_1, A_2 of

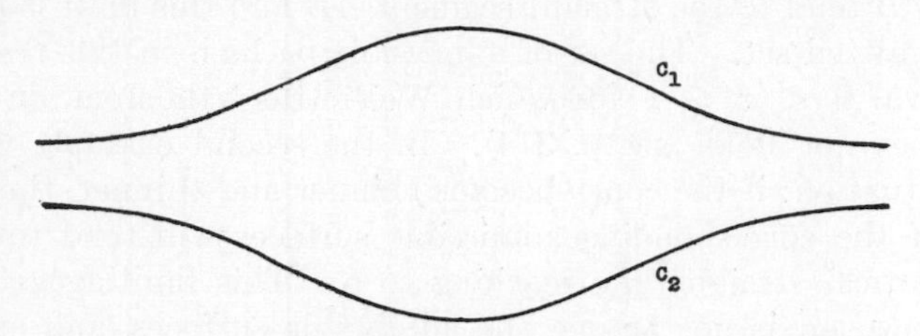

Fig. 224. Curves between which there is no longest or shortest distance.

points under consideration. However, if the two curves are not bounded but extend to infinity, then the problem may not have a solution. In the case shown in Figure 224 neither a smallest nor a largest distance between the curves is attained; the lower bound for the distance is zero, the upper bound is infinity, and neither is attained. In some cases a minimum but no maximum exists. For the case of two branches of a hyperbola (Fig. 17, p. 76) only a minimum distance is attained, by A and A', since obviously no two points exist with a maximum distance apart.

We can account for this difference in behavior by artificially restricting the domain of the variables. Select an arbitrary positive number R, and restrict x by the condition $|x| \leq R$. Then both a maximum and a minimum exist for each of the last two problems. In the first one, restricting the boundary in this way assures the existence of a maximum and a minimum distance, both of which are attained on the boundary. If R is increased, the points for which the extrema are attained are again on the boundary. Hence as R increases, these points disappear towards infinity. In the second case, the minimum distance is attained in the interior, and no matter how much R is increased the two points of minimum distance remain the same.

4. Difficulties in Higher Cases

While the existence question is not at all serious in the elementary problems involving one, two, or any finite number of independent variables, it is quite different with Dirichlet's principle or with even simpler problems of a similar type. The reason in these cases is either that the domain of the independent variable fails to be compact, or that the function fails to be continuous. In the first example of Article 2 we have a sequence of paths $AO'B$ where O' tends to the point A. Each path of the sequence satisfies the conditions of admissibility. But the paths $AO'B$ tend to the straight segment AB and this limit is no longer in the admitted set. The set of admissible paths is in this respect like the interval $0 < x \leq 1$ for which Weierstrass' theorem on extreme values does not hold (see p. 314). In the second example we find a similar situation: if the cones become thinner and thinner, then the sequence of the corresponding admissible surfaces will tend to the disk plus a vertical straight line reaching to S. This limiting geometrical entity, however, is not among the admissible surfaces, and again it is true that the set of admissible surfaces is not compact.

As an example of non-continuous dependence we may consider the length of a curve. This length is no longer a function of a finite number of numerical variables, since a whole curve cannot be characterized by a finite number of "coördinates," and it is not a continuous function of the curve. To see this let us join two points A and B at a distance d by a zigzag polygon P_n which together with the segment AB forms n equilateral triangles. It is clear from Figure 225 that the total length of P_n will be exactly $2d$ for every value of n. Now consider the sequence of polygons $P_1, P_2, \cdots$. The single waves of these polygons decrease in height as they increase in number, and it is clear that polygon P_n tends to the straight line AB, where, in the limit, the roughness has disappeared completely. The length of P_n is always $2d$, regardless of the index n, while the length of the limiting curve, the straight segment, is only d. Hence the length does not depend continuously on the curve.

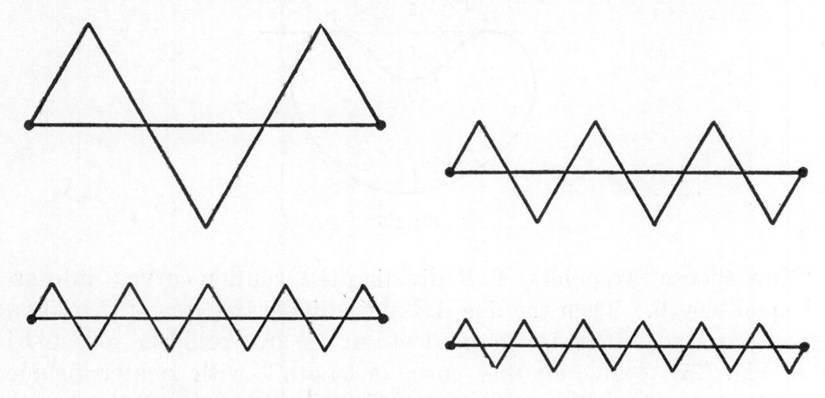

Fig. 225. Approximation to a segment by polygons of twice its length.

All these examples confirm the fact that caution as to the existence of a solution is really necessary in minimum problems of a more complex structure.

§8. THE ISOPERIMETRIC PROBLEM

That the circle encloses the largest area among all closed curves with a prescribed length is one of the "obvious" facts of mathematics for which only modern methods have yielded a rigorous proof. Steiner devised various ingenious ways of proving this theorem, of which we shall consider one.

Let us start with the assumption that a solution does exist. This granted, suppose the curve C is the required one with the prescribed length L and maximum area. Then we can easily show that C must be convex, in the sense that the straight segment joining any two points of C must lie entirely inside or on C. For if C were not convex, as in Figure 226, then a segment such as OP could be drawn between some pair of points O and P on C, such that OP lies outside of C. The arc $OQ'P$ which is the reflection of OQP in the line OP forms, together with the arc ORP, a curve of length L enclosing a larger area than the original curve C, since it includes the additional areas I and II. This contradicts the assumption that C contains the largest area for a closed curve of length L. Hence C must be convex.

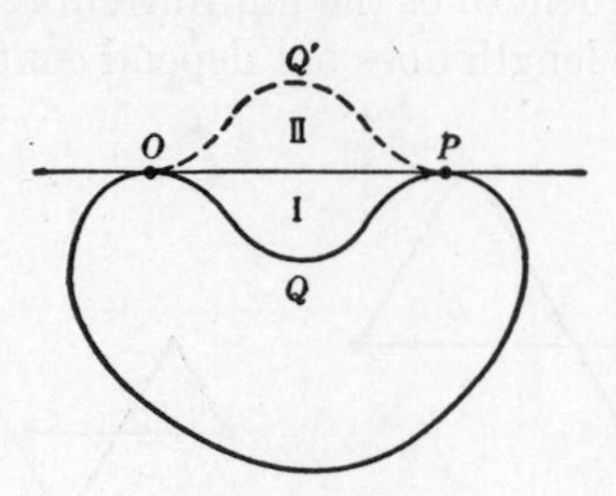

Fig. 226

Now choose two points, A, B, dividing the solution curve C into arcs of equal length. Then the line AB must divide the area of C into two equal parts, for otherwise the part of greater area could be reflected in AB (Fig. 227) to give another curve of length L with greater included area that C. It follows that half of the solution C must solve the following problem: To find the arc of length $L/2$ having its endpoints A, B on a straight line and enclosing a maximum area between it and

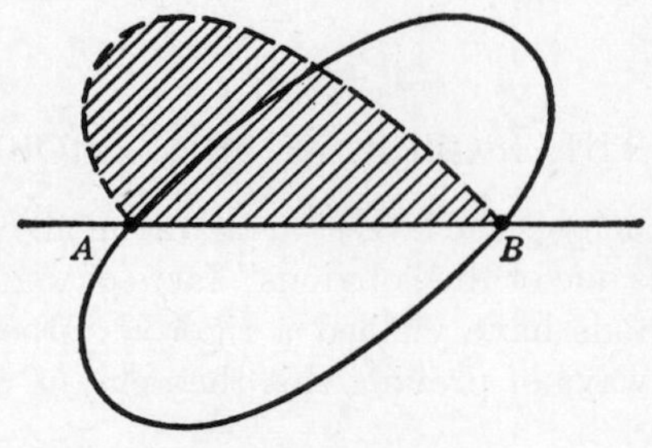

Fig. 227

this straight line. Now we shall show that the solution to this new problem is a semicircle, so that the whole curve C solving the isoperimetric problem is a circle. Let the arc AOB solve the new problem. It is sufficient to show that every inscribed angle such as $\measuredangle AOB$ in Figure 228 is a right angle, for this will prove that AOB is a semicircle. Suppose, on the contrary, that the angle AOB is not 90°. Then we can replace Figure 228 by another one, 229, in which the shaded areas and the length of the arc AOB are not changed, while the triangular area is increased by making $\sphericalangle AOB$ equal to or at least nearer to 90°. Thus Figure 229 gives a larger area than the original (see page 330). But we started with the assumption that Figure 228 solves the problem so that Figure 229 could not possibly yield a larger area. This contradiction shows that for every point O, $\sphericalangle AOB$ must be a right angle, and this completes the proof.

The isoperimetric property of the circle can be expressed in the form of an inequality. If L is the circumference of the circle, its area is $L^2/4\pi$, and therefore we must have the *isoperimetric inequality*, $A \leq L^2/4\pi$, between the area A and length L of any closed curve the equality sign holding only for the circle.

*As is apparent from the discussion in §7, Steiner's proof has only a conditional value: "If there is a curve of length L within maximal area

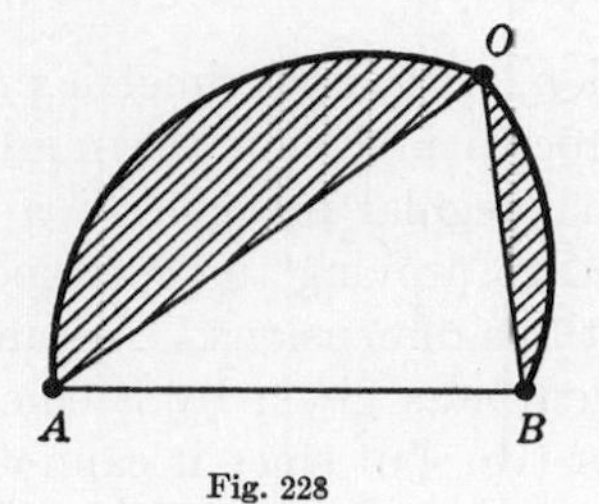

Fig. 228

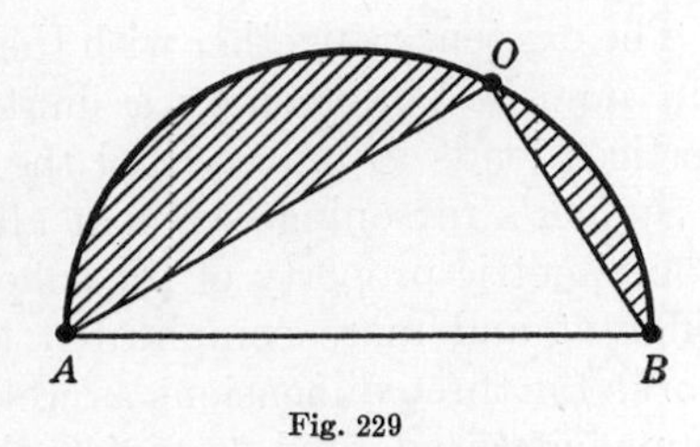

Fig. 229

then it must be a circle." To establish the hypothetical premise an essentially new argument is needed. First we prove an elementary theorem concerning closed polygons P_n with an even number $2n$ of edges: Among all such $2n$-gons with the same length, the regular $2n$-gon has the largest area. The proof follows the pattern of Steiner's reasoning with the following modifications. There is no difficulty about the question of existence here, since a $2n$-gon, together with its length and area, depends continuously on the $4n$ coördinates of its vertices, which may without loss of generality be restricted to a compact set of points in $4n$-dimentional space. Accordingly, in this problem for polygons we may safely begin with the assumption that some polygon P *is* the solution, and on this basis analyze the properties of P. Exactly as in Steiner's proof, it follows that P must be convex. We prove now that all the $2n$ edges of P must have the same length. For assume that two adjacent edges AB and BC had different lengths; then we could cut off triangle ABC from P and replace it by an isosceles triangle $AB'C$, in which $AB' + B'C = AB + BC$, and which has a larger area (see §1). Thus we would obtain a polygon P' with the same perimeter and a larger area, contrary to the assumption that P was the optimal polygon of $2n$ edges. Therefore all the edges of P must have equal length, and what remains to be shown is that P is regular; for this it suffices to know that all the vertices of P lie on a circle. The reasoning follows Steiner's pattern. First we show that any diagonal joining opposite vertices, e.g. the first with the $(n + 1)$-st, cuts the area in two equal parts. Then we prove that all the vertices of one of these parts lie on a semicircle. The details, which follow exactly the previous pattern, are left to the reader as an exercise.

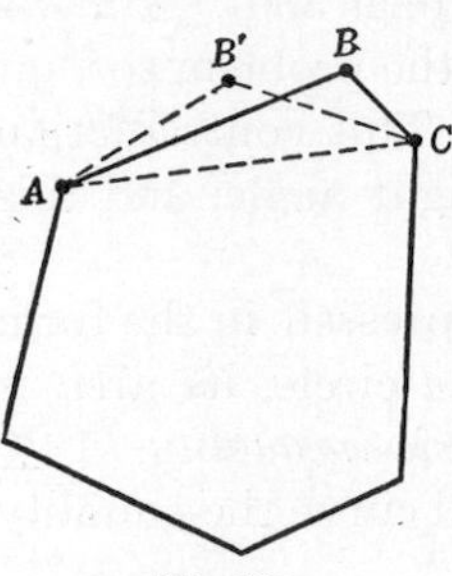

Fig. 230

The existence, together with the solution, of the isoperimetric problem can now be obtained by a limiting process in which the number of vertices tends to infinity and the optimal regular polygon to a circle.

Steiner's reasoning is not at all suited to proving the corresponding isoperimetric property of the sphere in three dimensions. A somewhat different and more complicated treatment was given by Steiner that works for three dimensions as well as for two, but since it cannot be so immediately adapted to giving the existence proof it is omitted here. As a matter of fact, proving the isoperimetric property of the sphere is a much harder task than for the circle; indeed, a complete and rigorous proof was first given much later, in a rather difficult paper by H. A. Schwarz. The three-dimensional isoperimetric property can be expressed by the inequality

$$36\pi V^2 \leq A^3$$

between the surface area A and the volume V of any closed three-dimensional body, the equality holding only for the sphere.

*§9. EXTREMUM PROBLEMS WITH BOUNDARY CONDITIONS. CONNECTION BETWEEN STEINER'S PROBLEM AND THE ISOPERIMETRIC PROBLEM

Interesting results arise in extremum problems when the domain of the variable is restricted by boundary conditions. The theorem of Weierstrass that in a compact domain a continuous function attains a largest and smallest value does not exclude the possibility that the extreme values are attained at the boundary of the domain. A simple, almost trivial, example is afforded by the function $u = x$. If x is not restricted and may range from $-\infty$ to $+\infty$, then the domain B of the independent variable is the entire number axis; and hence it is understandable that the function $u = x$ has no largest or smallest value anywhere. But if the domain B is limited by boundaries, say $0 \leq x \leq 1$, then there exists a largest value, 1, attained at the right endpoint, and a smallest value, 0, attained at the left endpoint. However, these extreme values are not represented by a summit or a depression in the curve of the function; they are not extrema relative to a full two-sided neighborhood. They change as soon as the interval is extended, because they remain at the endpoints. For a genuine peak or depression of a function, the extremal character always refers to a full neighborhood of the point where the value is attained; it is not affected by slight changes of the boundary. Such an extremum persists even under a free variation of the independent variable in the domain B, at least in a sufficiently small neighborhood. The distinction between such "free" extrema and those assumed at the boundary is illuminating in many apparently quite different contexts. For functions of one variable, of course, the distinction is simply that between monotone and non-monotone functions, and thus does not lead to particularly interesting observations. But there are many significant instances of extrema attained at the boundary of the domain of variability by functions of several variables.

This may occur, for example, in Schwarz's triangle problem. There the domain of variability of the three independent variables consists of all triples of points, one on each of the three sides of the triangle ABC. The solution of the problem involved two alternatives: either the minimum is attained when all three of the independently variable points P, Q, R lie inside the respective sides of the triangle, in which case the minimum is given by the altitude triangle, or the minimum is attained for the boundary position when two of the points P, Q, R coincide with the common endpoint of their respective intervals, in which case the minimum inscribed "triangle" is the altitude from this vertex, counted twice. Thus the character of the solution is quite different according to which of the alternatives occurs.

In Steiner's problem of the three villages the domain of variability of the point P is the whole plane, of which the three given points A, B, C may be considered as boundary points. Again there are two alternatives yielding two entirely different types of solutions: either the minimum is attained in the interior of the triangle ABC, which is the case of the three equal angles, or it is attained at a boundary point C. A similar pair of alternatives exists for the complementary problem.

As a last example we may consider the isoperimetric problem modified by restrictive boundary conditions. We shall thus obtain a surprising connection between the isoperimetric problem and Steiner's problem and at the same time what is perhaps the simplest instance of a new type of extremum problem. In the original problem the independent variable, the closed curve of given length, can be arbitrarily varied from the circular shape, and any such deformed curve is admissible into the competition, so that we have a genuine free minimum. Now let us consider the following modified problem: the curves C under consideration shall include in their interior, or pass through, three given points, P, Q, R, the area A is prescribed, and the length L is to be made a minimum. This represents a genuine boundary condition.

It is clear that, if A is prescribed sufficiently large, the three points P, Q, R will not affect the problem at all. Whenever the circle circumscribed about the triangle PQR has an area less than or equal to A, the solution will simply be a circle of area A including the three points. But what if A is smaller? We state the answer here but omit the somewhat detailed proof, although it would not be beyond our reach. Let us characterize the solutions for a sequence of values of A which decreases to zero. As soon as A falls below the area of the circumscribed circle, the original isoperimetric circle breaks up into three arcs, all having the same radius, which form a convex circular triangle with

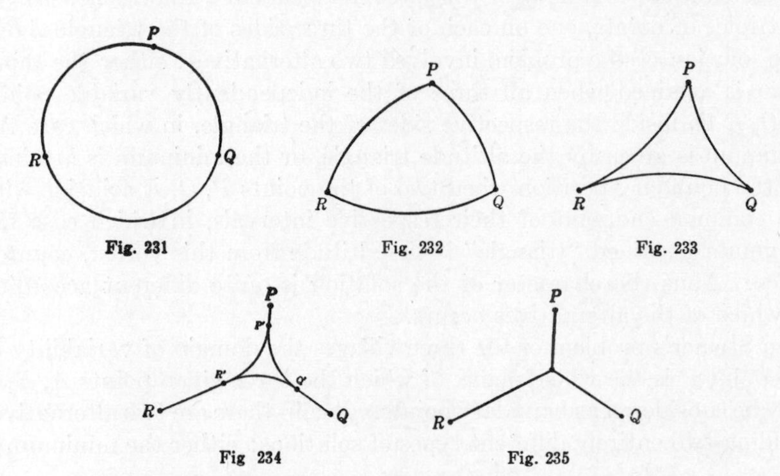

Fig. 231 Fig. 232 Fig. 233

Fig 234 Fig. 235

Figs 231-5 Isoperimetric figures tending to the solution of Steiner's problem.

P, Q, R as vertices (Fig. 232). This triangle is the solution; its dimensions can be determined from the given value of A. If A decreases further, the radius of these arcs will increase, and the arcs will become more and more nearly straight, until when A is exactly the area of the triangle PQR the solution is the triangle itself. If A now becomes even smaller, then the solution will again consist of three circular arcs having the same radius and forming a triangle with corners at P, Q, R. This time, however, the triangle is concave and the arcs are inside the triangle PQR (Fig. 233). As A continues to decrease, there will come a moment when, for a certain value of A, two of the concave arcs become tangent to each other in a corner R. With an additional decrease of A, it is no longer possible to construct a circular triangle of the previous type. A new phenomenon occurs: the solution is still given by a concave circular triangle, but one of its corners R' has become detached from the corresponding corner R, and the solution now consists of a circular triangle PQR' plus the straight line RR' counted twice (because it travels from R' to R and back). This straight segment is tangent to the two arcs tangent to each other at R'. If A decreases further, the separation process will also set in at the other vertices. Eventually we obtain as solution a circular triangle consisting of three arcs of equal radius tangent to each other and forming an equilateral circular triangle $P'Q'R'$, and in addition three doubly counted straight segments $P'P$, $Q'Q$, $R'R$ (Fig. 234). If, finally, A shrinks to zero, then the circular triangle reduces to a point, and we return to the solution of Steiner's problem; the latter is thus seen to be a limiting case of the modified isoperimetric problem.

If P, Q, R form an obtuse triangle with an angle of more than 120°, then the shrinking process leads to the corresponding solution of Steiner's problem, for then the circular arcs shrink toward the obtuse vertex. The solutions of the generalized Steiner problem (see Figs. 216–8 on p. 360) may be obtained by limiting processes of a similar nature.

§10. THE CALCULUS OF VARIATIONS

1. Introduction

The isoperimetric problem is one example, probably the oldest, of a large class of important problems to which attention was called in 1696 by Johann Bernoulli. In *Acta Eruditorum*, the great scientific journal of the time, he posed the following "brachistochrone" problem: Imagine a particle constrained to slide without friction along a certain curve joining a point A to a lower point B. If the particle is allowed to fall under the influence of gravity alone, along which such curve will the time required for the descent be least? It is easy to see that the falling particle will require different lengths of time for different paths. The straight line by no means affords the quickest journey, nor is the circular arc or any other elementary curve the answer. Bernoulli boasted of having a wonderful solution which he would not immediately publish in order to incite the greatest mathematicians of the time to try their skill at this new type of mathematical question. In particular, he challenged his elder brother Jacob, with whom he was at the time engaged in a bitter feud, and whom he publicly described as incompetent, to solve the problem. Mathematicians immediately recognized the different character of the brachistochrone problem. While heretofore, in problems treated by the differential calculus, the quantity to be minimized depended only on one or more numerical variables, in this problem the quantity under consideration, the time of descent, depends on the

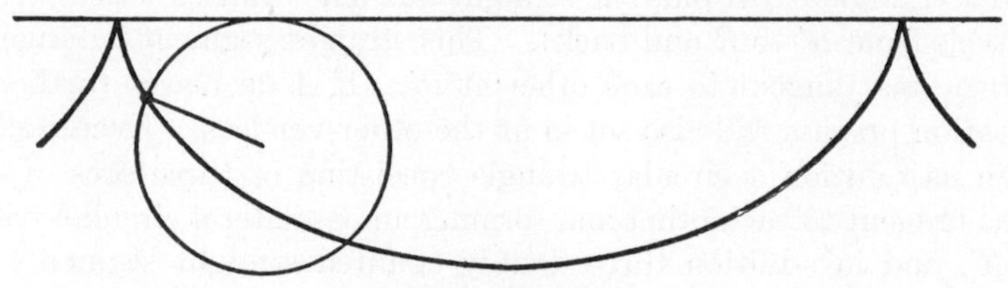

Fig. 236. The cycloid.

whole curve, and this makes for an essential difference, taking the problem out of the reach of the differential calculus or any other method known at the time.

The novelty of the problem—apparently the isoperimetric property of the circle was not clearly recognized as of the same nature—fascinated the contemporary mathematicians all the more when the solution turned out to be the cycloid, a curve that had just been discovered. (We recall the definition of the cycloid: it is the locus of a point on the circumference of a circle that rolls without slipping along a straight line, as shown in Fig. 236.) This curve had been brought into connection with interesting mechanical problems, especially with the construction of an ideal pendulum. Huygens had discovered that an ideal mass point which oscillates without friction under the influence of gravity on a vertical cycloid has a period of oscillation independent of the amplitude of the motion. On a circular path, such as is provided by an ordinary pendulum, this independence is only approximately true, and this was considered a drawback to the use of pendulums for precision clocks. The cycloid was honored by the name of tautochrone; now it acquired the new title of brachistochrone.

2. The Calculus of Variations. Fermat's Principle in Optics

Of the different ways in which the solution to the brachistochrone problem was found by the Bernoullis and others we shall presently explain one of the most original. The first methods were of a more or less special character, adapted to the specific problem. But it did not take long before Euler and Lagrange (1736–1813) evolved more general methods for solving extremum problems in which the independent element was not a single numerical variable or a finite number of such variables, but a whole curve or function or even a system of functions. The new

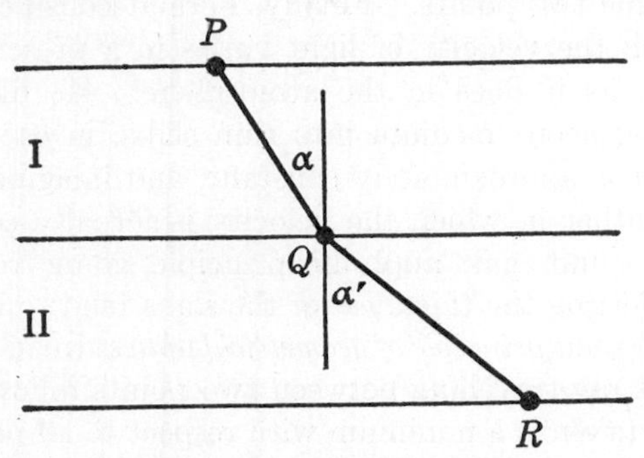

Fig. 237. Refraction of a light ray.

method for solving such problems was called the *calculus of variations.*

It is not possible to describe here the technical aspects of this branch of mathematics or to go deeper into the discussion of specific problems. The calculus of variations has many applications in physics. It was observed long ago that natural phenomena often follow some pattern of maxima and minima. As we have seen, Heron of Alexandria recognized that the reflection of a light ray in a plane mirror can be described by a minimum principle. Fermat, in the seventeenth century, took the next step: he observed that the law of refraction of light can also be stated in terms of a minimum principle. It is well known that the path of light travelling from one homogeneous medium into another is bent at the boundary. Thus in Figure 237, a light ray going from P in the upper medium where the velocity is v to R in the lower medium where the velocity is w will follow a path PQR. The empirical law found by Snell (1591–1626) states that the path consists of two straight segments, PQ and QR, forming angles α, α' with the normal determined by the conditions $\sin \alpha / \sin \alpha' = v/w$. By means of the calculus Fermat proved that this path is such that the time taken for the light ray to go from P to R is a minimum, i.e. smaller than it would be along any other connecting path. Thus Heron's law of reflection was supplemented sixteen hundred years later by a similar and equally important law of refraction.

Fermat generalized the statement of this law so as to include curved surfaces of discontinuity between media, such as the spherical surfaces used in lenses. In this case the statement still holds that light follows a path along which the time taken is a minimum relative to the time that would be required for the light to describe any other possible path between the same two points. Finally, Fermat considered any optical system in which the velocity of light varies in a prescribed way from point to point, as it does in the atmosphere. He divided the continuous inhomogeneous medium into thin slabs, in each of which the velocity of light is approximately constant, and imagined this medium replaced by another in which the velocity is actually constant in each slab. Then he could again apply his principle, going from each slab to the next. By letting the thickness of the slabs tend to zero, he arrived at the general *Fermat principle of geometrical optics*: In an inhomogeneous medium, a light ray travelling between two points follows a path along which the time taken is a minimum with respect to all paths joining the two points. This principle has been of the utmost importance, not only theoretically, but in practical geometrical optics. The technique of the calculus of variations applied to this principle provides the basis for calculating lens systems.

Minimum principles have also become dominant in other branches of physics. It was observed that stable equilibrium of a mechanical system is attained if the system is arranged in such a way that its "potential energy" is a minimum. As an example, let us consider a flexible homogeneous chain suspended at its two ends and allowing full play to the force of gravity. The chain will then assume a form in which its potential energy is a minimum. In this case the potential energy is determined by the height of the center of gravity above some fixed axis. The curve in which the chain hangs is called a catenary, and resembles superficially a parabola.

Not only the laws of equilibrium, but also those of motion, are dominated by maximum and minimum principles. It was Euler who obtained the first clear ideas about these principles, while philosophically and mystically inclined speculators, such as Maupertuis (1698–1759), were not able to separate the mathematical statements from hazy ideas about "God's intention to regulate physical phenomena by a general principle of highest perfection." Euler's variational principles of physics, rediscovered and extended by the Irish mathematician W. R. Hamilton (1805–1865), have proved to be among the most powerful tools in mechanics, optics, and electrodynamics, with many applications to engineering. Recent developments in physics—relativity and quantum theory—are full of examples revealing the power of the calculus of variations.

3. Bernoulli's Treatment of the Brachistochrone Problem

The early method developed for the brachistochrone problem by Jacob Bernoulli can be understood with comparatively little technical knowl-

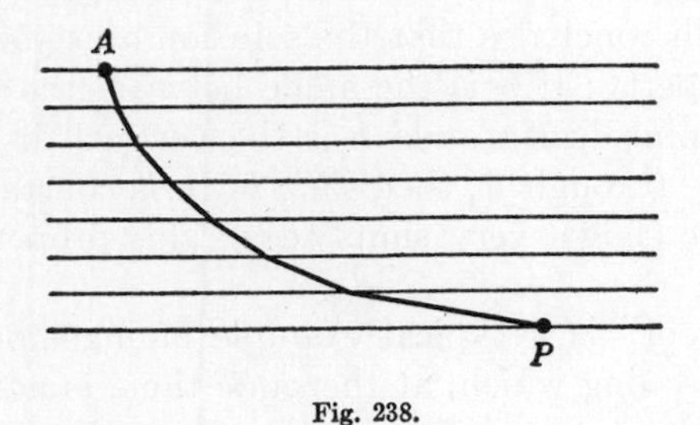

Fig. 238.

edge. We start with the fact, taken from mechanics, that a mass point falling from rest at A along any curve C will have at any point P a velocity proportional to $\sqrt{h}$, where h is the vertical distance from A to P; that is, $v = c\sqrt{h}$, where c is a constant. Now we replace the given problem by a slightly different one. We dissect the space into many thin horizontal slabs, each of thickness d, and assume for the moment that the velocity of the moving particle changes, not continuously, but in little jumps from slab to slab, so that in the first slab adjacent to A the velocity is $c\sqrt{d}$, in the second $c\sqrt{2d}$, and in the nth slab $c\sqrt{nd} = c\sqrt{h}$, where h is the vertical distance from A to P (see Fig. 238). If this problem is considered, then there are really only a finite number of variables. In each slab the path must be a straight segment, no existence problem arises, the solution must be a polygon, and the only question is how to determine its corners. According to the minimum principle for the law of simple refraction, in each pair of successive slabs the motion from P to R by way of Q must be such that, with P and R fixed, Q provides the shortest possible time. Hence the following "refraction law" must hold:

$$\frac{\sin \alpha}{\sqrt{nd}} = \frac{\sin \alpha'}{\sqrt{(n+1)d}}.$$

Repeated application of this reasoning yields the succession of equalities

$$\frac{\sin \alpha_1}{\sqrt{d}} = \frac{\sin \alpha_2}{\sqrt{2d}} = \cdots, \tag{1}$$

where α_n is the angle between the polygon in the nth slab and the vertical.

Now Bernoulli imagines the thickness d to become smaller and smaller, tending to zero, so that the polygon just obtained as the solution of the approximate problem tends to the desired solution of the original problem. In this passage to the limit the equalities (1) are not affected, and therefore Bernoulli concludes that the solution must be a curve C with the following property: If α is the angle between the tangent and the vertical at any point P of C, and h is the vertical distance of P from the horizontal line through A, then $\sin \alpha/\sqrt{h}$ is constant for all points P of C. It can be shown very simply that this property characterizes the cycloid.

Bernoulli's "proof" is a typical example of ingenious and valuable mathematical reasoning which, at the same time, is not at all rigorous. There are several tacit assumptions in the argument, and their justification would be more complicated and lengthy than the argument itself. For example, the existence of a solution C, and the fact that the solution of the approximate problem approximates the actual solution, were both assumed. The question as to the intrinsic value of heuristic considerations of this type certainly deserves discussion, but would lead us too far astray.

4. Geodesics on a Sphere. Geodesics and Maxi-Minima

In the introduction to this chapter we mentioned the problem of finding the shortest arcs joining two given points of a surface. On a sphere, as is shown in elementary geometry, these "geodesics" are arcs of great circles. Let P and Q be two (not diametrically opposite) points on a sphere, and c the shorter connecting arc of the great circle through P and Q. Then the question presents itself, what is the longer arc c' of the same great circle? Certainly it does not give the minimum length, nor can it give the maximum length for curves joining P and Q, since arbitrarily long curves between P and Q can be drawn. The answer is that c' solves a maxi-minimum problem. Consider a point S on a fixed great circle separating P and Q; we ask for the shortest connection between P and Q on the sphere passing through S. Of course, the minimum is given by a curve consisting of two small arcs of great circles PS and QS. Now we seek a position of the point S for which this smallest distance PSQ becomes as large as possible. The solution is: S must be such that PSQ is the longer arc c' of the great

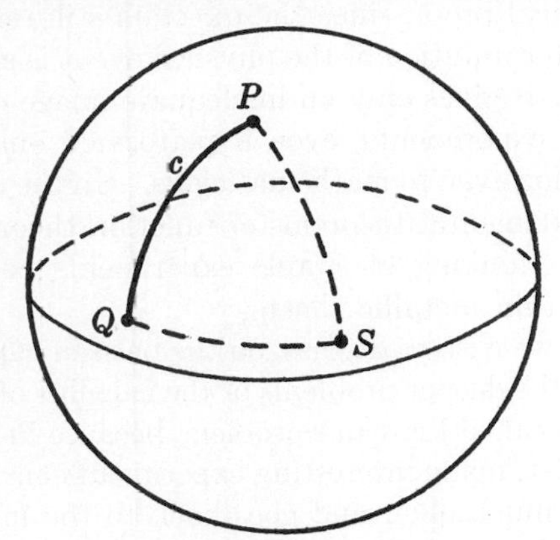

Fig. 239. Geodesics on a sphere.

circle PQ. We may modify the problem by first seeking the path of shortest length from P to Q passing through n prescribed points, S_1, $S_2, \cdots, S_n$, on the sphere, and then seeking to determine the points $S_1, \cdots, S_n$ so that this minimum length becomes as large as possible. The solution is given by a path on the great circle joining P and Q, but this path winds around the sphere so often that it passes through the points diametrically opposite P and Q exactly n times.

This example of a maximum-minimum problem is typical of a wide class of questions in the calculus of variations that have been studied with great success by methods developed by Morse and others.

§11. EXPERIMENTAL SOLUTIONS OF MINIMUM PROBLEMS. SOAP FILM EXPERIMENTS

1. Introduction

It is usually very difficult, and sometimes impossible, to solve variational problems explicitly in terms of formulas or geometrical constructions involving known simple elements. Instead, one is often satisfied with merely proving the existence of a solution under certain conditions and afterwards investigating properties of the solution. In many cases, when such an existence proof turns out to be more or less difficult, it is stimulating to realize the mathematical conditions of the problem by corresponding physical devices, or rather, to consider the mathematical problem as an interpretation of a physical phenomenon. The existence of the physical phenomenon then represents the solution of the mathematical problem. Of course, this is only a plausibility consideration and not a mathematical proof, since the question still remains whether the mathematical interpretation of the physical event is adequate in a strict sense, or whether it gives only an inadequate image of physical reality. Sometimes such experiments, even if performed only in the imagination, are convincing even to mathematicians. In the nineteenth century many of the fundamental theorems of function theory were discovered by Riemann by thinking of simple experiments concerning the flow of electricity in thin metallic sheets.

In this section we wish to discuss, on the basis of experimental demonstrations, one of the deeper problems of the calculus of variations. This problem has been called Plateau's problem, because Plateau (1801–1883), a Belgian physicist, made interesting experiments on this subject. The problem itself is much older and goes back to the initial phases of the calculus of variations. In its simplest form it is the following: to find the surface of smallest area bounded by a given closed contour in space. We shall also discuss experiments connected with some related questions, and it will turn out that much light can thus be thrown on some of our previous results as well as on certain mathematical problems of a new type.

2. Soap Film Experiments

Mathematically, Plateau's problem is connected with the solution of a "partial differential equation," or a system of such equations. Euler showed that all (non-plane) minimal surfaces must be saddle-shaped and that the mean curvature† at every point must be zero. The solution was shown to exist for many special cases during the last century, but

† The mean curvature of a surface at a point P is defined in the following way: Consider the perpendicular to the surface at P, and all planes containing it. These planes will intersect the surface in curves which in general have different curvatures at P. Now consider the curves of minimum and maximum curvature respectively. (In general, the planes containing these curves will be perpen-

the existence of the solution for the general case was proved only recently, by J. Douglas and by T. Radò.

Plateau's experiments immediately yield physical solutions for very general contours. If one dips any closed contour made of wire into a liquid of low surface tension and then withdraws it, a film in the form of a minimal surface of least area will span the contour. (We assume that we may neglect gravity and other forces which interfere with the tendency of the film to assume a position of stable equilibrium by attaining the smallest possible area and thus the least possible value of the potential energy due to surface tension.) A good recipe for such a liquid is the following: Dissolve 10 grams of pure dry sodium oleate in

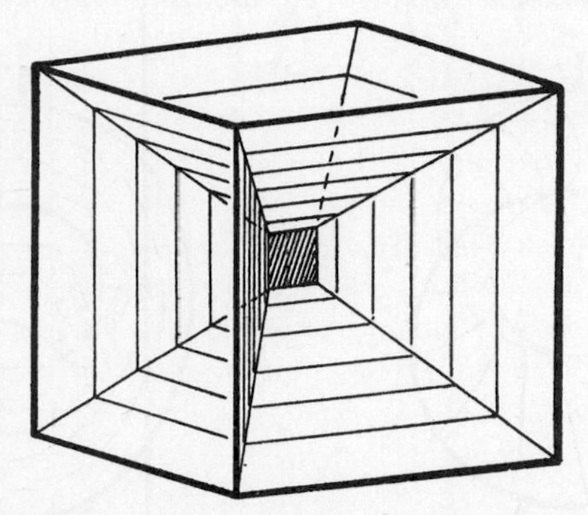

Fig. 240. Cubic frame spanning a soap film system of 13 nearly plane surfaces.

500 grams of distilled water, and mix 15 cubic units of the solution with 11 cubic units of glycerin. Films obtained with this solution and with frames of brass wire are relatively stable. The frames should not exceed five or six inches in diameter.

With this method it is very easy to "solve" Plateau's problem simply by shaping the wire into the desired form. Beautiful models are obtained in polygonal wire frames formed by a sequence of edges of a regular polyhedron. In particular, it is interesting to dip the whole frame of a cube into such a solution. The result is first a system of different surfaces meeting each other at angles of 120° along lines of intersection. (If the cube is withdrawn carefully, there will be thirteen nearly plane surfaces.) Then we may pierce and destroy enough of

---

dicular to each other.) One-half the sum of these two curvatures is the mean curvature of the surface at P.

these different surfaces so that only one surface bounded by a closed polygon remains. Several beautiful surfaces may be formed in this way. The same experiment can also be performed with a tetrahedron.

3. New Experiments on Plateau's Problem

The scope of soap film experiments with minimal surfaces is wider than these original demonstrations by Plateau. In recent years the problem of minimal surfaces has been studied when not only one but any number of contours is prescribed, and when, in addition, the topological structure of the surface is more complicated. For example, the surface might be one-sided or of genus different from zero. These more general problems produce an amazing variety of geometrical

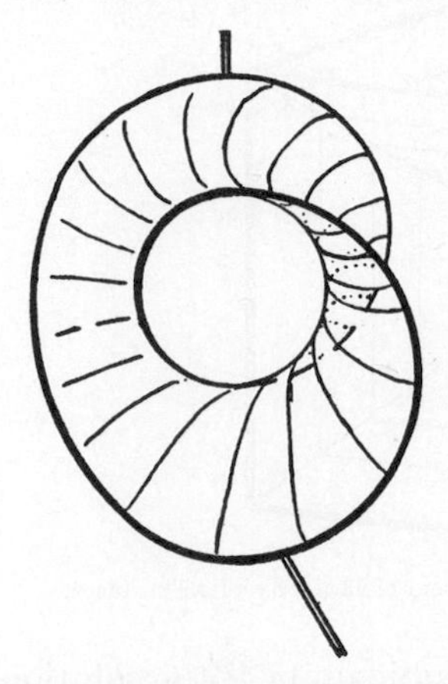

Fig. 241. One-sided surface (Moebius strip).

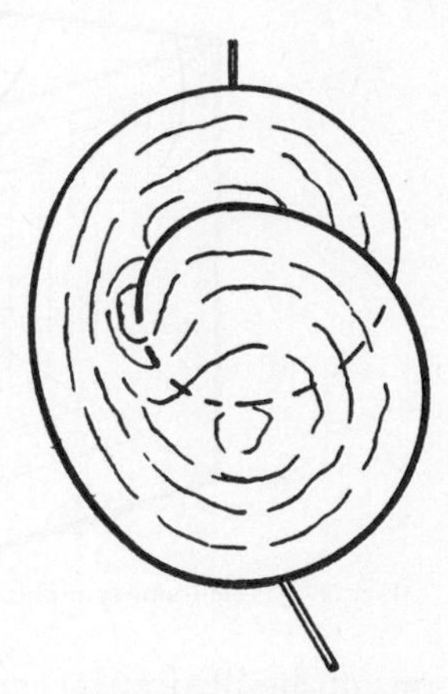

Fig. 242. Two-sided surface.

phenomena that can be exhibited by soap film experiments. In this connection it is very useful to make the wire frames flexible, and to study the effect of deformations of the prescribed boundaries on the solution.

We shall describe several examples:

1) If the contour is a circle we obtain a plane circular disk. If we continuously deform the boundary circle we might expect that the minimal surface would always retain the topological character of a disk. This is not the case. If the boundary is deformed into the shape indicated by Figure 241, we obtain a minimal surface that is no longer simply connected, like the disk, but is a one-sided Moebius strip. Conversely, we might start with this frame and with a soap film in the shape

of a Moebius strip. We may deform the wire frame by pulling handles soldered to it (Fig. 241). In this process we shall reach a moment when suddenly the topological character of the film changes, so that the surface is again of the type of a simply connected disk (Fig. 242). Reversing the deformation we again obtain a Moebius strip. In this alternating deformation process the mutation of the simply connected surface into the Moebius strip takes place at a later stage. This shows that there must be a range of shapes of the contour for which both the Moebius strip and the simply connected surface are stable, i.e. furnish relative minima. But when the Moebius strip has a much smaller area than the other surface, the latter is too unstable to be formed.

2) We may span a minimal surface of revolution between two circles. After the withdrawal of the wire frames from the solution we find, not one simple surface, but a structure of three surfaces meeting at angles of 120°, one of which is a simple circular disk parallel to the prescribed boundary circles (Figure 243). By destroying this intermediate surface the classical catenoid is produced (the catenoid is the surface obtained by revolving the catenary of page 382 about a line perpendicular to its axis of symmetry). If the two boundary circles are pulled apart, there is a moment when the doubly connected minimal surface (the catenoid) becomes unstable. At this moment the catenoid jumps discontinuously into two separated disks. This process is, of course, not reversible.

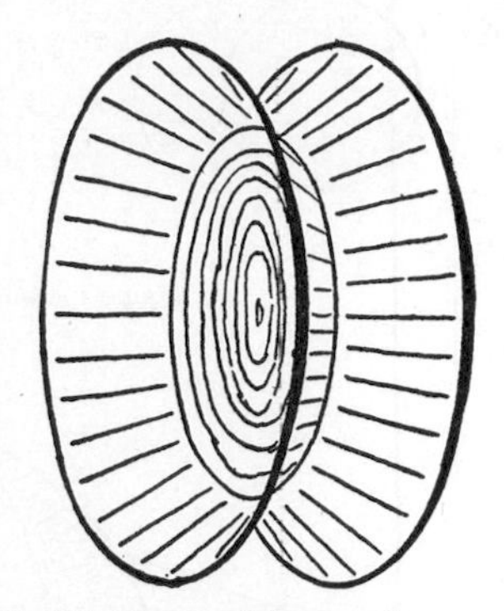

Fig. 243. System of three surfaces.

3) Another significant example is provided by the frame of Figures 244–6 in which can be spanned three different minimal surfaces. Each is bounded by the same simple closed curve; one (Figure 244) has the genus 1, while the other two are simply connected, and in a way symmetrical to each other. The latter have the same area if the contour is completely symmetrical. But if this is not the case then only one gives the absolute minimum of the area while the other will give a relative minimum, provided that the minimum is sought among simply connected surfaces. The possibility of the solution of genus 1 depends on the fact that by admitting surfaces of genus 1 one may obtain a smaller area than by requiring that the surface be simply connected. By deforming the frame we must, if the deformation is radical enough, come to a point where this is no longer true. At that moment the surface of genus 1

becomes more and more unstable and suddenly jumps discontinuously into the simply connected stable solution represented by Figure 245 or 246. If we start with one of these simply connected solutions, such as Figure 246, we may deform it in such a way that the other simply connected solution of Figure 245 becomes much more stable. The consequence is that at a certain moment a discontinuous transition from one to the other will take place. By slowly reversing the deforma-

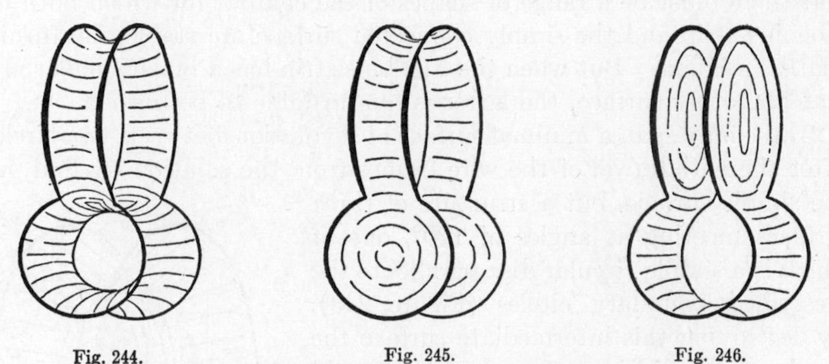

Fig. 244. Fig. 245. Fig. 246.
Frame spanning three different surfaces of genus 0 and 1.

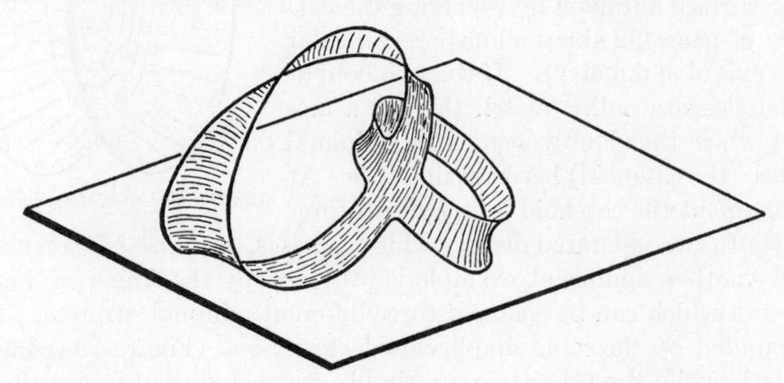

Fig. 247.
One-sided minimal surface of higher topological structure in a single contour.

tion, we return to the initial position of the frame, but now with the other solution in it. We can repeat the process in the opposite direction, and in this way swing back and forth by discontinuous transitions between the two types. By careful handling, one may also transform discontinuously either one of the simply connected solutions into that of genus 1. For this purpose we have to bring the disk-like parts very close to each other, so that the surface of genus 1 becomes markedly more stable. Sometimes in this process intermediate pieces of film appear first and have to be destroyed before the surface of genus 1 is obtained.

This example shows not only the possibility of different solutions of the same topological type, but also of another and different type in one and the same frame; moreover, it again illustrates the possibility of discontinuous transitions from one solution to another while the conditions of the problem are changed continuously. It is easy to construct more complicated models of the same sort and to study their behavior experimentally.

An interesting phenomenon is the appearance of minimal surfaces bounded by two or more interlocked closed curves. For two circles we obtain the surface shown in Figure 248. If, in this example, the circles are perpendicular to each other and the line of intersection of their planes is a diameter of both circles, there will be two symmetrically opposite forms of this surface with equal area. If the circles are now moved slightly with respect to each other, the form will be altered continuously, although for each position only one form is an absolute minimum, and the other one a relative minimum. If the circles are moved so that the relative minimum is formed, it will jump over into the absolute minimum at some point. Here both of the possible minimal surfaces have the same topological character, as do the surfaces of Figures 245–6 one of which can be made to jump into the other by a slight deformation of the frame.

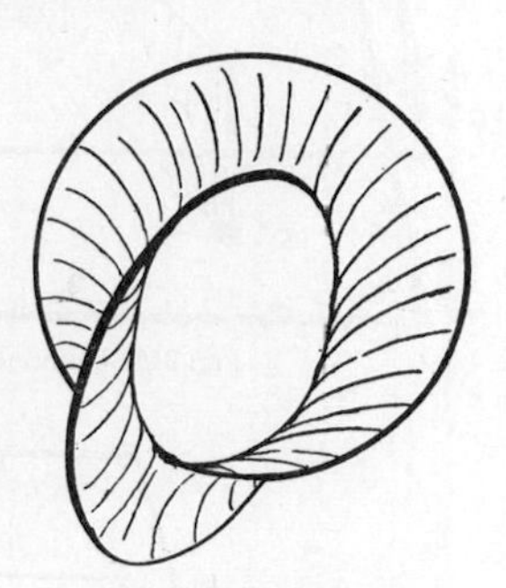

Fig. 248. Interlocked curves.

4. Experimental Solutions of Other Mathematical Problems

Owing to the action of surface tension, a film of liquid is in stable equilibrium only if its area is a minimum. This is an inexhaustible source of mathematically significant experiments. If parts of the boundary of a film are left free to move on given surfaces such as planes, then on these boundaries the film will be perpendicular to the prescribed surface.

We can use this fact for striking demonstrations of Steiner's problem and its generalizations (see §5). Two parallel glass or transparent plastic plates are joined by three or more perpendicular bars. If we immerse this object in a soap solution and withdraw it, the film forms a system of vertical planes between the plates and joining the fixed bars. The projection appearing on the glass plates is the solution of the problem discussed on page 359.

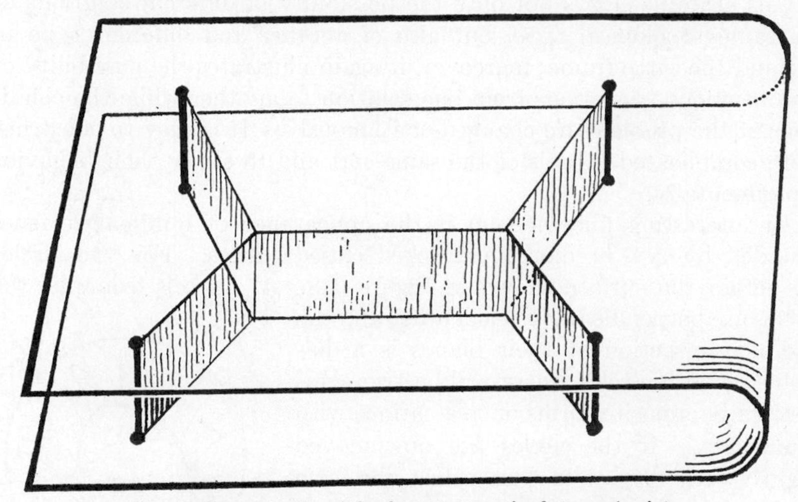

Fig. 249. Demonstration of the shortest connection between 4 points.

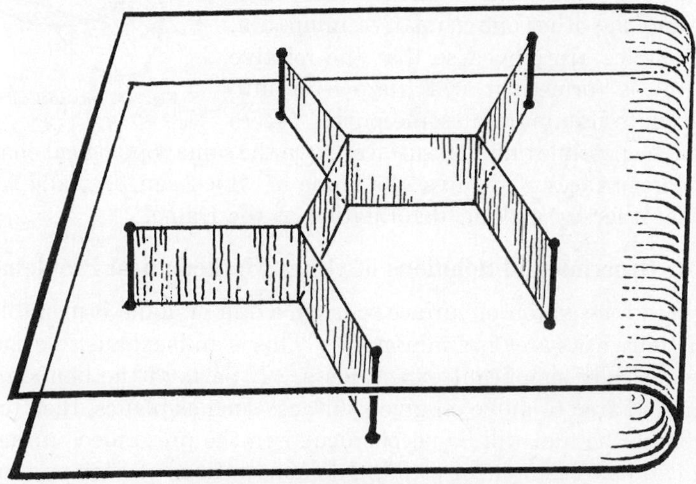

Fig. 250. Shortest connection between 5 points.

If the plates are not parallel, the bars not perpendicular to them, or the plates curved, then the curves formed by the film on the plates will not be straight, but will illustrate new variational problems.

The appearance of lines where three sheets of a minimal surface meet at angles of 120° may be regarded as the generalization to more dimensions of the phenomena connected with Steiner's problem. This becomes clear e.g. if we join two points A, B in space by three curves, and study the corresponding stable system of soap films. As the simplest case we take for one curve the straight segment AB, and for the

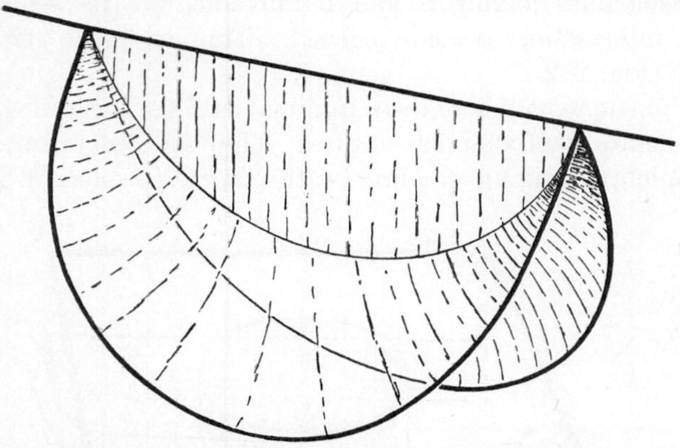

Fig. 251. Three surfaces meeting at 120° spanned between three wires joining two points.

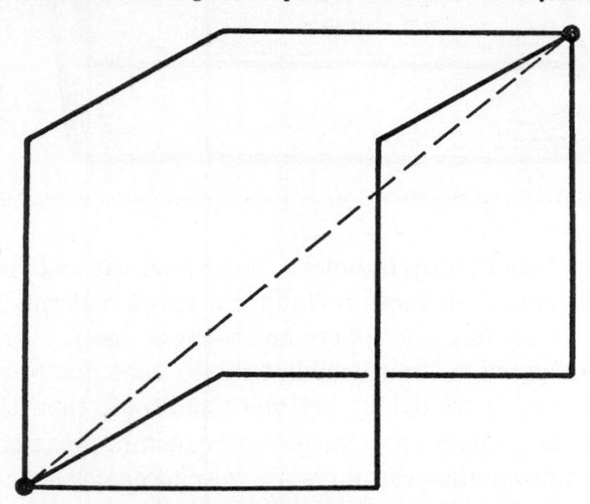

Fig. 252. Three broken lines joining two points.

others two congruent circular arcs. The result is shown in Figure 251. If the planes of the arcs form an angle of less than 120°, we obtain three surfaces meeting at angles of 120°; if we turn the two arcs, increasing the included angle, the solution changes continuously into two plane circular segments.

Now let us join A and B by three more complicated curves. As an example we may take three broken lines each consisting of three edges of the same cube that join two diagonally opposite vertices: we obtain three congruent surfaces meeting in the diagonal of the cube. (We obtain this system of surfaces from that depicted in Fig. 240 by destroying the films adjacent to three properly selected edges.) If we make the three broken lines joining A and B movable, we can see the line of threefold intersection become curved. The angles of 120° will be preserved (Fig. 252).

All the phenomena where three minimal surfaces meet in certain lines are fundamentally of a similar nature. They are generalizations of the plane problem of joining n points by the shortest system of lines.

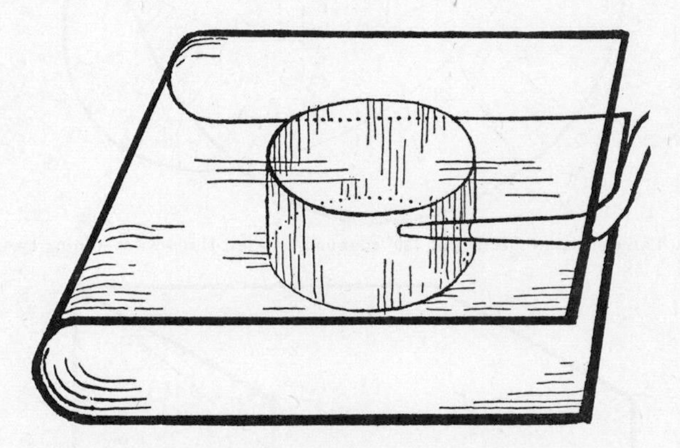

Fig. 253. Demonstration that the circle has least perimeter for a given area.

Finally, a word about soap bubbles. The spherical soap bubble shows that among all closed surfaces including a given volume (defined by the amount of air inside), the sphere has the least area. If we consider soap bubbles of given volume which tend to contract to a minimum area but which are restricted by certain conditions, then the resulting surfaces will be not spheres, but surfaces of constant mean curvature, of which spheres and circular cylinders are special examples.

For example, we blow a soap bubble between two parallel glass plates which have previously been wetted by the soap solution. When the bubble touches one plate, it suddenly assumes the shape of a hemisphere; as soon as it also touches the other plate, it jumps into the shape of a circular cylinder, thus demonstrating the isoperimetric property of the circle in a most striking way. The fact that the soap film adjusts itself vertically to the bounding surface is the key to this experiment. By blowing soap bubbles between two plates with perpendicular connecting rods, we can illustrate the problems discussed on pp. 378–9.

We can study the behavior of the solution of the isoperimetric problem by increasing or decreasing the amount of air in the bubble, using a tube with a fine point. By sucking out air, however, we do not obtain the figures of page 378 consisting of circular arcs tangent to each other. As the volume of air included decreases, the angles of the circular triangle will (theoretically) not decrease below 120°; we obtain the shapes shown in Figures 254–5, which again tend to straight segments as in Figure 235 as the area tends to zero. The mathematical reason for the failure of

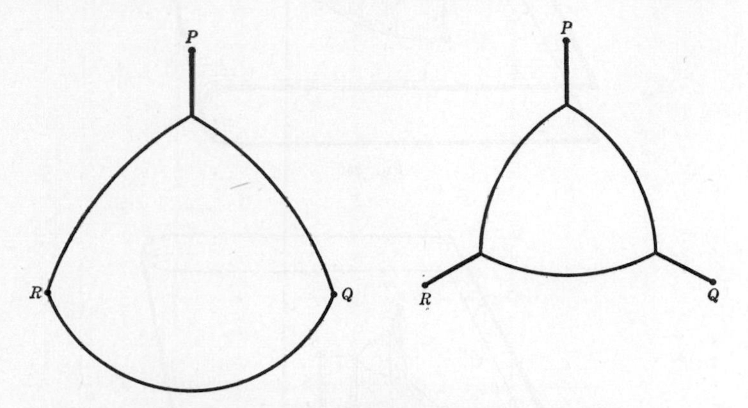

Figs. 254–5. Isoperimetric figures with boundary restrictions.

soap films to form tangent arcs is the fact that as soon as the bubble separates from the vertices, the connecting lines must no longer be counted twice. The corresponding experiments are illustrated by Figures 256 and 257.

* *Exercise:* Study the corresponding mathematical problem: a circular triangle is to be found including a given area and such that its perimeter plus three segments joining the vertices to the given points has a minimum length.

A cubic frame inside of which we blow a bubble will provide surfaces of constant mean curvature with a quadratic base, if the bubble bulges out of the frame. As we remove air from the bubble by sucking through a straw, we obtain a sequence of beautiful structures which result in that of Figure 258. The phenomena of stability and transition between different states of equilibria are a source of experiments that are very illuminating from the mathematical point of view. The experiments illustrate the theory of stationary values, since the transitions can be made to take place so as to lead through an unstable equilibrium which is a "stationary state."

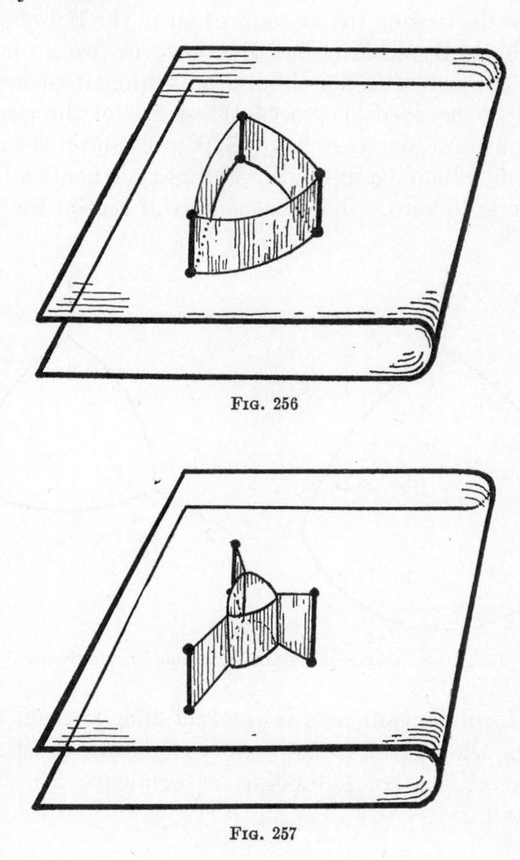

Fig. 256

Fig. 257

For example, the cubical structure of Figure 240 exhibits asymmetry insofar as a vertical plane in the center connects the twelve surfaces issuing from the edges. Hence there must be at least two other positions of equilibrium, one with a vertical and one with a horizontal central square. As a matter of fact, by blowing through a fine tube against the edges of this square, one can force the structure into a position where the square reduces to a point, the center of the cube; this position of unstable equilibrium will immediately go into one of the other stable positions obtained from the original by a rotation through 90°.

A similar experiment can be performed on the soap film that demonstrates Steiner's problem for four points forming a square (Figs. 219–20).

If we want to obtain the solutions of such problems as limiting cases of isoperimetric problems—for example, if we want to obtain Figure 240 from Figure 258—we must suck the air out of the bubble. Now Figure 258 is completely symmetric, and its limit for vanishing content of the bubble would be a symmetric system of 12 planes meeting at the center.

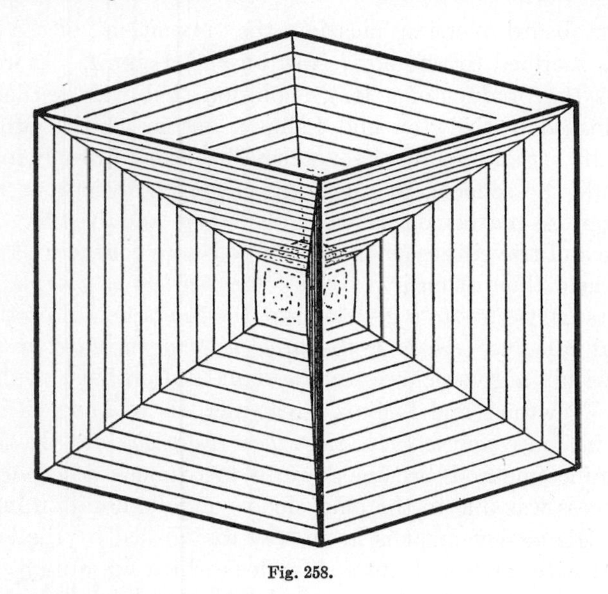

Fig. 258.

This can really be observed. But the position obtained as a limit is not in stable equilibrium; instead, it will change over into one of the positions of Figure 240. By using a somewhat more viscous liquid than that described above the whole phenomenon can be observed very easily. It exemplifies the fact that even in physical problems the solution of a problem need not depend continuously on the data; for in the limiting case for volume zero the solution, given by Figure 240, is not the limit of the solution, given by Figure 258, for volume ϵ as ϵ tends to zero.

CHAPTER VIII

THE CALCULUS

INTRODUCTION

With an absurd oversimplification, the "invention" of the calculus is sometimes ascribed to two men, Newton and Leibniz. In reality, the calculus is the product of a long evolution that was neither initiated nor terminated by Newton and Leibniz, but in which both played a decisive part. Scattered over seventeenth century Europe, for the most part outside the schools, was a group of spirited scientists who strove to continue the mathematical work of Galileo and Kepler. By correspondence and travel these men maintained close contact. Two central problems held their attention. First, the *problem of tangents*: to determine the tangent lines to a given curve, the fundamental problem of the differential calculus. Second, the *problem of quadrature*: to determine the area within a given curve, the fundamental problem of the integral calculus. Newton's and Leibniz' great merit is to have clearly recognized the intimate *connection between these two problems*. In their hands the new unified methods became powerful instruments of science. Much of the success was due to the marvelous symbolic notation invented by Leibniz. His achievement is in no way diminished by the fact that it was linked with hazy and untenable ideas which are apt to perpetuate a lack of precise understanding in minds that prefer mysticism to clarity. Newton, by far the greater scientist, appears to have been mainly inspired by Barrow (1630–1677), his teacher and predecessor at Cambridge. Leibniz was more of an outsider. A brilliant lawyer, diplomat, and philosopher, one of the most active and versatile minds of his century, he learned the new mathematics in an incredibly short time from the physicist Huygens while visiting Paris on a diplomatic mission. Soon afterwards he published results that contain the nucleus of the modern calculus. Newton, whose discoveries had been made much earlier, was averse to publication. Moreover, although he had originally found many of the results in his masterpiece, the *Principia*, by the methods of the calculus, he preferred a presentation in the style of classical geometry, and almost no trace of the calculus appears explicitly in the

Principia. Only later were his papers on the method of "fluxions" published. Soon his admirers started a bitter feud over priority with the friends of Leibniz. They accused the latter of plagiarism, although in an atmosphere saturated with the elements of a new theory, nothing is more natural than simultaneous and independent discovery. The resulting quarrel over priority in the "invention" of the calculus set an unfortunate example for the overemphasis on questions of precedence and claims to intellectual property that is apt to poison the atmosphere of natural scientific contact.

In the mathematical analysis of the seventeenth and most of the eighteenth centuries, the Greek ideal of clear and rigorous reasoning seemed to have been discarded. "Intuition" and "instinct" replaced reason in many important instances. This only encouraged an uncritical belief in the superhuman power of the new methods. It was generally thought that a clear presentation of the results of the calculus was not only unnecessary but impossible. Had not the new science been in the hands of a small group of extremely competent men, serious errors and even debacle might have resulted. These pioneers were guided by a strong instinctive feeling that kept them from going far astray. But when the French Revolution opened the way to an immense extension of higher learning, when increasingly large numbers of men wished to participate in scientific activity, the critical revision of the new analysis could no longer be postponed. This challenge was successfully met in the nineteenth century, and today the calculus can be taught without a trace of mystery and with complete rigor. There is no longer any reason why this basic instrument of the sciences should not be understood by every educated person.

This chapter is intended to serve as an elementary introduction in which the emphasis is on understanding the basic concepts rather than on formal manipulation. Intuitive language will be used throughout, but always in a manner consistent with precise concepts and clear procedure.

§1. THE INTEGRAL

1. Area as a Limit

In order to calculate the area of a plane figure we choose as the *unit of area* a square whose sides are of unit length. If the unit of length is the inch, the corresponding unit of area will be the square inch; i.e. the square whose sides are of length one inch. On the basis of this definition it is very easy to calculate the area of a rectangle. If p and q are the lengths of two adjacent sides measured in terms of the unit of length, then the area of the rectangle is pq square units, or, briefly, the area is equal to the product pq. This is true for arbitrary p and q, rational or not. For rational p and q we obtain this result by writing $p = m/n$, $q = m'/n'$, with integers m, n, m', n'. Then we find the common measure $1/N = 1/nn'$ of the two edges, so that $p = mn' \cdot 1/N$, $q = nm' \cdot 1/N$. Finally, we subdivide the rectangle into small squares of side $1/N$ and area $1/N^2$. The number of such squares is $nm' \cdot mn'$ and the total area is $nm'mn' \cdot 1/N^2 = nm'mn'/n^2n'^2 = m/n \cdot m'/n' = pq$. If p and q are irrational, the same result is obtained by first replacing p and q by approximate rational numbers p_r and q_r respectively, and then letting p_r and q_r tend to p and q.

It is geometrically obvious that the area of a triangle is equal to half the area of a rectangle with the same base b and altitude h; hence the area of a triangle is given by the familiar expression $\frac{1}{2}bh$. Any domain in the plane bounded by one or more polygonal lines can be decomposed into triangles; its area, therefore, can be obtained as the sum of the areas of these triangles.

The need for a more general method of computing areas arises when we ask for the area of a figure bounded, not by polygons, but by *curves*. How shall we determine, for example, the area of a circular disk or of a segment of a parabola? This crucial question, which is at the base of the integral calculus, was treated as early as the third century B.C. by Arch medes, who calculated such areas by a process of "exhaustion." With Archimedes and the great mathematicians until the time of Gauss, we may take the "naive" attitude that curvilinear areas are intuitively given entities, and that the question is not to *define*, but to *compute* them (see, however, the discussion on p. 464). We inscribe in the domain an approximating domain with a polygonal boundary and a well defined area. By choosing another polygonal domain which includes the former we obtain a better approximation to the given domain. Proceeding in this way, we can gradually "exhaust" the whole area, and we obtain the area of the given domain as the limit of the areas of a properly chosen sequence of inscribed polygonal domains with an increasing number of sides. The area of a circle of radius 1 may be computed in this way; its numerical value is denoted by the symbol π.

Archimedes carried out this general scheme for the circle and for the parabolic segment. During the seventeenth century many more cases were successfully treated. In each case, the actual calculation of the limit was made to depend on an ingenious device specially suited to the particular problem. One of the main achievements of the calculus was to replace these special and restricted procedures for the calculation of areas by a general and powerful method.

2. The Integral

The first basic concept of the calculus is that of integral. In this article we shall understand the integral as an expression of the *area under a curve* by means of a limit. If a positive continuous function $y = f(x)$ is given, e.g. $y = x^2$ or $y = 1 + \cos x$, then we consider the domain bounded below by the segment on the x-axis from a coördinate a to a greater coördinate b, on the sides by the perpendiculars to the x-axis at these points, and above by the curve $y = f(x)$. Our aim is to calculate the area A of this domain.

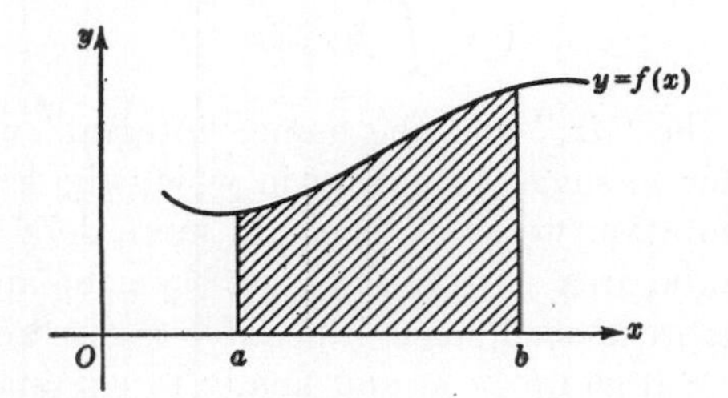

Fig. 259. The integral as an area.

Since such a domain cannot, in general, be decomposed into rectangles or triangles, no immediate expression of this area A is available for explicit calculation. But we can find an approximate value for A, and thus represent A as a limit, in the following way: We subdivide the interval from $x = a$ to $x = b$ into a number of small subintervals, erect perpendiculars at each point of subdivision, and replace each strip of the domain under the curve by a rectangle whose height is chosen somewhere between the greatest and the least height of the curve in that strip. The sum S of the areas of these rectangles gives an approximate value for the actual area A under the curve. The accuracy of this approximation will be better the larger the number of rectangles and the smaller the width of each individual rectangle. Thus we can characterize the exact area as a limit: If we form a sequence,

(1) $$S_1, S_2, S_3, \cdots,$$

of rectangular approximations to the area under the curve in such a manner that the width of the widest rectangle in S_n tends to 0 as n increases, then the sequence (1) approaches the limit A,

(2) $$S_n \to A,$$

and this limit A, the area under the curve, is independent of the particular way in which the sequence (1) is chosen, so long as the widths of the approximating rectangles tend to zero. (For example, S_n can arise from S_{n-1} by adding one or more new points of subdivision to those defining S_{n-1}, or the choice of points of subdivision for S_n can be entirely independent of the choice for S_{n-1}.) The area A of the domain, expressed by this limiting process, we call by definition *the integral of the function* f(x) *from* a *to* b. With a special symbol, the "integral sign," it is written

(3) $$A = \int_a^b f(x)\,dx.$$

The symbol $\int$, the "dx," and the name "integral" were introduced by Leibniz in order to suggest the way in which the limit is obtained. To explain this notation we shall repeat in more detail the process of approximation to the area A. At the same time the analytic formulation of the limiting process will make it possible to discard the restrictive assumptions $f(x) \geq 0$ and $b > a$, and finally to eliminate the prior intuitive concept of area as the basis of our definition of integral (the latter will be done in the supplement, §1).

Let us subdivide the interval from a to b into n small subintervals, which, for simplicity only, we shall assume to be of equal width, $(b - a)/n$. We denote the points of subdivision by

$$x_0 = a, \quad x_1 = a + \frac{b-a}{n},$$

$$x_2 = a + \frac{2(b-a)}{n}, \cdots, x_n = a + \frac{n(b-a)}{n} = b.$$

We introduce for the quantity $(b - a)/n$, the difference between consecutive x-values, the notation Δx (read, "delta x"),

$$\Delta x = \frac{b-a}{n} = x_{j+1} - x_j,$$

where the symbol Δ means simply "difference" (it is an "operator" symbol, and must not be mistaken for a number.) We may choose as

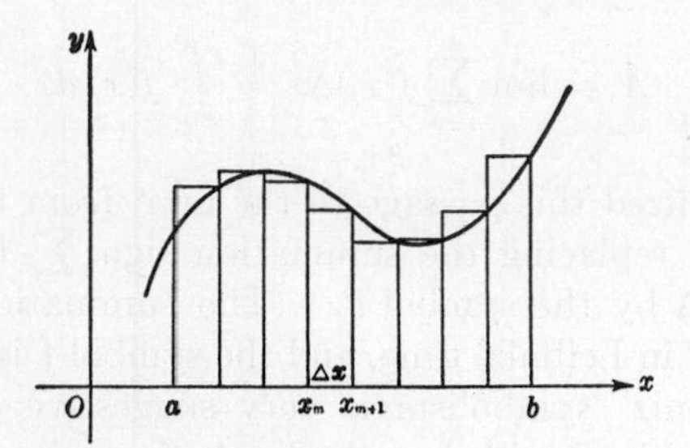

Fig. 260. Area approximated by small rectangles.

the height of each approximating rectangle the value of $y = f(x)$ at the right-hand endpoint of the subinterval. Then the sum of the areas of these rectangles will be

$$S_n = f(x_1)\cdot\Delta x + f(x_2)\cdot\Delta x + \cdots + f(x_n)\cdot\Delta x, \tag{4}$$

which is abbreviated as

$$S_n = \sum_{j=1}^{n} f(x_j)\cdot\Delta x. \tag{5}$$

Here the symbol $\sum_{j=1}^{n}$ (read "sigma from $j = 1$ to n") means the sum of all the expressions obtained by letting j assume in turn the values $1, 2, 3, \cdots, n$.

The use of the symbol $\sum$ to express in concise form the result of a summation may be illustrated by the following examples:

$$2 + 3 + 4 + \cdots + 10 = \sum_{j=2}^{10} j,$$

$$1 + 2 + 3 + \cdots + n = \sum_{j=1}^{n} j,$$

$$1^2 + 2^2 + 3^2 + \cdots + n^2 = \sum_{j=1}^{n} j^2,$$

$$aq + aq^2 + \cdots + aq^n = \sum_{j=1}^{n} aq^j,$$

$$a + (a + d) + (a + 2d) + \cdots + (a + nd) = \sum_{j=0}^{n} (a + jd).$$

Now we form a sequence of such approximations S_n in which n increases indefinitely, so that the number of terms in each sum (5) increases, while each single term $f(x_j)\Delta x$ approaches 0 because of the factor $\Delta x = (b - a)/n$. As n increases, this sum tends to the area A,

$$(6) \qquad A = \lim \sum_{j=1}^{n} f(x_j)\Delta x = \int_a^b f(x)\, dx.$$

Leibniz symbolized this passage to the limit from the approximating sum S_n to A by replacing the summation sign $\sum$ by $\int$ and the difference symbol Δ by the symbol d. (The summation symbol $\sum$ was usually written S in Leibniz' time, and the symbol $\int$ is merely a stylized S.) While Leibniz' symbolism is very suggestive of the manner in which the integral is obtained as the limit of a finite sum, one must be careful not to attach too much significance to what is, after all, a pure convention as to how the limit shall be denoted. In the early days of the calculus, when the concept of limit was not clearly understood and certainly not always kept in mind, one explained the meaning of the integral by saying that "the finite difference Δx is replaced by the infinitely small quantity dx, and the integral itself is the sum of infinitely many infinitely small quantities $f(x)\, dx$." Although the infinitely small has a certain attraction for speculative souls, it has no place in modern mathematics. No useful purpose is served by surrounding the clear notion of the integral with a fog of meaningless phrases. But even Leibniz was sometimes carried away by the suggestive power of his symbols; they work *as if* they denote a sum of "infinitely small" quantities with which one can nevertheless operate to a certain extent as with ordinary quantities. In fact, the word integral was coined to indicate that the whole or integral area A is composed of the "infinitesimal" parts $f(x)\, dx$. At any rate, it was almost a hundred years after Newton and Leibniz before it was clearly recognized that the limit concept and nothing else is the true basis for the definition of the integral. By firmly staying on this basis we may avoid all the haze, all the difficulties, and all the nonsense so disturbing in the early development of the calculus.

3. General Remarks on the Integral Concept. General Definition

In our geometrical definition of the integral as an area we assumed explicitly that $f(x)$ is never negative throughout the interval $[a, b]$ of integration, i.e. that no portion of the graph lies below the x-axis. But in our analytic definition of the integral as the limit of a sequence of sums S_n this assumption is superfluous. We simply take the small quantities $f(x_j) \cdot \Delta x$, form their sum, and pass to the limit; this procedure remains perfectly meaningful if some or all of the values $f(x_j)$ are negative. Interpreting this geometrically by means of areas (Fig. 261), we

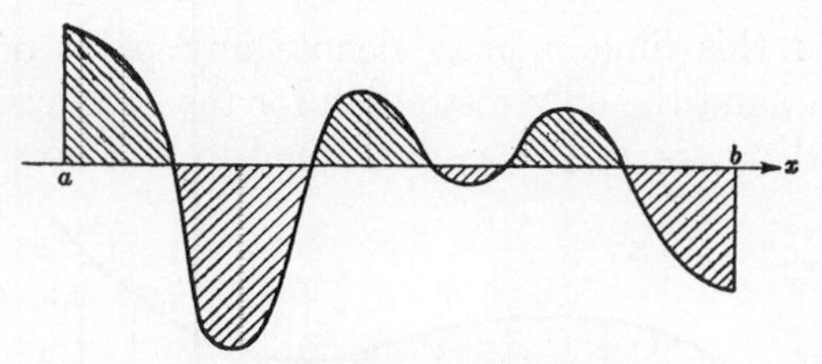

Fig. 261. Positive and negative areas.

find the integral of $f(x)$ to be the *algebraic* sum of the areas bounded by the graph and the x-axis, where areas lying below the x-axis are counted as negative and the others positive.

It may happen that in applications we are led to integrals $\int_a^b f(x)\,dx$ where b is less than a, so that $(b - a)/n = \Delta x$ is a negative number. In our analytic definition we have $f(x_j)\cdot\Delta x$ negative if $f(x_j)$ is positive and Δx negative, etc. In other words, the value of the integral will be the negative of the value of the integral from b to a. Thus we have the simple rule

$$\int_a^b f(x)\,dx = -\int_b^a f(x)\,dx.$$

We must emphasize that the value of the integral remains the same even if we do not restrict ourselves to equidistant points x_j of subdivision, or, what is the same, to equal x-differences $\Delta x = x_{j+1} - x_j$. We may choose the x_j in other ways, so that the differences $\Delta x_j = x_{j+1} - x_j$ are not equal (and must accordingly be distinguished by subscripts). Even then the sums

$$S_n = f(x_1)\Delta x_0 + f(x_2)\Delta x_1 + \cdots + f(x_n)\Delta x_{n-1}$$

and also the sums

$$S'_n = f(x_0)\Delta x_0 + f(x_1)\Delta x_1 + \cdots + f(x_{n-1})\Delta x_{n-1}$$

will tend to the same limit, the value of the integral $\int_a^b f(x)\,dx$, if only care is taken that with increasing n all the differences $\Delta x_j = x_{j+1} - x_j$ tend to zero in such a way that the largest such difference for a given value of n approaches zero as n increases.

Accordingly, the *final definition of the integral* is given by

(6a) $$\int_a^b f(x)\,dx = \lim \sum_{j=1}^{n} f(v_j)\Delta x_j$$

as $n \to \infty$. In this limit v_j may denote any point of the interval $x_j \leq v_j \leq x_{j+1}$, and the only restriction for the subdivision is that the longest interval $\Delta x_j = x_{j+1} - x_j$ must tend to zero as n increases.

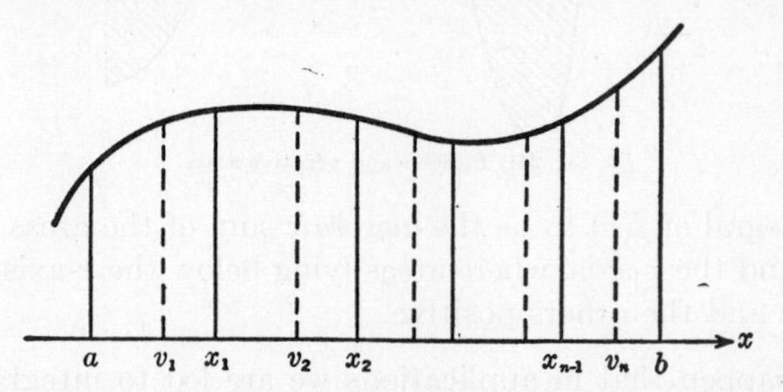

Fig. 262. Arbitrary subdivision in the general definition of integral.

The existence of the limit (6a) does not need a proof if we take for granted the concept of the area under a curve and the possibility of approximating this area by sums of rectangles. However, as will appear in a later discussion (p. 464), a closer analysis shows that it is desirable and even necessary for a logically complete presentation of the notion of integral to prove the existence of the limit for any continuous function $f(x)$ without reference to a prior geometrical concept of area.

4. Examples of Integration. Integration of x^r

Until now our discussion of the integral has been merely theoretical. The crucial question is whether the general pattern of forming a sum S_n and then passing to the limit actually leads to tangible results in concrete cases. Of course, this will require some additional reasoning adapted to the specific function $f(x)$ for which the integral is to be found. When Archimedes two thousand years ago found the area of the parabolic segment, he performed what we now call the integration of the function $f(x) = x^2$ by a very ingenious device; in the seventeenth century the forerunners of the modern calculus succeeded in solving problems of integration for simple functions such as x^n, again by special devices. Only after much experience with specific cases was a general approach to the problem of integration found in the systematic methods of the calculus, and thus the scope of solvable individual problems was greatly widened. In the present article we shall discuss a few of the instructive special problems belonging to the "pre-calculus" stage, for nothing can better illustrate integration as a limiting process.

a) We start with a quite trivial example. If $y = f(x)$ is a constant, for example $f(x) = 2$, then obviously the integral $\int_a^b 2\,dx$, understood as an area, is $2(b - a)$, since the area of a rectangle is equal to base times altitude. We shall compare this result with the definition of the integral (6) as a limit: If we substitute in (5) $f(x_j) = 2$ for all values of j, we find that

$$S_n = \sum_{j=1}^{n} f(x_j)\Delta x = \sum_{j=1}^{n} 2\Delta x = 2\sum_{j=1}^{n} \Delta x = 2(b - a)$$

for every n, since

$$\sum_{j=1}^{n} \Delta x = (x_1 - x_0) + (x_2 - x_1) + \cdots$$
$$+ (x_n - x_{n-1}) = x_n - x_0 = b - a.$$

b) Almost as simple is the integration of $f(x) = x$. Here $\int_a^b x\,dx$ is the area of a trapezoid (Fig. 263), and this, by elementary geometry, is

$$(b - a)\frac{b + a}{2} = \frac{b^2 - a^2}{2}.$$

This result again agrees with the definition (6) of the integral, as is seen by an actual passage to the limit without making use of the geometrical figure: If we substitute $f(x) = x$ in (5), then the sum S_n becomes

$$\begin{aligned} S_n &= \sum_{j=1}^{n} x_j\Delta x = \sum_{j=1}^{n} (a + j\Delta x)\Delta x \\ &= (na + \Delta x + 2\Delta x + 3\Delta x + \cdots + n\Delta x)\Delta x \\ &= na\Delta x + (\Delta x)^2(1 + 2 + 3 + \cdots + n). \end{aligned}$$

Using the formula (1) on page 12 for the arithmetical series $1 + 2 + 3 + \cdots + n$, we have

$$S_n = na\Delta x + \frac{n(n + 1)}{2}(\Delta x)^2.$$

Since $\Delta x = \dfrac{b - a}{n}$, this is equal to

$$S_n = a(b - a) + \tfrac{1}{2}(b - a)^2 + \frac{1}{2n}(b - a)^2.$$

If now we let n tend to infinity, the last term tends to zero, and we obtain

$$\lim S_n = \int_a^b x\,dx = a(b - a) + \tfrac{1}{2}(b - a)^2 = \tfrac{1}{2}(b^2 - a^2),$$

in conformity with the geometrical interpretation of the integral as an area.

c) Less trivial is the integration of the function $f(x) = x^2$. Archimedes used geometrical methods to solve the equivalent problem of finding the area of a segment of the parabola $y = x^2$. We shall proceed analytically on the basis of the definition (6a). To simplify the formal calculation we choose 0 as the "lower limit" a of the integral; then

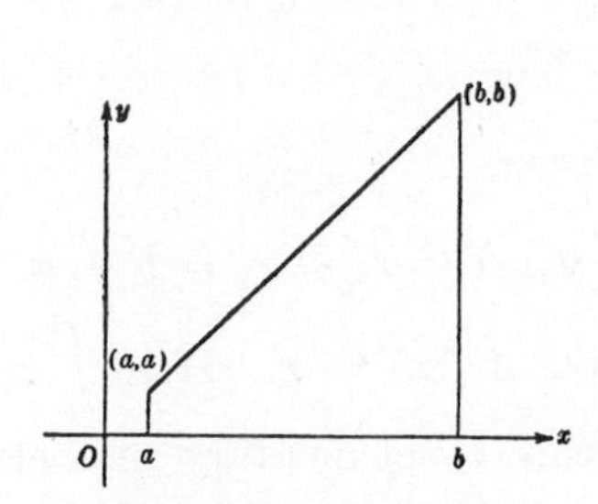

Fig. 263. Area of a trapezoid.

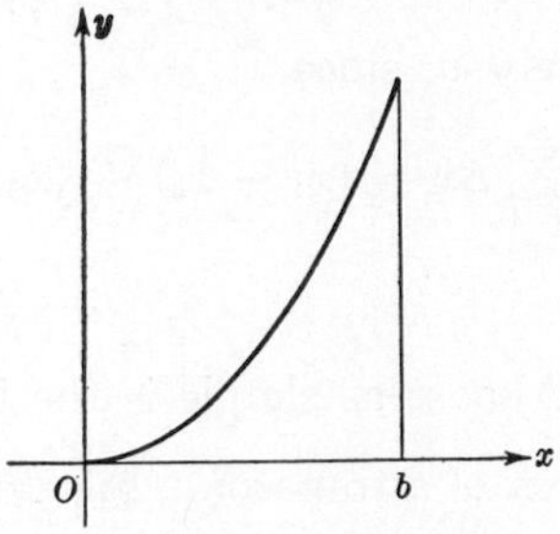

Fig. 264. Area under a parabola.

$\Delta x = b/n$. Since $x_j = j \cdot \Delta x$ and $f(x_j) = j^2(\Delta x)^2$, we obtain for S_n the expression

$$S_n = \sum_{j=1}^{n} f(j\Delta x)\Delta x = [1^2 \cdot (\Delta x)^2 + 2^2 \cdot (\Delta x)^2 + \cdots + h^2(\Delta x)^2] \cdot \Delta x$$

$$= (1^2 + 2^2 + \cdots + n^2)(\Delta x)^3.$$

Now we can actually calculate the limit. Using the formula

$$1^2 + 2^2 + \cdots + n^2 = \frac{n(n+1)(2n+1)}{6}$$

established on page 14, and making the substitution $\Delta x = b/n$, we obtain

$$S_n = \frac{n(n+1)(2n+1)}{6} \cdot \frac{b^3}{n^3} = \frac{b^3}{6}\left(1 + \frac{1}{n}\right)\left(2 + \frac{1}{n}\right).$$

This preliminary transformation makes the passage to the limit an easy matter, since $1/n$ tends to zero as n increases indefinitely. Thus we obtain as limit simply $\frac{b^3}{6} \cdot 1 \cdot 2 = \frac{b^3}{3}$, and thereby the result

$$\int_0^b x^2\, dx = b^3/3.$$

Applying this result to the area from 0 to a we have

$$\int_0^a x^2\,dx = a^3/3,$$

and by subtraction of the areas,

$$\int_a^b x^2\,dx = \frac{b^3 - a^3}{3}.$$

Exercise: Prove in the same way, using formula (5) on page 15 that

$$\int_a^b x^3\,dx = \frac{b^4 - a^4}{4}.$$

By developing general formulas for the sum $1^k + 2^k + \cdots + n^k$ of the kth powers of the integers from 1 to n, one can obtain the result

$$(7) \qquad \int_a^b x^k\,dx = \frac{b^{k+1} - a^{k+1}}{k+1}, \qquad k \text{ any positive integer.}$$

* Instead of proceeding in this way, we can obtain more simply an even more general result by utilizing our previous remark that we may calculate the integral by means of non-equidistant points of subdivision. We shall establish formula (7) not only for any positive integer k but for an arbitrary positive or negative rational number

$$k = u/v,$$

where u is a positive integer and v is a positive or negative integer. Only the value $k = -1$, for which formula (7) becomes meaningless, is excluded. We shall also suppose that $0 < a < b$.

To obtain the integral formula (7), we form S_n by choosing the points of subdivision $x_0 = a, x_1, x_2, \cdots, x_n = b$ in *geometrical progression.* We set $\sqrt[n]{\frac{b}{a}} = q$, so that $b/a = q^n$, and define $x_0 = a$, $x_1 = aq$, $x_2 = aq^2$, ..., $x_n = aq^n = b$. By this device, as we shall see, the passage to the limit becomes very easy. For the "rectangle sum" S_n we find, since $f(x_j) = x_j^k = a^k q^{jk}$ and $\Delta x_j = x_{j+1} - x_j = aq^{j+1} - aq^j$,

$$S_n = a^k(aq - a) + a^k q^k(aq^2 - aq) + a^k q^{2k}(aq^3 - aq^2) + \cdots + a^k q^{(n-1)k}(aq^n - aq^{n-1}).$$

Since each term contains the factor $a^k(aq - a)$, we may write

$$S_n = a^{k+1}(q - 1)\{1 + q^{k+1} + q^{2(k+1)} + \cdots + q^{(n-1)(k+1)}\}.$$

Substituting t for q^{k+1} we see that the expression in braces is the geometrical series

$1 + t + t^2 + \cdots + t^{n-1}$, whose sum, as shown on page 13, is $\dfrac{t^n - 1}{t - 1}$. But $t^n =$ $q^{n(k+1)} = \left(\dfrac{b}{a}\right)^{k+1} = \dfrac{b^{k+1}}{a^{k+1}}$. Hence

$$(8) \qquad S_n = (q-1)\left\{\frac{b^{k+1} - a^{k+1}}{q^{k+1} - 1}\right\} = \frac{b^{k+1} - a^{k+1}}{N},$$

where

$$N = \frac{q^{k+1} - 1}{q - 1}.$$

Thus far n has been a fixed number. Now we shall let n increase, and determine the limit of N. As n increases, the nth root $\sqrt[n]{\dfrac{b}{a}} = q$ will tend to 1 (see p. 323), and therefore both numerator and denominator of N will tend to zero, which makes caution necessary. Suppose first that k is a positive integer; then the division by $q - 1$ can be carried out, and we obtain (see p. 13) $N = q^k + q^{k-1} + \cdots + q + 1$. If now n increases, q tends to 1 and hence $q^2, q^3, \cdots, q^k$ will also tend to 1, so that N approaches $k + 1$. But this shows that S_n tends to $\dfrac{b^{k+1} - a^{k+1}}{k+1}$, as was to be proved.

Exercise: Prove that for any rational $k \neq -1$ the same limit formula, $N \to k + 1$, and therefore the result (7), remains valid. First give the proof, according to our model, for negative integers k. Then, if $k = u/v$, write $q^{1/v} = s$ and

$$N = \frac{s^{(k+1)v} - 1}{s^v - 1} = \frac{s^{u+v} - 1}{s^v - 1} = \frac{s^{u+v} - 1}{s - 1} \bigg/ \frac{s^v - 1}{s - 1}.$$

If n increases, both s and q tend to 1, and therefore the two quotients on the right hand side tend to $u + v$ and v respectively, which yields again $\dfrac{u+v}{v} = k + 1$ for the limit of N.

In §5 we shall see how this lengthy and somewhat artificial discussion may be replaced by the simpler and more powerful methods of the calculus.

Exercises: 1) Check the preceding integration of x^k for the cases $k = \frac{1}{2}, -\frac{1}{2}$, 2, −2, 3, −3.

2) Find the values of the integrals:

a) $\displaystyle\int_{-2}^{-1} x\,dx.$ b) $\displaystyle\int_{-1}^{+1} x\,dx.$ c) $\displaystyle\int_{1}^{2} x^2\,dx.$ d) $\displaystyle\int_{-1}^{-2} x^3\,dx.$ e) $\displaystyle\int_{0}^{n} x\,dx.$

3) Find the values of the integrals:

a) $\displaystyle\int_{-1}^{+1} x^3\,dx.$ b) $\displaystyle\int_{-2}^{2} x^3 \cos x\,dx.$ c) $\displaystyle\int_{-1}^{+1} x^4 \cos^2 x \sin^5 x\,dx.$ d) $\displaystyle\int_{-1}^{+1} \tan x\,dx.$

(Hint: Consider the graphs of the functions under the integral sign, take into account their symmetry with respect to $x = 0$, and interpret the integrals as areas.)

*4) Integrate $\sin x$ and $\cos x$ from 0 to b by substituting $\Delta x = h$ and using the formulas of page 488.

5) Integrate $f(x) = x$ and $f(x) = x^2$ from 0 to b by subdividing into equal parts and by using in (6a) the values $v_j = \frac{1}{2}(x_j + x_{j+1})$.

*6) Using the result (7) and the definition of the integral with equal values of Δx prove the limiting relation:

$$\frac{1^k + 2^k + \cdots + n^k}{n^{k+1}} \to \frac{1}{k+1} \text{ as } n \to \infty.$$

(Hint: Set $\frac{1}{n} = \Delta x$ and show that the limit is equal to $\int_0^1 x^k\,dx$.)

*7) Prove for $n \to \infty$:

$$\frac{1}{\sqrt{n}}\left(\frac{1}{\sqrt{1+n}} + \frac{1}{\sqrt{2+n}} + \cdots + \frac{1}{\sqrt{n+n}}\right) \to 2(\sqrt{2} - 1).$$

(Hint: Write this sum so that its limit appears as an integral.)

8) Calculate the area of a parabolic segment bounded by an arc P_1P_2 and the chord P_1P_2 of a parabola $y = ax^2$ in terms of the coördinates x_1 and x_2 of the two points.

5. Rules for the "Integral Calculus"

An important step in the development of the calculus was taken when certain general rules were formulated by means of which more involved problems could be reduced to simpler ones and thereby be solved by an almost mechanical procedure. This algorithmic feature is particularly emphasized by Leibniz' notation. Still, too much concentration on the mechanics of problem solving can degrade the teaching of the calculus into an empty drill.

Some simple rules for integrals follow at once either from the definition (6) or from the geometrical interpretation of integrals as areas.

The integral of the sum of two functions is equal to the sum of the integrals of the two functions. The integral of a constant c *times a function* f(x) *is* c *times the integral of* f(x). These two rules combined are expressed in the formula

$$\int_a^b [cf(x) + dg(x)]\,dx = c\int_a^b f(x)\,dx + d\int_a^b g(x)\,dx. \tag{9}$$

The proof follows immediately from the definition of the integral as the limit of the finite sum (5), since the corresponding formula for a sum S_n is obviously true. The rule extends immediately to sums of more than two functions.

As an example of the use of this rule we consider a polynomial,

$$f(x) = a_0 + a_1x + a_2x^2 + \cdots + a_nx^n,$$

where the coefficients $a_0, a_1, \cdots, a_n$ are constants. To form the integral of $f(x)$ from a to b, we proceed termwise, according to the rule. Using formula (7) we find

$$\int_a^b f(x)\,dx = a_0(b-a) + a_1\frac{b^2-a^2}{2} + \cdots + a_n\frac{b^{n+1}-a^{n+1}}{n+1}.$$

Another rule, obvious both from the analytic definition and the geometric interpretation, is given by the formula:

$$(10) \qquad \int_a^b f(x)\,dx + \int_b^c f(x)\,dx = \int_a^c f(x)\,dx.$$

Furthermore, it is clear that the integral becomes zero if b is equal to a. The rule of page 405,

$$(11) \qquad \int^b f(x)\,dx = -\int_b^a f(x)\,dx,$$

is in agreement with the last two rules, since it corresponds to (10) for $c = a$.

Sometimes it is convenient to use the fact that the value of the integral in no way depends upon the particular name x chosen for the independent variable in $f(x)$; for example

$$\int_a^b f(x)\,dx = \int_a^b f(u)\,du = \int_a^b f(t)\,dt, \text{ etc.}$$

For a mere change in the name of the coördinates in the system to which the graph of the function refers does not alter the area under the curve. The same remark applies even if we make certain changes in the coördinate system itself. For example, let us shift the origin to the right by one unit from O to O', as in Figure 265, so that x is replaced by a

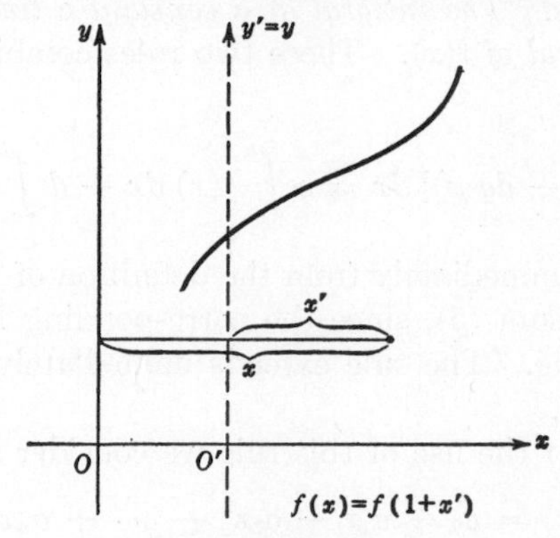

Fig. 265 Shifting of y-axis.

new coördinate x' such that $x = 1 + x'$. A curve with the equation $y = f(x)$ will have in the new coördinate system the equation $y = f(1 + x')$. (E.g. $y = 1/x = 1/(1+x')$.) A given area A under this curve, say between $x = 1$ and $x = b$, is, in the new coördinate system, the area under the arch between $x' = 0$ and $x' = b - 1$. Thus we have

$$\int_1^b f(x)\,dx = \int_0^{b-1} f(1 + x')\,dx',$$

or, changing the name x' to u,

$$\int_1^b f(x)\,dx = \int_0^{b-1} f(1 + u)\,du. \tag{12}$$

For example,

$$\int_1^b \frac{1}{x}\,dx = \int_0^{b-1} \frac{1}{1+u}\,du; \tag{12a}$$

and for the function $f(x) = x^k$,

$$\int_1^b x^k\,dx = \int_0^{b-1} (1 + u)^k\,du. \tag{12b}$$

Similarly,

$$\int_0^b x^k\,dx = \int_{-1}^{b-1} (1 + u)^k\,du \qquad (k \geq 0). \tag{12c}$$

Since the left side of (12c) is equal to $\dfrac{b^{k+1}}{k+1}$, we obtain

$$\int_{-1}^{b-1} (1 + u)^k\,du = \frac{b^{k+1}}{k+1}. \tag{12d}$$

Exercises: 1) Calculate the integral of $1 + x + x^2 + \cdots + x^n$ from 0 to b.

2) For $n > 0$ prove that the integral of $(1 + x)^n$ from -1 to z is equal to

$$\frac{(1 + z)^{n+1}}{(n + 1)}.$$

3) Show that the integral from 0 to 1 of $x^n \sin x$ is smaller than $1/(n + 1)$. (Hint: The latter value is the integral of x^n).

4) Prove directly and by use of the binomial theorem that the integral from -1 to z of $\dfrac{(1 + x)^n}{n}$ is $\dfrac{(1 + z)^{n+1}}{n(n + 1)}$.

Finally we mention two important rules which have the form of inequalities. These rules permit rough, but useful, appraisals of the values of integrals.

We suppose that $b > a$ and that the values of $f(x)$ in the interval nowhere exceed those of another function $g(x)$. Then we have

$$(13) \qquad \int_a^b f(x)\,dx \leq \int_a^b g(x)\,dx,$$

as is immediately clear either from Figure 266 or from the analytic defini-

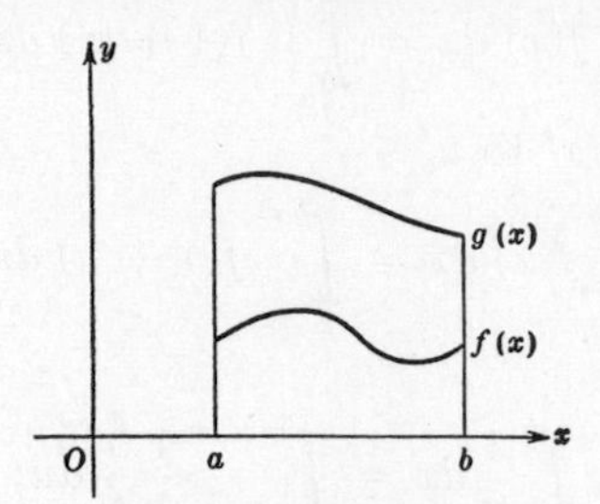

Fig. 266. Comparison of integrals.

tion of the integral. In particular, if $g(x) = M$ is a constant not exceeded by the values of $f(x)$, we have $\int_a^b g(x)\,dx = \int_a^b M\,dx = M(b - a)$. It follows that

$$(14) \qquad \int_a^b f(x)\,dx \leq M(b - a).$$

If $f(x)$ is not negative, then $f(x) = |f(x)|$. If $f(x) < 0$, then $|f(x)| > f(x)$. Hence, setting $g(x) = |f(x)|$ in (13), we obtain the useful formula

$$(15) \qquad \int_a^b f(x)\,dx \leq \int_a^b |f(x)|\,dx.$$

Since $|-f(x)| = |f(x)|$, we also have

$$-\int_a^b f(x)\,dx \leq \int_a^b |f(x)|\,dx,$$

which, together with (15), yields the somewhat stronger inequality

$$(16) \qquad \left|\int_a^b f(x)\,dx\right| \leq \int_a^b |f(x)|\,dx.$$

§2. THE DERIVATIVE

1. The Derivative as a Slope

While the concept of integral has its roots in antiquity, the other basic concept of the calculus, the derivative, was formulated only in the seventeenth century by Fermat and others. It was the discovery by Newton and Leibniz of the organic interrelation between these two seemingly quite diverse concepts that inaugurated an unparalleled development of mathematical science.

Fermat was interested in determining the maxima and minima of a function $y = f(x)$. In a graph of the function, a maximum corresponds to a summit higher than all other neighboring points, while a minimum corresponds to the bottom of a valley lower than all neighboring points. In Figure 191 on page 342 the point B is a maximum and the point C a minimum. To characterize the points of maximum and minimum it is natural to use the notion of *tangent* of a curve. We assume that the graph has no sharp corners or other singularities, and that at every point it possesses a definite direction given by a tangent line. At maximum or minimum points the tangent of the graph $y = f(x)$ must be parallel to the x-axis, since otherwise the curve would be rising or falling at these points. This remark suggests the idea of considering quite generally, at any point P of the graph $y = f(x)$, the direction of the tangent to the curve.

To characterize the direction of a straight line in the x, y-plane it is customary to give its *slope*, which is the trigonometrical tangent of the angle α from the direction of the positive x-axis to the line. If P is any point of the line L, we proceed to the right to a point R and then up or down to the point Q on the line; then slope of $L = \tan \alpha = \dfrac{RQ}{PR}$. The length PR is taken as positive, while RQ is taken as positive or negative according as the direction from R to Q is up or down, so that the slope gives the rise or fall per unit length along the horizontal when we proceed along the line from left to right. In Figure 267 the slope of the first line is $2/3$, while the slope of the second line is -1.

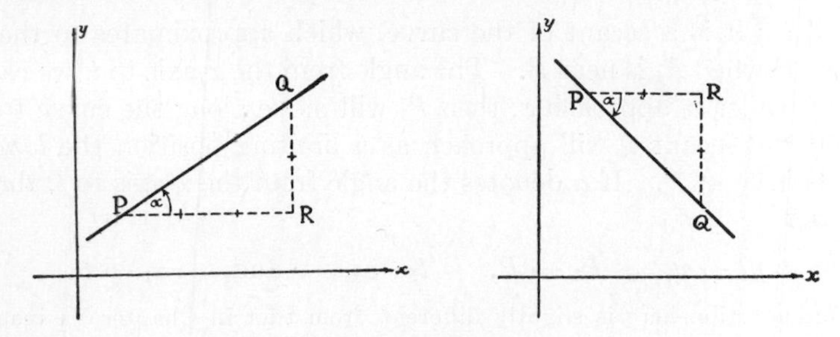

Fig. 267. Slopes of lines.

By the slope of a *curve* at a point P we mean the slope of the tangent to the curve at P. As long as we accept the tangent of a curve as an intuitively given mathematical concept there remains only the problem of *finding a procedure for calculating the slope.* For the moment we shall accept this point of view, postponing to the supplement a closer analysis of the problems involved.

2. The Derivative as a Limit

The slope of a curve $y = f(x)$ at the point P (x, y) cannot be calculated by referring to the curve at the point P alone. Instead, one must resort to a limiting process much like that involved in the calculation of the area under a curve. This limiting process is the basis of the differential calculus. We consider on the curve another point P_1, near P, with coördinates x_1, y_1. The straight line joining P to P_1

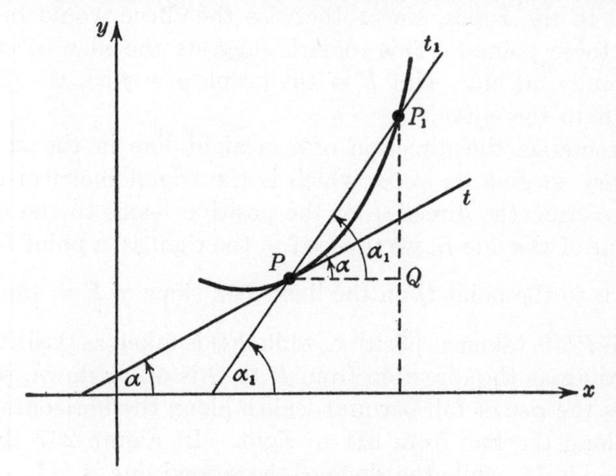

Fig. 268. The derivative as a limit.

we call t_1; it is a secant of the curve, which approximates to the tangent at P when P_1 is near P. The angle from the x-axis to t_1 we call α_1. Now if we let x_1 approach x, then P_1 will move along the curve toward P, and the secant t_1 will approach as a limiting position the *tangent* t to the curve at P. If α denotes the angle from the x-axis to t, then, as $x_1 \to x$,†

$$y_1 \to y, \qquad P_1 \to P, \qquad t_1 \to t, \qquad \text{and} \qquad \alpha_1 \to \alpha.$$

† Our notation here is slightly different from that in Chapter VI inasmuch as there we have $x \to x_1$, the latter value being fixed. No confusion should arise from this interchange of symbols.

The tangent is the limit of the secant, and the slope of the tangent is the limit of the slope of the secant.

Although we have no explicit expression for the slope of the tangent t itself, the slope of the secant t_1 is given by the formula

$$\text{slope of } t_1 = \frac{y_1 - y}{x_1 - x} = \frac{f(x_1) - f(x)}{x_1 - x},$$

or, if we again denote the operation of forming a difference by the symbol Δ,

$$\text{slope of } t_1 = \frac{\Delta y}{\Delta x} = \frac{\Delta f(x)}{\Delta x}.$$

The slope of the secant t_1 is a "difference quotient"—the difference Δy of the function values, divided by the difference Δx of the values of the independent variable. Moreover,

$$\text{slope of } t = \text{limit of slope of } t_1 = \lim \frac{f(x_1) - f(x)}{x_1 - x} = \lim \frac{\Delta y}{\Delta x},$$

where the limits are evaluated as $x_1 \to x$, i.e. as $\Delta x = x_1 - x \to 0$. *The slope of the tangent* t *to the curve is the limit of the difference quotient* $\Delta y/\Delta x$ *as* $\Delta x = x_1 - x$ *approaches zero.*

The original function $f(x)$ gave the *height* of the curve $y = f(x)$ for the value x. We may now consider the *slope* of the curve for a variable point P with the coördinates x and y [$= f(x)$] as a new function of x which we denote by $f'(x)$ and call the *derivative* of the function $f(x)$. The limiting process by which it is obtained is called *differentiation* of $f(x)$. This process is an operation which attaches to a given function $f(x)$ another function $f'(x)$ according to a definite rule, just as the function $f(x)$ is defined by a rule which attaches to any value of the variable x the value $f(x)$:

$$f(x) = \text{height of curve } y = f(x) \text{ at the point } x,$$

$$f'(x) = \text{slope of curve } y = f(x) \text{ at the point } x.$$

The word "differentiation" comes from the fact that $f'(x)$ is the limit of the difference $f(x_1) - f(x)$ divided by the difference $x_1 - x$:

$$(1) \qquad f'(x) = \lim \frac{f(x_1) - f(x)}{x_1 - x} \quad \text{as} \quad x_1 \to x.$$

Another notation, often useful, is

$$f'(x) = Df(x),$$

the "D" simply abbreviating "derivative of"; still different is Leibniz' notation for the derivative of $y = f(x)$,

$$\frac{dy}{dx} \quad \text{or} \quad \frac{df(x)}{dx},$$

which we shall discuss in §4 and which indicates the character of the derivative as limit of the difference quotient $\Delta y/\Delta x$ or $\Delta f(x)/\Delta x$.

If we describe the curve $y = f(x)$ in the direction of increasing values of x, then a *positive derivative*, $f'(x) > 0$, at a point means *ascending curve* (increasing values of y), a *negative derivative*, $f'(x) < 0$, means *descending curve*, while $f'(x) = 0$ means a horizontal direction of the curve for the value x. At a maximum or minimum, the slope must be zero (Fig. 269).

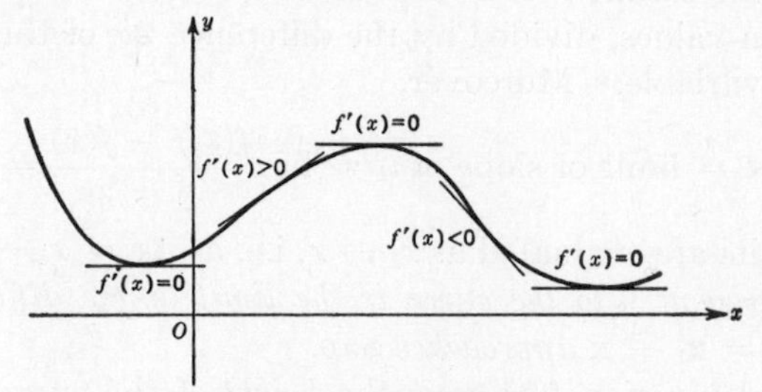

Fig. 269. The sign of the derivative.

Hence, by solving the equation

$$f'(x) = 0$$

for x we may find the positions of the maxima and minima, as was first done by Fermat.

3. Examples

The considerations leading to the definition (1) might seem to be without practical value. One problem has been replaced by another: instead of being asked to find the slope of the tangent to a curve $y = f(x)$ at a point, we are asked to evaluate a limit, (1), which at first sight appears equally difficult. But as soon as we leave the domain of generalities and consider specific functions $f(x)$ we shall obtain tangible results.

The simplest such function is $f(x) = c$, where c is a constant. The graph of the function $y = f(x) = c$ is a horizontal line coinciding with all its tangents, and it is obvious that

$$f'(x) = 0$$

for all values of x. This also follows from the definition (1), for

$$\frac{\Delta y}{\Delta x} = \frac{f(x_1) - f(x)}{x_1 - x} = \frac{c - c}{x_1 - x} = \frac{0}{x_1 - x} = 0,$$

so that, trivially,

$$\lim \frac{f(x_1) - f(x)}{x_1 - x} = 0 \quad \text{as} \quad x_1 \to x.$$

Next we consider the simple function $y = f(x) = x$, whose graph is a straight line through the origin bisecting the first quadrant. Geometrically it is clear that

$$f'(x) = 1$$

for all values of x, and the analytic definition (1) again yields

$$\frac{f(x_1) - f(x)}{x_1 - x} = \frac{x_1 - x}{x_1 - x} = 1,$$

so that

$$\lim \frac{f(x_1) - f(x)}{x_1 - x} = 1 \quad \text{as} \quad x_1 \to x.$$

The simplest non-trivial example is the differentiation of the function

$$y = f(x) = x^2,$$

which amounts to finding the slope of a parabola. This is the simplest case that teaches us how to carry out the passage to the limit when the result is not obvious from the outset. We have

$$\frac{\Delta y}{\Delta x} = \frac{f(x_1) - f(x)}{x_1 - x} = \frac{x_1^2 - x^2}{x_1 - x}.$$

If we should try to pass to the limit directly in numerator and denominator we should obtain the meaningless expression 0/0. But we can avoid this impasse by rewriting the difference quotient and cancelling, *before passing to the limit*, the disturbing factor $x_1 - x$. (In evaluating the limit of the difference quotient we consider only values $x_1 \neq x$, so that this is permissible; see p. 307.) Thus we obtain the expression:

$$\frac{x_1^2 - x^2}{x_1 - x} = \frac{(x_1 - x)(x_1 + x)}{x_1 - x} = x_1 + x.$$

Now, *after* the cancellation, there is no longer any difficulty with the limit as $x_1 \to x$. The limit is obtained "by substitution"; for the new form $x_1 + x$ of the difference quotient is continuous and the limit of a continuous function as $x_1 \to x$ is simply the value of the function for $x_1 = x$, in our case $x + x = 2x$, so that

$$f'(x) = 2x \quad \text{for} \quad f(x) = x^2.$$

In a similar way we can prove that for $f(x) = x^3$ we have $f'(x) = 3x^2$. For the difference quotient,

$$\frac{\Delta y}{\Delta x} = \frac{f(x_1) - f(x)}{x_1 - x} = \frac{x_1^3 - x^3}{x_1 - x},$$

can be simplified by the formula $x_1^3 - x^3 = (x_1 - x)(x_1^2 + x_1x + x^2)$; the denominator $\Delta x = x_1 - x$ cancels out, and we obtain the continuous expression

$$\frac{\Delta y}{\Delta x} = x_1^2 + x_1x + x^2.$$

Now if we let x_1 approach x, this expression simply approaches $x^2 + x^2 + x^2$, and we obtain as limit $f'(x) = 3x^2$.

In general, for

$$f(x) = x^n,$$

where n is any positive integer, we obtain the derivative

$$f'(x) = nx^{n-1}.$$

Exercise: Prove this result. (Use the algebraic formula

$$x_1^n - x^n = (x_1 - x)(x_1^{n-1} + x_1^{n-2}x + x_1^{n-3}x^2 + \cdots + x_1x^{n-2} + x^{n-1}).)$$

As a further example of simple devices that permit explicit determination of the derivative we consider the function

$$y = f(x) = \frac{1}{x}.$$

We have

$$\frac{\Delta y}{\Delta x} = \frac{y_1 - y}{x_1 - x} = \left(\frac{1}{x_1} - \frac{1}{x}\right)\cdot\frac{1}{x_1 - x} = \frac{x - x_1}{x_1x}\cdot\frac{1}{x_1 - x}.$$

Again we may cancel, and we find $\dfrac{\Delta y}{\Delta x} = -\dfrac{1}{x_1x}$, which is continuous at $x_1 = x$; hence we have in the limit

$$f'(x) = -\frac{1}{x^2}.$$

Of course, neither the derivative nor the function itself is defined for $x = 0$.

Exercises: Prove in a similar manner that for $f(x) = \frac{1}{x^2}$, $f'(x) = -\frac{2}{x^3}$; for $f(x) = \frac{1}{x^n}$, $f'(x) = -\frac{n}{x^{n+1}}$; for $f(x) = (1 + x)^n$, $f'(x) = n(1 + x)^{n-1}$.

We shall now carry out the differentiation of

$$y = f(x) = \sqrt{x}.$$

For the difference quotient we obtain

$$\frac{y_1 - y}{x_1 - x} = \frac{\sqrt{x_1} - \sqrt{x}}{x_1 - x}.$$

By the formula $x_1 - x = (\sqrt{x_1} - \sqrt{x})(\sqrt{x_1} + \sqrt{x})$ we can cancel a factor and get the continuous expression

$$\frac{y_1 - y}{x_1 - x} = \frac{1}{\sqrt{x_1} + \sqrt{x}}.$$

Passing to the limit yields

$$f'(x) = \frac{1}{2\sqrt{x}}.$$

Exercises: Prove that for $f(x) = \frac{1}{\sqrt{x}}$, $f'(x) = -\frac{1}{2(\sqrt{x})^3}$; for $f(x) = \sqrt[3]{x}$, $f'(x) = \frac{1}{3\sqrt[3]{x^2}}$; for $f(x) = \sqrt{1 - x^2}$, $f'(x) = \frac{-x}{\sqrt{1 - x^2}}$; for $f(x) = \sqrt[n]{x}$, $f'(x) = \frac{1}{n\sqrt[n]{x^{n-1}}}$.

4. Derivatives of Trigonometrical Functions

We now treat the very important question of the *differentiation of trigonometrical functions.* Here radian measure of angles will be used exclusively.

To differentiate the function $y = f(x) = \sin x$ we set $x_1 - x = h$, so that $x_1 = x + h$ and $f(x_1) = \sin x_1 = \sin (x + h)$. By the trigonometrical formula for $\sin (A + B)$,

$$f(x_1) = \sin (x + h) = \sin x \cos h + \cos x \sin h.$$

Hence

$$\begin{aligned}\frac{f(x_1) - f(x)}{x_1 - x} &= \frac{\sin (x + h) - \sin x}{h} \\ &= \cos x\left(\frac{\sin h}{h}\right) + \sin x\left(\frac{\cos h - 1}{h}\right).\end{aligned} \tag{2}$$

If now we let x_1 tend to x, then h tends to 0, sin h to 0, and cos h to 1. Moreover, by the results of page 308,

$$\lim \frac{\sin h}{h} = 1$$

and

$$\lim \frac{\cos h - 1}{h} = 0.$$

Hence the right side of (2) approaches cos x, giving the result:

The function $f(x) = \sin x$ *has the derivative* $f'(x) = \cos x$, or briefly,

$$D \sin x = \cos x.$$

Exercise: Prove that $D \cos x = -\sin x$.

To differentiate the function $f(x) = \tan x$, we write $\tan x = \dfrac{\sin x}{\cos x}$, and obtain

$$\begin{aligned}\frac{f(x+h) - f(x)}{h} &= \left(\frac{\sin (x+h)}{\cos (x+h)} - \frac{\sin x}{\cos x}\right)\frac{1}{h} \\ &= \frac{\sin (x+h) \cos x - \cos (x+h) \sin x}{h} \frac{1}{\cos (x+h) \cos x} \\ &= \frac{\sin h}{h} \frac{1}{\cos (x+h) \cos x}.\end{aligned}$$

(The last equality follows from the formula $\sin (A - B) = \sin A \cos B - \cos A \sin B$, with $A = x + h$ and $B = h$.) If now we let h approach zero, $\dfrac{\sin h}{h}$ approaches 1, $\cos (x + h)$ approaches $\cos x$, and we infer:

The derivative of the function $f(x) = \tan x$ *is* $f'(x) = \dfrac{1}{\cos^2 x}$, or

$$D \tan x = \frac{1}{\cos^2 x}.$$

Exercise: Prove that $D \cot x = -\dfrac{1}{\sin^2 x}$.

*5. Differentiation and Continuity

The differentiability of a function implies its continuity. For, if the limit of $\Delta y/\Delta x$ exists as Δx tends to zero, then it is easy to see that the change Δy of the function $f(x)$ must become arbitrarily small as the difference Δx tends to zero. Hence whenever we can differentiate a function, its continuity is automatically assured; we shall therefore dispense with explicitly mentioning or proving the continuity of the differentiable functions occurring in this chapter unless there is a particular reason for it.

6. Derivative and Velocity. Second Derivative and Acceleration

The preceding discussion of the derivative was carried out in connection with the geometrical concept of the graph of a function. But the significance of the derivative concept is by no means limited to the problem of finding the slope of the tangent to a curve. Even more important in the natural sciences is the problem of calculating the *rate of change* of some quantity $f(t)$ which varies with the time t. It was from this angle that Newton made his approach to the differential calculus. Newton wished in particular to analyze the phenomenon of velocity, where the time and the position of a moving particle are considered as the variable elements, or, as Newton expressed it, as the "fluent quantities."

If a particle moves along a straight line, the x-axis, its motion is completely described by giving the position x at any time t as a function $x = f(t)$. A "uniform motion" with constant velocity b along the x-axis is defined by a linear function $x = a + bt$, where a is the coördinate of the particle at the time $t = 0$.

In a plane the motion of a particle is described by two functions,

$$x = f(t), \qquad y = g(t),$$

characterizing the two coördinates as functions of the time. In particular, a uniform motion corresponds to a pair of linear functions,

$$x = a + bt, \qquad y = c + dt,$$

where b and d are the two "components" of the constant velocity, and a and c the coördinates of the particle at the moment $t = 0$; the path of the particle is a straight line with the equation $(x - a)d - (y - c)b = 0$, obtained by eliminating the time t from the two relations above.

If a particle moves in the vertical x, y-plane under the influence of gravity alone, then, as shown in elementary physics, the motion is described by two equations,

$$x = a + bt, \qquad y = c + dt - \tfrac{1}{2}gt^2,$$

where a, b, c, d are constants depending on the initial state of the particle and g the acceleration due to gravity, approximately equal to 32 if time is measured in seconds and distance in feet. The trajectory of the particle, obtained by eliminating t from the two equations, is now a parabola,

$$y = c + \frac{d}{b}(x - a) - \tfrac{1}{2}g\frac{(x - a)^2}{b^2},$$

if $b \neq 0$; otherwise it is a part of the vertical axis.

If a particle is confined to move along a given curve in the plane (like a train along a track), then its motion may be described by giving the arc length s, measured along the curve from a fixed initial point P_0 to the position P of the particle at the time t, as a function of t; $s = f(t)$. For example, on the unit circle $x^2 + y^2 = 1$ the function $s = ct$ describes a uniform rotation with the velocity c along the circle.

Exercises: *Draw the trajectories of the plane motion described by 1) $x = \sin t$, $y = \cos t$. 2) $x = \sin 2t$, $y = \sin 3t$. 3) $x = \sin 2t$, $y = 2 \sin 3t$.

4) In the parabolic motion described above, suppose the particle at the origin for $t = 0$, and $b > 0$, $d > 0$. Find the coördinates of the highest point of the trajectory. Find the time t and the value of x for the second intersection of the trajectory with the x-axis.

Newton's first aim was to determine the velocity of a non-uniform motion. For simplicity let us consider the motion of a particle along a straight line given by a function $x = f(t)$. If the motion were uniform, with constant velocity, then the velocity could be found by taking two values t and t_1 of the time, with corresponding values $x = f(t)$ and $x_1 = f(t_1)$ of the position, and forming the quotient

$$v = \text{velocity} = \frac{\text{distance}}{\text{time}} = \frac{x_1 - x}{t_1 - t} = \frac{f(t_1) - f(t)}{t_1 - t}.$$

For example, if t is measured in hours and x in miles, then, for $t_1 - t = 1$, $x_1 - x$ will be the number of miles covered in 1 hour and v will be the velocity in miles per hour. The statement that the velocity of the motion is constant simply means that the difference quotient

$$\frac{f(t_1) - f(t)}{t_1 - t} \tag{3}$$

is the same for all values of t and t_1. But when the motion is not uniform, as in the case of a freely falling body whose velocity increases as it falls, then the quotient (3) does not give the velocity *at the instant t*, but merely the *average velocity* during the time interval from t to t_1.

To obtain the velocity at the exact instant t we must take the limit of the average velocity as t_1 approaches t. Thus we define with Newton

$$(4) \qquad \text{velocity at the instant } t = \lim \frac{f(t_1) - f(t)}{t_1 - t} = f'(t).$$

In other words, the velocity is the derivative of the distance coördinate with respect to the time, or the "instantaneous rate of change" of the distance with respect to the time (as distinguished from the *average* rate of change given by (3)).

The *rate of change of the velocity* itself is called the *acceleration*. It is simply the derivative of the derivative, usually denoted by $f''(t)$, and called the *second derivative* of $f(t)$.

It was observed by Galileo that for a freely falling body the vertical distance x through which the body falls during the time t is given by the formula

$$(5) \qquad x = f(t) = \tfrac{1}{2}gt^2,$$

where g is the gravitational constant. It follows by differentiating (5) that the velocity v of the body at the time t is given by

$$(6) \qquad v = f'(t) = gt,$$

and the acceleration α by

$$\alpha = f''(t) = g,$$

which is constant.

Suppose it is required to find the velocity of the body 2 seconds after it has been released. The *average* velocity during the time interval from $t = 2$ to $t = 2.1$ is

$$\frac{\frac{1}{2}g(2.1)^2 - \frac{1}{2}g(2)^2}{2.1 - 2} = \frac{16(0.41)}{0.1} = 65.6 \text{ (feet per second)}.$$

But substituting $t = 2$ in (6) we find the *instantaneous* velocity at the end of two seconds to be $v = 64$.

Exercise: What is the average velocity of the body during the time interval from $t = 2$ to $t = 2.01$? from $t = 2$ to $t = 2.001$?

For motion in the plane the two derivatives $f'(t)$ and $g'(t)$ of the functions $x = f(t)$ and $y = g(t)$ define the components of the velocity. For motion along a fixed curve the velocity will be defined by the derivative of the function $s = f(t)$, where s is the arc length.

7. Geometrical Meaning of the Second Derivative

The second derivative is also important in analysis and geometry, for $f''(x)$, expressing the rate of change of the slope $f'(x)$ of the curve $y = f(x)$, gives an indication of the way the curve is bent. If $f''(x)$ is positive in an interval then the rate of change of $f'(x)$ is positive. A positive rate of change of a function means that the values of the function increase as x increases. Therefore $f''(x) > 0$ means that the slope $f'(x)$ increases as x increases, so that the curve becomes steeper where it has a positive slope and less steep where it has a negative slope. We say that the curve is *concave upward* (Fig. 270).

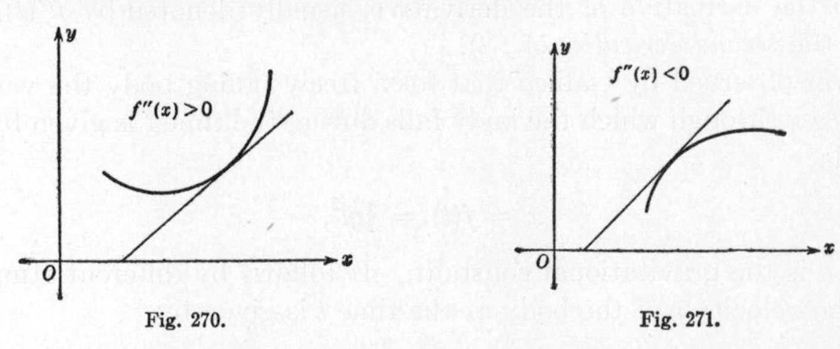

Fig. 270. Fig. 271.

Similarly, if $f''(x) < 0$, the curve $y = f(x)$ is *concave downward* (Fig. 271).

The parabola $y = f(x) = x^2$ is concave upward everywhere because $f''(x) = 2$ is always positive. The curve $y = f(x) = x^3$ is concave upward for $x > 0$ and concave downward for $x < 0$ (Fig. 153) because $f''(x) = 6x$, as the reader can easily prove. Incidentally, for $x = 0$ we have $f'(x) = 3x^2 = 0$ (but no maximum or minimum!); also $f''(x) = 0$ for $x = 0$. This point is called a *point of inflection.* At such a point the tangent, in this case the x-axis, crosses the curve.

If s denotes the arc-length along the curve, and α the slope-angle, then $\alpha = h(s)$ will be a function of s. As we travel along the curve $\alpha = h(s)$ will change. The rate of change $h'(s)$ is called the *curvature* of the curve at the point where the arc length is s. We mention without proof that the curvature κ can be expressed in terms of the first and second derivatives of the function $y = f(x)$ defining the curve:

$$\kappa = f''(x) \;/\; (1 + (f'(x))^2)^{3/2}.$$

8. Maxima and Minima

We can find the maxima and minima of a given function $f(x)$ by first forming $f'(x)$, then finding the values for which this derivative vanishes, and finally investigating which of these values furnish maxima and which minima. The latter question can be decided if we form the second derivative, $f''(x)$, whose sign indicates the convex or concave shape of the graph and whose vanishing usually indicates a point of inflection at which no extremum occurs. By observing the signs of $f'(x)$ and $f''(x)$ we can not only determine the extrema but also find the shape of the graph of the function. This method gives us the values of x for which extrema occur; to find the corresponding values of $y = f(x)$ itself we have to substitute these values of x in $f(x)$.

As an example we consider the polynomial

$$f(x) = 2x^3 - 9x^2 + 12x + 1,$$

and obtain

$$f'(x) = 6x^2 - 18x + 12, \qquad f''(x) = 12x - 18.$$

The roots of the quadratic equation $f'(x) = 0$ are $x_1 = 1$, $x_2 = 2$, and we have $f''(x_1) = -6 < 0$, $f''(x_2) = 6 > 0$. Hence $f(x)$ has a maximum, $f(x_1) = 6$, and a minimum, $f(x_2) = 5$.

Exercises: 1) Sketch the graph of the function considered above.

2) Discuss and sketch the graph of $f(x) = (x^2 - 1)(x^2 - 4)$.

3) Find the minimum of $x + 1/x$, of $x + a^2/x$, of $px + q/x$, where p and q are positive. Have these functions maxima?

4) Find the maxima and minima of $\sin x$ and $\sin(x^2)$.

§3. THE TECHNIQUE OF DIFFERENTIATION

Until now our efforts have been devoted to differentiating a variety of specific functions by transforming the difference quotients in preparation for passage to the limit. It was a decisive step when, through the work of Leibniz, Newton, and their successors, these individual devices were replaced by powerful general methods. By these methods one can differentiate almost automatically any function that normally occurs in mathematics, provided one has mastered a few simple rules and can recognize their applicability. Thus differentiation has acquired the character of an "algorithm" of calculation, and it is this aspect of the theory that is expressed by the term "calculus."

We cannot go far into the details of this technique. Only a few simple rules will be mentioned.

(a) *Differentiation of a sum.* If a and b are constants and the function $k(x)$ is given by

$$k(x) = af(x) + bg(x),$$

then, as the reader can easily verify,

$$k'(x) = af'(x) + bg'(x).$$

A similar rule holds for any number of terms.

(b) *Differentiation of a product.* For a product,

$$p(x) = f(x)g(x),$$

the derivative is

$$p'(x) = f(x)g'(x) + g(x)f'(x).$$

This is easily proved by the following device: we write, adding and subtracting the same term,

$$\begin{aligned} p(x + h) - p(x) &= f(x + h)g(x + h) - f(x)g(x) \\ &= f(x + h)g(x + h) - f(x + h)g(x) + f(x + h)g(x) - f(x)g(x), \end{aligned}$$

and obtain, by combining the first two and the second two terms,

$$\frac{p(x + h) - p(x)}{h} = f(x + h)\,\frac{g(x + h) - g(x)}{h} + g(x)\,\frac{f(x + h) - f(x)}{h}.$$

Now we let h approach zero; since $f(x + h)$ approaches $f(x)$, the statement to be proved follows immediately.

Exercise: Prove that the function $p(x) = x^n$ has the derivative $p'(x) = nx^{n-1}$. (Hint: Write $x^n = x \cdot x^{n-1}$ and use mathematical induction.)

Using rules (a) and (b) we can differentiate any polynomial

$$f(x) = a_0 + a_1x + \cdots + a_nx^n;$$

the derivative is

$$f'(x) = a_1 + 2a_2x + 3a_3x^2 + \cdots + na_nx^{n-1}.$$

As an application we may prove the *binomial theorem* (compare p. 17. This theorem concerns the expansion of $(1 + x)^n$ as a polynomial:

$$(1) \qquad f(x) = (1 + x)^n = 1 + a_1x + a_2x^2 + a_3x^3 + \cdots + a_nx^n,$$

and states that the coefficient a_k is given by the formula

$$(2) \qquad a_k = \frac{n(n-1)\cdots(n-k+1)}{k!}.$$

Of course, $a_n = 1$.

We have seen (Exercise, p. 421) that the left side of (1) differentiated yields $n(1 + x)^{n-1}$. Thus by the preceding paragraph we obtain

$$(3) \qquad n(1 + x)^{n-1} = a_1 + 2a_2x + 3a_3x^2 + \cdots + na_nx^{n-1}.$$

In this formula we now set $x = 0$ and find that $n = a_1$, which is (2) for $k = 1$. Then we differentiate (3) again, obtaining

$$n(n-1)(1+x)^{n-2} = 2a_2 + 3 \cdot 2a_3 x + \cdots + n(n-1)a_n x^{n-2}.$$

Substituting $x = 0$, we find $n(n-1) = 2a_2$ in agreement with (2) for $k = 2$.

Exercise: Prove (2) for $k = 3, 4$, and for general k by mathematical induction

(c) *Differentiation of a quotient.* If

$$q(x) = \frac{f(x)}{g(x)},$$

then

$$q'(x) = \frac{g(x)f'(x) - f(x)g'(x)}{(g(x))^2}.$$

The proof is left as an exercise. (Of course, we must assume $g(x) \neq 0$.)

Exercise: Derive by this rule the formulas of page 422 for the derivatives of $\tan x$ and $\cot x$ from those for $\sin x$ and $\cos x$. Prove that the derivatives of $\sec x = 1/\cos x$ and $\operatorname{cosec} x = 1/\sin x$ are $\sin x/\cos^2 x$ and $-\cos x/\sin^2 x$ respectively.

We are now able to differentiate any function that can be written as the quotient of two polynomials. For example,

$$f(x) = \frac{1-x}{1+x}$$

has the derivative

$$f'(x) = \frac{-(1+x)-(1-x)}{(1+x)^2} = -\frac{2}{(1+x)^2}.$$

Exercise: Differentiate

$$f(x) = \frac{1}{x^m} = x^{-m},$$

where m is a positive integer. The result is

$$f'(x) = -mx^{-m-1}.$$

(d) *Differentiation of inverse functions.* If

$$y = f(x) \quad \text{and} \quad x = g(y)$$

are inverse functions (e.g. $y = x^2$ and $x = \sqrt{y}$), then their derivatives are reciprocal:

$$g'(y) = \frac{1}{f'(x)} \quad \text{or} \quad Dg(y) \cdot Df(x) = 1.$$

This fact is easily proved by going back to the reciprocal difference quotients $\frac{\Delta y}{\Delta x}$ and $\frac{\Delta x}{\Delta y}$ respectively; it can also be seen from the geometrical interpretation of the inverse function given on page 281, if we refer the slope of the tangent to the y-axis instead of to the x-axis.

As an example we differentiate the function

$$y = f(x) = \sqrt[m]{x} = x^{\frac{1}{m}}$$

inverse to $x = y^m$. (See also the more direct treatment for $m = \frac{1}{2}$ on p. 421.) Since the latter function has as its derivative the expression my^{m-1}, we have

$$f'(x) = \frac{1}{my^{m-1}} = \frac{1}{m}\frac{y}{y^m} = \frac{1}{m}yy^{-m},$$

whence, after substituting $y = x^{\frac{1}{m}}$ and $y^{-m} = x^{-1}$, $f'(x) = \frac{1}{m}x^{\frac{1}{m}-1}$, or

$$D(x^{1/m}) = \frac{1}{m}x^{\frac{1}{m}-1}.$$

As a further example we differentiate the *inverse trigonometric function* (see page 281):

$$y = \text{arc tan } x, \text{ which means the same as } x = \tan y.$$

Here the variable y, denoting the radian measure, is restricted to the interval $-\frac{1}{2}\pi < y < \frac{1}{2}\pi$ so as to insure a unique definition of the inverse function.

Since we have (see page 422) $D \tan y = 1/\cos^2 y$ and since $1/\cos^2 y = (\sin^2 y + \cos^2 y)/\cos^2 y = 1 + \tan^2 y = 1 + x^2$, we find:

$$D \text{ arc tan } x = \frac{1}{1 + x^2}.$$

In the same way the reader may derive the following formulas:

$$D \text{ arc cot } x = -\frac{1}{1 + x^2}$$

$$D \text{ arc sin } x = \frac{1}{\sqrt{1 - x^2}}$$

$$D \text{ arc cos } x = -\frac{1}{\sqrt{1 - x^2}}.$$

Finally, we come to the important rule for

(e) *Differentiation of compound functions.* Such functions are compounded from two (or more) simpler ones (see p. 282). For example, $z = \sin(\sqrt{x})$ is compounded from $z = \sin y$ and $y = \sqrt{x}$; the function $z = \sqrt{x} + \sqrt{x^5}$ is compounded from $z = y + y^5$ and $y = \sqrt{x}$; $z =$

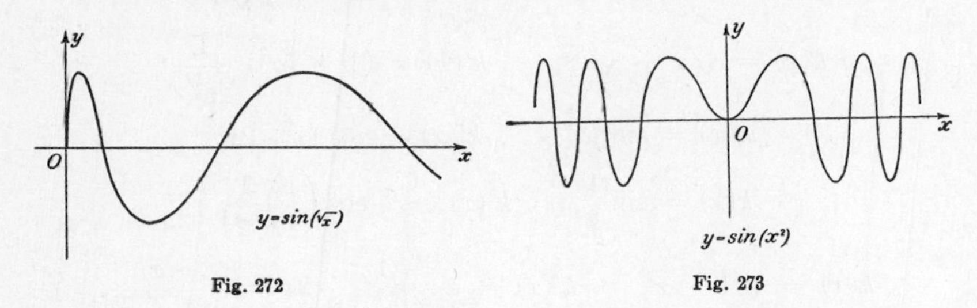

Fig. 272 Fig. 273

$\sin(x^2)$ is a compound of $z = \sin y$ and $y = x^2$; $z = \sin \frac{1}{x}$ is a compound of $z = \sin y$ and $y = \frac{1}{x}$.

If two functions

$$z = g(y) \quad \text{and} \quad y = f(x)$$

are given, and if the latter function is substituted in the former, we obtain the compound function

$$z = k(x) = g[f(x)].$$

We assert that

$$(4) \qquad k'(x) = g'(y)f'(x).$$

For if we write

$$\frac{k(x_1) - k(x)}{x_1 - x} = \frac{z_1 - z}{y_1 - y} \cdot \frac{y_1 - y}{x_1 - x},$$

where $y_1 = f(x_1)$ and $z_1 = g(y_1) = k(x_1)$, and then let x_1 approach x, the left side approaches $k'(x)$ and the two factors on the right hand side approach $g'(y)$ and $f'(x)$ respectively, thus proving (4).

In this proof the condition $y_1 - y \neq 0$ was necessary. For we divided by $\Delta y = y_1 - y$, and we cannot use values x_1 for which $y_1 - y = 0$. But the formula (4) remains valid even if $\Delta y = 0$ in an interval around x; y is then constant, $f'(x) = 0$, $k(x) = g(y)$ is constant with respect to x (since y does not change with x), and hence $k'(x) = 0$, as (4) states in this case.

The reader should verify the following examples:

$$k(x) = \sin \sqrt{x}, \qquad k'(x) = (\cos \sqrt{x}) \frac{1}{2\sqrt{x}},$$

$$k(x) = \sqrt{x} + \sqrt{x^5}, \qquad k'(x) = (1 + 5x^2) \frac{1}{2\sqrt{x}},$$

$$k(x) = \sin (x^2), \qquad k'(x) = \cos (x^2) \cdot 2x,$$

$$k(x) = \sin \frac{1}{x}, \qquad k'(x) = -\cos \left(\frac{1}{x}\right) \frac{1}{x^2},$$

$$k(x) = \sqrt{1 - x^2}, \qquad k'(x) = \frac{-1}{2\sqrt{1 - x^2}} \cdot 2x = \frac{-x}{\sqrt{1 - x^2}}.$$

Exercise: Combining the results of pages 420 and 430, show that the function

$$f(x) = \sqrt[m]{x^s} = x^{\frac{s}{m}}$$

has the derivative

$$f'(x) = \frac{s}{m} x^{\frac{s}{m} - 1}.$$

It should be noted that all our formulas concerning powers of x can now be combined into a single one:

If r *is any positive or negative rational number, then the function*

$$f(x) = x^r$$

has the derivative

$$f'(x) = rx^{r-1}.$$

Exercises: 1) Carry out the differentiations of the exercises on page 421 by using the rules of this section.

2) Differentiate the following functions: $x \sin x$, $\frac{1}{1 + x^2} \sin nx$, $(x^3 - 3x^2 - x + 1)^3$, $1 + \sin^2 x$, $x^2 \sin \frac{1}{x^2}$, $\arcsin (\cos nx)$, $\tan \frac{1 + x}{1 - x}$, $\arctan \frac{1 + x}{1 - x}$, $\sqrt[4]{1 - x^2}$, $\frac{1}{1 + x^2}$.

3) Find the second derivatives of some of the preceding functions and of $\frac{1 - x}{1 + x}$, $\arctan x$, $\sin^2 x$, $\tan x$.

4) Differentiate $c_1(x - x_1)^2 + y_1^2 + c_2(x - x_2)^2 + y_2^2$, *and prove the minimum properties of the light ray by reflection and by refraction stated in Chapter VII, pp. 330 and 382. The reflection or refraction is to be in the x-axis, and the coördinates of the endpoints of the path may be x_1, y_1 and x_2, y_2 respectively.

(Remark: The function possesses only one point with vanishing derivative; therefore, since a minimum but obviously no maximum occurs, there is no need to study the second derivative.)

More Problems on Maxima and Minima: 5) Find the extrema of the following functions, sketch their graphs, determine the intervals of increase, decrease, convexity, and concavity:

$$x^3 - 6x + 2,\ x/(1 + x^2),\ x^2/(1 + x^4),\ \cos^2 x.$$

6) Study the maxima and minima of the function $x^3 + 3ax + 1$ in their dependence on a.

7) Which point of the hyperbola $2y^2 - x^2 = 2$ is nearest to the point $x = 0$, $y = 3$?

8) Of all rectangles with given area find the one with the shortest diagonal.

9) Inscribe the rectangle of greatest area in the ellipse $x^2/a^2 + y^2/b^2 = 1$.

10) Of all circular cylinders with given volume find the one with the least area.

§4. LEIBNIZ' NOTATION AND THE "INFINITELY SMALL"

Newton and Leibniz knew how to obtain the integral and the derivative as limits. But the very foundations of the calculus were long obscured by an unwillingness to recognize the exclusive right of the limit concept as the source of the new methods. Neither Newton nor Leibniz could bring himself to such a clear-cut attitude, simple as it appears to us now that the limit concept has been completely clarified. Their example dominated more than a century of mathematical development during which the subject was shrouded by talk of "infinitely small quantities," "differentials," "ultimate ratios," etc. The reluctance with which these concepts were finally abandoned was deeply rooted in the philosophical attitude of the time and in the very nature of the human mind. One might have argued: "Of course integral and derivative can be and are calculated as limits. But what, after all, *are* these objects in themselves, irrespective of the particular way they are described by limiting processes? It seems obvious that intuitive concepts such as area or slope of a curve have an absolute meaning in themselves without any need for the auxiliary concepts of inscribed polygons or secants and their limits." Indeed, it is psychologically natural to search for adequate definitions of area and slope as "things in themselves." But to renounce this desire and rather to see in limiting processes their only scientifically relevant definitions, is in line with the mature attitude that has so often cleared the way for progress. In the seventeenth century there was no intellectual tradition to permit such philosophical radicalism.

Leibniz' attempt to "explain" the derivative started in a perfectly correct way with the difference quotient of a function $y = f(x)$,

$$\frac{\Delta y}{\Delta x} = \frac{f(x_1) - f(x)}{x_1 - x}.$$

For the limit, the derivative, which we called $f'(x)$ (following the usage introduced later by Lagrange), Leibniz wrote

$$\frac{dy}{dx},$$

replacing the difference symbol Δ by the "differential symbol" d. Provided we understand that this symbol is solely an indication that the limiting process $\Delta x \to 0$ and consequently $\Delta y \to 0$ is to be carried out, there is no difficulty and no mystery. *Before* passing to the limit, the denominator Δx in the quotient $\Delta y/\Delta x$ is cancelled out or transformed in such a way that the limiting process can be completed smoothly. This is always the crucial point in the actual process of differentiation. Had we tried to pass to the limit without such a previous reduction, we should have obtained the meaningless relation $\Delta y/\Delta x = 0/0$, in which we are not at all interested. Mystery and confusion only enter if we follow Leibniz and many of his successors by saying something like this:

"Δx does not approach zero. Instead, the 'last value' of Δx is not zero but an 'infinitely small quantity,' a 'differential' called dx; and similarly Δy has a 'last' infinitely small value dy. The actual quotient of these infinitely small differentials is again an ordinary number, $f'(x) = dy/dx$." Leibniz accordingly called the derivative the "*differential quotient*." Such infinitely small quantities were considered a new kind of number, not zero but smaller than any positive number of the real number system. Only those with a real mathematical sense could grasp this concept, and the calculus was thought to be genuinely difficult because not everybody has, or can develop, this sense. In the same way, the integral was considered to be a sum of infinitely many "infinitely small quantities" $f(x)\,dx$. Such a sum, people seemed to feel, *is* the integral or area, while the calculation of its value as the *limit of a finite sum of ordinary numbers* $f(x_i)\Delta x$ was regarded as something accessory. Today we simply discard the desire for a "direct" explanation and *define* the integral as the limit of a finite sum. In this way the difficulties are dispelled and everything of value in the calculus is secured on a sound basis.

In spite of this later development Leibniz' notation, dy/dx for $f'(x)$ and $\int f(x)\,dx$ for the integral, was retained and has proved extremely useful. There is no harm in it if we consider the symbols d only as symbols for a passage to the limit. Leibniz' notation has the advantage that limits of quotients and sums can in some ways be handled "as if" they were actual quotients or sums. The suggestive power of this symbolism has always tempted people to impute to these symbols some entirely unmathematical meaning. If we resist this temptation, then Leibniz' notation is at least an excellent abbreviation for the more cumbersome explicit notation of the limit process; as a matter of fact, it is almost indispensable in the more advanced parts of the theory.

For example, rule (d) of page 429 for differentiating the inverse function $x = g(y)$ of $y = f(x)$ was that $g'(y)f'(x) = 1$. In Leibniz' notation it reads simply

$$\frac{dx}{dy} \cdot \frac{dy}{dx} = 1,$$

"as if" the "differentials" may be cancelled out from something like an ordinary fraction. Likewise, rule (e) of page 431 for differentiating a compound function $z = k(x)$, where

$$z = g(y), \qquad y = f(x),$$

now reads

$$\frac{dz}{dx} = \frac{dz}{dy} \cdot \frac{dy}{dx}.$$

Leibniz' notation has the further advantage of emphasizing the *quantities* x, y, z rather than their explicit functional connection. The latter expresses a *procedure*, an *operation* producing one quantity y from another x, e.g. the function $y = f(x) = x^2$ produces a quantity y equal to the square of the quantity x. The operation (squaring) is the object of the mathematician's attention. But physicists and engineers are on the whole primarily interested in the quantities themselves. Hence the emphasis on quantities in Leibniz' notation has a particular appeal to people engaged in applied mathematics.

Another remark may be added. While "differentials" as infinitely small quantities are now definitely and dishonorably discarded, the same word "differential" has slipped in again through the back door—this time to denote a perfectly legitimate and useful concept. It now means simply a difference Δx when Δx is small in relation to the other quantities occurring. We cannot here go into a discussion of the value of this concept for approximate calculations. Nor can we discuss other legitimate mathematical notions for which the name "differential" has been adopted, some of which have proved quite useful in the calculus and in its applications to geometry.

§5. THE FUNDAMENTAL THEOREM OF THE CALCULUS

1. The Fundamental Theorem

The notion of integration, and to some extent that of differentiation, had been fairly well developed before the work of Newton and Leibniz. To start the tremendous evolution of the new mathematical analysis but one more simple discovery was needed. The two apparently unconnected limiting processes involved in the differentiation and integration of a function are intimately related. They are, in fact, inverse to one another, like the operations of addition and subtraction, or multiplication and division. There is no separate differential calculus and integral calculus, but only one *calculus*.

It was the great achievement of Leibniz and Newton to have first clearly recognized and exploited this *fundamental theorem of the calculus*. Of course, their discovery lay on the straight path of scientific development and it is only natural that several men should have arrived at a clear understanding of the situation independently and at almost the same time.

To formulate the fundamental theorem we consider the integral of a function $y = f(x)$ from the fixed lower limit a to the variable upper limit x. To avoid confusion between the upper limit of integration x and the variable x that appears in the symbol $f(x)$, we write this integral in the form (see p. 412)

$$F(x) = \int_a^x f(u)\,du, \tag{1}$$

indicating that we wish to study the integral as a function $F(x)$ of the upper limit x (Fig. 274). This function $F(x)$ is the area under the curve

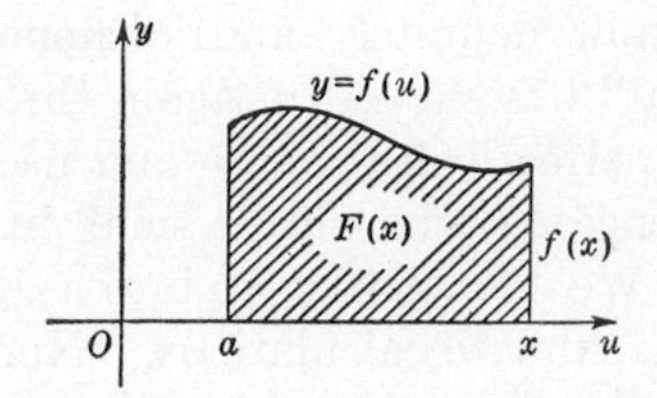

Fig. 274. The integral as function of upper limit.

$y = f(u)$ from the point $u = a$ to the point $u = x$. Sometimes the integral $F(x)$ with a variable upper limit is called an "indefinite" integral.

Now the fundamental theorem of the calculus is:

The derivative of the indefinite integral (1) *as a function of* x *is equal to the value of* f(u) *at the point* x:

$$F'(x) = f(x).$$

In other words, the process of integration, leading from the function f(x) *to* F(x), *is undone, inverted, by the process of differentiation, applied to* F(x).

On an intuitive basis the proof is very easy. It depends on the interpretation of the integral $F(x)$ as an area, and would be obscured if one tried to represent $F(x)$ by a graph and the derivative $F'(x)$ by its slope. Instead of this original geometrical interpretation of the derivative we retain the geometrical explanation of the integral $F(x)$ but proceed in an analytical way with the differentiation of $F(x)$. The difference

$$F(x_1) - F(x)$$

is simply the area between x and x_1 in Figure 275, and we see that this

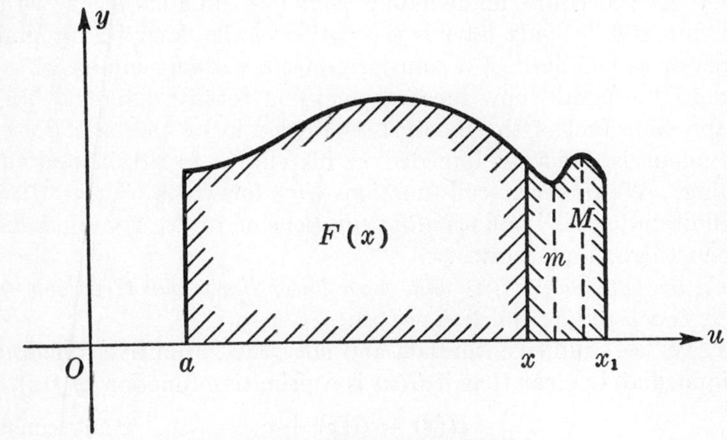

Fig. 275. Proof of the fundamental theorem.

area lies between the values $(x_1 - x)m$ and $(x_1 - x)M$,

$$(x_1 - x)m \leq F(x_1) - F(x) \leq (x_1 - x)M,$$

where M and m are respectively the greatest and least values of $f(u)$ in the interval between x and x_1. For these two products are the areas of rectangles including the curved area and included in it, respectively. Therefore

$$m \leq \frac{F(x_1) - F(x)}{x_1 - x} \leq M.$$

We shall assume that the function $f(u)$ is continuous, so that if x_1 approaches x, then M and m both approach $f(x)$. Hence we have

$$(2) \qquad F'(x) = \lim \frac{F(x_1) - F(x)}{x_1 - x} = f(x),$$

as stated. Intuitively, this expresses the fact that the rate of change of the area under the curve $y = f(x)$ as x increases is equal to the height of the curve at the point x.

In certain textbooks the salient point in the fundamental theorem is obscured by poorly chosen nomenclature. Many authors first introduce the derivative and then define the "indefinite integral" simply as the inverse of the derivative, saying that $G(x)$ is an indefinite integral of $f(x)$ if

$$G'(x) = f(x).$$

Thus their procedure immediately combines differentiation with the word "integral." Only later is the notion of the "definite integral" as an area or as the limit of a sum introduced, without emphasizing that the word "integral" now means something totally different. In this way the main fact of the theory is smuggled in by the back door, and the student is seriously impeded in his efforts to attain real understanding. We prefer to call functions $G(x)$ for which $G'(x) = f(x)$ not "indefinite integrals" but *primitive functions* of $f(x)$. The fundamental theorem then simply states:

F(x), *the integral of* f(u) *with fixed lower limit and a variable upper limit* x, *is a primitive function of* f(x).

We say "a" primitive function and not "the" primitive function, for it is immediately clear that if $G(x)$ is a primitive function of $f(x)$, then

$$H(x) = G(x) + c \qquad (c \text{ any constant})$$

is also a primitive function, since $H'(x) = G'(x)$. The converse is also true. *Two primitive functions,* G(x) and H(x), *can differ only by a constant.* For the difference $U(x) = G(x) - H(x)$ has the derivative $U'(x) = G'(x) - H'(x) = f(x) - f(x) = 0$, and is therefore constant, since a function represented by an everywhere horizontal graph must be constant.

This leads to a most important rule for finding the value of an integral between a and b, provided we know a primitive function $G(x)$ of $f(x)$. According to our main theorem,

$$F(x) = \int_a^x f(u)\,du$$

is also a primitive function of $f(x)$. Hence $F(x) = G(x) + c$, where c is a constant. The constant c is determined if we remember that $F(a) = \int_a^a f(u)\,du = 0$. This gives $0 = G(a) + c$, so that $c = -G(a)$. Then the definite integral between the limits a and x will be $F(x) = \int_a^x f(u)\,du = G(x) - G(a)$, or, if we write b instead of x,

$$(3) \qquad \int_a^b f(u)\,du = G(b) - G(a),$$

irrespective of what particular primitive function $G(x)$ we have chosen. In other words,

To evaluate the definite integral $\int_a^b \mathrm{f(x)\,dx}$ *we need only find a function* $\mathrm{G(x)}$ *such that* $\mathrm{G'(x) = f(x)}$, *and then form the difference* $\mathrm{G(b) - G(a)}$.

2. First Applications. Integration of x^r, cos x, sin x. Arc tan x

It is not possible here to give an adequate idea of the scope of the fundamental theorem, but the following illustrations may give some indication. In actual problems encountered in mechanics, physics, or pure mathematics, it is very often a definite integral whose value is wanted. The direct attempt to find the integral as the limit of a sum may be difficult. On the other hand, as we saw in §3, it is comparatively easy to perform any kind of differentiation and to accumulate a great wealth of information in this field. Each differentiation formula, $G'(x) = f(x)$, can be read inversely as providing a primitive function $G(x)$ for $f(x)$. By means of the formula (3), this can be exploited for calculating the integral of $f(x)$ between any two limits.

For example, if we want to find the integral of x^2 or x^3 or x^n we can now proceed much more simply than in §1. We know from our differentiation formula for x^n that the derivative of x^n is nx^{n-1}, so that the derivative of

$$G(x) = \frac{x^{n+1}}{n+1} \qquad (n \neq -1)$$

is

$$G'(x) = \frac{n+1}{n+1} x^n = x^n.$$

Therefore $x^{n+1}/(n + 1)$ is a primitive function of $f(x) = x^n$, and hence we have immediately

$$\int_a^b x^n\, dx = G(b) - G(a) = \frac{b^{n+1} - a^{n+1}}{n+1}.$$

This process is much simpler than the laborious procedure of finding the integral directly as the limit of a sum.

More generally, we found in §3 that for any rational s, positive or negative, the function x^s has the derivative sx^{s-1}, and therefore, for $s = r + 1$, the function

$$G(x) = \frac{1}{r+1} x^{r+1}$$

has the derivative $f(x) = G'(x) = x^r$. (We assume $r \neq -1$, i.e. $s \neq 0$.) Hence $x^{r+1}/(r + 1)$ is a primitive function or "indefinite integral" of x^r, and we have (for a, b positive and $r \neq -1$)

$$\int_a^b x^r\, dx = \frac{1}{r+1}(b^{r+1} - a^{r+1}). \tag{4}$$

In (4) we suppose that in the interval of integration the integrand x^r is defined and continuous, which excludes $x = 0$ if $r < 0$. We therefore make the assumption that in this case a and b are positive.

For $G(x) = -\cos x$ we have $G'(x) = \sin x$, hence

$$\int_0^a \sin x\, dx = -(\cos a - \cos 0) = 1 - \cos a.$$

Likewise, since for $G(x) = \sin x$ we have $G'(x) = \cos x$, it follows that

$$\int_0^a \cos x\, dx = \sin a - \sin 0 = \sin a.$$

A particularly interesting result is obtained from the formula for the differentiation of the inverse tangent, $D \arctan x = 1/(1 + x^2)$. It follows that the function $\arctan x$ is a primitive function of $1/(1 + x^2)$, and we obtain from formula (3) the result

$$\arctan b - \arctan 0 = \int_0^b \frac{1}{1+x^2}\, dx.$$

Now we have arc tan 0 = 0 because to the value 0 of the tangent the value 0 of the angle is attached. Hence we find

$$\text{arc tan } b = \int_0^b \frac{1}{1+x^2}\,dx. \tag{5}$$

If in particular $b = 1$, then arc tan b will be equal to $\pi/4$, because to the value 1 of the tangent corresponds an angle of 45°, or in radian measure $\pi/4$. Thus we obtain the remarkable formula

$$\pi/4 = \int_0^1 \frac{1}{1+x^2}\,dx. \tag{6}$$

This shows that the area under the graph of the function $y = 1/(1 + x^2)$ from $x = 0$ to $x = 1$ is one-fourth of the area of a circle of radius 1.

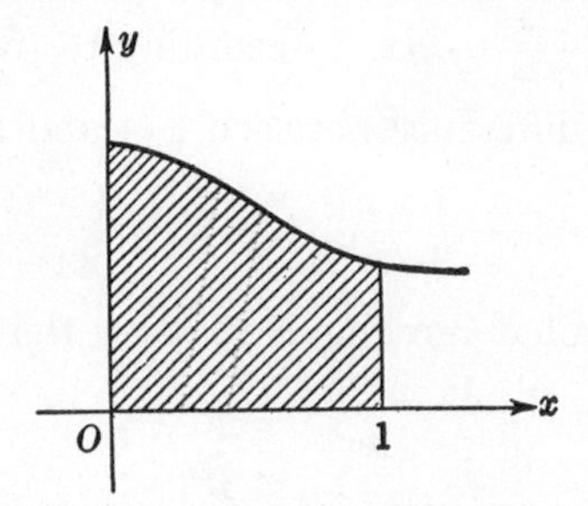

Fig. 276. $\pi/4$ as area under $y = 1/(1 + x^2)$ from 0 to 1.

3. Leibniz' Formula for π

The last result leads to one of the most beautiful mathematical discoveries of the seventeenth century—Leibniz' alternating series for π,

$$\frac{\pi}{4} = \frac{1}{1} - \frac{1}{3} + \frac{1}{5} - \frac{1}{7} + \frac{1}{9} - \frac{1}{11} + \cdots. \tag{7}$$

By the symbol $+ \cdots$ we mean that the sequence of finite "partial sums", formed by breaking off the expression on the right after n terms, converges to the limit $\pi/4$ as n increases.

To prove this famous formula, we have only to recall the finite geometrical series $\frac{1 - q^n}{1 - q} = 1 + q + q^2 + \cdots + q^{n-1}$, or

$$\frac{1}{1-q} = 1 + q + q^2 + \cdots + q^{n-1} + \frac{q^n}{1-q}.$$

In this algebraic identity we substitute $q = -x^2$ and obtain

$$\frac{1}{1+x^2} = 1 - x^2 + x^4 - x^6 + \cdots + (-1)^{n-1}x^{2n-2} + R_n, \tag{8}$$

where the "remainder" R_n is

$$R_n = (-1)^n \frac{x^{2n}}{1 + x^2}.$$

Equation (8) can now be integrated between the limits 0 and 1. By rule (a) of §3, we have to take on the right the sum of the integrals of the single terms. Since, by (4), $\int_a^b x^m\,dx = (b^{m+1} - a^{m+1})/(m + 1)$, we find $\int_0^1 x^m\,dx = 1/(m + 1)$, and therefore

$$(9) \qquad \int_0^1 \frac{dx}{1 + x^2} = 1 - \frac{1}{3} + \frac{1}{5} - \frac{1}{7} + \cdots + (-1)^{n-1} \frac{1}{2n - 1} + T_n,$$

where $T_n = (-1)^n \int_0^1 \frac{x^{2n}}{1 + x^2}\,dx$. According to (5), the left side of (9) is equal to $\pi/4$. The difference between $\pi/4$ and the partial sum

$$S_n = 1 - \frac{1}{3} + \frac{1}{5} + \cdots + \frac{(-1)^{n-1}}{2n - 1}$$

is $\pi/4 - S_n = T_n$. What remains is to show that T_n approaches zero as n increases. Now

$$\frac{x^{2n}}{1 + x^2} \leq x^{2n} \qquad \text{for } 0 \leq x \leq 1.$$

Recalling formula (13) of §1, which states that $\int_a^b f(x)\,dx \leq \int_a^b g(x)\,dx$ if $f(x) \leq g(x)$ and $a < b$, we see that

$$|T_n| = \int_0^1 \frac{x^{2n}}{1 + x^2}\,dx \leq \int_0^1 x^{2n}\,dx;$$

since the right side is equal to $1/(2n + 1)$, as we saw before (formula (4)), we find $|T_n| < 1/(2n + 1)$. Hence

$$\left| \frac{\pi}{4} - S_n \right| < \frac{1}{2n + 1}.$$

But this shows that S_n tends with increasing n to $\pi/4$, since $1/(2n + 1)$ tends to zero. Thus Leibniz' formula is proved.

§6. THE EXPONENTIAL FUNCTION AND THE LOGARITHM

The basic concepts of the calculus furnish a much more adequate theory of the logarithm and the exponential function than does the "elementary" procedure that underlies the usual instruction in school.

There one usually begins with the integral powers a^n of a positive number a, and then defines $a^{1/m} = \sqrt[m]{a}$, thus obtaining the value of a^r for every rational $r = n/m$. The value of a^x for any irrational x is next defined so as to make a^x a continuous function of x, a delicate point which is omitted in elementary instruction. Finally, the logarithm of y to the base a,

$$x = \log_a y,$$

is defined as the inverse function of $y = a^x$.

In the following theory of these functions on the basis of the calculus the order in which they are considered is reversed. We begin with the logarithm and then obtain the exponential function.

1. Definition and Properties of the Logarithm. Euler's Number e

We define the logarithm, or more specifically the "natural logarithm," $F(x) = \log x$ (its relation to the ordinary logarithm to the base 10 will be shown in Article 2), as the area under the curve $y = 1/u$ from $u = 1$ to $u = x$, or, what amounts to the same thing, as the integral

$$F(x) = \log x = \int_1^x \frac{1}{u}\, du \tag{1}$$

(see Fig. 5, p. 29). The variable x may be any positive number. Zero is excluded because the integrand $1/u$ becomes infinite as u tends to 0.

It is quite natural to study the function $F(x)$. For we know that the primitive function of any power x^n is a function $x^{n+1}/(n+1)$ of the same type, except for $n = -1$. In the latter case the denominator $n + 1$ would vanish and formula (4), p. 440 would be meaningless. Thus we might expect that the integration of $1/x$ or $1/u$ would lead to some new—and interesting—type of function.

Although we consider (1) the definition of the function $\log x$, we do not "know" the function until we have derived its properties and have found means for its numerical computation. It is quite typical of the modern approach that we start with general concepts such as area and integral, establish definitions such as (1) on this basis, then deduce properties of the objects defined and, only at the very end, arrive at explicit expressions for numerical calculation.

The first important property of $\log x$ is an immediate consequence of the fundamental theorem of §5. This theorem yields the equation

$$F'(x) = 1/x. \tag{2}$$

From (2) it follows that the derivative is always positive, which confirms the obvious fact that function $\log x$ is a monotone increasing function as we travel in the direction of increasing values of x.

The principal property of the logarithm is expressed by the formula

$$\log a + \log b = \log (ab). \tag{3}$$

The importance of this formula in the practical application of logarithms to numerical computations is well known. Intuitively, formula (3) could be obtained by looking at the areas defining the three quantities $\log a$, $\log b$, and $\log (ab)$. But we prefer to derive it by a reasoning typical of the calculus: Together with the function $F(x) = \log x$ we consider the second function

$$k(x) = \log (ax) = \log w = F(w),$$

setting $w = f(x) = ax$, where a is any positive constant. We can easily differentiate $k(x)$ by rule (e) of §3: $k'(x) = F'(w)f'(x)$. By (2), and since $f'(x) = a$, this becomes

$$k'(x) = a/w = a/ax = 1/x.$$

Therefore $k(x)$ has the same derivative as $F(x)$; hence, according to page 438, we have

$$\log (ax) = k(x) = F(x) + c,$$

where c is a constant not depending on the particular value of x. The constant c is determined by the simple procedure of substituting for x the specific number 1. We know from the definition (1) that

$$F(1) = \log 1 = 0,$$

because the defining integral has for $x = 1$ equal upper and lower limits. Hence we obtain

$$k(1) = \log (a \cdot 1) = \log a = \log 1 + c = c,$$

which gives $c = \log a$, and therefore for every x the formula

$$\log (ax) = \log a + \log x. \tag{3a}$$

Setting $x = b$ we obtain the desired formula (3).

In particular (for $a = x$), we now find in succession

$$\log (x^2) = 2 \log x$$

$$\log (x^3) = 3 \log x.$$

$$\dots\dots\dots\dots\dots\dots$$

$$\log (x^n) = n \log x. \tag{4}$$

Equation (4) shows that for increasing values of x the values of $\log x$ tend to infinity. For the logarithm is a monotone increasing function and we have, for example

$$\log (2^n) = n \log 2,$$

which tends to infinity with n. Furthermore we have

$$0 = \log 1 = \log \left(x \cdot \frac{1}{x}\right) = \log x + \log \frac{1}{x},$$

so that

$$\log \frac{1}{x} = -\log x. \tag{5}$$

Finally,

$$\log x^r = r \log x \tag{6}$$

for any rational number $r = \frac{m}{n}$. For, setting $x^r = u$, we have

$$n \log u = \log u^n = \log x^{\frac{m}{n} \cdot n} = \log x^m = m \log x,$$

so that

$$\log x^{\frac{m}{n}} = \frac{m}{n} \log x.$$

Since $\log x$ is a continuous monotone function of x, having the value 0 for $x = 1$ and tending to infinity as x increases, there must be some number greater than 1 such that for this value we have $\log x = 1$.

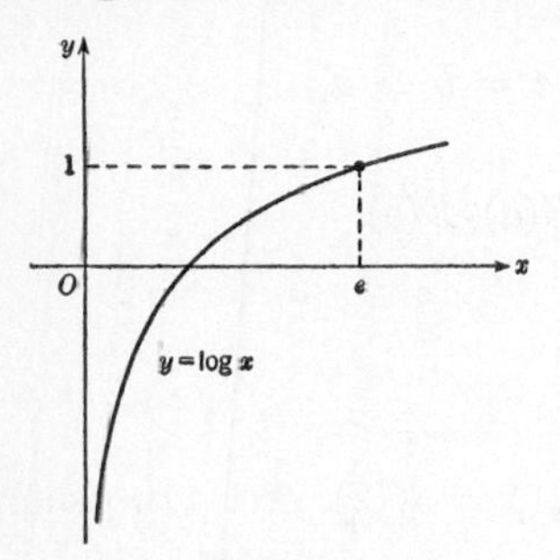

Fig. 277.

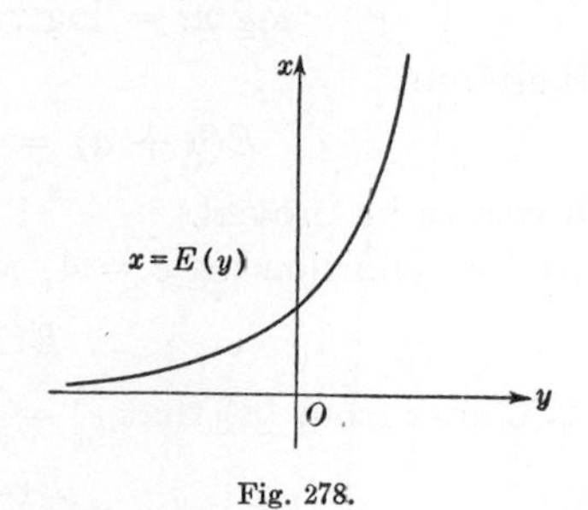

Fig. 278.

Following Euler, we call this number e. (The equivalence with the definition of p. 298 will be shown later.) Thus e is defined by the equation

$$\log e = 1. \tag{7}$$

We have introduced the number e by an intrinsic property which assures its *existence*. Presently we shall carry our analysis further, obtaining as a *consequence* explicit formulas giving arbitrarily exact approximations to the numerical value of e.

2. The Exponential Function

Summarizing our previous results, we see that the function $F(x) = \log x$ has the value zero for $x = 1$, increases monotonically to infinity but with decreasing slope $1/x$, and for positive values of x less than 1 is given by the negative of $\log 1/x$, so that $\log x$ becomes negatively infinite as $x \to 0$.

Because of the monotone character of $y = \log x$ we may consider the inverse function

$$x = E(y),$$

whose graph (Fig. 278) is obtained in the usual way from that of $y = \log x$ (Fig. 277), and which is defined for all values of y between $-\infty$ and $+\infty$. As y tends to $-\infty$ the value $E(y)$ tends to zero, and as y tends to $+\infty$ $E(y)$ tends to $+\infty$.

The E-function has the following fundamental property:

$$E(a) \cdot E(b) = E(a + b) \tag{8}$$

for any pair of values a and b. This law is merely another form of the law (3) for the logarithm. For if we set

$$E(b) = x, \qquad E(a) = z \qquad \text{(i.e. } b = \log x,\ a = \log z),$$

we have

$$\log xz = \log x + \log z = b + a,$$

and therefore

$$E(b + a) = xz = E(a) \cdot E(b),$$

which was to be proved.

Since by definition $\log e = 1$, we have

$$E(1) = e,$$

and it follows from (8) that $e^2 = E(1)E(1) = E(2)$, etc. In general,

$$E(n) = e^n$$

for any integer n. Likewise $E(1/n) = e^{\frac{1}{n}}$, so that $E(p/q) = E(1/q) \cdots E(1/q) = [e^{\frac{1}{q}}]^p$; hence, setting $p/q = r$, we have

$$E(r) = e^r$$

for any rational r. Therefore it is appropriate to *define* the operation of raising the number e to an irrational power by setting

$$e^y = E(y)$$

for any real number y, since the E-function is continuous for all values of y, and identical with the value of e^y for rational y. We can now express the fundamental law (8) of the E-function, or *exponential function*, as it is called, by the equation

$$e^a e^b = e^{a+b}, \tag{9}$$

which is thereby established for arbitrary rational or irrational a and b.

In all these discussions we have been referring the logarithm and exponential function to the number e as a "base," the "natural base" for the logarithm. The transition from the base e to any other positive number is easily made. We begin by considering the (natural) logarithm

$$\alpha = \log a,$$

so that

$$a = e^\alpha = e^{\log a}.$$

Now we define a^x by the compound expression

$$z = a^x = e^{\alpha x} = e^{x \log a}. \tag{10}$$

For example,

$$10^x = e^{x \log 10}.$$

We call the inverse function of a^x the *logarithm to the base* a, and we see immediately that the *natural logarithm* of z is x times α; in other words, the logarithm of a number z to the base a is obtained by dividing the natural logarithm of z by the fixed natural logarithm of a. For $a = 10$ this is (to four significant figures)

$$\log 10 = 2.303.$$

3. Formulas for Differentiation of e , a^x, x^s

Since we have defined the exponential function $E(y)$ as the inverse of $y = \log x$, it follows from the rule concerning differentiation of inverse functions (§3) that

$$E'(y) = \frac{dx}{dy} = \frac{1}{\frac{dy}{dx}} = \frac{1}{1/x} = x = E(y),$$

i.e.

$$E'(y) = E(y). \tag{11}$$

The natural exponential function is identical with its derivative.

This is really the source of all the properties of the exponential function and the basic reason for its importance in applications, as will become apparent in subsequent sections. Using the notation introduced in Section 2 we may write (11) as follows:

$$\frac{d}{dx} e^x = e^x. \tag{11a}$$

More generally, differentiating the compound function

$$f(x) = e^{\alpha x},$$

we obtain by the rule of §3

$$f'(x) = \alpha e^{\alpha x} = \alpha f(x).$$

Hence, for $\alpha = \log a$, we find that the function

$$f(x) = a^x$$

has the derivative

$$f'(x) = a^x \log a.$$

We may now define the function

$$f(x) = x^s$$

for any real exponent s and positive variable x by setting

$$x^s = e^{s \log x}.$$

Again applying the rule for differentiation of the compound functions, $f(x) = e^{sz}$, $z = \log x$, we find $f'(x) = se^{sz} \cdot \frac{1}{x} = sx^s \cdot \frac{1}{x}$ and therefore

$$f'(x) = sx^{s-1},$$

in accordance with our previous result for rational s.

4. Explicit Expressions for e, e^x, and log x as Limits

To find explicit formulas for these functions we shall exploit the differentiation formulas for the exponential function and the logarithm. Since the derivative of the function $\log x$ is $1/x$, by the definition of the derivative we obtain the relation

$$\frac{1}{x} = \lim \frac{\log x_1 - \log x}{x_1 - x} \quad \text{as} \quad x_1 \to x.$$

If we set $x_1 = x + h$ and let h tend to zero by running through the sequence

$$h = 1/2, 1/3, 1/4, \cdots, 1/n, \cdots,$$

then, on applying the rules of logarithms, we find

$$\frac{\log\left(x + \frac{1}{n}\right) - \log x}{1/n} = n \log \frac{x + \frac{1}{n}}{x} = \log\left[\left(1 + \frac{1}{nx}\right)^n\right] \to \frac{1}{x}.$$

By writing $z = 1/x$ and using again the laws for the logarithm we obtain

$$z = \lim \log\left[\left(1 + \frac{z}{n}\right)^n\right] \quad \text{as} \quad n \to \infty.$$

In terms of the exponential function,

$$(12) \qquad e^z = \lim\left(1 + \frac{z}{n}\right)^n \quad \text{as} \quad n \to \infty.$$

Here we have the famous formula defining the exponential function as a simple limit. In particular, for $z = 1$ we find

$$(13) \qquad e = \lim\,(1 + 1/n)^n,$$

and for $z = -1$,

$$(13a) \qquad \frac{1}{e} = \lim\,(1 - 1/n)^n.$$

These expressions lead at once to expansions in the form of infinite series. By the binomial theorem we find that

$$\left(1 + \frac{x}{n}\right)^n = 1 + n\frac{x}{n} + \frac{n(n-1)}{2!}\frac{x^2}{n^2} + \frac{n(n-1)(n-2)}{3!}\frac{x^3}{n^3} + \cdots + \frac{x^n}{n^n},$$

or

$$\left(1 + \frac{x}{n}\right)^n = 1 + \frac{x}{1!} + \frac{x^2}{2!}\left(1 - \frac{1}{n}\right) + \frac{x^3}{3!}\left(1 - \frac{1}{n}\right)\left(1 - \frac{2}{n}\right) + \cdots$$
$$+ \frac{x^n}{n!}\left(1 - \frac{1}{n}\right)\left(1 - \frac{2}{n}\right)\cdots\left(1 - \frac{n-2}{n}\right)\left(1 - \frac{n-1}{n}\right).$$

It is plausible and not difficult to justify completely (the details are omitted here) that we can perform the passage to the limit as $n \to \infty$ by replacing $\frac{1}{n}$ by 0 in each term. This gives the famous infinite series for e^x,

$$(14) \qquad e^x = 1 + \frac{x}{1!} + \frac{x^2}{2!} + \frac{x^3}{3!} + \cdots,$$

and in particular the series for e,

$$e = 1 + \frac{1}{1!} + \frac{1}{2!} + \frac{1}{3!} + \frac{1}{4!} + \cdots,$$

which establishes the identity of e with the number defined on page 298. For $x = -1$ we obtain the series

$$\frac{1}{e} = \frac{1}{2!} - \frac{1}{3!} + \frac{1}{4!} - \frac{1}{5!} + \cdots,$$

which gives an excellent numerical approximation with very few terms, the total error involved in breaking off the series at the nth term being less than the magnitude of the $(n + 1)$st term.

By exploiting the differentiation formula for the exponential function we can obtain an interesting expression for the logarithm. We have

$$\lim \frac{e^h - 1}{h} = \lim \frac{e^h - e^0}{h} = 1$$

as h tends to 0, because this limit is the derivative of e^y for $y = 0$, and this is equal to $e^0 = 1$. In this formula we substitute for h the values z/n, where z is an arbitrary number and n ranges over the sequence of positive integers. This gives

$$n \frac{e^{z/n} - 1}{z} \rightarrow 1,$$

or

$$n(\sqrt[n]{e^z} - 1) \rightarrow z$$

as n tends to infinity. Writing $z = \log x$ or $e^z = x$, we finally obtain

$$\text{(15)} \qquad \log x = \lim n(\sqrt[n]{x} - 1) \quad \text{as} \quad n \rightarrow \infty.$$

Since $\sqrt[n]{x} \rightarrow 1$ as $n \rightarrow \infty$ (see p. 323), this represents the logarithm as the limit of a product, one of whose factors tends to zero and the other to infinity.

Miscellaneous Examples and Exercises. By including the exponential function and the logarithm we now master a large class of functions and have access to many applications.

Differentiate: 1) $x(\log x - 1)$. 2) $\log(\log x)$. 3) $\log(x + \sqrt{1 + x^2})$. 4) $\log(x + \sqrt{1 - x^2})$. 5) e^{-x^2}. 6) e^{e^x} (a compound function e^z with $z = e^x$). 7) x^x (Hint: $x^x = e^{x \log x}$). 8) $\log \tan x$. 9) $\log \sin x$; $\log \cos x$. 10) $x/\log x$.

Find the maxima and minima of 11) xe^{-x}, 12) x^2e^{-x}, 13) xe^{-ax}.

*14) Find the locus of the maximum point of the curve $y = xe^{-ax}$ as a varies.

15) Show that all the successive derivatives of e^{-x^2} have the form e^{-x^2} multiplied by a polynomial in x.

*16) Show that the nth derivative of e^{-1/x^2} has the form $e^{-1/x^2} \cdot 1/x^{3n}$ multiplied by a polynomial of degree $2n - 2$.

*17) *Logarithmic differentiation.* By using the fundamental property of the logarithm, the differentiation of products can sometimes be effected in a simplified manner. We have for a product of the form

$$p(x) = f_1(x)f_2(x) \cdots f_n(x),$$

$$D(\log p(x)) = D(\log f_1(x)) + D(\log f_2(x)) + \cdots + D(\log f_n(x)),$$

and hence, by the rule for differentiating compound functions,

$$\frac{p'(x)}{p(x)} = \frac{f_1'(x)}{f_1(x)} + \frac{f_2'(x)}{f_2(x)} + \cdots + \frac{f_n'(x)}{f_n(x)}.$$

Use this for differentiating

a) $x(x + 1)(x + 2) \cdots (x + n)$, b) xe^{-ax^2}.

5. Infinite Series for the Logarithm. Numerical Calculation

It is not formula (15) that serves as the basis for numerical calculation of the logarithm. A quite different and more useful explicit expression of great theoretical importance is far better suited to this purpose. We shall obtain this expression by the method used on page 441 for finding π, exploiting the definition of the logarithm by formula (1). One small preparatory step is needed; instead of aiming at $\log x$, we shall try to express $y = \log(1 + x)$, composed of the functions $y = \log z$ and $z = 1 + x$. We have $\frac{dy}{dx} = \frac{dy}{dz} \cdot \frac{dz}{dx} = \frac{1}{z} \cdot 1 = \frac{1}{1+x}$. Hence $\log(1 + x)$ is a primitive function of $1/(1 + x)$, and we infer by the fundamental theorem that the integral of $1/(1 + u)$ from 0 to x is equal to $\log(1 + x) - \log 1 = \log(1 + x)$; in symbols,

$$\log(1 + x) = \int_0^x \frac{1}{1+u}\, du. \tag{16}$$

(Of course, this formula could just as well have been obtained intuitively from the geometrical interpretation of the logarithm as an area. Compare p. 413.)

In formula (16) we insert, as on page 442, the geometrical series for $(1 + u)^{-1}$, writing

$$\frac{1}{1+u} = 1 - u + u^2 - u^3 + \cdots + (-1)^{n-1}u^{n-1} + (-1)^n \frac{u^n}{1+u},$$

where, cautiously, we choose to write down not an infinite series, but rather a finite series with the remainder

$$R_n = (-1)^n \frac{u^n}{1+u}.$$

Substituting this series in (16) we may use the rule that such a (finite) sum can be integrated term by term. The integral of u^s from 0 to x yields $\frac{x^{s+1}}{s+1}$, and thus we obtain immediately

$$\log(1+x) = x - \frac{x^2}{2} + \frac{x^3}{3} - \frac{x^4}{4} + \cdots + (-1)^{n-1}\frac{x^n}{n} + T_n,$$

where the remainder T_n is given by

$$T_n = (-1)^n \int_0^x \frac{u^n}{1+u}\, du.$$

We shall now show that T_n tends to zero for increasing n provided that x is chosen greater than -1 and not greater than $+1$, in other words, for

$$-1 < x \leq 1,$$

where it is to be noted that $x = +1$ is included, while $x = -1$ is not. According to our assumption, in the interval of integration u is greater than a number $-\alpha$, which may be near to -1 but is at any rate greater than -1, so that $0 < 1 - \alpha < 1 + u$. Hence in the interval from 0 to x we have

$$\left|\frac{u^n}{1+u}\right| \leq \frac{|u|^n}{1-\alpha},$$

and therefore

$$|T_n| \leq \frac{1}{1-\alpha}\left|\int_0^x u^n\, du\right|,$$

or

$$|T_n| \leq \frac{1}{1-\alpha}\frac{|x|^{n+1}}{n+1} \leq \frac{1}{1-\alpha}\frac{1}{n+1}.$$

Since $1 - \alpha$ is a fixed factor, we see that for increasing n this expression tends to 0 so that from

$$\text{(17)} \qquad \left|\log(1+x) - \left\{x - \frac{x^2}{2} + \frac{x^3}{3} - \cdots + (-1)^n\frac{x^n}{n}\right\}\right| \leq \frac{1}{1-\alpha}\frac{1}{n+1},$$

we obtain the infinite series

$$\text{(18)} \qquad \log(1+x) = x - \frac{x^2}{2} + \frac{x^3}{3} - \frac{x^4}{4} + \cdots,$$

which is valid for $-1 < x \leq 1$.

If, in particular, we choose $x = 1$, we obtain the interesting result

$$\log 2 = 1 - \frac{1}{2} + \frac{1}{3} - \frac{1}{4} + \cdots. \tag{19}$$

This formula has a structure similar to that of the series for $\pi/4$.

The series (18) is not a very practical means for finding numerical values for the logarithm, since its range is limited to values of $1 + x$ between 0 and 2, and since its convergence is so slow that one must include many terms before obtaining a reasonably accurate result. By the following device we can obtain a more convenient expression. Replacing x by $-x$ in (18) we find

$$\log (1 - x) = -x - \frac{x^2}{2} - \frac{x^3}{3} - \frac{x^4}{4} - \cdots. \tag{20}$$

Subtracting (20) from (18) and using the fact that $\log a - \log b = \log a + \log (1/b) = \log (a/b)$, we obtain

$$\log \frac{1 + x}{1 - x} = 2\left(x + \frac{x^3}{3} + \frac{x^5}{5} + \cdots\right). \tag{21}$$

Not only does this series converge much faster, but now the left side can express the logarithm of any positive number z, since $\dfrac{1 + x}{1 - x} = z$ always has a solution x between -1 and $+1$. Thus, if we want to calculate $\log 3$ we set $x = \frac{1}{2}$ and obtain

$$\log 3 = \log \frac{1 + \frac{1}{2}}{1 - \frac{1}{2}} = 2\left(\frac{1}{1\cdot 2} + \frac{1}{3\cdot 2^3} + \frac{1}{5\cdot 2^5} + \cdots\right).$$

With only 6 terms, up to $\dfrac{2}{11\cdot 2^{11}} = \dfrac{1}{11{,}264}$, we find the value

$$\log 3 = 1.0986,$$

which is accurate to five digits.

§7. DIFFERENTIAL EQUATIONS

1. Definition

The dominating rôle of the exponential and trigonometrical functions in mathematical analysis and its applications to physical problems is rooted in the fact that these functions solve the simplest "differential equations."

A differential equation for an unknown function $u = f(x)$ with derivative $u' = f'(x)$—the notation u' is a very useful abbreviation for $f'(x)$ as long as the quantity u and its dependence on x as the function $f(x)$ need not be sharply distinguished—is an equation involving u, u', and possibly thc independent variable x, as for example

$$u' = u + \sin (xu)$$

or

$$u' + 3u = x^2.$$

More generally, a differential equation may involve the second derivative, $u'' = f''(x)$, or higher derivatives, as in the example

$$u'' + 2u' - 3u = 0.$$

In any case the problem is to find a function $u = f(x)$ that satisfies the given equation. Solving a differential equation is a wide generalization of the problem of integration in the sense of finding the primitive function of a given function $g(x)$, which amounts to solving the simple differential equation

$$u' = g(x).$$

For example, the solutions of the differential equation

$$u' = x^2$$

are the functions $u = x^3/3 + c$, where c is any constant.

2. The Differential Equation of the Exponential Function. Radioactive Disintegration. Law of Growth. Compound Interest

The differential equation

$$u' = u \tag{1}$$

has as a solution the exponential function $u = e^x$, since the exponential function is its own derivative. More generally, the function $u = ce^x$, where c is any constant, is a solution of (1). Similarly, the function

$$u = ce^{kx}, \tag{2}$$

where c and k are any two constants, is a solution of the differential equation

$$u' = ku. \tag{3}$$

Conversely, any function $u = f(x)$ satisfying equation (3) must be of the form ce^{kx}. For if $x = h(u)$ is the inverse function of $u = f(x)$, then according to the rule for finding the derivative of an inverse function we have

$$h' = \frac{1}{u'} = \frac{1}{ku}.$$

But $\frac{\log u}{k}$ is a primitive function of $\frac{1}{ku}$, so that $x = h(u) = \frac{\log u}{k} + b$, where b is some constant. Hence

$$\log u = kx - bk,$$

and

$$u = e^{kx} \cdot e^{-bk}.$$

Setting e^{-bk} (which is a constant) equal to c, we have

$$u = ce^{kx},$$

as was to be proved.

The great significance of the differential equation (3) lies in the fact that it governs physical processes in which a quantity u of some substance is a function of the time t,

$$u = f(t),$$

and in which the quantity u is changing at each instant at a rate proportional to the value of u at that instant. In such a case, the *rate of change* at the instant t,

$$u' = f'(t) = \lim \frac{f(t_1) - f(t)}{t_1 - t},$$

is equal to ku, where k is a constant, k being positive if u is increasing and negative if u is decreasing. In either case, u satisfies the differential equation (3); hence

$$u = ce^{kt}.$$

The constant c is determined if we know the amount u_0 which was

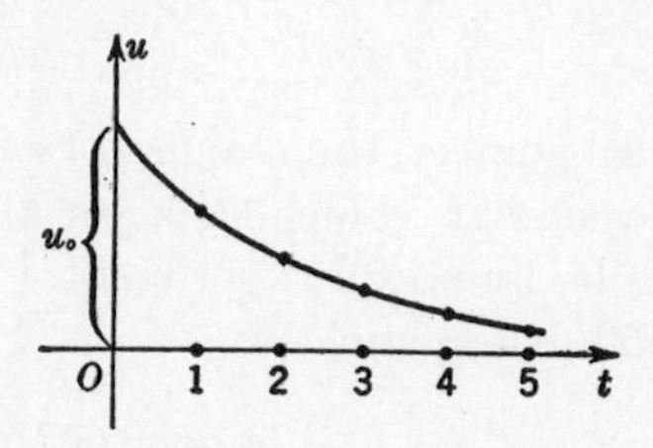

Fig. 279. Exponential decay. $u = u_0e^{kt}$, $k < 0$.

present at the time $t = 0$. We must obtain this amount if we set $t = 0$,

$$u_0 = ce^0 = c,$$

so that

$$u = u_0 e^{kt}. \tag{4}$$

Note that we start with a knowledge of the *rate of change* of u and deduce the law (4) which gives the actual *amount* of u at any time t. This is just the inverse of the problem of finding the derivative of a function.

A typical example is that of radioactive disintegration. Let $u = f(t)$ be the amount of some radioactive substance at the time t; then on the hypothesis that each individual particle of the substance has a certain probability of disintegrating in a given time, and that the probability is unaffected by the presence of other such particles, the rate at which u is disintegrating at a given time t will be proportional to u, i.e. to the total amount present at that time. Hence u will satisfy (3) with a negative constant k that measures the speed of the disintegration process, and therefore

$$u = u_0 e^{kt}.$$

It follows that the fraction of u which disintegrates in two equal time intervals is the same; for if u_1 is the amount present at time t_1 and u_2 the amount present at some later time t_2, then

$$\frac{u_2}{u_1} = \frac{u_0 e^{kt_2}}{u_0 e^{kt_1}} = e^{k(t_2 - t_1)},$$

which depends only on $t_2 - t_1$. To find out how long it will take for a given amount of the substance to disintegrate until only half of it is left, we must determine $s = t_2 - t_1$ so that

$$\frac{u_2}{u_1} = \frac{1}{2} = e^{ks},$$

from which we find

$$ks = \log \tfrac{1}{2}, \qquad s = (-\log 2)/k, \qquad \text{or} \qquad k = (-\log 2)/s. \tag{5}$$

For any radioactive substance, the value of s is called the half-life period, and s or some similar value (such as the value r for which $u_2/u_1 = 999/1000$) can be found by experiment. For radium, the half-life period is about 1550 years, and

$$k = \frac{\log \frac{1}{2}}{1550} = -0.0000447.$$

It follows that

$$u = u_0 \cdot e^{-0.0000447t}.$$

An example of a law of growth that is approximately exponential is provided by the phenomenon of compound interest. A given amount of money, u_0 dollars, is placed at 3% compound interest, which is to be compounded yearly. After 1 year, the amount of money will be

$$u_1 = u_0(1 + 0.03),$$

after 2 years it will be

$$u_2 = u_1(1 + 0.03) = u_0(1 + 0.03)^2,$$

and after t years it will be

$$(6) \qquad u_t = u_0(1 + 0.03)^t.$$

Now if, instead of being compounded at yearly intervals, the interest is compounded after each month or after each nth part of a year, then after t years the amount will be

$$u_0\left(1 + \frac{.03}{n}\right)^{nt} = u_0\left[\left(1 + \frac{.03}{n}\right)^n\right]^t.$$

If n is taken very large, so that the interest is compounded every day or even every hour, then as n tends to infinity the quantity in the brackets, according to §6, approaches $e^{0.03}$, and in the limit the amount after t years would be

$$(7) \qquad u_0 \cdot e^{0.03t},$$

which corresponds to a continuous process of compounding interest. We may also calculate the time s taken for the original capital to double at 3% continuous compound interest. We have $\frac{u_0 \cdot e^{.03s}}{u_0} = 2$, so that $s = \frac{100}{3} \log 2 = 23.10$. Thus the money will have doubled after about twenty-three years.

Instead of following this step-by-step procedure and then passing to the limit, we could have derived the formula (7) simply by saying that the rate of increase u' of the capital is proportional to u with the factor $k = .03$, so that

$$u' = ku, \qquad \text{where} \qquad k = .03.$$

The formula (7) then follows from the general result (4).

3. Other Examples. Simplest Vibrations

The exponential function often occurs in more complicated combinations. For example, the function

(8) $$u = e^{-kx^2},$$

where k is a positive constant, is a solution of the differential equation

$$u' = -2kxu.$$

The function (8) is of fundamental importance in probability and statistics, since it defines the "normal" frequency distributions.

The trigonometric functions $u = \cos t$, $v = \sin t$ also satisfy a very simple differential equation. We have first

$$u' = -\sin t = -v,$$

$$v' = \cos t = u,$$

which is a "system of two differential equations for two functions." By differentiating again, we find

$$u'' = -v' = -u,$$

$$v'' = u' = -v,$$

so that both functions u and v of the time variable t can be considered as solutions of the same differential equation

(9) $$z'' + z = 0,$$

which is a very simple differential equation of the "second order," i.e. involving the second derivative of z. This equation and its generalization with a positive constant k^2,

(10) $$z'' + k^2z = 0,$$

for which $z = \cos kt$ and $z = \sin kt$ are solutions, occur in the study of vibrations. This is why the oscillating curves $u = \sin kt$ and $u = \cos kt$

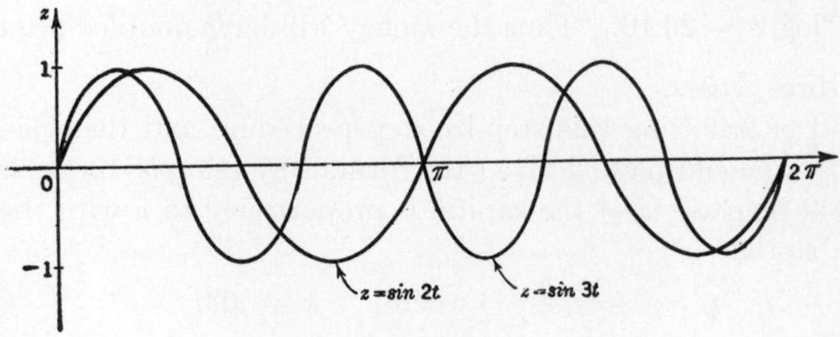

Fig. 280.

(Fig. 280) form the backbone of the theory of vibrating mechanisms. It should be stated that the differential equation (10) represents the ideal case, where there is no friction or resistance. Resistance is expressed in the differential equation of vibrating mechanisms by another term rz',

$$z'' + rz' + k^2z = 0, \tag{11}$$

and the solutions now are "damped" vibrations, mathematically expressed by the formula

$$e^{-rt/2} \cos \omega t,\ e^{-rt/2} \sin \omega t; \qquad \omega = \sqrt{k^2 - \left(\frac{r}{2}\right)^2},$$

and graphically represented by Figure 281. (As an exercise the reader

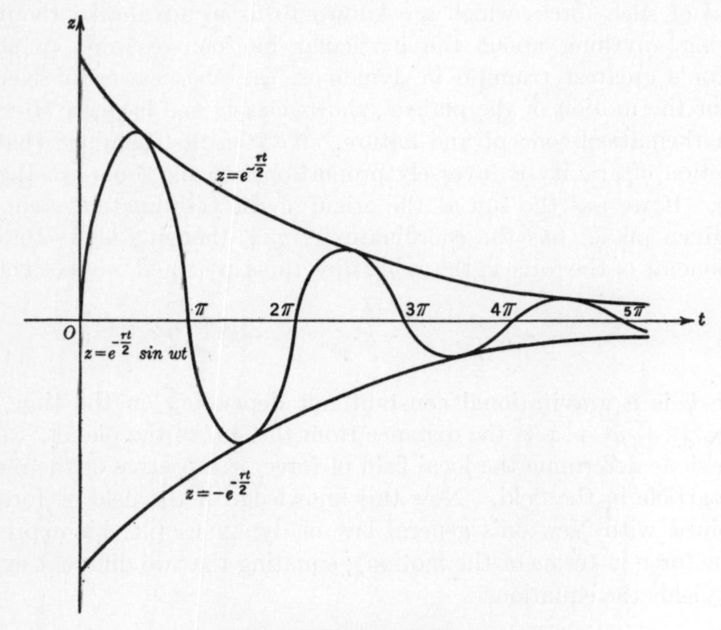

Fig. 281. Damped vibrations.

may verify these solutions by performing the differentiations.) The oscillations here are of the same type as those of the pure sine or cosine, but they are cut down in their intensity by an exponential factor, decreasing more or less rapidly according to the size of the friction coefficient r.

4. Newton's Law of Dynamics

Although a more detailed analysis of these facts is beyond our scope, we wish to bring them under the general aspect of the fundamental concepts with which Newton revolutionized mechanics and physics. He considered the motion of a particle with mass m and space coördinates $x(t)$, $y(t)$, $z(t)$ which are functions of the time t, so that the components of the acceleration are the second derivatives, $x''(t)$, $y''(t)$, $z''(t)$. The all-important step was Newton's realization that the quantities mx'', my'', mz'' can be considered as the components of force acting on the particle. At first sight this might appear to be only a formal definition of the word "force" in physics. But Newton's great achievement was to have shaped this definition in accordance with the actual phenomena of nature, inasmuch as nature very often provides a field of such forces which are known to us in advance without our knowing anything about the particular motion we want to study. Newton's greatest triumph in dynamics, the justification of Kepler's law for the motion of the planets, shows clearly the harmony between his mathematical concept and nature. Newton first assumed that the attraction of gravity is inversely proportional to the square of the distance. If we put the sun at the origin of the coördinate system, and if a given planet has the coördinates x, y, z, then it follows that the components of the force in the x, y, z directions are equal, respectively, to

$$-k\cdot\frac{x}{r^3}, \qquad -k\cdot\frac{y}{r^3}, \qquad -k\cdot\frac{z}{r^3},$$

where k is a gravitational constant not depending on the time, and $r = \sqrt{x^2 + y^2 + z^2}$ is the distance from the sun to the planet. These expressions determine the local field of force, irrespective of the motion of a particle in the field. Now this knowledge of the field of forces is combined with Newton's general law of dynamics (i.e. his expression for the force in terms of the motion); equating the two different expressions yields the equations

$$mx'' = \frac{-kx}{(x^2 + y^2 + z^2)^{3/2}},$$

$$my'' = \frac{-ky}{(x^2 + y^2 + z^2)^{3/2}},$$

$$mz'' = \frac{-kz}{(x^2 + y^2 + z^2)^{3/2}},$$

a system of three differential equations for three unknown functions $x(t)$, $y(t)$, $z(t)$. This system can be solved, and it turns out that, in accordance with Kepler's empirical observations, the orbit of the planet is a conic section with the sun at one focus, the areas swept out by a line joining the sun to the planet are equal for equal time intervals, and the squares of the periods of complete revolution for two planets are proportional to the cubes of their distances from the sun. We must omit the proof.

The problem of vibrations provides a more elementary illustration of Newton's method. Suppose that we have a particle moving along a straight line, the x-axis, and tied to the origin by an elastic force, such as a spring or a rubber band. If the particle is removed from its position of equilibrium at the origin to a position given by the coördinate x, the force will pull it back with an intensity that we assume proportional to the extension x; since the force is directed towards the origin, it will be represented by $-k^2x$, where $-k^2$ is a negative factor of proportionality expressing the strength of the elastic spring or rubber band. Furthermore, we assume that there is friction retarding the motion, and that this friction is proportional to the velocity x' of the particle, with a factor of proportionality $-r$. Then the total force at any moment will be given by $-k^2x - rx'$, and according to Newton's general principle we find $mx'' = -k^2x - rx'$ or

$$mx'' + rx' + k^2x = 0.$$

This is exactly the differential equation (11) of damped vibrations mentioned above.

This simple example is of great importance, since many types of vibrating mechanical and electrical systems can be described mathematically by exactly this differential equation. Here we have a typical instance where an abstract mathematical formulation bares with one stroke the innermost structure of many apparently quite different and unconnected individual phenomena. This abstraction from the particular nature of a given phenomenon to a formulation of the general law which governs the whole class of phenomena is one of the characteristic features of the mathematical treatment of physical problems.

SUPPLEMENT TO CHAPTER VIII

§1. MATTERS OF PRINCIPLE

1. Differentiability

We have linked the concept of derivative of a function $y = f(x)$ with the intuitive idea of tangent to the graph of the function. Since the general concept of function is so wide, it is necessary in the interests of logical completeness to do away with this dependence on geometrical intuition. For we have no guarantee that the intuitive facts familiar from the consideration of simple curves such as circles and ellipses will necessarily subsist for the graphs of more complicated functions. Consider, for example, the function in Figure 282, whose graph has a corner.

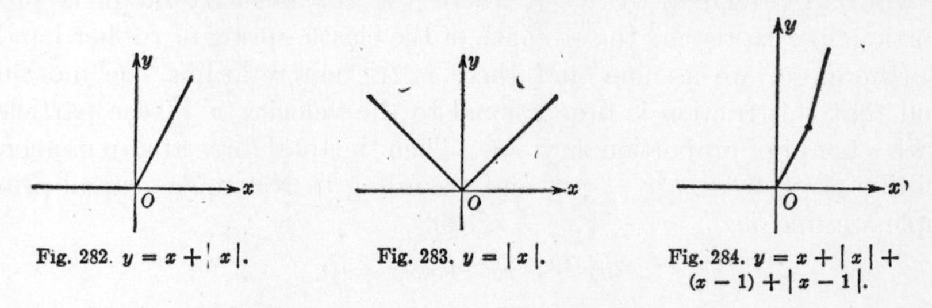

Fig. 282. $y = x + |x|$.

Fig. 283. $y = |x|$.

Fig. 284. $y = x + |x| + (x - 1) + |x - 1|$.

This function is defined by the equation $y = x + |x|$, where $|x|$ is the absolute value of x, i.e.

$$y = x + x = 2x \qquad \text{for} \quad x \geq 0,$$

$$y = x - x = 0 \qquad \text{for} \quad x < 0.$$

Another such example is the function $y = |x|$; still another is the function $y = x + |x| + (x - 1) + |x - 1|$. The graphs of these functions fail to have a definite tangent or direction at certain points; this means that the functions do not possess derivatives for the corresponding values of x.

Exercises: 1) Form the function $f(x)$ whose graph is one-half of a regular hexagon.

2) Where are the corners of the graph of

$$f(x) = (x + |x|) + \tfrac{1}{2}\{(x - \tfrac{1}{2}) + |x - \tfrac{1}{2}|\} + \tfrac{1}{4}\{(x - \tfrac{1}{4}) + |x - \tfrac{1}{4}|\}?$$

What are the discontinuities of $f'(x)$?

For another simple example of non-differentiability, we consider the function

$$y = f(x) = x \sin \frac{1}{x},$$

which is obtained from the function $\sin 1/x$ (see p. 283) by multiplication by the factor x; we define $f(x)$ to be zero for $x = 0$. This function, whose graph for positive values of x is shown in Figure 285, is con-

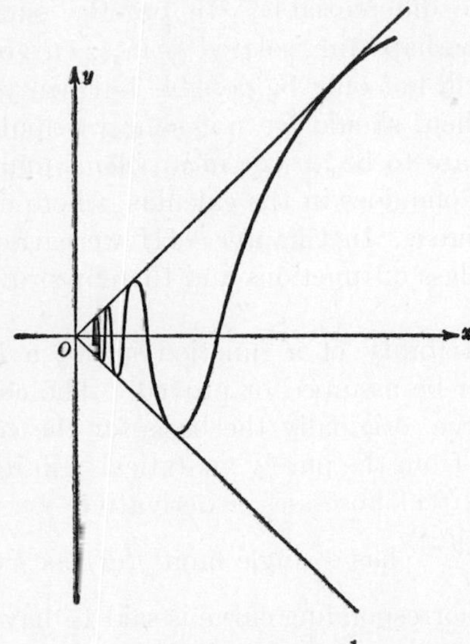

Fig. 285. $y = x \sin \frac{1}{x}$.

tinuous everywhere. The graph oscillates infinitely often in the neighborhood of $x = 0$, the "waves" becoming very small as we approach $x = 0$. The slope of these waves is given by

$$f'(x) = \sin \frac{1}{x} - \frac{1}{x} \cos \frac{1}{x}$$

(the reader may verify this as an exercise); as x tends to 0 this slope oscillates between ever-increasing positive and negative bounds. For $x = 0$ we may try to find the derivative as the limit for $h \to 0$ of the difference quotient

$$\frac{f(0 + h) - f(0)}{h} = \frac{h \sin \frac{1}{h}}{h} = \sin \frac{1}{h}.$$

But as $h \to 0$ this difference quotient oscillates between -1 and $+1$ and does not approach a limit; hence the function cannot be differentiated at $x = 0$.

These examples indicate a difficulty inherent in the subject. Weierstrass has most strikingly illustrated the situation by constructing a continuous function whose graph does not have a tangent at any point. While differentiability implies continuity, this shows that continuity does not imply differentiability, since Weierstrass' function is continuous and nowhere differentiable. In practice such difficulties will not arise. Except perhaps for isolated points, curves will be smooth and differentiation will not only be possible but will yield a continuous derivative. Why, then, should we not simply stipulate that "pathological" phenomena are to be absent in problems under consideration? This is exactly what one does in the calculus, where only differentiable functions are considered. In Chapter VIII we carried out the differentiation of a large class of functions and thereby proved their differentiability.

Since the differentiability of a function is not a logical matter of course, it must either be assumed or proved. The concept of tangent or direction of a curve, originally the basis for the concept of derivative, is then derived from the purely analytical definition of derivative: If the function $y = f(x)$ posessses a derivative, i.e. if the difference quotient $\frac{f(x+h) - f(x)}{h}$ has a single limit $f'(x)$ as h tends to 0 from either side, then the corresponding curve is said to have a tangent with the slope $f'(x)$. Thus the naive attitude of Fermat, Leibniz, and Newton is reversed in the interests of logical cogency.

Exercises: 1) Show that the continuous function defined by $x^2 \sin (1/x)$ has a derivative at $x = 0$.

2) Show that the function $\arctan (1/x)$ is discontinuous for $x = 0$, that $x \arctan (1/x)$ is continuous there but has no derivative, and that $x^2 \arctan (1/x)$ has a derivative at $x = 0$.

2. The Integral

The situation is similar with respect to the integral of a continuous function $f(x)$. Instead of considering the "area under the curve" $y = f(x)$ as a quantity which obviously exists and which can be expressed *a posteriori* as the limit of a sum, we *define* the integral by this limit, and consider the concept of integral as the primary basis from which the general concept of area is afterward derived. This attitude is forced upon us by a realization of the vagueness of geometrical intuition when applied to analytical concepts as general as that of continuous function. We start by forming a sum

$$(1) \qquad S_n = \sum_{j=1}^{n} f(v_j)(x_j - x_{j-1}) = \sum_{j=1}^{n} f(v_j)\Delta x_j,$$

where $x_0 = a, x_1, \cdots, x_n = b$ is a subdivision of the interval of integration, $\Delta x_j = x_j - x_{j-1}$ is the x-difference or length of the jth subinterval, and v_j is an arbitrary value of x in this subinterval, i.e. $x_{j-1} \leq v_j \leq x_j$. (We may take, for example, $v_j = x_j$ or $v_j = x_{j-1}$.) Now we form a sequence of such sums in which the number n of subintervals increases and at the same time the maximum length of the subintervals decreases to zero. Then the main fact is: The sum S_n for a given continuous function $f(x)$ tends to a definite limit A, which is independent of the specific way in which the subintervals and points v_j are chosen. By definition, this limit is the integral $A = \int_a^b f(x)\,dx$. Of course, the existence of this limit requires analytical proof if we do not wish to rely on an intuitive geometrical notion of area. This proof is given in every rigorous textbook on the calculus.

Comparing differentiation and integration, we are confronted with the following antithetical situation. Differentiability is definitely a restrictive condition on a continuous function, but the actual carrying out of the differentiation, i.e. the algorithm of the differential calculus, is in practice a straightforward procedure based on a few simple rules. On the other hand, every continuous function without exception possesses an integral between any two given limits. But the explicit calculation of such integrals, even for quite simple functions, is in general a very difficult task. At this point the fundamental theorem of the calculus becomes in many cases the decisive instrument for carrying out the integration. However, for most functions, even for very elementary ones, integration does not yield simple explicit expressions, and the numerical computation of integrals requires advanced methods.

3. Other Applications of the Concept of Integral. Work. Length

Dissociating the analytical notion of integral from its original geometrical interpretation, we meet a number of other, equally important, interpretations and applications. For example, the integral can be interpreted in mechanics as expressing the concept of work. The following simplest case will suffice for our explanation. Suppose a mass moves along the x-axis under the influence of a force directed along the axis. This mass is thought of as concentrated at the point with the coördinate x, and the force is given as a function $f(x)$ of the position, the sign of $f(x)$ indicating whether it points in the positive or negative x-direction. If the force is constant and moves the mass from a to b, then the work done is given by the product, $(b - a)f$, of the intensity f of the force and the distance traversed by the mass. But if the intensity varies with x, we shall have to define the amount of work done by a limiting process (as we defined velocity). To this end we divide the interval from a to b as before into small subintervals by the points $x_0 = a, x_1, \cdots, x_n = b$; then we imagine that in each subinterval the force is constant and equal, say, to $f(x_\nu)$, the actual value at the end-point, and calculate the work that would correspond to this stepwise varying force:

$$S_n = \sum_{\nu=1}^{n} f(x_\nu)\Delta x_\nu.$$

If we now refine the subdivision as before and let n increase, we see that the sum tends to the integral

$$\int_a^b f(x)\,dx.$$

Thus the work done by a continuously varying force is defined by an integral.

As an example let us consider a mass m fastened by an elastic spring to the origin $x = 0$. The force $f(x)$ will, in line with the discussion on page 461, be proportional to x,

$$f(x) = -k^2x,$$

where k^2 is a positive constant. Then the work done by this force if the mass moves from the origin to the position $x = b$ will be

$$\int_0^b - k^2x\,dx = -k^2\frac{b^2}{2},$$

and the work we must do against this force, if we want to pull out the spring to this position, is $+\,k^2\frac{b^2}{2}$.

A second application of the general notion of integral is to the concept of arc length of a curve. Let us suppose that the portion of the curve under consideration is represented by a function $y = f(x)$ whose derivative $f'(x) = \frac{dy}{dx}$ is also a continuous function. To define length we proceed exactly as though we had to measure a curve for practical purposes with a straight yardstick. We inscribe in the arc AB a polygon with n small edges, measure the total length L_n of this polygon, and consider the length L_n as an approximation; letting n increase and the maximum length of the edges of the polygon decrease toward zero, we define

$$L = \lim L_n$$

as the length of the arc AB. (In Chapter VI the length of a circle was obtained in this way as the limit of the perimeters of inscribed regular n-gons.) It can be shown that for sufficiently smooth curves this limit exists and is independent of the specific way in which the sequence of inscribed polygons is chosen. Curves for which this holds are said to be *rectifiable*. Any "reasonable" curve that arises in theory or applications will be rectifiable, and we shall not dwell on the investigation of pathological cases. It will suffice to show that the arc AB, for a function $y = f(x)$ with a continuous derivative $f'(x)$, has a length L in this sense, and that L can be expressed by an integral.

To this end, let us denote the x-coördinates of A and B by a and b respectively, then subdivide the x-interval from a to b as before by the points $x_0 = a, x_1, \cdots, x_j, \cdots, x_n = b$, with the differences $\Delta x_j = x_j - x_{j-1}$, and consider the polygon with the vertices $x_j, y_j = f(x_j)$ above these points of subdivision. A single edge of the polygon will

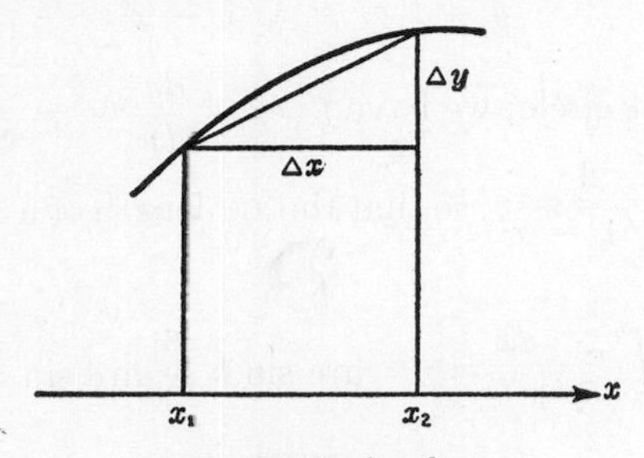

Fig. 286. Arc length.

have the length $\sqrt{(x_j - x_{j-1})^2 + (y_j - y_{j-1})^2} = \sqrt{\Delta x_j^2 + \Delta y_j^2} = \Delta x_j \sqrt{1 + \left(\frac{\Delta y_j}{\Delta x_j}\right)^2}$. Hence we have for the total length of the polygon

$$L_n = \sum_{j=1}^{n} \sqrt{1 + \left(\frac{\Delta y_j}{\Delta x_j}\right)^2}\, \Delta x_j.$$

If now n tends to infinity, the difference quotients $\frac{\Delta y_j}{\Delta x_j}$ will tend to the derivative $\frac{dy}{dx} = f'(x)$ and we obtain for the length L the integral expression

$$(2) \qquad L = \int_a^b \sqrt{1 + (f'(x))^2}\, dx.$$

Without going into further details of this theoretical discussion we make two supplementary remarks. First, if B is considered as a variable point on the curve with the coördinate x, then $L = L(x)$ becomes a function of x, and we have by the fundamental theorem,

$$L'(x) = \frac{dL}{dx} = \sqrt{1 + [f'(x)]^2},$$

a frequently used formula. Second, while formula (2) gives the "general" solution of the problem, it hardly yields an explicit expression for arc length in particular cases. For this we have to substitute the specific function $f(x)$, or rather $f'(x)$, in (2), and then to undertake the actual integration of the expression obtained. Here the difficulty is in general insurmountable if we restrict ourselves to the realm of the elementary functions considered in this book. We shall mention a few cases in which the integration is possible. The function

$$y = f(x) = \sqrt{1 - x^2}$$

represents the unit circle; we have $f'(x) = \frac{dy}{dx} = -\frac{x}{\sqrt{1 - x^2}}$, whence $\sqrt{1 + f'(x)^2} = \frac{1}{\sqrt{1 - x^2}}$, so that the arc length of a circular arc is given by the integral

$$\int_a^b \frac{dx}{\sqrt{1 - x^2}} = \text{arc sin } b - \text{arc sin } a.$$

For the parabola $y = x^2$ we have $f'(x) = 2x$ and the arc length from $x = 0$ to $x = b$ is

$$\int_0^b \sqrt{1 + 4x^2}\, dx.$$

For the curve $y = \log \sin x$ we have $f'(x) = \cot x$ and the arc length is expressed by

$$\int_a^b \sqrt{1 + \cot^2 x}\, dx.$$

We shall be content with merely writing down these integral expressions. They could be evaluated with a little more technique than we have at our command, but we shall go no farther in this direction.

§2. ORDERS OF MAGNITUDE

1. The Exponential Function and Powers of x

Frequently in mathematics we encounter sequences a_n which tend to infinity. Often we need to compare such a sequence with another sequence, b_n, also tending to infinity, but perhaps "faster" than a_n. To make this concept precise, we shall say that b_n tends to infinity faster than a_n, or has a *higher order of magnitude* than a_n, if the *ratio* a_n/b_n (numerator and denominator of which both tend to infinity) tends to zero as n increases. Thus the sequence $b_n = n^2$ tends to infinity faster than the sequence $a_n = n$, and the latter in turn faster than $c_n = \sqrt{n}$, for

$$\frac{a_n}{b_n} = \frac{n}{n^2} = \frac{1}{n} \to 0, \qquad \frac{c_n}{a_n} = \frac{\sqrt{n}}{n} = \frac{1}{\sqrt{n}} \to 0.$$

It is clear that n^s tends to infinity faster than n^r whenever $s > r > 0$, since then $n^r/n^s = 1/n^{(s-r)} \to 0$.

If the ratio a_n/b_n approaches a finite constant c, different from zero, we say that the two sequences a_n and b_n approach infinity at the same rate or have the *same order of magnitude*. Thus $a_n = n^2$ and $b_n = 2n^2 + n$ have the same order of magnitude, since

$$\frac{a_n}{b_n} = \frac{n^2}{2n^2 + n} = \frac{1}{2 + \frac{1}{n}} \to \frac{1}{2}.$$

One might think that with the powers of n as a yardstick one could measure the different degrees of becoming infinite for any sequence a_n that tends to infinity. To do this one would have to find a suitable power n^s with the same order of magnitude as a_n; i.e. such that a_n/n^s tends to a finite constant different from zero. It is a remarkable fact that this is by no means always possible, since *the exponential function* a^{n} *with* $\mathrm{a} > 1$ (e.g. e^{n}) *tends to infinity faster than any power* n^{s}, *however large we choose* s, *while* log n *tends to infinity slower than any power* n^{s}, *however small the positive exponent* s. In other words, we have the relations

$$\frac{n^s}{a^n} \to 0 \tag{1}$$

and

$$\text{(2)} \qquad \frac{\log n}{n^s} \to 0$$

as $n \to \infty$. The exponent s here need not be an integer, but may be any fixed positive number.

To prove (1) we first simplify the statement by taking the sth root of the ratio; if the root tends to zero, the original ratio does also. Hence we need only prove that

$$\frac{n}{a^{n/s}} \to 0$$

as n increases. Let $b = a^{1/s}$; since a is assumed to be greater than 1, b and also $\sqrt{b} = b^{\frac{1}{2}}$ will be greater than 1. We may write

$$b^{\frac{1}{2}} = 1 + q,$$

where q is positive. Now by the inequality (6) on page 15,

$$b^{n/2} = (1 + q)^n \geq 1 + nq > nq,$$

so that

$$a^{n/s} = b^n > n^2 q^2$$

and

$$\frac{n}{a^{n/s}} < \frac{n}{n^2 q^2} = \frac{1}{nq^2}.$$

Since the latter quantity tends to zero as n increases, the proof is complete.

As a matter of fact, the relation

$$\text{(3)} \qquad \frac{x^s}{a^x} \to 0$$

holds when x becomes infinite in any manner by running through a sequence $x_1, x_2, \cdots$, which need not coincide with the sequence 1, 2, 3, $\cdots$ of positive integers. For if $n - 1 \leq x \leq n$, then

$$\frac{x^s}{a^x} < \frac{n^s}{a^{n-1}} = a \cdot \frac{n^s}{a^n} \to 0.$$

This remark may be used to prove (2). Setting $x = \log n$ and $e^s = a$, so that $n = e^x$ and $n^s = (e^s)^x$, the ratio in (2) becomes

$$\frac{x}{a^x},$$

which is the special case of (3) for $s = 1$.

Exercises: 1) Prove that for $x \to \infty$ the function $\log \log x$ tends to infinity more slowly than $\log x$. 2) The derivative of $x/\log x$ is $1/\log x - 1/(\log x)^2$. Prove that for large x it is "asymptotically" equivalent to the first term, $1/\log x$, i.e. that their ratio tends to 1 as $x \to \infty$.

2. Order of Magnitude of log (n!)

In many applications, e.g. in the theory of probability, it is important to know the order of magnitude or "asymptotic behavior" of $n!$ for large values of n. We shall here be content with studying the logarithm of $n!$, i.e. the expression

$$P_n = \log 2 + \log 3 + \log 4 + \cdots + \log n.$$

We shall show that the "asymptotic value" of P_n is given by $n \log n$; i.e. that

$$\frac{\log (n!)}{n \log n} \to 1$$

as $n \to \infty$.

The proof is typical of a much used method of comparing a sum with an integral. In Figure 287 the sum P_n is equal to the sum of the areas

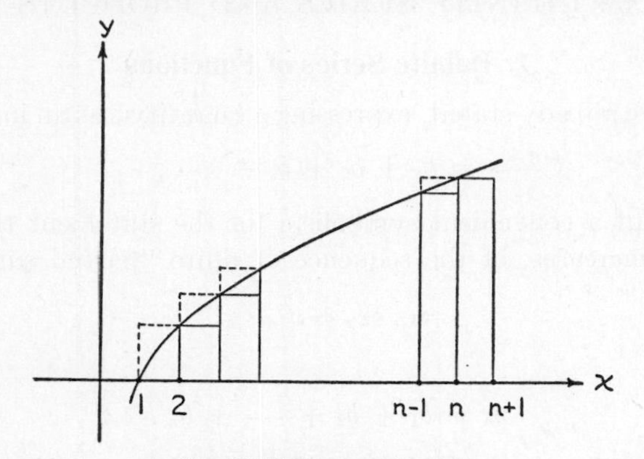

Fig. 287. Estimation of log (n!).

of the rectangles whose tops are marked by solid lines, and which together do not exceed the area

$$\int_1^{n+1} \log x \, dx = (n + 1) \log (n + 1) - (n + 1) + 1$$

under the logarithmic curve from 1 to $n + 1$ (see p. 450, Exercise 1)). But the sum P_n is likewise equal to the total area of the rectangles whose tops are marked by broken lines and which together exceed the area under the curve from 1 to n, given by

$$\int_1^n \log x\,dx = n \log n - n + 1.$$

Thus we have

$$n \log n - n + 1 < P_n < (n + 1) \log (n + 1) - n,$$

and dividing by $n \log n$,

$$1 - \frac{1}{\log n} + \frac{1}{n \log n} < \frac{P_n}{n \log n} < (1 + 1/n) \frac{\log (n + 1)}{\log n} - \frac{1}{\log n}$$

$$= (1 + 1/n) \frac{\log n + \log (1 + 1/n)}{\log n} - \frac{1}{\log n}.$$

Obviously the two bounds tend to 1 as n tends to infinity, and our statement is proved.

Exercise: Prove that the two bounds are greater than $1 - 1/n$ and less than $1 + 1/n$ respectively.

§3. INFINITE SERIES AND PRODUCTS

1. Infinite Series of Functions

As we have already stated, expressing a quantity s as an infinite series,

$$s = b_1 + b_2 + b_3 + \cdots, \tag{1}$$

is nothing but a convenient symbolism for the statement that s is the limit, as n increases, of the sequence of finite "partial sums",

$$s_1, s_2, s_3, \cdots,$$

where

$$s_n = b_1 + b_2 + \cdots + b_n. \tag{2}$$

Thus the equation (1) is equivalent to the limiting relation

$$\lim s_n = s \text{ as } n \to \infty, \tag{3}$$

where s_n is defined by (2). When the limit (3) exists we say that the series (1) *converges* to the value s, while if the limit (3) does not exist, we say that the series *diverges.*

Thus, the series

$$1 - \tfrac{1}{3} + \tfrac{1}{5} - \tfrac{1}{7} + \cdots$$

converges to the value $\pi/4$, and the series

$$1 - \tfrac{1}{2} + \tfrac{1}{3} - \tfrac{1}{4} + \cdots$$

converges to the value log 2; on the other hand, the series

$$1 - 1 + 1 - 1 + \cdots$$

diverges (since the partial sums alternate between 1 and 0), and the series

$$1 + 1 + 1 + 1 + \cdot$$

diverges because the partial sums tend to infinity.

We have already encountered series whose terms b_i are functions of x of the form

$$b_i = c_i x^i,$$

with constant factors c_i. Such series are called *power series*; they are limits of polynomials representing the partial sums

$$S_n = c_0 + c_1 x + c_2 x^2 + \cdots + c_n x^n$$

(the addition of the constant term c_0 requires an unessential change in the notation (2)). An expansion

$$f(x) = c_0 + c_1 x + c_2 x^2 + \cdots$$

of a function $f(x)$ in a power series is thus a way of expressing an approximation of $f(x)$ by polynomials, the simplest functions. Summarizing and supplementing previous results, we list the following power series expansions:

$$\frac{1}{1+x} = 1 - x + x^2 - x^3 + \cdots, \qquad \text{valid for } -1 < x < +1 \tag{4}$$

$$\tan^{-1} x = x - \frac{x^3}{3} + \frac{x^5}{5} - \cdots, \qquad \text{valid for } -1 \leq x \leq +1 \tag{5}$$

$$\log(1 + x) = x - \frac{x^2}{2} + \frac{x^3}{3} - \cdots, \qquad \text{valid for } -1 < x \leq +1 \tag{6}$$

$$\tfrac{1}{2}\log\frac{1+x}{1-x} = x + \frac{x^3}{3} + \frac{x^5}{5} + \cdots, \qquad \text{valid for } -1 < x < +1 \tag{7}$$

$$e^x = 1 + x + \frac{x^2}{2!} + \frac{x^3}{3!} + \frac{x^4}{4!} + \cdots, \qquad \text{valid for all } x. \tag{8}$$

To this collection we now add the following important expansions:

$$\sin x = x - \frac{x^3}{3!} + \frac{x^5}{5!} - \cdots, \qquad \text{valid for all } x, \tag{9}$$

$$\text{(10)} \qquad \cos x = 1 - \frac{x^2}{2!} + \frac{x^4}{4!} - \cdots, \qquad \text{valid for all } x.$$

The proof is a simple consequence of the formulas (see p. 440)

$$\text{(a)} \qquad \int_0^x \sin u\, du = 1 - \cos x,$$

$$\text{(b)} \qquad \int_0^x \cos u\, du = \sin x.$$

We start with the inequality

$$\cos x \leq 1.$$

Integrating from 0 to x, where x is any fixed positive number, we find (see formula (13), p. 414)

$$\sin x \leq x;$$

integrating this again,

$$1 - \cos x \leq \frac{x^2}{2},$$

which is the same thing as

$$\cos x \geq 1 - \frac{x^2}{2}.$$

Integrating once more, we obtain

$$\sin x \geq x - \frac{x^3}{2 \cdot 3} = x - \frac{x^3}{3!}.$$

Proceeding indefinitely in this manner, we get the two sets of inequalities

$$\begin{array}{ll}
\sin x \leq x & \cos x \leq 1 \\
\sin x \geq x - \dfrac{x^3}{3!} & \cos x \geq 1 - \dfrac{x^2}{2!} \\
\sin x \leq x - \dfrac{x^3}{3!} + \dfrac{x^5}{5!} & \cos x \leq 1 - \dfrac{x^2}{2!} + \dfrac{x^4}{4!} \\
\sin x \geq x - \dfrac{x^3}{3!} + \dfrac{x^5}{5!} - \dfrac{x^7}{7!} & \cos x \geq 1 - \dfrac{x^2}{2!} + \dfrac{x^4}{4!} - \dfrac{x^6}{6!} \\
\cdots\cdots & \cdots\cdots
\end{array}$$

Now $x^n/n! \to 0$ as n tends to infinity. To show this we choose a fixed integer m such that $x/m < \frac{1}{2}$, and write $c = x^m/m!$. For any integer $n > m$ let us set $n = m + r$; then

$$0 < \frac{x^n}{n!} = c \cdot \frac{x}{m+1} \cdot \frac{x}{m+2} \cdots \frac{x}{m+r} < c(\tfrac{1}{2})^r,$$

and as $n \to \infty$, r also $\to \infty$ and hence $c(\frac{1}{2})^r \to 0$. It follows that

$$\begin{cases} \sin x = x - \dfrac{x^3}{3!} + \dfrac{x^5}{5!} - \dfrac{x^7}{7!} + \cdots \\ \cos x = 1 - \dfrac{x^2}{2!} + \dfrac{x^4}{4!} - \dfrac{x^6}{6!} + \cdots . \end{cases}$$

Since the terms of the series are of alternating sign and decreasing magnitude (at least for $|x| \leq 1$), it follows that *the error committed by breaking off either series at any term will not exceed in magnitude the value of the first term dropped.*

Remarks. These series can be used for the computation of tables. Example: What is $\sin 1^0$? 1^0 is $\pi/180$ in radian measure; hence

$$\sin \frac{\pi}{180} = \frac{\pi}{180} - \frac{1}{6}\left(\frac{\pi}{180}\right)^3 + \cdots .$$

The error committed by breaking off here is not greater than $\dfrac{1}{120}\left(\dfrac{\pi}{180}\right)^5$, which is less than 0.000 000 000 02. Hence $\sin 1^0 = 0.017\,452\,406\,4$, to 10 places of decimals.

Finally, we mention without proof the "binomial series"

$$(1 + x)^a = 1 + ax + C_2^a x^2 + C_3^a x^3 + \cdots , \tag{11}$$

where C_s^a is the "binomial coefficient"

$$C_s^a = \frac{a(a-1)(a-2)\cdots(a-s+1)}{s!}.$$

If $a = n$ is a positive integer, then we have $C_n^a = 1$, and for $s > n$ all the coefficients C_s^a in (11) are zero, so that we simply retain the finite formula of the ordinary binomial theorem. It was one of Newton's great discoveries, made at the beginning of his career, that the elementary binomial theorem can be extended from positive integral exponents n to arbitrary positive or negative, rational or irrational exponents a. When a is not an integer the right side of (11) yields an infinite series, valid for $-1 < x < +1$. For $|x| > 1$ the series (11) is divergent and thus the equality sign is meaningless.

In particular, we find, by substituting $a = \frac{1}{2}$ in (11), the expansion

$$\sqrt{1 + x} = 1 + \tfrac{1}{2}x - \frac{1}{2^2 \cdot 2!}x^2 + \frac{1 \cdot 3}{2^3 \cdot 3!}x^2 - \frac{1 \cdot 3 \cdot 5}{2^4 \cdot 4!}x^4 + \cdots . \tag{12}$$

Like the other mathematicians of the eighteenth century, Newton did not give a real proof for the validity of his formula. A satisfactory analysis of the convergence and range of validity of such infinite series was not given until the nineteenth century.

Exercise: Write the power series for $\sqrt{1 - x^2}$ and for $1/\sqrt{1 - x}$.

The expansions (4) – (11) are special cases of the general formula of Brook Taylor (1685–1731), which aims at expanding any one of a large class of functions $f(x)$ in the form of a power series,

$$f(x) = c_0 + c_1x + c_2x^2 + c_3x^3 + \cdots, \tag{13}$$

by finding a law that expresses the coefficients c_i in terms of the function f and its derivatives. It is not possible here to give a precise proof of Taylor's formula by formulating and establishing the conditions for its validity. But the following plausibility considerations will illuminate the interconnections of the relevant mathematical facts.

Let us tentatively assume that an expansion (13) is possible. Let us further assume that $f(x)$ can be differentiated, that $f'(x)$ can be differentiated, and so on, so that the unending succession of derivatives

$$f'(x), f''(x), \cdots, f^{(n)}(x), \cdots$$

actually exists. Finally, we take as granted that an infinite power series may be differentiated term by term just like a finite polynomial. Under these assumptions, we can determine the coefficients c_n from a knowledge of the behavior of $f(x)$ in the neighborhood of $x = 0$. First, by substituting $x = 0$ in (13), we find

$$c_0 = f(0),$$

since all terms of the series containing x disappear. Now we differentiate (13) and obtain

$$f'(x) = c_1 + 2c_2x + 3c_3x^2 + \cdots + nc_nx^{n-1} + \cdots. \tag{13'}$$

Again substituting $x = 0$, but this time in (13′) and not in (13), we find

$$c_1 = f'(0).$$

By differentiating (13) we obtain

$$f''(x) = 2c_2 + 2\cdot3\cdot c_3x + \cdots + (n - 1)\cdot n\cdot c_nx^{n-2} + \cdots; \tag{13''}$$

then substituting $x = 0$ in (13″), we see that

$$2!\, c_2 = f''(0).$$

Similarly, diff rentiating (13″) and then substituting $x = 0$,

$$3!\, c_3 = f'''(0),$$

and by continuing this procedure, we get the general formula

$$c_n = \frac{1}{n!} f^{(n)}(0),$$

where $f^{(n)}(0)$ is the value of the nth derivative of $f(x)$ at $x = 0$. The result is the *Taylor series*

$$(14) \qquad f(x) = f(0) + xf'(0) + \frac{x^2}{2!} f''(0) + \frac{x^3}{3!} f'''(0) + \cdots.$$

As an exercise in differentiation the reader may verify that, in examples (4)–(11), the law of formation of the coefficients of a Taylor series is satisfied.

2. Euler's Formula, cos x + i sin x = e^ix

One of the most fascinating results of Euler's formalistic manipulations is an intimate connection in the domain of complex numbers between the sine and cosine functions on the one hand, and the exponential function on the other. It should be stated in advance that Euler's "proof" and our subsequent argument have in no sense a rigorous character; they are typically eighteenth century examples of formal manipulation.

Let us start with De Moivre's formula proved in Chapter II,

$$(\cos n\varphi + i \sin n\varphi) = (\cos \varphi + i \sin \varphi)^n.$$

In this we substitute $\varphi = x/n$, obtaining the formula

$$(\cos x + i \sin x) = \left(\cos \frac{x}{n} + i \sin \frac{x}{n}\right)^n.$$

Now if x is given, then $\cos \frac{x}{n}$ will differ but slightly from $\cos 0 = 1$ for large n; moreover, since

$$\frac{\sin \frac{x}{n}}{\frac{x}{n}} \to 1 \quad \text{as} \quad \frac{x}{n} \to 0$$

(see p. 307), we see that $\sin \frac{x}{n}$ is asymptotically equal to $\frac{x}{n}$. We may therefore find it plausible to proceed to the limit formula

$$(14) \qquad \cos x + i \sin x = \lim \left(1 + \frac{ix}{n}\right)^n \text{ as } n \to \infty.$$

Comparing the right side of this equation with the formula (p. 449)

$$e^z = \lim \left(1 + \frac{z}{n}\right)^n \text{ as } n \to \infty ,$$

we have

$$\text{(15)} \qquad \cos x + i \sin x = e^{ix},$$

which is Euler's result.

We may obtain the same result in another formalistic way from the expansion of e^z,

$$e^z = 1 + \frac{z}{1!} + \frac{z^2}{2!} + \frac{z^3}{3!} + \cdots,$$

by substituting in it $z = ix$, x being a real number. If we recall that the successive powers of i are i, -1, $-i$, $+1$, and so on periodically, then by collecting real and imaginary parts we find

$$e^{ix} = \left(1 - \frac{x^2}{2!} + \frac{x^4}{4!} - \frac{x^6}{6!} + \cdots\right) + i\left(x - \frac{x^3}{3!} + \frac{x^5}{5!} - \frac{x^7}{7!} + \cdots\right);$$

comparing the right hand side with the series for $\sin x$ and $\cos x$ we again obtain Euler's formula.

Such reasoning is by no means an actual proof of the relation (15). The objection to our second argument is that the series expansion for e^z was derived under the assumption that z is a real number; therefore the substitution $z = ix$ requires justification. Likewise the validity of the first argument is destroyed by the fact that the formula

$$e^z = \lim (1 + z/n)^n \text{ as } n \to \infty$$

was derived for real values of z only.

To remove Euler's formula from the sphere of mere formalism to that of rigorous mathematical truth required the development of the theory of functions of a complex variable, one of the great mathematical achievements of the nineteenth century. Many other problems stimulated this far-reaching development. We have seen, for example, that the expansions of functions in power series converge for different x-intervals. Why do some expansions converge always, i.e. for all x, while others become meaningless for $|x| > 1$?

Consider, for example, the geometrical series (4), page 473, which converges for $|x| < 1$. The left side of this equation is perfectly meaningful when $x = 1$, taking the value $\frac{1}{1+1} = \frac{1}{2}$, while the series on the right behaves most strangely, becoming

$$1 - 1 + 1 - 1 + \cdots.$$

This series does not converge, since its partial sums oscillate between 1 and 0. This indicates that functions may give rise to divergent series even when the functions themselves do not show any irregularity. Of course, the function $\frac{1}{1+x}$ becomes infinite when $x \to -1$. Since it can easily be shown that convergence of a power series for $x = a > 0$ always implies convergence for $-a < x < a$, we might find an "explanation" of the queer behavior of the expansion in the discontinuity of $\frac{1}{1+x}$ for $x = -1$. But the function $\frac{1}{1+x^2}$ may be expanded into the series

$$\frac{1}{1+x^2} = 1 - x^2 + x^4 - x^6 + \cdots$$

by substituting x^2 for x in (4). This series will also converge for $|x| < 1$, while for $x = 1$ it again leads to the divergent series $1 - 1 + 1 - 1 + \cdots$, and for $|x| > 1$ it diverges explosively, although the function itself is everywhere regular.

It has turned out that a complete explanation of such phenomena is possible only when the functions are studied for *complex* values of the variable x, as well as for real values. For example, the series for $\frac{1}{1+x^2}$ must diverge for $x = i$ because the denominator of the fraction becomes zero. It follows that the series must also diverge for all x such that $|x| > |i| = 1$, since it can be shown that its convergence for any such x would imply its convergence for $x = i$. Thus the question of the convergence of series, completely neglected in the early period of the calculus, became one of the main factors in the creation of the theory of functions of a complex variable.

3. The Harmonic Series and the Zeta Function. Euler's Product for the Sine

Series whose terms are simple combinations of the integers are particularly interesting. As an example we consider the "harmonic series"

$$1 + \frac{1}{2} + \frac{1}{3} + \frac{1}{4} + \cdots + \frac{1}{n} + \cdots, \tag{16}$$

which differs from that for log 2 by the signs of the even-numbered terms only.

To ask whether this series converges is to ask whether the sequence

$$s_1, s_2, s_3, \cdots,$$

where

$$s_n = 1 + \frac{1}{2} + \frac{1}{3} + \cdots + \frac{1}{n}, \tag{17}$$

tends to a finite limit. Although the terms of the series (16) approach 0 as we go out farther and farther, it is easy to see that the series does not converge. For by taking enough terms we can exceed any positive number whatever, so that s_n increases without limit and hence the series (16) "diverges to infinity." To see this we observe that

$$s_2 = 1 + \tfrac{1}{2},$$

$$s_4 = s_2 + (\tfrac{1}{3} + \tfrac{1}{4}) > s_2 + (\tfrac{1}{4} + \tfrac{1}{4}) = 1 + \tfrac{2}{2},$$

$$s_8 = s_4 + (\tfrac{1}{5} + \cdots + \tfrac{1}{8}) > s_4 + (\tfrac{1}{8} + \cdots + \tfrac{1}{8}) = s_4 + \tfrac{1}{2} > 1 + \tfrac{3}{2},$$

and in general

$$s_{2^m} > 1 + \frac{m}{2}. \tag{18}$$

Thus, for example, the partial sums s_{2^m} exceed 100 as soon as $m \geq 200$.

Although the harmonic series does not converge, the series

$$1 + \frac{1}{2^s} + \frac{1}{3^s} + \frac{1}{4^s} + \cdots + \frac{1}{n^s} + \cdots \tag{19}$$

may be shown to converge for any value of s greater than 1, and defines for all $s > 1$ the so-called zeta function,

$$\zeta(s) = \lim \left(1 + \frac{1}{2^s} + \frac{1}{3^s} + \frac{1}{4^s} + \cdots + \frac{1}{n^s}\right) \textit{ as } n \to \infty, \tag{20}$$

as a function of the variable s. There is an important relation between the zeta-function and the prime numbers, which we may derive by using our knowledge of the geometrical series. Let $p = 2, 3, 5, 7, \cdots$ be any prime; then for $s \geq 1$,

$$0 < \frac{1}{p^s} < 1,$$

so that

$$\frac{1}{1 - 1/p^s} = 1 + \frac{1}{p^s} + \frac{1}{p^{2s}} + \frac{1}{p^{3s}} + \cdots.$$

Let us multiply together these expressions for all the primes $p = 2, 3, 5, 7, \cdots$ without concerning ourselves with the validity of such an operation. On the left we obtain the infinite "product"

$$\left(\frac{1}{1 - 1/2^s}\right)\cdot\left(\frac{1}{1 - 1/3^s}\right)\cdot\left(\frac{1}{1 - 1/5^s}\right)\cdots$$

$$= \text{limit as } n \to \infty \text{ of } \left[\frac{1}{1 - 1/p_1^s}\cdots\frac{1}{1 - 1/p_n^s}\right],$$

while on the other side we obtain the series

$$1 + \frac{1}{2^s} + \frac{1}{3^s} + \cdots = \zeta(s),$$

by virtue of the fact that every integer greater than 1 can be expressed uniquely as the product of powers of distinct primes. Thus we have represented the zeta-function as a product:

$$\zeta(s) = \left(\frac{1}{1 - 1/2^s}\right)\left(\frac{1}{1 - 1/3^s}\right)\left(\frac{1}{1 - 1/5^s}\right)\cdots. \tag{21}$$

If there were but a finite number of distinct primes, say $p_1, p_2, p_3, \cdots, p_r$, then the product on the right side of (21) would be an ordinary finite product and would therefore have a finite value, even for $s = 1$. But as we have seen, the zeta series for $s = 1$

$$\zeta(1) = 1 + \tfrac{1}{2} + \tfrac{1}{3} + \cdots$$

diverges to infinity. This argument, which can easily be made into a rigorous proof, shows that there are infinitely many primes. Of course, this is much more involved and sophisticated than the proof given by Euclid (see p. 22). But it has the fascination of a difficult ascent of a mountain peak which could be reached from the other side by a comfortable road.

Infinite products such as (21) are sometimes just as useful as infinite series for representing functions. Another infinite product, whose discovery was one more of Euler's achievements, concerns the trigonometric function $\sin x$. To understand this formula we start with a remark on polynomials. If $f(x) = a_0 + a_1x + \cdots + a_nx^n$ is a polynomial of degree n and has n distinct zeros, $x_1, \cdots, x_n$, then it is known from algebra that $f(x)$ can be decomposed into linear factors:

$$f(x) = a_n(x - x_1)\cdots(x - x_n)$$

(see p. 101). By factoring out the product $x_1x_2\cdots x_n$ we can write

$$f(x) = C\left(1 - \frac{x}{x_1}\right)\left(1 - \frac{x}{x_2}\right)\cdots\left(1 - \frac{x}{x_n}\right),$$

where C is a constant which, by setting $x = 0$, is recognized as $C = a_0$. Now, if instead of polynomials we consider more complicated functions $f(x)$, the question arises whether a product decomposition by means of the zeros of $f(x)$ is still possible. (In general this cannot be true, as shown by the example of the exponential function which has no zeros at all, since $e^x \neq 0$ for every value of x.) Euler discovered that for the sine function such a decomposition *is* possible. To write the formula in the simplest way, we consider not $\sin x$, but $\sin \pi x$. This function has the zeros $x = 0, \pm 1, \pm 2, \pm 3, \cdots$, since $\sin \pi n = 0$ for all integers n and for no other numbers. Euler's formula now states that

$$\sin \pi x = \pi x \left(1 - \frac{x^2}{1^2}\right)\left(1 - \frac{x^2}{2^2}\right)\left(1 - \frac{x^2}{3^2}\right)\left(1 - \frac{x^2}{4^2}\right) \cdots. \tag{22}$$

This infinite product converges for all values of x, and is one of the most beautiful formulas in mathematics. For $x = \frac{1}{2}$ it yields

$$\sin \frac{\pi}{2} = 1 = \frac{\pi}{2}\left(1 - \frac{1}{2^2 \cdot 1^2}\right)\left(1 - \frac{1}{2^2 \cdot 2^2}\right)\left(1 - \frac{1}{2^2 \cdot 3^2}\right)\left(1 - \frac{1}{2^2 \cdot 4^2}\right) \cdots.$$

If we write

$$1 - \frac{1}{2^2 \cdot n^2} = \frac{(2n - 1)(2n + 1)}{2n \cdot 2n},$$

we obtain Wallis' product,

$$\frac{\pi}{2} = \frac{2}{1} \cdot \frac{2}{3} \cdot \frac{4}{3} \cdot \frac{4}{5} \cdot \frac{6}{5} \cdot \frac{6}{7} \cdot \frac{8}{7} \cdot \frac{8}{9} \cdots,$$

mentioned on page 300.

For proofs of all these facts we must refer the reader to textbooks on the calculus (see also pp. 509–510).

**§4. THE PRIME NUMBER THEOREM OBTAINED BY STATISTICAL METHODS

When mathematical methods are applied to the study of natural phenomena one is usually satisfied with arguments in the course of which the chain of strict logical reasoning is interrupted by more or less plausible assumptions. Even in pure mathematics one encounters reasoning which, while it does not provide a rigorous proof, nevertheless suggests the correct solution and points the direction in which a rigorous proof may be sought. Bernoulli's solution of the brachistochrone problem (see p. 383) has this character, as does most of the early work in analysis.

By a procedure typical of applied mathematics and particularly of statistical mechanics we shall here present an argument that at least makes plausible the truth of Gauss's famous law of the distribution of the primes. (A related procedure was suggested to one of the authors by the experimental physicist Gustav Hertz.) This theorem, discussed empirically in the supplement to Chapter I, states that the number $A(n)$ of primes not exceeding n is asymptotically equivalent to the quantity $n/\log n$:

$$A(n) \sim \frac{n}{\log n}.$$

By this is meant that the ratio of $A(n)$ to $n/\log n$ tends to the limit 1 as n tends to infinity.

We start by making the assumption that there *exists* a mathematical law which describes the distribution of the primes in the following sense: for large values of n the function $A(n)$ is approximately equal to the integral $\int_2^n W(x)\,dx$, where $W(x)$ is a function which measures the "density" of the primes. (We choose 2 as lower limit of the integral because for $x < 2$ clearly $A(x) = 0$.) More precisely, let x be a large number and Δx another large number but such that the order of magnitude of x is greater than that of Δx. (For example, we might agree to set $\Delta x = \sqrt{x}$.) Then we are assuming that the distribution of the primes is so smooth that the number of primes in the interval from x to $x + \Delta x$ is approximately equal to $W(x)\cdot\Delta x$, and moreover that $W(x)$ as a function of x changes so slowly that the integral $\int_2^n W(x)\,dx$ may be replaced by a subsequent rectangular approximation without changing its asymptotic value. With these preliminary remarks we are ready to begin the argument.

We have proved (p. 471) that for large integers $\log n!$ is asymptotically equal to $n\cdot\log n$,

$$\log n! \sim n\cdot\log n.$$

Now we proceed by giving a second formula for $\log n!$ involving the primes and comparing the two expressions. Let us count how often an arbitrary prime p less than n is contained as a factor in the integer $n! = 1\cdot2\cdot3 \cdots n$. We shall denote by $[a]_p$ the largest integer k such that p^k divides a. Since the prime decomposition of every integer is unique, it follows that $[ab]_p = [a]_p + [b]_p$ for any two integers a and b. Hence

$$[n!]_p = [1]_p + [2]_p + [3]_p + \cdots + [n]_p.$$

The terms in the sequence 1, 2, 3, $\cdots$, n which are divisible by p^k are p^k, $2p^k$, $3p^k$, $\cdots$; their number N_k for large n is approximately n/p^k. The number M_k of these terms which are divisible by p^k and no higher power of p is equal to $N_k - N_{k+1}$. Hence

$$\begin{aligned}[n!]_p &= M_1 + 2M_2 + 3M_3 + \cdots \\ &= (N_1 - N_2) + 2(N_2 - N_3) + 3(N_3 - N_4) + \cdots \\ &= N_1 + N_2 + N_3 + \cdots \\ &= \frac{n}{p} + \frac{n}{p^2} + \frac{n}{p^3} + \cdots = \frac{n}{p-1}.\end{aligned}$$

(These equalities are, of course, only approximate.)

It follows that for large n the number $n!$ is given approximately by the product of all the expressions $p^{\frac{n}{p-1}}$ for all primes $p < n$. Thus we have the formula

$$\log n! \sim \sum_{p<n} \frac{n}{p-1} \log p.$$

Comparing this with our previous asymptotic relation for $\log n!$ we find, writing x instead of n,

$$\log x \sim \sum_{p<x} \frac{\log p}{p-1}. \tag{1}$$

The next and decisive step is to obtain an asymptotic expression in terms of $W(x)$ for the right side of (1). When x is very large we may subdivide the interval from 2 to $x = n$ into a large number r of large subintervals by choosing points $2 = \xi_1, \xi_2, \cdots, \xi_r, \xi_{r+1} = x$, with corresponding increments $\Delta\xi_j = \xi_{j+1} - \xi_j$. In each subinterval there may be primes, and all the primes in the jth subinterval will have approximately the value ξ_j. By our assumption on $W(x)$ there are approximately $W(\xi_j)\cdot\Delta\xi_j$ primes in the jth subinterval; hence the sum on the right side of (1) is approximately equal to

$$\sum_{j=1}^{r+1} W(\xi_j)\,\frac{\log \xi_j}{\xi_j - 1}\cdot\Delta\xi_j.$$

Replacing this finite sum by the integral which it approximates, we have as a plausible consequence of (1) the relation

$$\log x \sim \int_2^x W(\xi) \frac{\log \xi}{\xi - 1} \, d\xi. \tag{2}$$

From this we shall determine the unknown function $W(x)$. If we replace the sign $\sim$ by ordinary equality and differentiate both sides with respect to x, then by the fundamental theorem of the calculus

$$\begin{aligned} \frac{1}{x} &= W(x) \frac{\log x}{x - 1}, \\ W(x) &= \frac{x - 1}{x \log x}. \end{aligned} \tag{3}$$

We assumed at the beginning of our discussion that $A(x)$ is approximately equal to $\int_2^x W(x)\, dx$; hence $A(x)$ is approximately given by the integral

$$\int_2^x \frac{x - 1}{x \log x} \, dx. \tag{4}$$

In order to evaluate this integral we observe that the function $f(x) = x/\log x$ has the derivative

$$f'(x) = \frac{1}{\log x} - \frac{1}{(\log x)^2}.$$

For large values of x the two expressions

$$\frac{1}{\log x} - \frac{1}{(\log x)^2}, \qquad \frac{1}{\log x} - \frac{1}{x \log x}$$

are approximately equal, since for large x the second term in both cases will be much smaller than the first. Hence the integral (4) will be asymptotically equal to the integral

$$\int_2^x f'(x)\, dx = f(x) - f(2) = \frac{x}{\log x} - \frac{2}{\log 2},$$

since the integrands will be almost equal over most of the range of integration. The term $2/\log 2$ can be neglected for large x since it is a constant, and thus we obtain the final result

$$A(x) \sim \frac{x}{\log x},$$

which is the prime number theorem.

We cannot pretend that the preceding argument has more than a suggestive value. But on closer analysis the following fact emerges. It is not difficult to give a complete justification for all the steps that we have so boldly made; in particular, for equation (1), for the asymptotic equivalence between this sum and the integral in (2), and for the step leading from (2) to (3). It is far more difficult to prove the *existence* of a smooth density function $W(x)$, which we assumed at the beginning. Once we accept this, the *evaluation* of the function is a comparatively simple matter; from this point of view the proof of the existence of such a function is the central difficulty of the prime number problem.

APPENDIX

SUPPLEMENTARY REMARKS, PROBLEMS, AND EXERCISES

Many of the following problems are intended for the somewhat advanced reader. They are designed not so much to develop routine technique as to stimulate inventive ability.

Arithmetic and Algebra

(1) How do we know that 3 does not divide any power of 10, as stated on page 61? (See p. 47.)

(2) Prove that the principle of the smallest integer is a consequence of the theorem of mathematical induction. (See p. 19.)

(3) By the binomial theorem applied to the expansion of $(1 + 1)^n$, show that $C_0^n + C_1^n + C_2^n + \cdots + C_n^n = 2^n$.

(*4) Take any integer, $z = abc \cdots$, form the sum of its digits, $a + b + c + \cdots$, subtract this from z, cross out any one digit from the result, and denote the sum of the remaining digits by w. From a knowledge of w alone, can a rule be found for determining the value of the digit crossed out? (There will be one ambiguous case, when $w = 0$.) Like many other simple facts about congruences, this can be used as the basis for a parlor trick.

(5) An arithmetical progression of first order is a sequence of numbers, $a, a + d, a + 2d, a + 3d, \cdots$, such that the difference between successive members of the sequence is a constant. An arithmetical progression of second order is a sequence of numbers, $a_1, a_2, a_3, \cdots$ such that the differences $a_{i+1} - a_i$ form an arithmetical progression of first order. Similarly, an arithmetical progression of kth order is a sequence such that the differences form an arithmetical progression of order $k - 1$. Prove that the squares of the integers form an arithmetical progression of second order, and prove by induction that the kth powers of the integers form an arithmetical progression of order k. Prove that any sequence whose nth term, a_n, is given by the expression $c_0 + c_1n + c_2n^2 + \cdots + c_kn^k$, where the c's are constants, is an arithmetical progression of order k. *Prove the converse of this statement for $k = 2$; $k = 3$; for general k.

(6) Prove that the sum of the first n terms of an arithmetical progression of order k is an arithmetical progression of order $k + 1$.

(7) How many divisors has 10,296? (See p. 25.)

(8) From the algebraic formula $(a^2 + b^2)(c^2 + d^2) = (ac - bd)^2 + (ad + bc)^2$, prove by induction that any integer $r = a_1 a_2 \cdots a_n$, where all the a's are sums of two squares, is itself a sum of two squares. Check this with $2 = 1^2 + 1^2$, $5 = 1^2 + 2^2$, $8 = 2^2 + 2^2$, etc. for $r = 160$, $r = 1600$, $r = 1300$, $r = 625$. If possible, give several different representations of these numbers as sums of two squares.

(9) Apply the result of Exercise 8 to construct new Pythagorean number triples from given ones.

(10) Set up rules for divisibility similar to those on page 35 for number systems with the bases 7, 11, 12.

(11) Show that for two positive rational numbers, $r = a/b$ and $s = c/d$, the inequality $r > s$ is equivalent to $ac - bd > 0$.

(12) Show that for positive r and s, with $r < s$, we always have

$$r < \frac{r + s}{2} < s \quad \text{and} \quad \frac{2}{[(1/r) + (1/s)]^2} < 2rs < (r + s)^2.$$

(13) If z is any complex number, prove by induction that $z^n + 1/z^n$ can be expressed as a polynomial of degree n in the quantity $w = z + 1/z$. (See p. 100.)

(*14) Introducing the abbreviation $\cos \varphi + i \sin \varphi = E(\varphi)$, we have $[E(\varphi)]^m = E(m\varphi)$. Use this and the formulas of page 13 on geometrical series, which remain correct for complex quantities, in order to prove that

$$\sin \varphi + \sin 2\varphi + \sin 3\varphi + \cdots + \sin n\varphi = \frac{\cos \frac{\varphi}{2} - \cos (n + \frac{1}{2})\varphi}{2 \sin \frac{\varphi}{2}},$$

$$\tfrac{1}{2} + \cos \varphi + \cos 2\varphi + \cos 3\varphi + \cdots + \cos n\varphi = \frac{\sin (n + \frac{1}{2})\varphi}{2 \sin \frac{1}{2}\varphi}.$$

(15) Find what the formula of Exercise 3 on page 18 yields, if we substitute $q = E(\varphi)$.

Analytic Geometry

A careful study of the following exercises, supplemented by drawings and numerical examples, will help in mastering the elements of analytic geometry. The definitions and simplest facts of trigonometry are presupposed.

It will often be useful to think of a line or a segment as being directed from one of its points toward another. By the *directed* line PQ (or the *directed* segment PQ) we shall mean the line (or segment) having the direction from P toward Q. In the absence of explicit specification a directed line l will be supposed to have a fixed but arbitrary direction; except that the directed x-axis will be taken to be directed from O toward a point on it with positive x-coördinate, and similarly for the directed y-axis. Directed lines (or directed segments) will then be said to be parallel if and only if they have the same direction. The direction of a directed segment on a directed line can be indicated by attaching a plus or a minus sign to the distance between the endpoints of the segment, according as the segment has the same direction as the line or the opposite direction. It will be desirable to extend the terminology "segment PQ" to the case in which P and Q coincide; to such a "segment" we must clearly assign length zero, but no direction.

(16) Prove: If $P_1(x_1, y_1)$ and $P_2(x_2, y_2)$ are any two points, the coördinates of the midpoint, P_0 (x_0, y_0), of the segment P_1P_2 are $x_0 = (x_1 + x_2)/2$, $y_0 = (y_1 + y_2)/2$. More generally, show that, if P_1 and P_2 are distinct, then the point P_0 on the directed line P_1P_2 for which the ratio $P_1P_0 : P_1P_2$ of the directed lengths has the value k, has the coördinates

$$x_0 = (1 - k)x_1 + kx_2, \qquad y_0 = (1 - k)y_1 + ky_2.$$

(Hint: Parallel lines cut two transversals in proportional segments.)

Thus the points on the line P_1P_2 have coördinates of the form $x = \lambda_1 x_1 + \lambda_2 x_2$, $y = \lambda_1 y_1 + \lambda_2 y_2$, with $\lambda_1 + \lambda_2 = 1$. The values $\lambda_1 = 1$ and $\lambda_1 = 0$ characterize the points P_1 and P_2 respectively. Negative values of λ_1 characterize points beyond P_2, and negative values of λ_2 characterize points before P_1.

(17) Characterize the position of points on the line in a similar manner by means of the values of k.

It is just as important to use positive and negative numbers to indicate the directions of rotations as of distances. By definition, the direction of rotation that brings the directed x-axis into coincidence with the directed y-axis after a rotation of 90° is taken as positive. In the usual coördinate system, with the positive x-axis directed to the right and the positive y-axis upward, this is the counterclockwise sense of rotation. We now define the angle from a directed line l_1 to a directed line l_2 as the angle through which l_1 must be rotated in order to become parallel to l_2. Of course, this angle is determined only up to integral multiples of a complete revolution of 360°. Thus the angle from the directed x-axis to the directed y-axis is 90° or $-270°$, etc.

(18) If α is the angle from the directed x-axis to the directed line l, if P_1, P_2 are any two points on l, and if d denotes the directed distance from P_1 to P_2, show that

$$\cos \alpha = \frac{x_2 - x_1}{d}, \quad \sin \alpha = \frac{y_2 - y_1}{d}, \quad (x_2 - x_1) \sin \alpha = (y_2 - y_1) \cos \alpha.$$

If the line l is not perpendicular to the x-axis, the *slope* of l is defined as

$$m = \tan \alpha = \frac{y_2 - y_1}{x_2 - x_1}.$$

The value of m does not depend on the choice of direction on the line, since $\tan \alpha = \tan (\alpha + 180°)$, or, equivalently, $(y_1 - y_2)/(x_1 - x_2) = (y_2 - y_1)/(x_2 - x_1)$.

(19) Prove: The slope of a line is zero, positive, or negative, according as a parallel to it through the origin lies on the x-axis, in the first and third quadrants, or in the second and fourth quadrants, respectively.

We distinguish a positive and a negative side of a directed line l as follows. Let P be any point not on l, and let Q be the foot of the perpendicular to l through P. Then P is on the positive or negative side of l according as the angle from l to the directed line QP is 90° or $-90°$.

We shall now determine the equation of a directed line l. We draw through the origin O a line m perpendicular to l, and direct m so that the angle from it to l is 90°. The angle from the directed x-axis to m will be called β. Then $\alpha = 90° + \beta$, $\sin \alpha = \cos \beta$, $\cos \alpha = -\sin \beta$. Let R with coördinates x_1, y_1 be the point where m meets l. We shall denote by d the directed distance OR on directed m.

(20) Show that d is positive if and only if O lies on the negative side of l.

We have $x_1 = d \cos \beta$, $y_1 = d \sin \beta$ (compare Ex. 18). Hence, $(x - x_1) \sin \alpha = (y - y_1) \cos \alpha$, or $(x - d \cos \beta) \cos \beta = -(y - \mathrm{d} \sin \beta) \sin \beta$, which gives the equation

$$x \cos \beta + y \sin \beta - d = 0.$$

This is the *normal form* of the equation of the line l. Note that this equation does not depend on the direction assigned to l, for a change in direction would change the sign of every term on the left side, and hence would leave the equation unchanged.

By multiplying the normal equation with an arbitrary factor, we obtain the general form of the equation of the line:

$$ax + by + c = 0.$$

To retrieve from this general form the geometrically significant normal form we must multiply by a factor which will reduce the first two coefficients to $\cos\beta$ and $\sin\beta$, whose squares add up to 1. This may be done by the factor $1/\sqrt{a^2 + b^2}$, which yields the normal form

$$\frac{a}{\sqrt{a^2 + b^2}}\, x + \frac{b}{\sqrt{a^2 + b^2}}\, y + \frac{c}{\sqrt{a^2 + b^2}} = 0,$$

so that we have

$$\frac{a}{\sqrt{a^2 + b^2}} = \cos\beta, \qquad \frac{b}{\sqrt{a^2 + b^2}} = \sin\beta, \qquad -\frac{c}{\sqrt{a^2 + b^2}} = d.$$

(21) Show: (a) that the only factors that will reduce the general form to the normal form are $1/\sqrt{a^2 + b^2}$ and $-1/\sqrt{a^2 + b^2}$; (b) that the choice of the one or the other of these factors determines which direction is assigned to the line; and (c) that, when one of these factors has been used, the origin is on the positive or negative side of the resulting directed line, or is on the line, according as d is negative, positive, or zero.

(22) Prove directly that the line with slope m through a given point $P_0(x_0, y_0)$ is given by the equation

$$y - y_0 = m(x - x_0), \qquad \text{or} \qquad y = mx + y_0 - mx_0.$$

Prove that the line through two given points, $P_1(x_1, y_1)$, $P_2(x_2, y_2)$, has an equation

$$(y_2 - y_1)(x - x_1) = (x_2 - x_1)(y - y_1).$$

The x-coördinate of a point in which a line or curve cuts the x-axis is called an x-*intercept* of the curve; similarly for y-*intercept*.

(23) By dividing the general equation of Exercise 20 by an appropriate

factor, show that the equation of a line may be written in the *intercept form*,

$$\frac{x}{a} + \frac{y}{b} = 1,$$

where a and b are the x- and y-intercepts. What exceptions are there?

(24) By a similar procedure show that the equation of a line not parallel to the y-axis may be written in the *slope-intercept* form,

$$y = mx + b.$$

(If the line is parallel to the y-axis, its equation may be written as $x = a$.)

(25) Let $ax + by + c = 0$ and $a'x + b'y + c' = 0$ be equations of undirected lines l and l', with slopes m and m' respectively. Show that l and l' are parallel or perpendicular according as: (a) $m = m'$ or $mm' = -1$. (b) $ab' - a'b = 0$ or $aa' + bb' = 0$. (Note that (b) holds even when a line has no slope, i.e. is parallel to the y-axis.)

(26) Show that the equation of a line through a given point $P_0(x_0, y_0)$ and parallel to a given line l with equation $ax + by + c = 0$ has the equation $ax + by = ax_0 + by_0$. Show that a similar formula, $bx - ay = bx_0 - ay_0$, holds for the equation of the line through P_0 and perpendicular to l. (Note that if the equation of l is in the normal form, so also will the new equation be, in each case.)

(27) Let $x \cos \beta + y \sin \beta - d = 0$ and $ax + by + c = 0$ be the normal and general forms of the equation of a line l. Show that the directed distance h from l to any point $Q(u, v)$ is given by

$$h = u \cos \beta + v \sin \beta - d,$$

or by

$$h = \frac{au + bv + c}{\pm\sqrt{a^2 + b^2}};$$

and that h is positive or negative according as Q is on the positive or negative side of the directed line l (the direction having been determined by β, or by the choice of the sign before $\sqrt{a^2 + b^2}$). (Hint: Write the normal form of the equation of the line m through Q parallel to l, and find the distance from l to m.)

(28) Let $l(x, y) = 0$ represent the equation $ax + by + c = 0$ of a line l; similarly for $l'(x, y) = 0$. Let λ and λ' be constants, with

$\lambda + \lambda' = 1$. Show that, if l and l' intersect in $P_0(x_0, y_0)$, then every line through P_0 has an equation

$$\lambda l(x, y) + \lambda' l'(x, y) = 0,$$

and conversely; and that every such line is uniquely determined by the choice of a pair of values for λ and λ'. (Hint: P_0 lies on l if and only if $l(x_0, y_0) = ax_0 + by_0 + c = 0$.) What lines are represented if l and l' are parallel? Note that the condition $\lambda + \lambda' = 1$ is unnecessary, but serves to determine a unique equation for each line through P_0.

(29) Use the result of the previous exercise to find the equation of a line through the intersection P_0 of l and l' and through another point, $P_1(x_1, y_1)$, without finding the coördinates of P_0. (Hint: Find λ and λ' from the conditions $\lambda l(x_1, y_1) + \lambda' l'(x_1, y_1) = 0, \lambda + \lambda' = 1$.) Check by finding the coördinates of P_0 (see pp. 76–77) and showing that P_0 lies on the line whose equation you have found.

(30) Prove that the equations of the bisectors of the angles formed by intersecting lines l and l' are

$$\sqrt{a'^2 + b'^2}\, l(x, y) = \pm\sqrt{a^2 + b^2}\, l'(x, y).$$

(Hint: See Ex. 27.) What do these equations represent if l and l' are parallel?

(31) Find the equation of the perpendicular bisector of the segment P_1P_2 by each of the following methods: (a) Find the equation of line P_1P_2; find the coördinates of the midpoint P_0 of segment P_1P_2; find the equation of the line through P_0 perpendicular to P_1P_2. (b) Write the equation expressing the fact that the distance (p. 74) between P_1 and any point $P(x, y)$ on the perpendicular bisector is equal to the distance between P_2 and P; square both sides of the equation and simplify.

(32) Find the equation of the circle through three non-collinear points, P_1, P_2, P_3, by each of the following methods: (a) Find the equation of the bisectors of the segments P_1P_2 and P_2P_3; find the coordinates of the center as the point of intersection of these lines; find the radius as the distance between the center and P_1. (b) The equation must be of the form $x^2 + y^2 - 2ax - 2by = k$ (see p. 74). Since each of the given points lies on the circle we must have

$$x_1^2 + y_1^2 - 2ax_1 - 2by_1 = k,$$
$$x_2^2 + y_2^2 - 2ax_2 - 2by_2 = k,$$
$$x_3^2 + y_3^2 - 2ax_3 - 2by_3 = k,$$

for a point lies on a curve if and only if its coördinates satisfy the equation of the curve. Solve these simultaneous equations for a, b, k.

(33) To find the equation of the ellipse with major axis $2p$, minor axis $2q$, and foci at $F(e, 0)$ and $F(-e, 0)$, where $e^2 = p^2 - q^2$, use the distances r and r' from F and F' to any point on the curve. By definition of the ellipse, $r + r' = 2p$. By using the distance formula on page 74, show that

$$r'^2 - r^2 = (x + e)^2 - (x - e)^2 = 4ex.$$

Since

$$r'^2 - r^2 = (r' + r)(r' - r) = 2p(r' - r),$$

show that $r' - r = 2ex/p$. Solve this relation and $r' + r = 2p$ to find the important formulas

$$r = -\frac{e}{p}x + p, \qquad r' = \frac{e}{p}x + p.$$

Since (again by the distance formula) $r^2 = (x - e)^2 + y^2$, equate this expression for r^2 to the expression $\left(-\frac{e}{p}x + p\right)^2$ just above,

$$(x - e)^2 + y^2 = \left(-\frac{e}{p}x + p\right)^2.$$

Expand, collect terms, substitute $p^2 - q^2$ for e^2, and simplify. Show that the result may be expressed in the form

$$\frac{x^2}{p^2} + \frac{y^2}{q^2} = 1.$$

Carry out the same procedure for the hyperbola, defined as the locus of all points P for which the absolute value of the difference $r - r'$ is equal to a given quantity $2p$. Here $e^2 = p^2 + q^2$.

(34) The parabola is defined as the locus of a point whose distance from a fixed line (the directrix) is equal to its distance from a fixed point (the focus). If we choose the line $x = -a$ as directrix and the point $F(a, 0)$ as focus, show that the equation of the parabola may be written in the form $y^2 = 4ax$.

Geometrical Constructions

(35) Prove the impossibility of constructing with ruler and compass the numbers $\sqrt[3]{3}$, $\sqrt[3]{4}$, $\sqrt[3]{5}$. Prove that the construction of $\sqrt[3]{a}$

is only possible if a is the cube of a rational number. (See p. 134, ff.)

(36) Find the sides of the regular $3 \cdot 2^n$-gon and of the $5 \cdot 2^n$-gon and characterize the corresponding sequences of extension fields.

(37) Prove the impossibility of trisecting with ruler and compass an angle of 120 or 30 degrees. (Hint for the case of 30°: The equation to be discussed is $4z^3 - 3z = \cos 30° = \frac{1}{2}\sqrt{3}$. Introduce a new unknown, $u = z\sqrt{3}$, and obtain an equation for z from which the non-constructibility of z follows as in the text, p. 139.)

(38) Prove that the regular 9-gon is not constructible.

(39) Prove that the inversion of a point $P(x, y)$ into the point $P'(x', y')$ in the circle with the radius r about the origin is given by the equations

$$x' = \frac{xr}{x^2 + y^2}, \qquad y' = \frac{yr}{x^2 + y^2}.$$

Find algebraically the equations giving x, y in terms of x', y'.

(*40) Prove analytically by using Exercise 39 that by inversion the totality of circles and straight lines is transformed into itself. Check the properties a) – d) on page 142 separately, and likewise the transformations corresponding to Figure 61.

(41) What becomes of the two families of lines, x = const. and y = const., parallel to the coördinate axes, after inversion in the unit circle about the origin? Find the answer without and with analytic geometry. (See p. 160.)

(42) Carry out the Apollonius constructions in simple cases of your own selection. Try the solution analytically according to the method of page 125.

Projective and Non-Euclidean Geometry

(43) Find all the values of the cross ratio λ of four harmonic points, if the points are subjected to permutations. (Answer: $\lambda = -1, 2, \frac{1}{2}$).

(44) For what configurations of four points do some of the six values of the cross-ratio on page 176 coincide? (Answer: Only for $\lambda = -1$ or $\lambda = 1$; there is also one imaginary value of λ for which $\lambda = 1/(1 - \lambda)$, the "equianharmonic" cross-ratio.)

(45) Show that a cross-ratio $(ABCD) = 1$ means coincidence of the points C and D.

(46) Prove the statements about the cross-ratio of planes, page 176.

(47) Prove that if P and P' are inverse with respect to a circle and if the diameter AB is collinear with P, P', then the points A, B, P, P'

form a harmonic quadruple. (Hint: Use the analytic expression (2) on p. 178, take the circle as the unit circle and AB as the axis.)

(48) Find the coördinates of the fourth harmonic point to three points P_1, P_2, P_3. What happens if P_3 moves to the midpoint of P_1P_2? (See p. 178.)

(*49) Use Dandelin's spheres to develop the theory of conic sections. In particular prove that they are all (except for the circle) geometrical loci of points whose distances from a fixed point F and a fixed line l have a constant ratio k. For $k > 1$ we have a hyperbola, for $k = 1$ a parabola, for $k < 1$ an ellipse. The line l is obtained by intersecting the plane of the conic with the plane through the circle in which the Dandelin sphere touches the cone. (Since the circle does not come under this characterization except as a limiting case, it is not entirely appropriate to choose this property as a definition of the conics, although this is sometimes done.)

(50) Discuss: "A conic, regarded as both a set of points *and* a set of lines, is self-dual." (See p. 209.)

(*51) Try to prove Desargues's theorem in the plane by carrying out the passage to the limit from the three-dimensional configuration of Figure 73. (See p. 172.)

(*52) How many lines intersecting four given skew lines can be drawn? How can they be characterized? (Hint: Draw a hyperboloid through three of the given lines, see p. 212.)

(*53) If the Poincaré circle is the unit circle of the complex plane, then two points z_1 and z_2 and the z-values w_1, w_2 of the two points of intersection of the "straight line" through these two points with the unit circle define a cross-ratio $\frac{z_1 - w_1}{z_1 - w_2} : \frac{z_2 - w_1}{z_2 - w_2}$ which, according to Exercise 8 on page 97, is real. Its logarithm is by definition the hyperbolic distance between z_1 and z_2.

(*54) By an inversion transform the Poincaré circle into the upper half plane. Develop the Poincaré model and its properties for this half plane directly and by means of this inversion. (See p. 224.)

Topology

(55) Verify Euler's formula for the five regular polyhedra and for other polyhedra. Carry out the corresponding network reductions.

(56) In the proof of Euler's formula (p. 239) we are required to reduce any plane network of triangles, by successive application of two fundamental operations, to a network consisting of a single triangle, for

which $V - E + F = 3 - 3 + 1 = 1$. How can we be sure that the final result will not be a *pair* of triangles with no vertices in common, so that $V - E + F = 6 - 6 + 2 = 2$? (Hint: We can assume that the original network is *connected*, i.e. that one can pass from any vertex to any other along edges of the network. Show that this property cannot be destroyed by the two fundamental operations.)

(57) We have admitted only two fundamental operations in the reduction of the network. Might it not happen at some stage that a triangle appears having only one vertex in common with the other triangles of the network? (Construct an example.) This would require a third operation: Removal of two vertices, three edges, and a face. Would this affect the proof?

(58) Can a wide rubber band be wrapped three times around a broomstick so as to lie flat (i.e. untwisted) on the broomstick? (Of course, the rubber band must cross itself somewhere.)

(59) Show that a circular disk from which the point at the center has been removed admits a continuous, fixed-point-free transformation into itself.

(*60) The transformation which shifts each point of a disk one unit in a fixed direction obviously has no fixed points. Of course, this is not a transformation of the disk into *itself*, since some points will be taken into points outside the disk. Why does not the argument of page 255, based on the transformation $P \rightarrow P^*$, hold in this case?

(61) Suppose we have a rubber inner tube, the inside of which is painted white and the outside black. Is it possible, by cutting a small hole, deforming the tube, and then sealing up the hole, to turn the tube inside out, so that the inside will be black and the outside white?

(*62) Show that there is no "four color problem" in three dimensions by proving that for any desired number n, n bodies can be placed in space so that each touches all the others.

(*63) Using either an actual torus surface (inner tube, anchor ring) or a plane region with boundary identification (Fig. 143), construct a map consisting of seven regions, each of which touches all the others. (See p. 248.)

(64) The 4-dimensional tetrahedron of Figure 118 consists of five points, a, b, c, d, e, each of which is joined to the other four. Even if the connecting lines are allowed to be curved, the figure cannot be drawn in the plane in such a way that no two of the connections cross. Another configuration, containing ten connections, that cannot be drawn in the plane without crossings consists of six points, a, b, c, a', b', c',

such that each of the points a, b, c, is connected to each of the points a', b', c'. Verify these facts by experiment, * and try to devise a proof, using the Jordan curve theorem as a basis. (It has been proved that any configuration of points and lines that cannot be represented in the plane without crossings must contain one of these two configurations as a part.)

(65) A configuration is formed by taking the six sides of a 3-dimensional tetrahedron and adding one line joining the midpoints of two opposite sides. (Two sides of a tetrahedron are opposite if they have no common endpoint.) Show that this configuration is equivalent to one described in the preceding exercise.

(*66) Let p, q, r be the three tips of the symbol E. The symbol is shifted some distance away, giving another E, with tips p', q', r'. Can one join p to p', q to q', and r to r' by three curves which do not cross each other or the E's?

If we go around a square, we change our direction four times, each time by an amount 90°, giving a total change of $\Delta = 360°$. If we go around a triangle, it is known from elementary geometry that $\Delta = 360°$.

(67) Prove that if C is any simple closed polygon, then $\Delta = 360°$. (Hint: Cut the interior of C into triangles, then remove boundary segments, as on p. 239. Let the successive boundaries be B_1, B_2, B_3, $\cdots$, B_n. Then $B_1 = C$, and B_n is a triangle. Show that, if Δ_i corresponds to B_i, then $\Delta_i = \Delta_{i-1}$.)

(*68) Let C be any simple closed curve with a continuously turning tangent vector. If Δ denotes the total change in the angle of the tangent as we traverse the curve once, show that here also $\Delta = 360°$. (Hint: Let p_0, p_1, p_2, $\cdots$, p_n, p_0 be points cutting C into small, nearly straight segments. Let C_i be the curve with the segments p_0p_1, p_1p_2, $\cdots$, $p_{i-1}p_i$, and the original arcs p_ip_{i+1}, $\cdots$, p_np_0. Then $C_0 = C$, and C_n is composed of line segments. Show that $\Delta_i = \Delta_{i+1}$, and use the result of the preceding exercise). Does this apply to the hypocycloid of Figure 55?

(68) Show that if in the diagram of the Klein bottle on page 263 all four arrows are drawn clockwise, a surface is formed that is equivalent to a sphere with one disk replaced by a cross-cap. (This surface is topologically equivalent to the extended plane of projective geometry.)

(70) The Klein bottle of Figure 142 may be cut into two symmetrical halves by a plane. Show that the result consists of two Moebius strips.

(*71) In the Moebius strip of Figure 139 the two endpoints of each transversal segment are identified. Show that the result is topologically equivalent to a Klein bottle.

All possible ordered pairs of points on a line segment (the two points coinciding or not) form a square, in the following sense. If the points of the segment are designated by their distances x, y from one end A, the ordered pairs of numbers (x, y) may be regarded as the Cartesian coördinates of a point of the square.

All possible pairs of points without regard to order (i.e. with (x, y) regarded as the same as (y, x)) form a surface S which is topologically equivalent to a square. To see this choose that representation which has the first point nearest the end A of the segment, if $x \neq y$. Thus S is the set of all pairs (x, y) where either x is less than y or $x = y$. Using Cartesian coördinates, this gives the triangle in the plane with vertices (0, 0), (0, 1), (1, 1).

(*72) What surface is formed by the set of all ordered pairs of points of which the first belongs to a line and the second to the circumference of a circle? (Answer: A cylinder.)

(73) What surface is formed by the set of all ordered pairs of points on a circle? (Answer: A torus.)

(*74) What surface is formed by the set of all *unordered* pairs of points on a circle? (Answer: A Moebius strip.)

(75) Here are the rules of a game, played with pennies on a large circular table: A and B in turn place pennies on the table. The pennies need not touch each other, and a penny may be placed anywhere on the table, so long as it does not extend over the edge or overlap a penny already on the table. Once placed, a penny may not be moved. In time, the table will be covered with pennies in such a way that no space large enough for another penny remains. The player who is able to place the last penny on the table wins. If A plays first, prove that no matter how B plays, A can be sure of winning, provided that he plays correctly.

(76) If, in the game of Exercise 75 the table has the form of Figure 125, b, prove that B can always win.

Functions, Limits and Continuity

(77) Find the continued fraction expansion for the ratio $OB{:}AB$ of page 123.

(78) Show that the sequence $a_0 = \sqrt{2}$, $a_{n+1} = \sqrt{2 + a_n}$ is mono-

tone increasing, bounded by $B = 2$, and hence has a limit. Show that this limit must be the number 2. (See pp. 125 and 326.)

(*79) Try to prove, by methods similar to those used on pages 318 and following, that given any smooth closed curve, a square may always be drawn whose sides are tangent to the curve.

The function $u = f(x)$ is called *convex* if the midpoint of the segment joining any two points of the graph of the function lies above the graph. For example, $u = e^x$ (Fig. 278) is convex, while $u = \log x$ (Fig. 277) is not.

(80) Prove that the function $u = f(x)$ is convex if, and only if,

$$\frac{f(x_1) + f(x_2)}{2} \geq f\left(\frac{x_1 + x_2}{2}\right),$$

with equality only for $x_1 = x_2$.

(*81) Prove that for convex functions the more general inequality

$$\lambda_1 f(x_1) + \lambda_2 f(x_2) \geq f(\lambda_1 x_1 + \lambda_2 x_2)$$

holds, where λ_1, λ_2 are any two constants such that $\lambda_1 + \lambda_2 = 1$ and $\lambda_1 \geq 0$, $\lambda_2 \geq 0$. This is equivalent to the statement that no point of the segment joining two points of the graph lies below the graph.

(82) Using the condition of Exercise 80 prove that the functions $u = \sqrt{1 + x^2}$ and $u = 1/x$ (for $x > 0$) are convex, i.e. that

$$\frac{\sqrt{1 + x_1^2} + \sqrt{1 + x_2^2}}{2} \geq \sqrt{1 + \left(\frac{x_1 + x_2}{2}\right)^2},$$

$$\frac{1}{2}\left(\frac{1}{x_1} + \frac{1}{x_2}\right) \geq \frac{2}{x_1 + x_2} \quad \text{for positive } x_1 \text{ and } x_2.$$

(83) The same for $u = x^2$, $u = x^n$ for $x > 0$, $u = \sin x$ for $\pi \leq x \leq 2\pi$, $u = \tan x$ for $0 \leq x \leq \pi/2$, $u = -\sqrt{1 - x^2}$ for $|x| \leq 1$.

Maxima and Minima

(84) Find the path of shortest length between P and Q as in Figure 178, if the path is supposed to meet the two given lines alternately n times. (See p. 333.)

(85) Find the shortest connection between two points P and Q within a triangle with acute angles if the path is required to meet the sides of the triangle in a given order. (See p. 334.)

(86) Draw the level lines and check the existence of at least two saddle points in a surface over a triply connected domain whose boundary is on the same level. (See p. 345.) Again we must exclude the case

where the tangent plane to the surface is horizontal along a whole closed curve.

(87) Starting with two arbitrary positive rational numbers, a and b, form, step by step, the pairs $a_{n+1} = \sqrt{a_n b_n}$, $b_{n+1} = \frac{1}{2}(a_n + b_n)$. Prove that they define a sequence of nested intervals. (The limit point as $n \to \infty$, the so-called arithmetical-geometrical mean of a_0 and b_0, played a great rôle in the early researches of Gauss).

(88) Find the length of the whole graph in Figure 219, and compare this with the total length of the two diagonals.

(*89) Investigate conditions on four points, A_1, A_2, A_3, A_4, that show whether they lead to the case of Figure 216 or 218.

(*90) Find systems of five points for which different street nets satisfying the angular conditions exist. Only some of them will yield relative minima. (See p. 361.)

(91) Prove Schwarz's inequality,

$$(a_1b_1 + \cdots + a_nb_n)^2 \leq (a_1^2 + \cdots + a_n^2)(b_1^2 + \cdots + b_n^2),$$

valid for any set of pairs of numbers a_i, b_i; prove that the inequality sign holds only if the a_i are proportional to the b_i. (Hint: Generalize the algebraic formula of Ex. 8.)

(*92) With n positive numbers $x_1, \cdots, x_n$ we form the expressions s_k defined by

$$s_k = (x_1x_2 \cdots x_k + \cdots)/C_k^n,$$

where the symbol "$+ \cdots$" means that all the C_k^n products of combinations of k of these quantities are to be added. Then prove that

$$\sqrt[k+1]{s_{k+1}} \leq \sqrt[k]{s_k},$$

where the equality sign holds only if all the quantities x_i are equal.

(93) For $n = 3$ these inequalities state that for three positive numbers a, b, c

$$\sqrt[3]{abc} \leq \sqrt{\frac{ab + ac + bc}{3}} \leq \frac{a + b + c}{3}.$$

What extremal properties of the cube are implied by these inequalities?

(*94) Find an arc of a curve of shortest length joining two points A, B and including with the segment AB a prescribed area. (Answer: The arc must be circular.)

(*95) Given two segments AB and $A'B'$, find an arc joining A to B and one joining A' to B' such that the two arcs include with the two segments a prescribed area and have a minimum total length. (Answer: The arcs are circular with the same radius.)

(*96) The same for any number of segments, AB, $A'B'$, etc.

(*97) On two lines intersecting at O find two points A and B, respectively, and join A with B by an arc of minimal length such that the area included by it and the lines is prescribed. (Answer: The arc is circular and perpendicular to the lines.)

(*98) The same problem, but now the total perimeter of the domain included, i.e. the arc plus OA plus OB is to be a minimum. (Answer: The solution is given by an arc of a circle which bulges outward and touches the two lines.)

(*99) The same problem for several angular sectors.

(*100) Prove that the nearly plane surfaces in Figure 240 are not plane except for the stabilizing surface in the center. Remark: To find or characterize these curved surfaces analytically is a challenging unsolved problem. The same is true for the surfaces in Figure 251. In Figure 258 we actually have twelve symmetric planes meeting at 120° in the diagonals.

Advice for some additional soap film experiments. Carry out experiments indicated by Figures 256 and 257 for more than three connecting rods. Study the limiting cases for volume of air tending to zero. Experiment with non-parallel planes or other surfaces. Blow up the cubic bubble of Figure 258 until it fills the whole cube and bulges over the edges. Then suck the air out again, reversing the process.

(*101) Find two equilateral triangles with given total perimeter and minimum area. (Answer: The triangles must be congruent (use calculus).)

*(102) Find two triangles with given total perimeter and maximum area. (Answer: One triangle degenerates into a point; the other one must be equilateral.)

*(103) Find two triangles with given total area and minimum perimeter.

(*104) Find two equilateral triangles with given total area and maximum perimeter.

The Calculus

(105) Differentiate the functions $\sqrt{1+x}$, $\sqrt{1+x^2}$, $\sqrt{\dfrac{x+1}{x-1}}$ by applying directly the definition of derivative, forming and transforming the difference quotient until the limit can be obtained easily by substituting $x_1 = x$. (See p. 421.)

(106) Prove that the function $y = e^{-1/x^2}$, with $y = 0$ for $x = 0$, has all its derivatives zero at $x = 0$.

(107) Show that the function of Exercise 106 cannot be expanded in a Taylor series. (See p. 477.)

(108) Find the points of inflection ($f''(x) = 0$) of the curves $y = e^{-x^2}$ and $y = xe^{-x^2}$.

(109) Prove that for a polynomial $f(x)$ with all n roots $x_1, \cdots, x_n$ distinct we have

$$\frac{f'(x)}{f(x)} = \sum_{i=1}^{n} \frac{1}{x - x_i}$$

*(110) Using the direct definition of the integral as limit of a sum, prove that for $n \to \infty$ we have

$$n\left(\frac{1}{1^2 + n^2} + \frac{1}{2^2 + n^2} + \cdots + \frac{1}{n^2 + n^2}\right) \to \frac{\pi}{4}.$$

(*111) Prove in a similar way that

$$\frac{b}{n}\left(\sin\frac{b}{n} + \sin\frac{2b}{n} + \cdots + \sin\frac{nb}{n}\right) \to \cos b - 1.$$

(112) By drawing Figure 276 in large scale on coördinate paper and counting the small squares in the shaded area, find an approximate value for π.

(113) Use formula (7), page 441 for the numerical calculation of π with a guaranteed accuracy of at least 1/100.

(114) Prove: $e^{\pi i} = -1$. (See p. 478.)

(115) A curve of given shape is expanded in the ratio $1:x$. $L(x)$ and $A(x)$ denote the length and area of the expanded curve. Show that $L(x)/A(x) \to 0$ as $x \to \infty$ and, more generally, $L(x)/A(x)^k \to 0$ as $x \to \infty$, if $k > \frac{1}{2}$. Check for circle, square and, * ellipse. (Area is of a higher order of magnitude than circumference. See p. 472.)

(116) Often the exponential function occurs in combinations given and denoted as follows:

$$u = \sinh x = \tfrac{1}{2}(e^x - e^{-x}), \qquad v = \cosh x = \tfrac{1}{2}(e^x + e^{-x})$$

$$w = \tanh x = \frac{e^x - e^{-x}}{e^x + e^{-x}},$$

called *hyperbolic sine*, *hyperbolic cosine*, and *hyperbolic tangent* respectively. These functions have many properties analogous to those of the trigonometric functions; they are linked to the hyperbola $u^2 - v^2 = 1$ much as the functions $u = \cos x$ and $v = \sin x$ are linked to the circle $u^2 + v^2 = 1$. The following facts should be proved by the reader and compared with the corresponding facts concerning trigonometric functions:

$$D \cosh x = \sinh x, \qquad D \sinh x = \cosh x, \qquad D \tanh x = 1/\cosh^2 x,$$

$$\sinh (x + x') = \sinh x \cdot \cosh x' + \cosh x \sinh x',$$

$$\cosh (x + x') = \cosh x \cdot \cosh x' + \sinh x \cdot \sinh x'.$$

The inverse functions are called $x = \text{arc sinh } u = \log (u + \sqrt{u^2 + 1})$; $x = \text{arc cosh } v = \log (v + \sqrt{v^2 - 1})$ $(v \geq 1)$.

Their derivatives are given by

$$D \text{ arc sinh } u = \frac{1}{\sqrt{1 + u^2}}; \qquad D \text{ arc cosh } v = \frac{1}{\sqrt{v^2 - 1}}$$

$$D \text{ arc tanh } w = \frac{1}{1 - w^2}, \qquad (|w| > 1).$$

(117) On the basis of Euler's formula check the analogy between hyperbolic and trigonometric functions.

(*118) Find simple summation formulas for

$$\sinh x + \sinh 2x + \cdots + \sinh nx$$

and

$$\tfrac{1}{2} + \cosh x + \cosh 2x + \cdots + \cosh nx$$

analogous to those in Exercise 14 for trigonometric functions.

Technique of Integration

The theorem of p. 439 reduces the problem of integrating a function $f(x)$ between the limits a and b to that of finding a primitive function $G(x)$ for $f(x)$, i.e. one for which $G'(x) = f(x)$. The integral is then simply the difference $G(b) - G(a)$. For these primitive functions, which are determined by $f(x)$ (except for an arbitrary additive constant), the name "indefinite integral" and the suggestive notation

$$G(x) = \int f(x)\, dx,$$

without limits of integration, is customary. (This notation may be misleading for the beginner; see the remark on p. 438.)

Every formula of differentiation contains the solution of a problem of indefinite integration simply by interpreting it inversely as a formula of integration. We can extend this somewhat empirical procedure by two important rules, which are nothing but the equivalent of the rules of differentiation of a compound function and of a product of functions. In their integral form these are called the rules of *integration by substitution* and *integration by parts.*

A) The first rule results from the formula for the differentiation of a compound function,

$$H(u) = G(x),$$

where

$$x = \psi(u) \qquad \text{and} \qquad u = \varphi(x)$$

are supposed to be functions of each other, uniquely determined in the interval under consideration. Then we have

$$H'(u) = G'(x)\psi'(u).$$

If

$$G'(x) = f(x),$$

we can write

$$G(x) = \int f(x)\,dx$$

and also

$$G'(x)\psi'(u) = f(x)\psi'(u),$$

which, in consequence of the formula above for $H'(u)$, is equivalent to

$$H(u) = \int f(\psi(u))\psi'(u)\,du.$$

Hence, since $H(u) = G(x)$,

$$\text{(I)} \qquad \int f(x)\,dx = \int f(\psi(u))\psi'(u)\,du.$$

Written in Leibniz' notation (see p. 434) this rule takes the very suggestive form

$$\int f(x)\,dx = \int f(x)\frac{dx}{du}\,du,$$

which means that the symbol dx may be replaced by the symbol $\frac{dx}{du}\,du$, just as if dx and du were numbers and $\frac{dx}{du}$ a fraction.

The usefulness of formula (I) will be illustrated by a few examples.

a) $J = \int \frac{1}{u \log u}\,du$. Here we start with the right hand side of (I), substituting $x = \log u = \psi(u)$. We then have $\psi'(u) = \frac{1}{u}$, $f(x) = \frac{1}{x}$; hence

$$J = \int \frac{dx}{x} = \log x,$$

or

$$\int \frac{du}{u \log u} = \log \log u.$$

We can verify this result by differentiating both sides. We find $\frac{1}{u \log u} = \frac{d}{du}(\log \log u)$, which is easily shown to be correct.

b) $J = \int \cot u\,du = \int \frac{\cos u}{\sin u}\,du$. Setting $x = \sin u = \psi(u)$ we find

$$\psi'(u) = \cos u, \qquad f(x) = x,$$

hence,

$$J = \int \frac{dx}{x} = \log x$$

or

$$\int \cot u\,du = \log \sin u.$$

This result can again be verified by differentiation.

c) In general, if we have an integral of the form

$$J = \int \frac{\psi'(u)}{\psi(u)}\,du,$$

we set $x = \psi(u)$, $f(x) = x$ and find

$$J = \int \frac{dx}{x} = \log x = \log \psi(u).$$

d) $J = \int \sin x \cos x \, dx$. We put $\sin x = u$, $\cos x = \dfrac{du}{dx}$. Then

$$J = \int u \frac{du}{dx} \, dx = \int u \, du = \frac{u^2}{2} = \tfrac{1}{2} \sin^2 x.$$

e) $J = \int \dfrac{\log u}{u} \, du$. Set $\log u = x$, $\dfrac{1}{u} = \dfrac{dx}{du}$. Then

$$J = \int x \frac{dx}{du} \, du = \int x \, dx = \frac{x^2}{2} = \tfrac{1}{2}(\log u)^2.$$

In the examples below (I) is used, starting from the left side.

f) $J = \int \dfrac{dx}{\sqrt{x}}$. Set $\sqrt{x} = u$. Then $x = u^2$ and $\dfrac{dx}{du} = 2u$. Therefore

$$J = \int \frac{1}{u} \cdot 2u \, du = 2u = 2\sqrt{x}.$$

g) By the substitution $x = au$, where a is a constant, we find

$$\int \frac{dx}{a^2 + x^2} = \int \frac{dx}{du} \cdot \frac{1}{a^2} \cdot \frac{1}{1 + u^2} \, du = \int \frac{1}{a} \frac{du}{1 + u^2} = \frac{1}{a} \cdot \operatorname{arc\,tan} \frac{x}{a}.$$

h) $J = \int \sqrt{1 - x^2} \, dx$. Set $x = \cos u$, $\dfrac{dx}{du} = -\sin u$. Then

$$J = -\int \sin^2 u \, du = -\int \frac{1 - \cos 2u}{2} \, du = -\frac{u}{2} + \frac{\sin 2u}{4}.$$

Using $\sin 2u = 2 \sin u \cos u = 2 \cos u \sqrt{1 - \cos^2 u}$, we have

$$J = -\tfrac{1}{2} \operatorname{arc\,cos} x + \tfrac{1}{2} x \sqrt{1 - x^2}.$$

Evaluate the following indefinite integrals and verify the results by differentiation:

119) $\displaystyle\int \frac{u \, du}{u^2 - u + 1}$.

120) $\displaystyle\int u e^{u^2} \, du$.

121) $\displaystyle\int \frac{du}{u(\log u)^n}$.

124) $\displaystyle\int \frac{dx}{x^2 + 2ax + b}$.

125) $\displaystyle\int t^2 \sqrt{1 + t^3} \, dt$.

126) $\displaystyle\int \frac{t + 1}{\sqrt{1 - t^2}} \, dt$.

122) $\int \frac{8x}{3+4x}\,dx.$

123) $\int \frac{dx}{x^2+x+1}.$

127) $\int \frac{t^4}{1-t}\,dt.$

128) $\int \cos^n t \cdot \sin t \cdot dt.$

129) Prove that $\int \frac{dx}{a^2-x^2} = \frac{1}{a}\,\text{arc tanh}\,\frac{x}{a}$; $\int \frac{dx}{\sqrt{a^2-x^2}} = \text{arc sinh}\,\frac{x}{a}$. (Compare examples g, h.)

B) The rule (p. 428) for the differentiation of a product,

$$(p(x)\cdot q(x))' = p(x)\cdot q'(x) + p'(x)\cdot q(x),$$

can be written as an integral formula:

$$p(x)\cdot q(x) = \int p(x)\cdot q'(x)\,dx + \int p'(x)\cdot q(x)\,dx$$

or

$$\text{(II)} \qquad \int p(x)\cdot q'(x)\,dx = p(x)q(x) - \int p'(x)\cdot q(x)\,dx.$$

In this form it is called the rule of *integration by parts.* This rule is useful when the function to be integrated can be written as a product of the form $p(x)q'(x)$, where the primitive function $q(x)$ of $q'(x)$ is known. In that case formula (II) reduces the problem of finding the indefinite integral of $p(x)q'(x)$ to that of the integration of the function $p'(x)q(x)$, which is often much simpler to solve.

Examples:

a) $J = \int \log x\,dx$. Set $p(x) = \log x$, $q'(x) = 1$, so that $q(x) = x$. Then (II) leads to

$$\int \log x\,dx = x\log x - \int \frac{x}{x}\,dx = x\log x - x.$$

b) $J = \int x\log x\,dx$. Set $p(x) = \log x$, $q'(x) = x$. Then

$$J = \frac{x^2}{2}\log x - \int \frac{x^2}{2x}\,dx = \frac{x^2}{2}\log x - \frac{x^2}{4}.$$

c) $J = \int x\sin x\,dx$. Here we set $p(x) = x$, $q(x) = -\cos x$ and find

$$\int x\sin x\,dx = -x\cos x + \sin x.$$

Evaluate the following integrals using integration by parts.

130) $\int xe^x\,dx.$

131) $\int x^2 \cos x\,dx.$ (Hint: Apply (II) twice.)

132) $\int x^a \log x\,dx \quad (a \neq -1).$

133) $\int x^2 e^x\,dx.$ (Hint: Use Ex. 130.)

Integration by parts of the integral $\int \sin^m x\,dx$ leads to a remarkable expression for the number π as an infinite product. To derive it we write the function $\sin^m x$ in the form $\sin^{m-1} x \cdot \sin x$ and integrate by parts between the limits 0 and $\pi/2$. This leads to the formula

$$\int_0^{\pi/2} \sin^m x\,dx = (m-1)\int_0^{\pi/2} \sin^{m-2} x \cos^2 x\,dx$$

$$= -(m-1)\int_0^{\pi/2} \sin^m x\,dx + (m-1)\int_0^{\pi/2} \sin^{m-2} x\,dx,$$

or

$$\int_0^{\pi/2} \sin^m x\,dx = \frac{m-1}{m}\int_0^{\pi/2} \sin^{m-2} x\,dx,$$

because the first term on the right side of (II), pq, is equal to zero for the values 0 and $\pi/2$. By repeated application of the last formula we find the following value for $I_m = \int_0^{\pi/2} \sin^m x\,dx$ (the formulas differ according as m is even or odd):

$$I_{2n} = \frac{2n-1}{2n}\cdot\frac{2n-3}{2n-2}\cdots\frac{1}{2}\cdot\frac{\pi}{2},$$

$$I_{2n+1} = \frac{2n}{2n+1}\cdot\frac{2n-2}{2n-1}\cdots\frac{2}{3}.$$

Since $0 < \sin x < 1$ for $0 < x < \pi/2$, we have $\sin^{2n-1} x > \sin^{2n} x > \sin^{2n+1} x$, so that

$$I_{2n-1} > I_{2n} > I_{2n+1}$$ (see p. 414)

or

$$\frac{I_{2n-1}}{I_{2n+1}} > \frac{I_{2n}}{I_{2n+1}} > 1.$$

Substituting the values calculated above for I_{2n-1}, etc. in the last inequalities, we find

$$\frac{2n+1}{2n} > \frac{1\cdot 3\cdot 3\cdot 5\cdot 5\cdot 7 \cdots (2n-1)(2n-1)(2n+1)}{2\cdot 2\cdot 4\cdot 4\cdot 6\cdot 6 \cdots (2n)(2n)}\cdot\frac{\pi}{2} > 1.$$

If we now pass to the limit as $n \to \infty$ we see that the middle term tends to 1, hence we obtain Wallis' product representation for $\pi/2$:

$$\frac{\pi}{2} = \frac{2\cdot 2\cdot 4\cdot 4\cdot 6\cdot 6 \cdots 2n\cdot 2n \cdots}{1\cdot 3\cdot 3\cdot 5\cdot 5\cdot 7 \cdots (2n-1)(2n-1)\cdot(2n+1) \cdots}$$

$$= \lim \frac{2^{4n}(n!)^4}{[(2n)!]^2(2n+1)} \quad \text{as } n \to \infty.$$

SUGGESTIONS FOR FURTHER READING

GENERAL REFERENCES

W. Ahrens. *Mathematische Unterhaltungen und Spiele*, 2nd edition, 2 vols. Leipzig: Teubner, 1910.

W. W. Rouse Ball. *Mathematical Recreations and Essays*, 11th edition, revised by H. S. M. Coxeter. New York: Macmillan, 1939.

E. T. Bell. *The Development of Mathematics*. New York: McGraw-Hill, 1940.

———. *Men of Mathematics*. New York: Simon and Schuster, 1937.

T. Dantzig. *Aspects of Science*. New York: Macmillan, 1937.

A. Dresden. *An Invitation to Mathematics*. New York: Holt, 1936.

F. Enriques. *Questioni riguardanti le matematiche elementari*, 3rd edition, 2 vols. Bologna: Zanichelli, 1924 and 1926.

E. Kasner and J. Newman. *Mathematics and the Imagination*. New York: Simon and Schuster, 1940.

F. Klein. *Elementary Mathematics from an Advanced Standpoint*, translated by E. R. Hedrick and C. A. Noble, 2 vols. New York: Macmillan, 1932 and 1939.

M. Kraitchik. *La Mathématique des Jeux*. Brussels: Stevens, 1930.

O. Neugebauer. *Vorlesungen über Geschichte der antiken mathematischen Wissenschaften*. Erster Band: *Vorgriechische Mathematik*. Berlin: Springer, 1934.

H. Poincaré. *The Foundations of Science*. Lancaster, Pa.: Science Press, 1913.

H. Rademacher und O. Toeplitz. *Von Zahlen und Figuren*, 2nd edition. Berlin: Springer, 1933.

B. Russell. *Introduction to Mathematical Philosophy*. London: Allen and Unwin, 1924.

———. *The Principles of Mathematics*, 2nd edition. New York: Norton, 1938.

D. E. Smith. *A Source Book in Mathematics*. New York: McGraw-Hill, 1929.

H. Steinhaus. *Mathematical Snapshots*. New York: Stechert, 1938.

H. Weyl. "The Mathematical Way of Thinking," *Science*, XCII (1940), p. 437 ff.

H. Weyl. *Philosophie der Mathematik und Naturwissenschaft,* Handbuch der Philosophie, Bd. II. Munich: Oldenbourg, 1926, pp. 3–162.

CHAPTER I

L. E. Dickson. *Introduction to the Theory of Numbers.* Chicago: University of Chicago Press, 1931.

———. *Modern Elementary Theory of Numbers.* Chicago: University of Chicago Press, 1939.

G. H. Hardy. "An Introduction to the Theory of Numbers," *Bulletin of the American Mathematical Society,* XXXV (1929), p. 789 ff.

G. H. Hardy and E. M. Wright. *An Introduction to the Theory of Numbers.* Oxford: Clarendon Press, 1938.

J. V. Uspensky and M. H. Heaslet. *Elementary Number Theory.* New York: McGraw-Hill, 1939.

CHAPTER II

G. Birkhoff and S. MacLane. *A Survey of Modern Algebra.* New York: Macmillan, 1941.

M. Black. *The Nature of Mathematics.* New York: Harcourt, Brace, 1935.

T. Dantzig. *Number, the Language of Science,* 3rd edition. New York: Macmillan, 1939.

G. H. Hardy. *A Course of Pure Mathematics,* 7th edition. Cambridge: University Press, 1938.

K. Knopp. *Theory and Application of Infinite Series,* translated by Miss R. C. Young. London: Blackie, 1928.

A. Tarski. *Introduction to Logic.* New York: Oxford University Press, 1939.

F. Enriques. *The Historic Development of Logic,* translated by J. Rosenthal. New York: Holt, 1929.

CHAPTER III

J. L. Coolidge. *A History of Geometrical Methods.* Oxford: Clarendon Press, 1940.

A. De Morgan. *A Budget of Paradoxes,* 2 vols. Chicago: Open Court, 1915.

L. E. Dickson. *New First Course in the Theory of Equations.* New York: Wiley, 1939.

F. Enriques (editor). *Fragen der Elementargeometrie,* 2nd edition, 2 vols. Leipzig: Teubner, 1923.

E. W. Hobson. *"Squaring the Circle," a History of the Problem.* Cambridge: University Press, 1913.

A. B. Kempe. *How to Draw a Straight Line.* London: Macmillan, 1877.

F. Klein. *Famous Problems of Geometry*, translated by W. W. Beman and D. E. Smith, 2nd edition. New York: Stechert, 1930.

L. Mascheroni. *La geometria del compasso.* Palermo: Reber, 1901.

G. Mohr. *Euclides Danicus.* Copenhagen: Høst, 1928.

J. M. Thomas. *Theory of Equations.* New York: McGraw-Hill, 1938.

L. Weisner. *Introduction to the Theory of Equations.* New York: Wiley, 1939.

CHAPTER IV

W. C. Graustein. *Introduction to Higher Geometry.* New York: Macmillan, 1930.

D. Hilbert. *The Foundations of Geometry*, translated by E. J. Townsend, 3rd edition. La Salle, Ill.: Open Court, 1938.

C. W. O'Hara and D. R. Ward. *An Introduction to Projective Geometry.* Oxford: Clarendon Press, 1937.

G. de B. Robinson. *The Foundations of Geometry.* Toronto: University of Toronto Press, 1940.

Girolamo Saccheri. *Euclides ab omni naevo vindicatus*, translated by G. B. Halsted. Chicago: Open Court, 1920.

R. G. Sanger. *Synthetic Projective Geometry.* New York: McGraw-Hill, 1939.

O. Veblen and J. W. Young. *Projective Geometry*, 2 vols. Boston: Ginn, 1910 and 1918.

J. W. Young. *Projective Geometry.* Chicago: Open Court, 1930.

CHAPTER V

P. Alexandroff. *Einfachste Grundbegriffe der Topologie.* Berlin: Springer, 1932.

D. Hilbert und S. Cohn-Vossen. *Anschauliche Geometrie.* Berlin: Springer, 1932.

M. H. A. Newman. *Elements of the Topology of Plane Sets of Points.* Cambridge: University Press, 1939.

H. Seifert und W. Threlfall. *Lehrbuch der Topologie.* Leipzig: Teubner, 1934.

CHAPTER VI

R. Courant. *Differential and Integral Calculus*, translated by E. J. McShane, revised edition, 2 vols. New York: Nordemann, 1940.

G. H. Hardy. *A Course of Pure Mathematics*, 7th edition. Cambridge: University Press, 1938.

W. L. Ferrar. *A Text-book of Convergence*. Oxford: Clarendon Press, 1938.

For the theory of continued fractions see, e.g.

S. Barnard and J. M. Child. *Advanced Algebra*. London: Macmillan, 1939.

CHAPTER VII

R. Courant. "Soap Film Experiments with Minimal Surfaces," *American Mathematical Monthly*, XLVII (1940), pp. 167–174.

J. Plateau. "Sur les figures d'équilibre d'une masse liquide sans pésanteur," *Mémoires de l'Académie Royale de Belgique*, nouvelle série, XXIII (1849).

———. *Statique expérimentale et théoretique des Liquides*. Paris: 1873.

CHAPTER VIII

C. B. Boyer. *The Concepts of the Calculus*. New York: Columbia University Press, 1939.

R. Courant. *Differential and Integral Calculus*, translated by E. J. McShane, revised edition, 2 vols. New York: Nordemann, 1940.

G. H. Hardy. *A Course of Pure Mathematics*, 7th edition. Cambridge: University Press, 1938.

INDEX